THE KINETIC THEORY OF GASES
An Anthology of Classic Papers with Historical Commentary

History of Modern Physical Sciences Series

Aims and Scope

The series will include a variety of books dealing with the development of physics, astronomy, chemistry and geology during the past two centuries (1800–2000). During this period there were many important discoveries and new theories in the physical sciences which radically changed our understanding of the natural world, at the same time stimulating technological advances and providing a model for the growth of scientific understanding in the biological and behavioral sciences.

While there is no shortage of popular or journalistic writing on these subjects, there is a need for more accurate and comprehensive treatments by professional historians of science who are qualified to discuss the substance of scientific research. The books in the series will include new historical monographs, editions and translations of original sources, and reprints of older (but still valuable) histories. Efforts to understand the worldwide growth and impact of physical science, not restricted to the traditional focus on Europe and the United States, will be encouraged. The books should be authoritative and readable, useful to scientists, graduate students and anyone else with a serious interest in the history, philosophy and social studies of science.

Series Board

Professor Igor Aleksander
Imperial College, London, UK

Professor Tom Kibble
Imperial College, London, UK

Professor Stephen G Brush
University of Maryland, USA

Dr Katherine Russell Sopka
Consultant and Historian of Science

Professor Richard H Dalitz
Oxford University, UK

Professor Roger H Stuewer
University of Minnesota – Twin Cities, USA

Professor Freeman J Dyson
Princeton University, USA

Dr Andrew C Warwick
Imperial College, London, UK

Professor Chris J Isham
Imperial College, London, UK

Professor John Archibald Wheeler
Princeton University, USA

Professor Maurice Jacob
CERN, France

Professor C N Yang
*State University of New York,
Stony Brook, USA*

Forthcoming Titles

Stalin's Great Science
by Dr Alexei Kojevnikov, University of Georgia at Athens, USA

Critical Edition of the Peierls Correspondence
edited by Dr Sabine Lee, Birmingham University, UK

HISTORY OF MODERN PHYSICAL SCIENCES – VOL. 1

THE KINETIC THEORY OF GASES

An Anthology of Classic Papers with Historical Commentary

by

Stephen G Brush
University of Maryland
USA

edited by

Nancy S Hall
University of Maryland & University of Delaware
USA

Imperial College Press

Published by

Imperial College Press
57 Shelton Street
Covent Garden
London WC2H 9HE

Distributed by

World Scientific Publishing Co. Pte. Ltd.
5 Toh Tuck Link, Singapore 596224
USA office: Suite 202, 1060 Main Street, River Edge, NJ 07661
UK office: 57 Shelton Street, Covent Garden, London WC2H 9HE

British Library Cataloguing-in-Publication Data
A catalogue record for this book is available from the British Library.

THE KINETIC THEORY OF GASES
An Anthology of Classic Papers with Historical Commentary

Copyright © 2003 by Imperial College Press

All rights reserved. This book, or parts thereof, may not be reproduced in any form or by any means, mechanical, including photocopying, recording or any information storage and retrieval system, be invented, without written permission from the Publisher.

For photocopying of material in this volume, please pay a copying fee through the Copyright Inc., 222 Rosewood Drive, Danvers, MA 01923, USA. In this case permission to photocopy is the publisher.

ISBN 1-86094-347-0
ISBN 1-86094-348-9 (pbk)

Printed by FuIsland Offset Printing (S) Pte Ltd, Singapore

CONTENTS

PREFACE ix
ACKNOWLEDGMENTS xi

PART I   *The Nature of Gases and of Heat*

Introduction 1
Bibliography 34

1. Robert Boyle: The Spring of the Air (from *New Experimentas Physical-Mechanical, etc.* 1660) 43

2. Isaac Newton: The Repulsion Theory (from *Philosophiae Naturalis Principia Mathematica*, 1687) 52

3. Daniel Bernoulli: On the Properties and Motions of Elastic Fluids, especially Air (from *Hydrodynamica*, 1738) 57

4. George Gregory: The Existence of Fire (from *The Economy of Nature*, 1798) 66

5. Robert Mayer: The Forces of Inorganic Nature (from *Annalen der Chemie und Pharmacie*, 1842) 71

6. James Joule: On Matter, Living Force, and Heat (from the *Manchester Courier*, 1847) 78

7. Hermann von Helmholtz: The Conservation of Force (from *Ueber die Erhaltung der Kraft*, 1847) 89

8. Rudolf Clausius: The Nature of the Motion which We Call Heat (from *Annalen der Physik*, 1857) 111

9. Rudolf Clausius: On the Mean Length of the Paths Described by the Separate Molecules of Gaseous Bodies (from *Annalen der Physik*, 1858) 135

10. James Clerk Maxwell: Illustrations of the Dynamical   148
    Theory of Gases (from *Philosophical Magazine*, 1858)

11. Rudolf Clausius: On a Mechanical Theorem Applicable   172
    to Heat (from *Sitzungsberichte der Niedderrheinischen
    Gesellschaft*, Bonn, 1870)

PART II *Irreversible Processes*

Introduction   179
Bibliography   195

1. James Clerk Maxwell: On the Dynamical Theory of   197
   Gases (from *Philosophical Transactions of the
   Royal Society of London*, for 1866)

2. Ludwig Boltzmann: Further Studies on the Thermal   262
   Equilibrium of Gas Molecules (from *Sitzungs-
   berichte der kaiserlichen Akademie der
   Wissenschaften*, Vienna, 1872)

3. William Thomson: The Kinetic Theory of the Dissipation   350
   of Energy (from *Proceedings of the Royal Society
   of Edinburgh*, for 1874)

4. Ludwig Boltzmann: On the Relation of a General   362
   Mechanical Theorem to the Second Law of
   Thermodynamics (from *Sitzungsberichte der
   kaiserlichen Akademie der Wissenschaften*,
   Vienna, 1877)

5. Henri Poincaré: On the Three-body Problem and the   368
   the Equations of Dynamics (from *Acta
   Mathematica*, 1890)

6. Henri Poincaré: Mechanism and Experience (from   377
   Revue de Métaphysique et de Morale, 1893)

7. Ernst Zermelo: On a Theorem of Dynamics and the   382
   Mechanical Theory of Heat (from *Annalen der
   Physik*, 1860)

CONTENTS vii

8. Ludwig Boltzmann: Reply to Zermelo's Remarks on the 392
    Theory of Heat (from *Annalen der Physik*, 1896)

9. Ernst Zermelo: On the Mechanical Explanation of 403
    Irreversible Processes (from *Annalen der
    Physik*, 1896)

10. Ludwig Boltzmann: On Zermelo's Paper "On the 412
    Mechanical Explanation of Irreversible
    Processes" (from *Annalen der Physik*, 1897)

PART III *Historical Discussions by Stephen G. Brush*

1. Gadflies and Geniuses in the History of Gas Theory 421
    (from *Synthese*, 1999)

2. Interatomic Forces and Gas Theory from Newton 451
    to Lennard-Jones (from *Archive for Rational
    Mechanics and Analysis*, 1970)

3. Irreversibility and Indeterminism: Fourier to 480
    Heisenberg (from *Journal of the History
    of Ideas*, 1976)

4. Proof of the Impossibility of Ergodic Systems: the 505
    1913 Papers of Rosenthal and Plancherel (from
    Transport Theory and Statistical Physics, 1971)

5. Statistical Mechanics and the Philosophy of Science: 524
    Some Historical Notes (from *PSA 1976*, vol. 2, 1977)

PART IV *A Guide to Historical Commentaries: Kinetic* 555
    *Theory of Gases, Thermodynamics, and Related
    Topics*

INDEX TO PARTS I, II, III 637

Preface

The first two Parts of this book were originally prepared about 40 years ago at the suggestion of Dirk ter Haar for a series called "Selected Readings in Physics" which he edited for Pergamon Press. The purpose of the series was to encourage students and teachers of physics to learn how their subject developed, by reading the papers that recorded the development of important new ideas. I first encountered ter Haar in 1957 at Oxford University, where I was a graduate student and he was a lecturer. Although I had made good use of his recently-published book *Elements of Statistical Mechanics* in my research and had consulted him on some technical points, it was not until my oral dissertation defense, at which he was one of the examiners, that we discovered our mutual interest in the history of the kinetic theory of gases. My interest in that topic went back to 1954, when Thomas S. Kuhn had suggested it as a subject for my paper in his seminar on the history of 19th-century thermodynamics. The seminar paper grew into a series of journal articles and eventually an 800-page book, *The Kind of Motion We Call Heat* (1976). On the other hand, ter Haar's book on statistical mechanics, and his own research method, were based on lectures of H. A. Kramers, went "straight back to Boltzmann, via Kramers and Ehrenfest" as noted in the preface of the book (whose title used the word "Elements" in the same sense as Gibbs' classic work, *Elementary Principles in Statistical Mechanics*).

In planning the kinetic theory volumes of the "Selected Readings" series, ter Haar and I agreed that the selections should be in English; the books were not intended for professional historians of science, who would of course be expected to read the original sources in doing their own research, although this collection may be useful for graduate students and others who need an overview of the subject. We wanted the selections to be long enough to enable the reader to follow the author's train of reasoning to its conclusion, although in some cases they may be only portions of longer works. The criterion for selection is not that the work necessarily represents an important original contribution, but rather that reading it enables one to understand more clearly the development of the subject. Thus for example we present not only the theories that were "correct" from the modern viewpoint, but also some that seemed plausible at the time but were later rejected. If one wishes to understand the process by which scientific theories are developed, and accepted or rejected, one must be willing to look at the alternatives and objections to the theory that was ultimately successful. One's respect for the pioneer kinetic theorists cannot but be increased by the realization that they had to overcome another theory whose proponents could invoke the authority of such giants as Newton and Laplace. Nor should one forget that the triumph of the kinetic theory in the 1860s and 1870s was made possible in part by the general acceptance of the principle of conversion and conservation of energy, and therefore we decided to include the fundamental papers of Mayer, Joule, and Helmholtz on energy conservation.

The papers included in Part II will probably be of even greater value to physicists and philosophers of physics now than when this Part was originally published, because there has been since the mid-1960s a strong revival of research on the kinetic theory of gases, and of interest in the problems and paradoxes of irreversibility. Anyone participating in current research and controversies should be familiar with the original papers of Maxwell, Boltzmann, William Thomson (Kelvin), Poincaré and Zermelo.

In Part III, I have reprinted 5 of my articles that were not included in or go beyond the subject matter of *The Kind of Motion We Call Heat*. The first one, "Gadflies and Geniuses," presents an overview of the history of gas theory from a particular perspective. The second and third connect 19th-century research on interatomic forces and irreversibility to 20th-century problems. The fourth includes the first (and as far as I know the only) English translation of the two classic papers of Rosenthal and Plancherel, proving the mathematical impossibility of the ergodic hypothesis in its original form and opening the way to modern research in ergodic theory. The fifth article shows the relevance of the history of kinetic theory to some late-20th-century debates in the philosophy of science.

Since the early 1960s there has been a great flowering of research and writing on all aspects of the history of science, including the kinetic theory of gases, thermodynamics, and related subjects. While much has been learned, there has also been a certain amount of redundancy and waste of effort because many authors have not bothered to search out what other historians have already done before publishing their own articles and books. In the interests of promoting more effective historical research, I have presented in Part IV a comprehensive (although certainly not complete) bibliography of secondary sources (and reprints of primary sources) published since about 1964.

It is a pleasure to thank those who have, over several decades, assisted in many ways my work on the history of kinetic theory: John Blackmore, S. R. de Groot, C. W. F. Everitt, Elizabeth Garber, Frank Haber, Roger Hahn, Sir Harold Hartley, Erwin Hiebert, Gerald Holton, W. James King, Martin J. Klein, J. L. Lebowitz, P. E. Liley, E. A. Mason, Eric Mendoza, A. Michels, Richard G. Olson, N. H. Robinson, J. S. Rowlinson, Judah Schwartz, Harold I. Sharlin, C. Truesdell, J. D. Van der Waals Jr.,

Nancy Hall has done a marvelous job of transforming this varied collections of materials into the finished book you are now holding.

SGB

Acknowledgments

The author gratefully acknowledges permission to reprint the following of his publications.

Kinetic Theory, Vol. 1, *The Nature of Gases and of Heat.* Pergamon Press, Oxford (1965).

Kinetic Theory, Vol. 2, *IrreversibleProcesses.* Pergamon Press, Oxford (1966).

"Interatomic Forces and Gas Theory from Newton to Lennard-Smith," *Archive for Rational Mechanics and Analysis* 39: 1-29 (1970).

"Proof of the Impossibility of Ergodic Systems: The 1913 Papers of Rosenthal and Plancherel," *Transport Theory and Statistical Physics* 1: 287-311 (1971).

"Irreversibility and Indeterminism from Fourier to Heisenberg," *Journal of the History of Ideas* 37: 603-630 (1976).

"Statistical Mechanics and the Philosophy of Science: Some Historical Notes," *PSA 1976, Vol. 2*, Philosophy of Science Association, East Lansing, MI. (1977).

"Gadflies and Geniuses in the History of Gas Theory," *Synthese* 119: 11-43 (1999).

Part I

The Nature of Gases and of Heat

Introduction

THE history of atomism goes back to the ancient Greek and Roman philosophers, but the history of the kinetic theory of gases does not really begin until the 17th century when Torricelli, Pascal, and Boyle first established the physical nature of the air. By a combination of experiments and theoretical reasoning they persuaded other scientists that the earth is surrounded by a " sea " of air that exerts pressure in much the same way that water does, and that air pressure is responsible for many of the phenomena previously attributed to " nature's abhorrence of a vacuum ". We may view this development of the concept of air pressure as part of the change in scientific attitudes which led to the " mechanico-corpuscular " view of nature, associated with the names of Galileo, Boyle, Newton, and others. Instead of postulating " occult forces " or teleological principles to explain natural phenomena, scientists started to look for explanations based simply on matter and motion.

It was well known in the time of Galileo Galilei (1564–1642) that water will not rise more than 34 ft in a pump, although Galileo himself seems to have been the first to put this fact on record in 1638.†
A few years later (1643 or 1644) his student Evangelista Torricelli (1608–47) devised an experiment to illustrate the same effect in the laboratory. Since mercury is about fourteen times as dense as water, one might expect that it can be lifted only about 1/14 as far. This is indeed what is observed, and this fact tends to make plausible the

---

† *Dialogues concerning two new sciences*, English translation by Crew and de Salvio, pp. 12–17. Further details of this and other works mentioned in this Introduction may be found in the Bibliography.

mechanical explanation based on air pressure. Taking a glass tube about a yard long with one end closed, Torricelli filled it with mercury to the top; then, placing a finger over the open end, he inverted the tube so that the open end was immersed in an open dish of mercury. When he removed his finger from the open end, the mercury in the tube fell until the top of the mercury column was about 30 in. above the level of the mercury in the open dish. Between the top of the mercury column and the upper end of the tube was an open space, which became known as the " Torricellian vacuum ".

According to Torricelli, it is just the mechanical pressure of the air that raises the mercury in the tube. Blaise Pascal (1623–62), the celebrated philosopher and mathematician, then pointed out that— by analogy with the laws of hydrostatics—the pressure of air should be less on the top of a mountain than at sea level. An experiment to test this prediction was carried out by Pascal's brother-in-law, Florin Perier, in 1648, according to Pascal's instructions, and the results conformed to expectations. Further experiments with Torricelli's " barometer " were conducted by Otto von Guericke (1602–86), who also constructed a suction pump and performed the famous experiment of the Magdeburg hemispheres in 1654. In this experiment, two hollow bronze hemispheres were fitted carefully edge to edge, and the interior was evacuated. A team of eight horses was harnessed to each hemisphere and the two teams were driven in opposite directions, but they were unable to pull the hemispheres apart.

These early experiments in pneumatics were carried on at about the same time as the formation of the first scientific societies in Italy, England and France, and in many cases several scientists collaborated in the experiments. The Accademia del Cimento (Academy of Experiments) was founded in Florence in 1657; Torricelli himself had taught some of its charter members, and they in turn carried on his researches into the nature of air pressure. The Royal Society developed from an informal association of scientists in London; some of these men moved to Oxford and formed a separate group there about 1649, but after the Restoration the group again concentrated in London, and the Royal Society received its official charter in 1662. Several members of this group played important roles in the development of the theory of gases. Robert Boyle (1627–91), for example, was the seventh and last son of the Earl of Cork, a British nobleman who owned considerable property in Ireland; Boyle's fortune was put to good use in buying expensive scientific apparatus. The Oxford group met in his lodgings for a time. The best experimentalist of the group was Robert Hooke (1635–1703)—now remembered chiefly for his discovery of the relation between stress and strain—who constructed an improved air pump for Boyle about 1658. It was with this " pneumatical engine " that Boyle performed the experiments recorded in his book, *New Experiments Physico-Mechanicall, touching the Spring of the Air, and its effects* (Oxford, 1660). Boyle mentions that some of his experiments were done in the presence of such colleagues as Christopher Wren (1632–1723), the architect who designed St. Paul's Cathedral and many other buildings still standing in London and elsewhere, and John Wallis (1616–1703), mathematician and divine who made several important contributions to analytical geometry and algebraic analysis. It was Wallis, Wren, and Christian Huygens (1629–95), Dutch astronomer and physicist, who—in response to a request from the Royal Society in 1668—independently formulated the laws of impact. These laws are the basis of the kinetic theory of " billiard balls " (atoms represented by elastic spheres).

Isaac Newton (1642–1727) was elected to the Royal Society in 1671, after he had already invented the differential and integral calculus and discovered the unequal refrangibility of the rays of light and the binomial theorem (1665–66) and had become Lucasian Professor of Mathematics at Cambridge in 1669. His *Principia* was published in 1687, and he served as President of the Royal Society from 1703 until his death.

In all this illustrious company, perhaps we can even find a place for Richard Towneley (1628–1707), who though not a member of the Society was an active correspondent with many of its members; he performed a number of minor scientific experiments and meteorological observations at his estate in Towneley, Lancashire, but the chief reason for mentioning him here is that he was the person who first suggested Boyle's law.

Robert Boyle is generally credited with the discovery that the pressure exerted by a gas is inversely proportional to the volume of the space in which it is confined. From Boyle's point of view that discovery by itself was insignificant, and though he had provided the experimental evidence for it he readily admitted that he had not found any general quantitative relation between pressure and volume before Richard Towneley suggested his simple hypothesis. Robert Hooke also provided further experimental confirmation of the hypothesis. It was long known as " Mariotte's law " on the Continent, because Edmé Mariotte (1620–84), a French priest, proposed it in his *Essay de la nature de l'air* (1679). There are good reasons for believing that Mariotte was familiar with Boyle's work even though he does not mention it, so that he does not even deserve the credit for independent (much less simultaneous) discovery.†

Boyle's researches were carried out to illustrate not just a quantitative relation between pressure and volume, but rather the qualitative fact that *air has elasticity* ("spring ") and can exert a mechanical pressure of a magnitude sufficient to support a column of water or mercury. His achievement was to introduce a new variable—pressure —into physics; he could well afford to be generous about giving others the credit for perceiving the numerical relations between this variable and others. He considered that the crucial experiment was his No. 17,‡ in which he enclosed the lower part of the Torricellian barometer (a column of mercury in a glass tube sitting in a dish of mercury) in a container from which the air could be removed by means of his pump. As the air was exhausted, the mercury in the tube fell nearly to the level of that in the dish. This was interpreted to mean that the mercury had in fact been supported by air pressure, or rather by the difference between atmospheric pressure and the negligible pressure of the Torricellian vacuum at the top of the tube.

Boyle also proposed a theoretical explanation for the elasticity of air—he likened it to " a heap of little bodies, lying one upon another " and attributed the elasticity of the whole to the elasticity of the parts (Selection 1 in this volume). The atoms are said to behave like springs which resist compression. To a modern reader this explanation does not seem very satisfactory, for it does no more than attribute to atoms the observable properties of macroscopic objects. It is interesting to note that Boyle also tried the " crucial experiment " which was to help overthrow his own theory in favor of the kinetic theory two centuries later, though he does not realize its significance;

---

† See for example W. S. JAMES, *Science Progress* 23, 261 (1928); **24**, 57 (1929).

‡ See Boyle's *Works*, p. 33.

in his Experiment No. 26 he places a pendulum in the evacuated chamber and discovers, to his surprise, that the presence or absence of air makes hardly any difference to the period of the swings or the time needed for the pendulum to come to rest. In 1859, James Clerk Maxwell deduced from the kinetic theory that the viscosity of a gas should be independent of its density (Selection 10)—a property which would be very hard to explain on the basis of Boyle's theory.

These criticisms are irrelevant in a sense, since Boyle's theory should be compared with other ideas current at the time, rather than with modern views. Soon after the publication of Boyle's *New Experiments*, attacks on the experiments and their interpretation were advanced by Thomas Hobbes (1588–1679), the writer on political philosophy, and Franciscus Linus, alias Francis Hall (1595–1675), Jesuit scientist and sometime Professor of Hebrew and Mathematics at Liège. Hobbes, though a participant in the new scientific movement, was engaged in a mathematical dispute with Wallis and also resented his exclusion from the Royal Society. He believed that a "subtle matter" exists, filling all space; this was a view that hampered the development of the kinetic theory right up to the beginning of the 20th century. Linus asserted that the Torricellian vacuum contained an invisible cord or membrane (Latin *funiculus*, diminutive of *funis*, rope). When air is stretched or rarefied, the *funiculus* exerts a violent attraction on all surrounding bodies, and it is this attraction which pulls the mercury up the tube. Indeed, if you put your finger over the end of the tube from a suction pump (or vacuum cleaner) you can actually feel the *funiculus* pulling in the flesh of your finger!

Laugh if you like at this fantastic idea, but remember that the funicular hypothesis was an example of the type of pseudo-mechanical explanation of physical phenomena that used to be quite popular in the early days of science. Moreover, the idea that a vacuum contains an entity that sucks things into it is much closer to "common sense" than the theory that the suction is merely due to the absence of normal atmospheric pressure inside the vacuum. It takes a considerable degree of sophistication to accept the idea that we are living at the bottom of a sea of air which exerts the tremendous pressure of 14·7 lb on every square inch of our bodies.

Boyle published in 1662 a *Defence* against the objections of Linus and Hobbes; in the course of refuting the funicular hypothesis, he provides some new experimental evidence on the compression and rarefaction of air. He presents this evidence as confirmation of Towneley's hypothesis (" Boyle's law "), which is mentioned here for the first time.

Newton discusses very briefly in his *Principia* (1687) the consequences of various hypotheses about the forces between atoms for the relation between pressure and volume (Selection 2). One particular hypothesis, a repulsive force inversely proportional to distance, leads to Boyle's law. It seems plausible that Newton was trying to put Boyle's theory in mathematical language, and that he thought of the repulsive forces as being due to the action of atomic springs in contact with each other, but there seems to be no direct evidence for this.

Neither Boyle nor Newton claimed that the hypothesis of repulsive forces between atoms is the only correct explanation for gas pressure; both were willing to leave the question open. Boyle mentions the Descartes theory of vortices (1644), for example, which is somewhat closer in spirit to the kinetic theory since it relies more heavily on the rapid motion of the parts of the atom as a cause of repulsion.† (Though Descartes did not believe in " atoms " in the classical sense.)

---

† Incidentally, it is important to realize that there is more to the kinetic theory than just the statement that heat is atomic motion. That statement was frequently made, especially in the 17th century, but usually by scientists who did not make the important additional assumption that in gases the atoms move freely most of the time. It was quite possible to accept the " heat is motion " idea and still reject the kinetic theory of gases, as did Humphry Davy early in the 19th century.

Here is an example of a derivation of Boyle's law, which Tait (1885) mistakenly calls an " anticipation of the kinetic theory ", by Robert Hooke, in his *Lectures de Potentia Restitutiva* (1678) pp. 15–16: " The air then is a body consisting of particles so small as to be almost equal to the particles of the Heterogeneous fluid medium incompassing the earth. . . . If therefore a quantity of this body be inclosed by a solid body, and that be so contrived as to compress it into less room, the motion thereof (supposing the heat the same) will continue the same, and consequently the Vibrations and Occursions will be increased in reciprocal proportion, that is, if it be condensed into half the space the Vibrations and Occursions will be double in number. . . . Again, if the containing Vessel be so contrived as to leave it more space, the length of the Vibrations will be proportionably inlarged, and the number of Vibrations and Occursions will

Nevertheless, the Boyle–Newton theory of gases was apparently accepted by most scientists until about the middle of the 19th century, when the kinetic theory finally managed to overcome Newton's authority. Thus a theory which was decidedly a step forward from older ideas was able to retard further progress for a considerable time.

It is difficult to understand the relative lack of progress in gas theory during the 18th century as compared to the 17th. Several brilliant mathematicians refined and clarified Newton's principles of mechanics and applied them to the analysis of the motions of celestial bodies and of continuous solids and fluids, but there was little interest in the properties of systems of freely moving atoms. The atoms in a gas were still conceived as being suspended in the ether, although they could vibrate or rotate enough to keep other atoms from coming too close. This model was rather awkward to formulate mathematically, as may be seen from an unsuccessful attempt by Leonhard Euler in 1727.

However, one contribution from this period has been generally recognized as the first kinetic theory of gases. This is Daniel Bernoulli's derivation of the gas laws from a " billiard ball " model—much like the one still used in elementary textbooks today—in 1738 (Selection 3). Bernoulli (1700–82) was one of a famous family of mathematicians;† his kinetic theory is only a small part of a treatise on hydrodynamics, a subject to which he made important contributions. For example, he established the principle that when a fluid is flowing through a region of varying pressures, every increase in its velocity corresponds to a decrease in pressure. This fact may easily be understood if one accepts the principle of conservation of mechanical energy, since every change in the velocity of a fluid element

---

be reciprocally diminished. . . . These explanations will serve *mutatis mutandis* for explaining the spring of any other body whatsoever." See P. G. Tait, *Scientific Papers*, Cambridge University Press, 1900, Vol. 2, pp. 122–3.

† James (or Jakob) Bernoulli (1654–1705) developed the calculus and probability theory. His brother Johann Bernoulli (1667–1748) applied calculus to extremum problems and advanced analytical trigonometry. Daniel Bernoulli was the second son of Johann. Successive generations carried on the tradition of mathematical physics; for example, in 1912, Leonhard Bernoulli (1879–1939) was lecturing on the kinetic theory of matter at Basel (see *Physikalische Zeitschrift*, vol. 13, p. 335).

must be produced by work done on it by pressure forces. Thus when the fluid element enters a region in which a greater velocity of flow prevails (as in a part of a pipe with narrower cross-section) it can acquire this velocity only because a greater pressure acts on its rear surface than on its front surface; in other words, only when it moves from points of higher to points of lower pressure can its kinetic energy be increased. The same principle is involved in the " lift " of an airfoil.

From the viewpoint of the historical development of the kinetic theory, Bernoulli's formulation and successful applications of the principle of conservation of mechanical energy (usually known at that time as *vis viva*, Latin for " living force ") were really more important than the fact that he actually proposed a kinetic theory himself. Bernoulli's kinetic theory, while in accord with modern ideas, was a century ahead of its time, for scientists were not yet ready to accept the physical description of a gas that it employed. Heat was still generally regarded as a substance, even though it was also recognized that heat might have some relationship to atomic motions; Bernoulli's assumption that heat is *nothing but* atomic motion was unacceptable, especially to scientists who were interested in the phenomena of radiant heat. The assumption that atoms could move freely through space until they collided like billiard balls was probably regarded as too drastic an approximation, since it neglected the drag of the ether and oversimplified the interaction between atoms. Bernoulli's kinetic theory may even have been regarded as a step backwards, although we do not know much about how the scientific world reacted to it. The best that could be said for it was that it explained some of the properties of gases that were already well understood. When physics reached the stage of development at which the kinetic theory no longer conflicted with other established principles, Bernoulli's theory had been almost forgotten and had to be rediscovered.† Here we have one more illustration of the inade-

---

† The kinetic theory was kept alive to a certain extent by the writings of such scientists as M. V. Lomonosov (1711–65) in Russia, and J. A. De Luc (1727–1817) and George-Louis Le Sage (1724–1803) in Geneva. See for example De Luc's review of Bernoulli's theory in *Recherches sur les modifications de l'atmosphere*, Geneve, 1772, pp. 165–71. Nevertheless little was done to advance the mathematical development of the theory or to convince the majority of scientists of its validity.

quacy of the view of history of science which is concerned primarily with questions of *priority*. While we can recognize the genius of a man who solves a problem long before anyone else knew that the problem existed, we must also ask whether such a discovery has any effect on the progress of science. In a very real sense, the man who persuades the world to adopt a new idea has accomplished as much as the man who conceived that idea.

THE CALORIC THEORY

It was not possible for the kinetic theory to be fully accepted until the doctrine that heat is a substance—" caloric "—had been overthrown and replaced by the " mechanical theory of heat " (essentially what is now known as thermodynamics) which treats heat as simply one form of energy that can be converted into other forms such as mechanical work. Although the caloric theory is apparently no longer of interest in physics, it is necessary to describe it briefly here in order to show why it constituted a real obstacle to the adoption of the modern theory of heat. It is important to realize that, from the standpoint of the early 19th-century scientist, there were many perfectly valid reasons for retaining the caloric theory and rejecting the mechanical theory.

Calorique (or " caloric " in English) was a fluid composed of particles that repel each other but are attracted to the particles of ordinary matter. Caloric is able to diffuse into and penetrate the interstices of all matter; and each particle of matter is surrounded by an atmosphere of caloric whose density increases with temperature. These atmospheres cause two particles of matter to repel each other at small distances, though at larger distances the atmosphere becomes attenuated and the attractive force of gravity predominates; there is thus an intermediate equilibrium point at which there is no net attractive or repulsive force. As the temperature rises and more caloric is added to a substance, this equilibrium point is shifted outward and the average distance between particles becomes greater, thus leading to an expansion of the body. On the other hand, if a body is compressed, caloric is squeezed out and appears at the surface as evolved heat.

An important extension of this theory was made by Joseph Black (1728–99), a Scottish chemist, whose discoveries, though taught in lectures, were not published until 1803. (Black's lectures at Edinburgh were attended by many of the students at the world-renowned Edinburgh medical school, and some of these students later played important roles in 19th century science.) Black formulated the doctrine of latent and specific heats; he showed that various substances absorb different amounts of heat when their temperature is raised by a given amount, and that they require large amounts of heat to be added during the process of melting or vaporization though there may be little change in temperature. Thus a clear distinction was made between *heat* and *temperature*. According to the caloric theory, a definite amount of caloric combines with each particle of matter when it is melted or liquefied.

Opinions differed on the question: does caloric have weight? Benjamin Thompson (1753–1814), an American who was created Count Rumford by the Elector Palatine of Bavaria and established the Royal Institution of Great Britain, was one of the scientists who studied this question. He performed a series of experiments before 1798 from which he concluded that the weight of heat could never be detected; this was one of his arguments against the caloric theory. Rumford also pointed out that an indefinite amount of heat can be produced from matter by mechanical work, for instance by friction in boring cannon. Similar arguments and experiments were adduced in 1799 by Humphry Davy (1778–1829), British chemist who discovered several elements, and was President of the Royal Society of London from 1820 to 1827.

The experiments of Rumford and Davy, which were considered much later to have disproved the caloric theory, were not at the time regarded as constituting serious objections to it. One reason was that Rumford and Davy had no coherent alternative theory; they did not explain in detail how heat, if it were simply molecular motion, could be transferred from one substance to another. The phenomena of radiant heat were frequently cited in support of the materiality of heat; since heat could be transferred across a " vacuum " without any accompanying movement of matter, it must be a substance and not a property of matter.

The chemical system worked out by the French chemists in 1787

(Guyton de Morveau, Lavoisier, Berthollet and de Fourcroy) included *calorique* in the list of elements; it was adopted by most British chemists along with Black's ideas. In following the history of the kinetic theory we shall be concerned with British and German science rather than French, so rather than examine the details of the French formulations of the caloric theory we shall quote an example of a popular British exposition (Selection 4). The extract is from the treatise *The Economy of Nature* by the Rev. George Gregory (1754–1808), Rector of West Ham, Essex. This book went through three editions very quickly (1796, 1798, 1804) and may thus be considered a widely-read English version of contemporary doctrines. While Gregory himself has no permanent reputation as a scientist, it is interesting to note that another of his popularizations, *Lectures on Experimental Philosophy and Chemistry* (1808) is credited with arousing the scientific interests of the American scientist Joseph Henry (1799?–1878), who made important discoveries in electricity.†

Laplace

To illustrate how the caloric theory was used to explain the properties of gases, we shall summarize the derivation of the gas laws given by Pierre Simon, Marquis de Laplace (1749–1827) in his *Traité de Mécanique Celeste*, Book XII (Paris, 1825). Laplace is known for his mathematical developments of Newton's theory of planetary motion, and also for his work on the theory of probability. His statement on the mechanical state of the universe, as being the effect of its anterior state and the cause of all future states, is often quoted as an example of the deterministic philosophy of science:

> " Given for one instant an intelligence which could comprehend all the forces by which nature is animated and the respective situation of the beings who compose it—an intelligence sufficiently vast to submit these data to analysis—it would embrace in the same formula the movements of the greatest bodies of the universe and those of the lightest atom; for it, nothing would be uncertain and the future, as the past, would be present to its eyes."

It is frequently forgotten, however, that this statement is made in an essay on probability theory, and leads up to a justification of the use

---

† T. COULSON, *Joseph Henry, His Life and Work*, Princeton University Press, 1950, pp. 14–15.

of *statistical* methods rather than deterministic ones.† Indeed, the kinetic theory of gases and statistical mechanics owe much to Laplace's development of probability theory, though his caloric theory has been rejected.

Whereas Newton had not attempted to explain the cause of the repulsive force between gas molecules, Laplace attributed it to the atmospheres of caloric which were assumed to surround them. He also thought it more reasonable that forces between molecules, unlike gravitational forces, are significant only at short distances. Denoting by c the amount of caloric in a molecule, he assumed that two similar molecules at a distance r would repel each other with a force equal to

$$F = Hc^2\phi(r) \qquad (1)$$

where H is a constant, probably (but not necessarily) the same for all gases, and $\phi(r)$ is a function that decreases rapidly as r increases. The total pressure in the gas due to such forces is

$$P = 2\pi HK\rho^2 c^2 \qquad (2)$$

where ρ is the density and K is the sum of the forces exerted on a given molecule by all the others. (K is finite even if the sum is extended to infinity because of the short-range character of the forces, and moreover its value is independent of the volume and shape of the container, if the latter is large enough.)

It might appear from equation (2) that the pressure is proportional to the square of the density, and hence the theory disagrees with the observed linear variation of pressure with density. However, Laplace now argues that c, the amount of caloric in a molecule, must depend on density because of the nature of the radiative equilibrium. Each molecule is continually sending out and receiving rays of caloric; the quantity of caloric rays that a surface receives in each time interval depends only on its temperature u, and is denoted by $\Pi(u)$; it is independent of the nature of the surrounding bodies. The surface absorbs a fraction q of these, where q depends on the nature of the molecules of the gas. The radiation of heat by a molecule is regarded as being caused by the action of the caloric of the surrounding molecules on its own caloric, and thus the amount of radiation is

---

† *A Philosophical Essay on Probabilities*, Dover, New York, p. 4.

proportional to the density and caloric of the surrounding gas, i.e. to ρc, and to the caloric of the molecule considered, c. This quantity, proportional to ρc^2, will have to be equal to the extinction at that temperature, $q\Pi(u)$, since the system as a whole is supposed to be in thermal equilibrium, with the total caloric remaining constant. Laplace thus obtains the relation

$$\rho c^2 = q'\Pi(u) \qquad (3)$$

where q' is another constant depending on the nature of the gas, while $\Pi(u)$ is a universal function of temperature. Combining equations (2) and (3), he gets the result

$$P = (2\pi HKq')\rho\Pi(u) = i\rho\Pi(u) \qquad (4)$$

the constant i being independent of density and temperature. The ideal gas laws follow immediately from (4): P is proportional to density at constant temperature (Boyle-Towneley law); and if the temperature changes from u to u', the density changes in the ratio

$$\frac{\rho'}{\rho} = \frac{\Pi(u)}{\Pi(u')}$$

regardless of the nature of the gas (Gay-Lussac's law). Since $\Pi(u)$ measures the density of radiant caloric in space, it is natural (according to Laplace) to equate it to the temperature: $\Pi(u) = u$.

One of the outstanding accomplishments of Laplace's caloric theory (with some later contributions by Poisson) is its explanation of adiabatic compression and of the velocity of sound. Laplace was able to show that in adiabatic compressions or expansions, PV^γ remains constant, where γ is the ratio of specific heat at constant pressure to that at constant volume. By assuming that the propagation of sound involves adiabatic rather than isothermal compressions and rarefactions, he was able to show theoretically that Newton's value for the velocity of sound should be multiplied by a factor $\sqrt{\gamma}$, thus bringing it into good agreement with experiment. (On the other hand, the theory predicts that the specific heat should increase with decreasing pressure or density. If experimental techniques had been good enough to test this prediction, the caloric theory might have been deposed much earlier.)

Herapath

This is not the place to analyse Laplace's theory from the viewpoint of modern physics; instead, we shall mention some of the objections which were raised against it by a contemporary kinetic theorist, John Herapath. Herapath himself is a rather interesting figure, who might even be regarded today as the founder of the kinetic theory except for the fact that his work had hardly any influence on later developments. He was born at Bristol in 1790, the son of a maltster, and had a scanty formal education. In 1820 he submitted to the Royal Society a manuscript entitled "A Mathematical Inquiry into the Causes, Laws and Principal Phenomena of Heat, Gases, Gravitation, etc." It contained, among other things, a comprehensive (if somewhat faulty) exposition of the kinetic theory of gases. Humphry Davy was primarily responsible for the fate of Herapath's memoir; he considered it too speculative and would not support its publication in the *Philosophical Transactions*. Herapath then withdrew the article and sent it instead to the *Annals of Philosophy* where it appeared in 1821. Despite all of Herapath's attempts to publicize his theory—including a number of letters to *The Times* (London), and articles in the *Railway Magazine* which he later edited—it received little notice from scientists until 35 years later, when the kinetic theory was revived by Joule, Krönig, Clausius and Maxwell. Joule was the only one of these who was directly influenced by Herapath (in particular, by Herapath's book on *Mathematical Physics* published in 1847), and Joule's own paper on the kinetic theory attracted no attention until the theory was again revived by Krönig in 1856. Herapath's failure to put over his theory indicates that the caloric theory was still too firmly entrenched in the first half of the 19th century, and even Davy, who presumably should have welcomed it, was still under the sway of older ideas about the constitution of matter and heat. It must be emphasized again that what needed to be established was not simply a connection between heat and molecular motion, for that was already admitted by many scientists and was not considered incompatible with the caloric theory; it was the notion that heat is *nothing but* molecular motion, and the idea that molecules *move freely through space* in gases rather than simply vibrating around fixed positions, that could not yet be accepted.

Herapath had sent " an announcement of the publication and objects " of his paper to Laplace in June 1821; and " on the 10th of the following September, this nobleman communicated to the Royal Academy of Sciences " in Paris a paper containing the theory we have summarized above. According to Herapath, " . . . it was obvious from the perfect coincidence of the object of M. Laplace's paper with that of a part of mine, and its being presented to the Royal Academy so long after the printing and notice of my paper, that his communication was in consequence of mine and intended to supersede it."† Perhaps this allegation of Herapath's should not be taken too seriously; years later he maintained that Davy's resignation from the Presidency of the Royal Society had been forced by his defeat in a controversy with Herapath, though there seems to be no reason to doubt the conventional account that Davy resigned because of his illness.

Herapath's first objection is to the assumption that the law of repulsion can be expressed as $Hc^2\phi(r)$, the force at a given distance varying as the square of the amount of caloric.

> " Supposing therefore the distance r between the particles to remain the same, the function $\phi(r)$ must involve the dimensions of the spheres or shells (of caloric surrounding the particles); consequently, as these dimensions would vary with the quantity of caloric, the repulsion would not be as c^2, as M. Laplace conceives; unless when the particles of caloric mutually repel one another by a force reciprocally proportional to the square of the distance, which would give the gas a very different law to that which experiment requires."

The second objection was to the arguments that led to equation (3). Laplace assumes that the radiation of caloric by a molecule is proportional, first, to its own caloric, and, second, to the caloric in the surrounding molecules.

> " Philosophers will, I presume, hardly . . . grant the . . . conclusion, that the radiation of a molecule is proportional to $\rho c \times c$. We might easily show that the surrounding molecules tend rather by their repulsion to compress the caloric of an enclosed molecule closer towards the centre, than to disperse it."

Yet even granting Laplace's argument that the radiation increases as the repulsive force, it should still be as $\rho c^2 \phi(r)$ rather than just ρc^2,

---

† *Phil. Mag.* **62**, 61 (1823).

since it has already been assumed that the repulsion is proportional to $\phi(r)$. The appropriate value of r will be proportional to the average distance of separation of the particles: $r = (a/\rho)^{1/3}$, a being some constant. But in this case equation (3), which states that the quantity ρc^2 is independent of density, is wrong, and one cannot derive the ideal gas law (4) from it.

In the same year, 1823, the French mathematician S. D. Poisson (1781–1840) published a paper on the caloric theory in which he derived Laplace's results without relying on a molecular model, assuming simply that the quantity of caloric in a gas depends only on its density and pressure, and that at constant pressure the caloric content is proportional to volume. Herapath translated this paper for the *Philosophical Magazine*, adding an introduction and critical notes. In the introduction Herapath says:

> "From this paper and those of M. Laplace it is plain with what ardour the subject of gases and heat is pursued on the continent. Would our English philosophers but lend their aid in the securer course of deciding some of the more important and disputed points by experiments, it is manifest we should speedily come to decisive conclusions respecting the nature and laws of heat . . ."†

French supremacy in the theory of heat was soon to come to an end, however. In 1824, Sadi Carnot (1796–1824) published his famous essay, *Réflexions sur la puissance motrice du feu*, in which he deduced from the caloric theory and the impossibility of perpetual motion the principle now known as the Second Law of Thermodynamics. Carnot was the last of the eminent school of French writers on heat; he completed the caloric theory and then transcended it, leaving in his notebook (unpublished until 1878) what is essentially a calculation of the mechanical equivalent of heat as well as a recognition of the equivalence of heat and work. Except for a paper by Carnot's classmate from the École Polytechnique, Emile Clapeyron, in 1834, French scientists and engineers ignored Carnot's work. It was only in Germany and England that his achievement was appreciated and incorporated into the modern theory of heat. (The history of thermodynamics will be treated in more detail in a separate volume of this series.)

---

† *Phil. Mag.* **62**, 328 (1823).

Waterston

John James Waterston (1811–83) was another British scientist who, like Herapath, worked out the elementary kinetic theory and some of its consequences, but did not succeed in obtaining any recognition for his work in his own lifetime. His theory goes beyond that of Herapath in its treatment of specific heats and the effects of expansion and contraction on the kinetic energy of a system of molecules. He also gave the first statement of the law of partition of energy among molecules of different weights in mixed gases. He explicitly identified the absolute temperature of a gas with the square of the velocity of its molecules, though he emphasized that the identification was only based on the *analogy* between the behaviour of the real gas and that of his ideal system of elastic spheres.

Like Herapath, Waterston sent his paper to the Royal Society, hoping that it would be published in the *Philosophical Transactions*. At that time the custom of the Royal Society was that a paper submitted by someone who was not a Fellow of the Society could be " read " (officially presented) if it were sponsored by a Fellow, but it then became the property of the Society and could not be returned to the author even if it were not published. However, the author could get permission to have a copy made for his own use if he wished. Herapath, on being informed by Davy that his paper would not be published though it could be read, preferred to withdraw it in order to submit it to another journal. Waterston, however, was in India at the time his paper was submitted for him by a friend, and either did not know the procedure regarding rejected papers, or did not bother to have a copy made for himself. The two referees who read his paper recommended that it should not be published. One of them, Baden Powell, who was Professor of Geometry at Oxford, said that Waterston's basic principle—that the pressure of a gas is due to the impacts of molecules against the surface of the container—was " . . . very difficult to admit, and by no means a satisfactory basis for a mathematical theory." The other referee was the astronomer Sir John William Lubbock; he said that ". . . the paper is nothing but nonsense, unfit even for reading before the Society."†

---

† *Phil. Trans. Roy. Soc. London* **183A**, 1–4 (1893). I am indebted to Sir Harold Hartley and N. H. Robinson for identifying the referees. The original letters are still on the Royal Society's archives. [*Cont. on next page.*]

The outcome was that Waterston's paper was read to the Royal Society in March 1846, and an abstract was printed in the *Proceedings of the Royal Society*, but the paper itself remained in the Society's archives and was generally unknown until 1892, when Lord Rayleigh discovered it and had it printed in the *Philosophical Transactions*. In an introductory note to the paper, Rayleigh discussed its history and concluded that such " highly speculative investigations, especially by an unknown author, are best brought before the world through some other channel than a scientific society " and that a young scientist should first establish his reputation " by work whose scope is limited, and whose value is easily judged, before embarking on higher flights". This was perhaps sound advice, yet it was natural that young scientists confident of the importance of their own work should try to secure recognition for it from the leading scientific institution of their country. Rayleigh's remarks are a reflection of the sad state to which the Royal Society had sunk in the 19th century, after its promising beginnings. Instead of encouraging new scientific advances, it had become such a formidable obstacle to progress that a new organization—the British Association for the Advancement of Science—had to be founded in 1831 to provide a forum for discussions of scientific matters by those who were not part of the established London elite. The Royal Society reformed itself by restricting its membership to men of definite scientific attainments (as opposed to aristocratic dilettantes) and eventually regained part of its reputation in the scientific world, but with the vast increase in the

---

[*Cont. from previous page.*]

The Secretary of the Royal Society at that time, who handled the correspondence concerning Waterston's paper and prepared the abstract that was printed in the *Proceedings* (vol. 5, p. 604) was Peter Mark Roget (1779–1869), now best known for his *Thesaurus of English Words and Phrases* (London, 1852, and many later editions). Roget had earlier read Herapath's paper and reported unfavorably on it to Davy. It is ironic that Roget's name also appears on some lists of early statements of the conservation of energy principles, because of his remarks about the impossibility of perpetual motion in his article on Galvanism published in 1832 (*Library of Useful Knowledge*, Vol. 2, Chap. VI, p. 32, par. 113; Baldwin and Cradock, London).

It would be of some interest to know why the few people who knew about Waterston's paper, such as Powell, Lubbock, and Roget, never called attention to it when the same theory was revived by others a few years later.

number of scientists, journals, and societies during the last century it would be impossible for it ever again to recover the monopolistic position it enjoyed in the 17th century.

Waterston did send an account of his theory to the 1851 meeting of the British Association at Ipswich, where it should have found a more sympathetic reception. Unfortunately the only scientist who paid any attention to it there was W. J. M. Rankine. Rankine was one of the leaders in the development of thermodynamics, but he was also strongly attached to the hypothesis of rotating molecular vortices, and he considered the Herapath–Waterston theory inadequate because simple translational motion of gas molecules could not account for all their thermal energy.†

THE CONSERVATION OF ENERGY

The principle of conservation of energy (generally called " force " at that time) was proposed independently by several scientists between 1842 and 1847. The major share of the credit is usually divided between the German physician Julius Robert Mayer (1814–78) and the British experimental physicist James Prescott Joule (1818–89). Both stated the general proposition that heat, mechanical work, electricity, and other apparently distinct entities are merely different forms of the same thing, now called " energy ". They also computed approximate values for the conversion coefficient of heat and work. Mayer emphasized the philosophical generality of the principle, while Joule provided the experimental verification in particular cases. From Joule's popular lecture reprinted here (Selection 6) we can see that he had recognized that the next logical step is the kinetic theory of matter, as proposed by Herapath, but Joule himself did not have sufficient mathematical talents to develop this theory. On the other hand, it may be noted that Mayer, and later writers such

---

† Rankine mentioned Waterston's theory (without a specific reference to where it could be found) in three of his own published papers: *Transactions of the Royal Society of Edinburgh* **20**, 561 (1853) and **25**, 559 (1869) [I am indebted to Dr. C. W. F. Everitt for calling my attention to these two references] and *Proceedings of the Glasgow Philosophical Society* **5**, 128 (1864). In 1876 Preston told Maxwell about Waterston's theory [letter found by Dr. Everitt in the Royal Society archives].

as Wilhelm Ostwald (1853–1932) and Ernst Mach (1838–1916) would not concede this further step, and preferred to avoid the assumption that heat is molecular motion.

The connection between conservation of energy and the kinetic theory of matter was established very clearly by the memoir " On the conservation of force " (Selection 7) by Hermann von Helmholtz (1821–94). Helmholtz was trained as a physician and taught physiology and physics at several German universities; unlike several other physicians who made contributions to physics, he had a firm grounding in mathematics. Helmholtz made important contributions to the analysis of the faculties of hearing and seeing, as well as to mathematical physics. His results in the theory of vortex motions in fluids were later used by Lord Kelvin and other British scientists as a foundation for the " vortex atom " theory, which enjoyed considerable popularity in the latter half of the 19th century. His studies of " monocyclic " mechanical systems influenced the work of Ludwig Boltzmann, and through him Willard Gibbs, on statistical mechanics.

We shall merely list the other scientists who have been credited with more or less general statements of the principle of conservation of energy, without attempting to evaluate their contributions: (1) Ludvig August Colding (1815–88), Danish engineer; (2) Karl Friedrich Mohr (1806–79), German chemist; (3) Sadi Carnot, mentioned on page 16; (4) Marc Seguin (1786–1875), French engineer; (5) Karl Holtzmann (1811–65), German physicist; (6) Gustav Adolphe Hirn (1815–90), Belgian engineer; (7) William Robert Grove (1811–96), English judge and physicist; (8) Justus Liebig (1803–73), German chemist; (9) Michael Faraday (1791–1867), English physicist. (See the Bibliography for references.) Some of these statements give one the impression that the author does not believe himself to be proposing a new physical principle, but is merely making explicit the current scientific view that perpetual motion is impossible, and force cannot be created or destroyed unless some kind of conversion takes place. The list could probably be increased indefinitely by adding the names of writers who made this type of statement.†

---

† See T. S. KUHN, pp. 321–56 in *Critical Problems in the History of Science* (ed. M. Clagett), University of Wisconsin Press, 1959.

The reasons for such a case of " simultaneous discovery " must clearly be sought in the general scientific and intellectual climate common to all these men. Among the reasons that have been suggested are: (1) discoveries of various conversion processes showing that many electrical, magnetic, thermal, mechanical and chemical phenoma must be related to each other; (2) attempts by theoretical physicists to carry out Newton's program of explaining all natural phenomena by reducing them to the motions and interactions of atoms; (3) scientific analysis of engines, in particular the development of the concept of " work " (and its economic importance); (4) the influence of the doctrines of German romanticism and *Naturphilosophie*, especially the idea that there must be a single unifying principle underlying all natural phenomena.

A complete description of these factors would be out of place here, although it is hoped that the reader will find them interesting topics for further study. It will be sufficient to point out their relevance to the works reprinted in this volume. Joule is best known for his experimental work on the relations between thermal, electrical and mechanical energy, and these researches followed directly on Volta's invention of the battery (1800), Oersted's discovery of electromagnetism (1820), Seebeck's discovery of thermo-electricity (1822), Peltier's discovery of the production or adsorption of heat by passage of currents at bimetallic junctions (1834), and Faraday's many discoveries in electricity and magnetism. While Joule's experiments, strictly interpreted, show only that the various forms of energy are interconvertible but do not appear to single out any one of them as more fundamental than the others, Joule's popular lecture " On Matter, Living Force, and Heat " (Selection 6) and Helmholtz's memoir " On the Conservation of Force " (Selection 7) indicate quite clearly that *mechanical* energy is regarded as the basic entity. It is this prejudice toward mechanical explanations that makes the kinetic theory appear to be an obvious consequence of the principle of conservation of energy; for if heat and mechanical energy are interconvertible, what is more natural than to conclude that heat *is* mechanical energy?

While the selections from the writings of Helmholtz and Joule should be easily comprehensible for any reader familiar with elementary physics, Mayer's paper (Selection 5) and the contributions of some of the other scientists mentioned above need to be interpreted in the light of the ideas of Friedrich Schelling (1775–1854) and his followers. Schelling and other *Naturphilosophen* taught that all forces in nature must be interrelated, and it is well known that Oersted, for example, was led to look for a connection between electricity and magnetism mainly because he believed that there must be such a connection on philosophical grounds. Similarly, Mayer's arguments are almost entirely philosophical: every force in nature is inherently a causal relation; cause is transformed into effect and must hence be equivalent to it. Indeed, it was probably the dislike of *Naturphilosophie* by empirical scientists like J. C. Poggendorff, editor of *Annalen der Physik*, that was responsible for the difficulty that Mayer, Helmholtz, and other conservationists encountered in getting their works published. Later in the 19th century there were various attacks on the kinetic theory by scientists such as Mach and Ostwald, who argued that attempts to explain phenomena in terms of the motions of invisible atoms were simply metaphysical or materialist speculation, and were not properly part of science. They rejected the argument that the interconvertibility of heat and mechanical energy proves their identity. It was not until after the experiments of Perrin and others on Brownian movement, in the first decade of the 20th century, that these critics were willing to accept the existence of atoms.

REVIVAL OF THE KINETIC THEORY

It was the overthrow of the caloric theory, and its replacement by the " dynamical " theory of heat, that created a situation favourable to the revival of the kinetic theory of gases. As we have already noted, it was natural that the old idea, that heat is itself nothing but the motion of the component parts of a body, should be resurrected; and it was not surprising that two of the leaders in the establishment of thermodynamics, the experimentalist Joule and the theoretician Clausius, should independently propose a kinetic theory. August Karl Krönig (1822–79), a German chemist, is usually credited with reviving the kinetic theory after 1850 and stimulating the further developments of Clausius and Maxwell. His short paper, published in Poggendorff's *Annalen der Physik* in 1856, seems to have had an influence very much out of proportion to its actual substance. This may perhaps be accounted for by the fact that Krönig was a fairly well-known scientist in Germany at the time; he was a Professor at the Realschule (technical high school) in Berlin, and was editor of *Fortschritte der Physik* (an annual review of progress in physics) for several years. He was presumably influential in the Physikalische Gesellschaft of Berlin. His paper represented no real advance over the work of Bernoulli and Herapath, though it seemed to be independent of them. Krönig deduced the ideal-gas law from the simplest assumption of perfectly elastic spheres moving parallel to three perpendicular axes with a common velocity. He also gave a rather elementary discussion of molecular velocities and specific heats of gases, but did not make any calculations or comparisons with experiment. Joule, in a paper read to the Manchester Literary and Philosophical Society in October 1848, had used Herapath's theory to calculate the velocity of a hydrogen molecule; he also attempted to derive the specific heat at constant volume. The paper was not published until 1851, and apparently remained unknown to most physicists until after Clausius mentioned it in his paper (see the footnote on the second page of Selection 8). Likewise it is doubtful whether anything would have come of the suggestions of Krönig if Clausius, who had already conceived similar ideas but had not published them, had not developed them further.

Clausius

Rudolf Clausius (1822–88) was one of the outstanding physicists of the 19th century. His most important contributions were his formulation of the second law of thermodynamics, and his development of the mechanical theory of heat on the basis of the concept of entropy. His first paper on the kinetic theory was published in 1857 (Selection 8), and his contributions were made over a period of about twenty years. He introduced the concept of the mean free path (Selection 9), which he used to give one of the first mathematical treatments of the conduction of heat in gases. He introduced the "virial theorem" which was the basis of later calculations of the pressure–volume–temperature relations ("equation of state") for imperfect gases, taking into account intermolecular forces. He suggested the hypothesis that electrolytes are dissociated in solution. To a large degree it is to Clausius that we owe the establishment of the kinetic theory as a bridge between the atomic theory and thermodynamics; he perceived not only that the atomic theory could thus be used to explain macroscopic thermodynamics, but also that experimental data on thermal properties of matter could be translated into specific statements about the nature of atoms themselves.

Clausius imposed on himself very strict standards of mathematical rigour; yet he did not allow these standards to stifle imagination and conjecture. As a result, his qualitative description of atomic motion avoided many of the pitfalls of dogmatic theorizing which marred the work of other contemporary scientists, although it included many ideas then unproved but essentially correct. Clausius also attempted to maintain a separation between his works on thermodynamics (the general theory of heat) and on kinetic theory (the particular theory) so that the possible failure of the latter would not impair the validity of the former. One consequence of this attitude was the conclusion of some later writers that if there is any conflict between the second law of thermodynamics and the predictions of the kinetic theory, then the latter must be discarded.

The ideal-gas theory was confronted with a very practical objection in 1858 by the Dutch meteorologist C. H. D. Buys-Ballot (1817–90). Buys-Ballot pointed out that if it were really true, as Joule, Krönig and Clausius claimed, that the molecules of a gas move at speeds of the order of several hundred metres per second, one would expect gases to diffuse and mix with each other very rapidly. In fact, however, if hydrogen sulphide or chlorine is evolved in one corner of a room, it may be several minutes before it is noticed in another corner of the room, although according to the kinetic theory each of the molecules should have traversed the room hundreds of times by then. Other common phenomena also appear to contradict the assertion that molecules move at high velocities; for example, carbon dioxide may remain for a long time in an open vessel.

Clausius realized that this was a valid objection to the theory as thus far developed, and he attempted to answer it by showing that in real gases, where the intermolecular forces are not negligible, the molecules could not travel for great distances in a straight line. In order to prove this, he introduced the concept of " mean free path " which he related to the range of influence of the repulsive forces exerted by a molecule (Selection 9). Although this derivation was a valuable contribution to the theory, it did not completely answer the original objection, since Clausius had no way of estimating the size of the sphere of action. He could only guess that a value of 1000:1 for the ratio of the mean free path (l) to the radius of the sphere of action (ρ) might be reasonable, and he thought that ρ itself was probably so small that this would still make l small compared with macroscopic dimensions.

The concept of mean free path which Clausius introduced turned out to be very useful in investigating many properties of gases, such as the conduction of heat, viscosity, and diffusion; indeed it is still the basis for expositions of the kinetic theory given in most modern textbooks. It is not completely satisfactory, especially if one wishes to take account of the fact that not all molecules in the gas have the same velocity, but instead the velocities vary according to some statistical distribution law. The mean free path of a molecule will depend on its velocity, and the velocity distribution law itself will vary from place to place in the gas if some process like heat conduction, viscosity, or diffusion is taking place. A more rigorous treatment was worked out by Maxwell, Boltzmann, Chapman and Enskog (selections reprinted in later volumes of this series); in most cases it was found that the mean-free-path theory predicted correctly the order of magnitude of the transport coefficients and their variation with density. However, the important phenomenon of thermal diffusion (which has been used for separating isotopes of uranium and of uranium hexafluoride) was not predicted by the mean-free-path theory at all; its discovery was one of the most important accomplishments of the Chapman–Enskog theory.

Maxwell

James Clerk Maxwell (1831–79) is best known as the discoverer of the laws of electromagnetism. His mathematical theory of the behaviour and interrelations of electric and magnetic fields is still a fundamental part of modern physics; it was found to need very little modification on the advent of quantum mechanics, and none at all to make it consistent with special relativity theory. Maxwell's theoretical prediction of electromagnetic waves was confirmed by Hertz's experiments in 1887, and became the basis for a large part of modern communications and technology.

It is to Maxwell that we owe the introduction of the statistical approach in the kinetic theory. Most modern theories of gases and liquids are based on the assumption of random molecular motions. The early kinetic theorists tended to ignore this property; they based their mathematical proofs on the assumption that all molecules move with the same velocity, and sometimes in addition that they are all arranged in a regular array in space. It was thought plausible that the results would still be the same if the molecules moved randomly, although this assumption later turned out to be wrong in some important cases (for example, Brownian motion).

Maxwell's basic hypothesis was that the numerous collisions between molecules in a gas, instead of tending to equalize the velocities of all the molecules, as some scientists had expected, would instead produce a statistical distribution of velocities in which all velocities might occur, with a known probability. Ludwig Boltzmann (1844–1906) later attempted to justify this hypothesis by means of his H-theorem, and derived an integro-differential equation for the evolution of the velocity-distribution function. (Boltzmann's work is reprinted in a later volume of this series.)

Maxwell's first thoughts on the kinetic theory are recorded in a letter to Sir George Gabriel Stokes, on 30 May, 1859:

> "I saw in the Philosophical Magazine of February, '59, a paper by Clausius on the 'mean length of path of a particle of air or gas between consecutive collisions', on the hypothesis of the elasticity of gas being due to the velocity of its particles and of their paths being rectilinear except when they come into close proximity to each other, which event may be called a collison. ... I thought that it might be worth while examining the hypothesis of free particles acting by impact and comparing it with phenomena which seem to depend on this 'mean path'. I have therefore begun at the beginning and drawn up the theory of the motions and collisions of free particles acting only by impact, applying it to internal friction of gases, diffusion of gases, and conduction of heat through a gas (without radiation). Here is the theory of gaseous friction with its results ..."

Then follows an outline of his theory, as published later and reprinted in Selection 10.

> "... I do not know how far such speculations may be found to agree with facts, even if they do not it is well to know that Clausius' (or

rather Herapath's) theory is wrong [footnote: i.e. inadequate] and at any rate as I found myself able and willing to deduce the laws of motion of systems of particles acting on each other only by impact, I have done so as an exercise in mechanics. Now do you think there is any so complete a refutation of this theory of gases as would make it absurd to investigate it further so as to found arguments upon measurements of strictly 'molecular' quantities before we know whether there be any molecules? One curious result is that μ is independent of the density, for

$$\mu = MNlv = \frac{Mv}{\sqrt{(2)}\pi s^2}.$$

This is certainly very unexpected, that the friction should be as great in a rare as in a dense gas. The reason is, that in the rare gas the mean path is greater, so that the frictional action extends to greater distances.

Have you the means of refuting this result of the hypothesis?

Of course my particles have not all the same velocity, but the velocities are distributed according to the same formula as the errors are distributed in the theory of least squares.

If two sets of particles act on each other the mean *vis viva* of a particle will become the same for both, which implies, that equal volumes of gases at same press. and temp. have the same number of particles, that is, are chemical equivalents. This is one satisfactory result at least.

I have been rather diffuse on gases but I have taken to the subject for mathematical work lately and I am getting fond of it and require to be snubbed a little by experiments, and I have only a few of Prof. Graham's, quoted by Herapath, on diffusion so that I am tolerably high-minded still . . ."

[From *Memoir and Scientific Correspondence of the late Sir George Gabriel Stokes, Bart.*, Cambridge University Press, 1909, Vol. II, pp. 8–11.]

Thus Maxwell's researches on the kinetic theory, which not only helped to establish the theory but also laid the foundations for modern statistical mechanics, were thought by Maxwell himself to be only an "exercise in mechanics" whose probable consequence would be to refute the kinetic theory by showing that it led to absurd predictions. Maxwell read this paper at a meeting of the British Association in September 1859, and in October he wrote again to Stokes, saying that he intended to arrange his propositions about the motions of elastic spheres "in a manner independent of the speculations about gases" and submit them to the *Philosophical Magazine* (*Memoir*, ibid. p. 11). The paper appeared in 1860 (Selection 10) and it will be observed from the last paragraph that in the theory of specific heats Maxwell had found more evidence that the kinetic theory of gases must be wrong.

The prediction that the viscosity is independent of density suggested a clear-cut experimental test of the validity of the kinetic theory, since the alternative static theory would certainly lead one to expect that the viscosity should increase with density (as it in fact does for a liquid). At that time no accurate experiments had yet been done on the viscosity of gases, and Boyle's experiment on the swings of a pendulum in an evacuated chamber (see above) had been forgotten, so Maxwell designed and carried out his own experiment. He found that the viscosity of air, at a given temperature, remained constant when the pressure was varied between $\frac{1}{2}$ in. and 30 in. This result, confirmed independently by O. E. Meyer, probably converted most of the scientists who had not already accepted the kinetic theory.

Clausius (Selection 8) had derived a general formula for the ratio of specific heats, from which it follows that this ratio should be equal to 5/3 for a monatomic gas regarded as a system of elastic spheres. If the molecules can rotate or vibrate, or if they are composed of smaller particles which must share the total energy of motion, then the ratio should be smaller. The general formula for the ratio is

$$\gamma = 1 + \frac{2}{3 + n},$$

where n is the number of "degrees of freedom" of the internal motions of the molecule. According to Newtonian mechanics, a system of n particles will have $3n$ degrees of freedom, corresponding to the three independent directions in space in which each particle may move. Out of the total of $3n$, three are counted as "translational" degrees of freedom, representing the motion of the center of mass of the molecule, and the rest are considered "internal" degrees of freedom. Thus for a diatomic molecule such as hydrogen, nitrogen, or oxygen, regarded as a mechanical system of two spheres held together by some kind of attractive forces, one would expect to have $n = 3$ and $\gamma = 1 \cdot 333 \ldots$.

At the time when Clausius and Maxwell wrote their first papers on the kinetic theory, the only available experimental evidence indicated $\gamma \approx 1.4$, which would correspond to $n = 2$. It was not until 1875 that Kundt and Warburg determined the ratio of specific heats of a monatomic gas, mercury vapour, and found that $\gamma \approx 1\frac{2}{3}$ in agreement with the formula for no internal degrees of freedom. Later on, in the eighteen-nineties, Ramsay and Rayleigh discovered the rare gases helium, argon, and krypton, and it was found that these monatomic gases also had specific heat ratios of about $1\frac{2}{3}$.

Clausius did not originally make the assumption that there would be equipartition of energy among all degrees of freedom; he merely asserted that the ratio of translatory to rotatory motion in gases would always be constant. Maxwell and later writers could not find any convincing reasons why the equipartition theorem should not apply to the internal motions of gas molecules, and the discrepancy between theoretical and experimental specific heats remained as an unsolved problem until the advent of the quantum theory. Not only was it difficult to understand why diatomic molecules should appear to have only five rather than six degrees of freedom, but there were many reasons for expecting that diatomic molecules should have many more than six degrees of freedom. The spectroscope was revealing that all molecules, even monatomic ones, must have a complex internal structure. If the parts of a molecule can move in such a way as to absorb and emit light of various frequencies, then they should also—according to the kinetic theory—have their proper share of the total energy of the molecule. It was also believed that light and other types of electromagnetic waves were propagated by means of an ether, which might be composed of discrete particles or might be similar to a continuous fluid or solid. In any case if the ether interacts with gas molecules in some way—as it apparently must in order to transmit their vibrations—then it must also share their energy of motion. In other words, in order to heat up a gas one must also heat up the ether surrounding the gas molecules.

The only way to make progress in the kinetic theory was to ignore these difficulties and hope someone would eventually work out a better theory of atoms and of the propagation of electromagnetic waves which would be consistent with the assumption that gas molecules behave in certain respects as if they were elastic spheres or simple point-centres of force. The particular problem concerning diatomic molecules was tentatively resolved in 1876, when R. H. M. Bosanquet in Oxford and Ludwig Boltzmann in Vienna independently proposed a mechanical model with five degrees of freedom. A molecule consisting of two rigidly connected mass-points, or in general any "solid of rotation" constructed by rotating a plane figure about its axis of symmetry, would have five degrees of freedom in a restricted sense: its motion of rotation around its axis is not changed by collisions with other similar molecules. This degree of freedom, while present, will not contribute to the specific heat, since its energy cannot reach equilibrium with the other degrees of freedom. Maxwell† objected to this explanation, but his criticism was never published (in any case he had nothing better to suggest) and the Boltzmann–Bosanquet model remained the only accepted one until the advent of quantum theory.

Maxwell's derivation of his velocity distribution (Selection 10) was not a rigorous proof, since it assumes that the probability distribution for each component of the velocity is independent of the values of the other components. Maxwell himself recognized the weakness of this proof and gave better ones in his later papers. Nevertheless the introduction of a statistical distribution of velocities was a great step forward in the kinetic theory and provided one of the basic elements of modern statistical mechanics.‡

THE VIRIAL THEOREM

The last selection in this volume (11) is a paper by Clausius in which is derived the theorem now known as the "virial theorem". The theorem states that the mean value of the kinetic energy in a

---

† Unpublished review of a book by Watson (1876), MS. at Cambridge University Library.

‡ For the historical background see C. C. Gillispie, pp. 431–53 in *Scientific Change* (ed. A. C. Crombie), Basic Books, New York, 1963. Dr. C. W. F. Everitt has pointed out the remarkable similarity of Maxwell's first derivation to Herschel's proof of the law of errors in the *Edinburgh Review*, July 1850; a letter from Maxwell to Campbell indicates that Maxwell may have read Herschel's article (see p. 143 of Campbell and Garnett's life of Maxwell).

system of material points is equal to the mean value of a quantity called the virial:

$$\Sigma \tfrac{1}{2}m \langle V^2 \rangle_{av} = - \tfrac{1}{2} \Sigma \langle Xx + Yy + Zz \rangle_{av}$$

where x, y and z represent the rectangular coordinates of the molecules, and X, Y and Z represent the respective components of the force acting on each molecule; the average value (denoted by: $\langle \ \rangle_{av}$) is taken over a time, in the case of periodic motion, equal to a complete period; or, in the case of irregular motion, sufficiently long that the mean value becomes constant. It is further assumed that the system is in " stationary motion ", i.e. that the points move within a limited space and the velocities do not change continuously in any particular direction.

In the special case of a gas acted on by an external pressure p which confines it to a volume v, the theorem leads to the equation

$$E = \tfrac{1}{2} \Sigma\, rf(r) + \tfrac{3}{2} pv$$

where $f(r)$ represents the force between two molecules at a distance r, and E denotes the mean *vis viva* (kinetic energy) of the internal motions. From this equation one may calculate the " equation of state " (relation between p, v and temperature T) if the force law $f(r)$ is known.

ATOMIC MAGNITUDES

By means of the virial theorem—or by even simpler arguments such as those of J. D. van der Waals—one can deduce a direct relation between the deviation from the ideal gas law and the size of a molecule. By combining this relation with the Maxwell formula relating viscosity to molecular size, it is possible to determine explicitly both the molecular diameter and the number of molecules in unit volume (Loschmidt number). As soon as the latter is known, one can also estimate the mass of a molecule.

By the latter part of the 19th century there was plenty of evidence from other sources to attest to the atomic nature of matter, but what could be more persuasive—if one can't actually *see* an atom—than the physicists' ability to specify an atom's diameter in centimetres, and its weight in grammes?

Although there had been a number of guesses of atomic magnitudes—some of them remarkably accurate—the first convincing estimate was made in 1865 by Josef Loschmidt (1821–95). He used the mean free path as deduced from viscosity measurements, together with an estimate of the fraction of the total gas volume occupied by the molecules themselves, obtained by comparing the density of liquid and gas. He concluded that the diameter of an air molecule is about 0·000001 mm (10 Å in modern units), which is about four times too large. The corresponding value of " Loschmidt's number ", viz. the number of molecules in a cubic centimetre of an ideal gas at standard conditions (0° C, 1 atm pressure) would be $N_L \approx 2 \times 10^{18}$. (Although Loschmidt himself did not give this result explicitly, it can easily be deduced from his formula.) The presently accepted value is $N_L = 2 \cdot 687 \times 10^{19}$. (Those who have been educated in English-speaking countries will be more familiar with " Avogadro's number " which is defined as the number of molecules per gram-mole, and is equal to $N_L/V_0 = 6 \cdot 025 \times 10^{23}$; V_0 is the standard volume of a perfect gas, 22,420·7 cm^3 atm mole^{-1}.)

Similar estimates were published soon afterwards by G. Johnstone Stoney (1868), Lothar Meyer (1867), Louis Lorenz (1870) and William Thomson (1870).† Thomson showed that several independent methods beside the kinetic-theory arguments of Loschmidt led to similar lower limits for atomic magnitudes. It should be noted that Thomson (later Lord Kelvin) enjoyed a very high scientific reputation in Britain during his own lifetime (1824–1907), and the support he gave to the kinetic theory in 1870 probably helped to establish it, even though he later questioned the validity of certain parts of the theory.

The revival of the atomic theory in chemistry during the second half of the 19th century, combined with the establishment of the kinetic theory in physics, produced a profound modification of scientific ideas about the structure of matter. Belief in atoms is one of the oldest and most persistent components of philosophical

---

† G. J. STONEY, *Phil. Mag.* [4] **36**, 132 (1868); L. MEYER, *Ann. Chem. Pharm.* (*Suppl.*) **5**, 129 (1867); L. LORENZ, *Vid. Selsk. Forh.* 40 (1870), English translation in *Phil. Mag.* [4] **40**, 390 (1870); W. THOMSON, *Nature* **1**, 551 (1870).

speculation about the physical universe, but it had never before made any real contact with experimental science. As Thomson remarked (1870),

> "The idea of an atom has been so constantly associated with incredible assumptions of infinite strength, absolute rigidity, mystical actions at a distance and indivisibility, that chemists and many other reasonable naturalists of modern times, losing all patience with it, have dismissed it to the realms of metaphysics, and made it smaller than ' anything we can conceive '."

But now for the first time there seemed to be good reason to hope that theories based on the assumption that atoms are particles of small but finite size, moving and colliding according to the laws of mechanics, could provide not only reasonable interpretations of macroscopic phenomena and predictions of new phenomena that could be tested by experiment, but also consistent values for atomic properties. As we now know, this hope could not be completely realized as long as the " laws of mechanics " were thought to be those of Newton. Nevertheless, the kinetic theory was so successful in most respects that its failures could not be explained simply by denying the existence of atoms. Instead, by pushing the application of Newtonian mechanics to its ultimate limits in the atomic realm, the kinetic theorists both established the reality of atoms as physical objects, and prepared the way for the modern quantum theory by revealing some of the inadequacies of the classical doctrines of mechanics. Indeed, quite apart from its success in explaining the properties of gases, we may regard the kinetic theory's most significant achievements as being the " discovery " of the physical atom and the elucidation of its properties.

Addendum to Volume 1

After the first volume of this series was prepared, historical research was published which showed that the relation between pressure and volume of a gas, commonly known as Boyle's Law, should be credited to Henry Power as well as to Towneley and Boyle. See C. Webster, *Nature* **197**, 226 (1963), *Arch. Hist. Exact Sci.* **2**, 441 (1965), and I. B. Cohen, *Nature* **204**, 618 (1964).

Bibliography

IN ORDER to help the reader to locate the major works mentioned in the introduction, detailed citations have been collected here in alphabetical order by author. We have attempted in all cases to give complete details of the original publications, together with recent reprints and English translations where they are available. While the average university library may not have some of the older scientific books and periodicals cited here, it should possess the collected works of all of the major scientists, so that there should be no difficulty in finding most of the items listed here in some form.

Since the purpose of the *Selected Readings in Physics* series is to encourage the use of original documents, we have given very few references to secondary sources. Here are a few such sources which may be consulted:

On Boyle, Newton, and the 17th century: J. B. CONANT, *Robert Boyle's Experiments in Pneumatics*, Harvard University Press, Cambridge, 1950; M. BOAS, *Osiris* **10**, 412 (1952); J. F. FULTON, *Isis* **18**, 77 (1932); A. WOLF, *A History of Science, Technology and Philosophy in the 16th and 17th Centuries*, Macmillan, New York, 2nd ed. 1950.

On the gas laws: W. S. JAMES, *Science Progress* **23**, 261 (1928); **24**, 57 (1929).

On the caloric theory: T. S. KUHN, *Isis* **49**, 132 (1958); S. C. BROWN, *Amer. J. Phys.* **18**, 367 (1950); F. CAJORI, *Isis* **4**, 483 (1922); S. LILLEY, *Arch. Int. Hist. Sci.* 630 (1948).

On the atomic theory: *Foundations of the Molecular Theory* (comprising papers and extracts by JOHN DALTON, J.-L. GAY-LUSSAC, and AMADEO AVOGADRO, 1808–11), *Alembic Club reprints*, No. 4, Livingstone, Edinburgh, 1950; A. G. VAN MELSEN, *From Atomos to Atom*, Duquesne University Press, Pittsburgh, 1952.

On energy conservation: T. S. KUHN, p. 321 in *Critical Problems in the History of Science* (ed. M. CLAGETT), University of Wisconsin Press, Madison, 1959.

On the 19th century: J. T. MERZ, *A History of European Thought in the Nineteenth Century*, Blackwood, Edinburgh, 4 vols., 1904–14.

On the kinetic theory: L. BOLTZMANN, *Lectures on Gas Theory*, University of California Press, Berkeley, 1964; L. B. LOEB, *Kinetic Theory of Gases*, McGraw-Hill, New York, 2nd ed. 1934; E. H. KENNARD, *Kinetic Theory of Gases*, McGraw-Hill, New York, 1938; S. G. BRUSH, *Annals of Science* **13**, 188, 273 (1957); **14**, 185, 243 (1958); *Amer. Scientist* **49**, 202 (1961); *Notes and Records of the Royal Society of London* **18**, 161 (1963).

General reference: H. T. PLEDGE, *Science since 1500*, H.M.S.O., London, 1939, reprinted by Harper, New York, 1959; J. C. POGGENDORFF, *Biographisch-Literarisches Handworterbuch zur Geschichte der exacten Wissenschaften*, J. A. Barth, Leipzig, 1863–1904, and later volumes under similar title; Royal Society of London, *Catalogue of Scientific Papers*, 1800–63, Eyre and Spottiswoode, London, 1867–72, and later volumes; G. SARTON et al., Critical Bibliographies in *Isis*, vols. 1 *et seq.* (1912 to date).

Journals frequently cited are abbreviated as follows:

Ann. Phys. = *Annalen der Physik*. Series 2, indicated by [2] before the volume number, is also cited as *Poggendorff's Annalen* or just *Pogg. Ann.* by contemporary writers.

Phil. Mag. = *The London, Edinburgh, and Dublin Philosophical Magazine and Journal.*

Phil. Trans. Roy. Soc. London = *Philosophical Transactions of the Royal Society of London.*

Wien. Ber. = *Sitzungsberichte der kaiserlichen Akademie der Wissenschaften in Wien, Klasse IIa.*

BLACK, JOSEPH. *Lectures on the Elements of Chemistry* (ed. J. ROBINSON). Creech, Edinburgh, 1803.

BOLTZMANN, LUDWIG. Weitere Studien über das Warmegleichgewicht unter Gasmolekülen. *Wien. Ber.* **66**, 275 (1872); reprinted in Boltzmann's *Wissenschaftliche Abhandlungen*, Barth, Leipzig, 1909, Vol. 1, p. 316. English translation to be published in another volume of *Selected Readings in Physics*.

BOLTZMANN, LUDWIG. Über die Natur der Gasmolecüle. *Wien. Ber.* **74**, 553 (1877); *Ann. Phys.* [2] **160**, 175 (1877); *Abhandlungen*, Vol. 2, p. 103. English translation in *Phil Mag.* [5] **3**, 320 (1877).

BOSANQUET, R. H. M. Notes on the theory of sound. *Phil. Mag.* [5] **3**, 271 (1877).

BOYLE, ROBERT. *A Defence of the Doctrine touching the spring and weight of the air, proposed by Mr. R. Boyle in his New Physico-Mechanical Experiments; Against the objections of Franciscus Linus, wherewith the Objector's Funicular Hypothesis is also examined.* Oxford, 1662. Reprinted in Boyle's *Works* (ed. T. BIRCH) London, 1744, second edition 1772, Vol. I, p. 118. (Warning: the section of this work in which the evidence for "Boyle's law" is presented is frequently reprinted in anthologies, but the following paragraph in which Boyle gives Towneley the credit for the hypothesis is usually omitted.)

BUYS-BALLOT, CHRISTIAN HEINRICH DIEDERICH. Ueber die Art von Bewegung, welche wir Wärme und Electricität nennen. *Ann. Phys.* [2] **103**, 240 (1858).

CARNOT, SADI. *Reflexions sur la puissance motrice du feu et sur les machines propres à développer cette puissance.* Bachelier, Paris, 1824. Reprinted with additional material by Blanchard, Paris, 1878. English translation by R. H. THURSTON, together with other papers by Clapeyron and Clausius, in *Reflections on the Motive Power of Fire*, etc. (ed. E. MENDOZA), Dover, New York, 1960.

CLAPEYRON, ÉMILE. Mémoire sur la puissance motrice de la chaleur. *Journal de l'Ecole Royale Polytechnique, Paris*, **14**, Cahier 23, p. 153 (1834). English translation by E. MENDOZA in *Reflections on the Motive Power of Fire*, etc., Dover, New York, 1960.

CLAUSIUS, RUDOLF. Ueber die bewegende Kraft der Wärme und die Gesetze die sich daraus für die Wärmelehre selbst ableiten lassen. *Ann. Phys.* [2] **79**, 368, 500 (1850); reprinted with other writings on thermodynamics in Clausius' *Abhandlungen über die mechanische Warmetheorie*, Part I, Vieweg, Braunschweig, 1864. English translation in *Phil. Mag.* [4] **2**, 1, 102 (1851) and in *Reflections on the Motive Power of Fire*, etc., Dover, New York, 1960.

COLDING, LUDVIG. *Nogle saetninger om Kraefterne* ("Some remarks on the forces") Lunos, Copenhagen, 1856 (read 1843); published with the 1856 volume of Oversigt over det Kongelige Danske Videnskabernes Forhandlingar. Translation into the international language "Ido" in Progreso, Agosto 1913. (See also: *Phil. Mag.* [4] **27**, 56 (1864); **42**, 1 (1871); *Isis* **1**, 522 (1913).)

DAVY, HUMPHRY. Essay on Heat, Light, and the Combinations of Light. *Contributions to physical and medical knowledge, primarily from the West of England* (collected by T. BEDDOES) Bristol, 1799. Reprinted in Davy's *Collected Works*, Smith, Elder & Co., London, 1839, Vol. 2. p. 5.

DESCARTES, RENE. *Principia Philosophiae.* Elzevir, Amsterdam, 1644. Reprinted in Descartes' *Oeuvres,* Cerf, Paris, 1905, Vol. 8, first part. French translation (Paris, 1647) reprinted in *Oeuvres,* Vol. 9, second part.

EULER, LEONHARD. Tentamen explicationis phaenomenorum aeris. *Commentarii Academiae Scientiarum Imperialis Petropolitanae* **2**, 347 (1727). (See R. HOOYKAAS, *Archives Internationales d'Histoire des Sciences* **2**, 180 (1948).)

FARADAY, MICHAEL. On the source of power in the voltaic pile. *Phil. Trans. Roy. Soc. London* **130**, 93 (1840); reprinted with many other papers in Faraday's *Experimental Researches in Electricity,* Quaritch, London, 1844, Vol. II, p. 59. (See pp. 125–26 or pp. 102–3, respectively.)

FOURCROY, ANTOINE FRANCOIS DE: *See* MORVEAU.

GALILEI, GALILEO. *Discorsi e dimostrazioni matematiche, intorno a due nuoue scienze, attenenti alla mecanica & i movimenti locali.* Elsevir, Leiden, 1638. English translation by H. CREW and A. de SALVIO: *Dialogues Concerning Two New Sciences,* Macmillan, New York, 1914.

GROVE, WILLIAM. *On the correlations of Physical Forces: Being the substance of a course of lectures delivered in the London Institution in the year 1843.* Skipper and East, London, 1846. Reprinted, with papers by Mayer, Helmholtz, Faraday, Liebig and Carpenter, in *The Correlation and Conservation of Forces* (ed. E. L. YOUMANS). Appleton, New York, 1868.

GUERICKE, OTTO VON. *Experimenta nova (ut vocantur) Magdeburgica de vacuo spatio primùm à R. P. Gaspare Schotto.* Jansson, Amsterdam, 1672. German translation by F. Danneman: *Otto von Guericke's neue "Magdeburgische" Versuche über den leeren Raum (1672).* Engelmann, Leipzig, 1894.

GUYTON DE MORVEAU: *See* MORVEAU.

HERAPATH, JOHN. A mathematical inquiry into the causes, laws and principal phenomena of heat, gases, gravitation, etc. *Annals of Philosophy* [2] **1**, 273, 340, 401 (1821).

HERAPATH, JOHN. Tables of temperature, and a mathematical development of the causes and laws of phenomena which have been adduced in support of hypotheses of "Calorific Capacity," "Latent Heat," etc. *Annals of Philosophy* [2] **2**, 50, 89, 201, 256, 363, 434; **3**, 16 (1821).

HERAPATH, JOHN. Observations on M. Laplace's Communication to the Royal Academy of Sciences, "Sur l'Attraction des Sphères, et sur la Repulsion des Fluides Elastiques." *Phil. Mag.* **62**, 61, 136 (1823).

HERAPATH, JOHN. *Mathematical Physics; or the Mathematical Principles of Natural Philosophy: with a development of the causes of Heat, Gaseous Elasticity, Gravitation and other great phenomena of Nature.* Whittaker & Co. and Herapath's Railway Journal Office, London, 1847 (2 vols.).

HIRN, GUSTAV ADOLPHE. Études sur les principaux phénomènes que presentent les frottements médiats, et sur les diverses manières de déterminer la valeur mécanique des matières employées au graissage des machines. Notice sur les lois de la production de calorique par les frottements médiats. *Bulletin de la Societé Industrielle de Mulhouse* **26**, 188, 238 (1854).

HOBBES, THOMAS. *Dialogus Physicus, sive de Natura Aeris conjectura sumpta ab experimentis nuper Londini habitis in Collegio Greshamensi.* London, 1661.

HOLTZMANN, KARL. Über die Warme und Elasticität der Gase und Dampfe. Mannheim, 1845. English translation by W. FRANCIS in *Scientific Memoirs* (ed. R. TAYLOR), R. and J. Taylor, London, 1846, Vol. 4, p. 189.

HUYGENS, CHRISTIAAN. De motu corporum ex percussione. Published posthumously in his *Opera Reliqua*, Janssonio-Waesbergios, Amtelodami, 1728, p. 75; reprinted with French translation in *Oeuvres Complètes*, Nijhoff, La Haye, 1888–1950, Vol. 16, p. 30. The results were announced in *Journal des Sçavans* **2**, 531 (1669); *Phil. Trans. Roy. Soc. London* **4**, 925 (1669); Lowthrop's abridged *Philosophical Transactions* (London, 1749) Vol. I, p. 460.

JOULE, JAMES PRESCOTT. On the calorific effects of magneto-electricity and the mechanical value of heat. British Association reports, 13th meeting, p. 33 (1843), reprinted in *Isis* **13**, 41 (1929). A longer paper with the same title appeared in *Phil. Mag.* [3] **23**, 263, 347, 435 (1843); reprinted in *The Scientific Papers of James Prescott Joule*, The Physical Society, London, 1884, p. 123.

JOULE, JAMES PRESCOTT. On the existence of an equivalent relation between heat and the ordinary forms of mechanical power. *Phil. Mag.* [3] **27**, 205 (1845); *Scientific Papers* p. 203.

JOULE, JAMES PRESCOTT. Some remarks on heat, and the constitution of elastic fluids. *Memoirs of the Literary and Philosophical Society of Manchester* [2] **9**, 107 (1851) (read in 1848); *Phil. Mag.* [4] **14**, 211 (1857); *Scientific Papers* p. 290. A summary appeared in the British Association reports, 18th meeting, part 2, p. 21 (1848); *Scientific Papers* p. 288.

KELVIN, WILLIAM THOMSON (LORD). The size of atoms. *Nature* **1**, 551 (1870). Reprinted many times, including p. 495 in Thomson and Tait's *Principles of Mechanics and Dynamics* (originally titled *Treatise on Natural Philosophy*), Part 2, Dover, New York, 1962.

KRÖNIG, AUGUST KARL. *Grundzüge einer Theorie der Gase.* Hayn, Berlin, 1856. Reprinted in *Ann. Phys.* [2] **99**, 315 (1856).

KUNDT, AUGUST; WARBURG, EMIL. Ueber die specifische Wärme des Quecksilbergases. *Berichte der Deutschen Chemischen Gesellschaft (Berlin)* **8**, 945 (1875); *Ann. Phys.* [2] **157**, 353 (1876).

LAPLACE, PIERRE SIMON, MARQUIS DE. *Essai philosophique sur les probabilities.* Paris, 1814. English translation from the 6th French edition by F. W. TRUSCOTT and F. L. EMORY: *A Philosophical Essay on Probabilities.* Chapman and Hall, London, 1902; reprinted by Dover, New York, 1951.

LAPLACE, PIERRE SIMON, MARQUIS DE. Sur l'attraction des spheres et sur la répulsion des fluides élastiques. *Annales de Chimie* **18**, 181 (1821); revised version, including theory of velocity of sound, in Livre XII (1823) of his *Traité de Mecanique Celeste*, Duprat, Paris, 1798–1825; reprinted in *Oeuvres Complètes de Laplace*, Gauthier-Villars, Paris, **5**, 99 (1882); see also **13**, 272 (1904); **14**, 305 (1912). *Mécanique Céleste*, English translation with commentary by N. BOWDITCH; Hillard, Gray, Little & Wilkins, Boston, 1829–39.

LAVOISIER, ANTOINE. See MORVEAU.

LIEBIG, JUSTUS. Zehnter Briefe. *Chemische Briefe*, Winter, Heidelberg, 1844, p. 114.

LINUS, FRANCISCUS, alias Francis Hall. *Tractatus de corporum inseparabilitate, in quo experimenta de vacuo tam Torricelliana quam Magdeburgica et Boyliana examinantur.* Martin, London, 1661.

LOSCHMIDT, JOSEF. Zur Grösse der Luftmolecule. *Wien. Ber.* **52**, 395 (1865).

MACH, ERNST. *Die Geschichte und die Wurzel des Satzes von der Erhaltung der Arbeit.* Calve, Prague, 1872. English translation by P. E. B. JOURDAIN, *History and Root of the Principle of Conservation of Energy.* Open Court, Chicago, 1911. See also a later version: On the principle of the conservation of energy, *The Monist* **5**, 22 (1894); reprinted in *Popular Scientific Lectures*, Open Court, Chicago, 1894, 5th ed. 1943, p. 137.

MARIOTTE, EDMÉ. *Essay de la nature de l'air.* Paris, 1679. Reprinted in Mariotte's *Oeuvres*, P. Vander Aa, Leiden, 1717, and J. Neaulme, La Haye, 1740, p. 148.

MAXWELL, JAMES CLERK. Viscosity or internal friction of air and other gases. *Phil. Trans. Roy. Soc. London* **156**, 249 (1866); reprinted in *The Scientific Papers of James Clerk Maxwell*, Cambridge University Press, 1890, Vol. II, p. 1.

MOHR, KARL FRIEDRICH. Ueber die Natur der Wärme. *Zeitschrift für Physik, Mathematik, und verwandte Wissenschaften* (hrsg. A. Baumgartner, Wien) [2] **5**, 419 (1837); summary in *Annalen der Chemie und Pharmacie* **24**, 141 (1837). English translation, with anti-Mayer comments, by P. G. Tait, in *Phil. Mag.* [5] **2**, 110 (1876). (See J. T. MERZ, *A History of European Thought in the Nineteenth Century*, Blackwood, Edinburgh, 1912, Vol. 2, p. 107.)

MORVEAU, LOUIS BERNARD GUYTON DE; LAVOISIER, ANTOINE; BERTHOLLET, CLAUDE LOUIS; FOURCROY, ANTOINE FRANCOIS DE. *Méthode de Nomenclature Chimique.* Paris, 1787. English translation by J. ST

JOHN: *Method of Chymical Nomenclature.* Kearsley, London, 1788. Another translation by G. PEARSON, London, 1794; 2nd ed. 1799.

OERSTED, HANS CHRISTIAN. *Experimenta Circa Effectum Conflictus Electrici in Acum Magneticam.* Hafniae, 1820. English translation in *Annals of Philosophy* **16**, 273 (1820). (See B. DIBNER, *Oersted and the Discovery of Electromagnetism*, Burndy Library, Norwalk, Conn., 1961.)

OSTWALD, WILHELM. Die Ueberwindung des wissenschaftlichen Materialismus. Verhandlungen der Gesellschaft deutscher Naturforscher und Aerzte (Leipzig) Th. 1, p. 155 (1895); reprinted in his *Abhandlungen und Vorträge allgemeinen Inhaltes* (1887–1903), Veit, Leipzig, 1904, p. 220. English translation in *Science Progress* **4**, 419 (1896).

PASCAL, BLAISE; PERIER, FLORIN. *Récit de la grande expérience de l'equilibre des liqueurs, projetée par le sieur B. P. pour l'accomplissement du traité qu'il a promis dans son abrégé touchant le vide et faite par le sieur F. P. en une des plus hautes montagnes d'Auvergne.* Savreuz, Paris, 1648. Reprinted, with Pascal's other works on hydrostatics, in his *Oeuvres* (ed. L. BRUNSCHVICG and P. BOUTROUX), Hachette, Paris, 1908–21, Vol. 2, pp. 147, 349; also in *Oeuvres Complètes* (ed. J. CHEVALIER), Gallimard, Paris, 1954, p. 392. (See also: *The Physical Treatises of Pascal: the equilibrium of liquids and the weight of the mass of the air*, translated by I. H. B. and A. G. H. SPIERS, with introduction and notes by F. BARRY; Columbia University Press, New York, 1937. This includes related writings by Simon Stevin, Galileo Galilei, and Torricelli.)

PELTIER, JEAN CHARLES ATHANASE. Nouvelles expériences sur la caloricité des courants electriques. *Annales de Chimie* **56**, 371 (1834). (See also: *Annales de Chimie* **60**, 261 (1835); *Comptes Rendus de l'Academie des Sciences (Paris)* **1**, 470 (1835).)

PERIER, FLORIN: *See* PASCAL.

POISSON, SIMON DENIS. Sur la chaleur des gaz et des vapeurs. *Annales de Chimie* **23**, 337 (1823); English translation, with notes, by J. HERAPATH, *Phil. Mag.* **62**, 328 (1823).

RAMSAY, WILLIAM. Sur l'argon et l'helium. *Comptes Rendus de l'Academie des Sciences (Paris)* **120**, 1049 (1895).

RAMSAY, WILLIAM; TRAVERS, MORRIS WILLIAM. On a new constituent of atmospheric air; On the companions of argon. *Proceedings of the Royal Society (London)* **63**, 405, 437 (1898).

RAYLEIGH, JOHN WILLIAM STRUTT (LORD); RAMSAY, WILLIAM. Argon, a new constituent of the atmosphere. *Phil. Trans. Roy. Soc. London* **186**, 187 (1896); reprinted in Rayleigh's *Scientific Papers*, Cambridge University Press, 1899–1920, Vol. 4, p. 130.

RUMFORD, BENJAMIN THOMPSON (COUNT). An inquiry concerning the source of the heat which is excited by friction. *Phil. Trans. Roy. Soc.*

London **80** (1798); reprinted in his *Essays, Political, Economical, and Philosophical*, new edition, Cadell and Davies, London, 1800, vol. 2, p. 469; and in his *Complete Works*, American Academy of Arts and Sciences, Boston, 1870–75, Vol. 2, p. 469. Reprinted in part in *Classics of Modern Science* (ed. W. S. KNICKERBOCKER), Beacon Boston, 1962, p. 157.

SCHELLING, FRIEDRICH WILLIAM JOSEPH VON. *Sämmtliche Werke*. Cotta, Stuttgart and Augsburg, 1856–61, 14 vols. *Schellings Werke, nach der Original-ausgabe in neuer Anordnung*. Beck, München, 1927, reprinted 1958–59, 6 vols. The chief works relevant to science, which may be found along with many others in both these collections, are: *Ideen zu einer Philosophie der Natur* (1797); *Von der Weltseele* (1798); *Erster Entwurf eines Systems der Naturphilosophie* (1799); *Allgemeine Deduktion des dynamischen Processes* (1800); *Darstellung meines Systems der Philosophie* (1801). (See E. BREHIER, *Schelling*, Alcan, Paris, 1912, pp. 26–118.)

SEEBECK, THOMAS JOHANN. Magnetische Polarisation der Metalle und Erze durch Temperatur-Differenz. *Abhandlungen der Königlichen Akademie der Wissenschaften (Berlin)* 265 (1822–23); reprinted as *Ostwald's Klassiker der exakten Wissenschaften*, Nr. 70; Engelmann, Leipzig, 1895. See also *Ann. Phys.* [2] 6, 1, 133, 253 (1826).

SEEBECK, THOMAS JOHANN. Ueber den Magnetismus der galvanischen Kette. *Abhandlungen der Königlichen Akademie der Wissenschaften (Berlin)* 289 (1820–21); reprinted in *Zur Entdeckung des Electromagnetismus* (hrsg. A. J. V. OETTINGEN), Engelmann, Leipzig, 1895.

SEGUIN, MARC. On the effects of heat and motion. *Edinburgh Journal of Science* 3, 276 (1825).

SEGUIN, MARC. *De l'Influence des Chemins der Fer et de l'Art de les tracer et de les construire*. Carilian-Goeury, Paris, and Société Typographique Belge, Bruxelles, 1839. (In the Paris edition the remarks on heat are on pp. xvi–xviii and 378–422; in the Bruxelles edition, pp. ix–x and 243–71.)

THOMPSON, BENJAMIN: *See* RUMFORD.

TORRICELLI, EVANGELISTA. Esperienza dell'Argento vivo (Correspondence with Ricci: June 11, 18 and 28, 1644). See his *Opere*, Montanari, Faenza, 1919, Vol. 3, pp. 186–8, 193–5, 198–201, where letters to other scientists announcing Torricelli's discoveries may also be found. English translation by V. CIOFFARI in *The Physical Treatises of Pascal*, etc., Columbia University Press, New York, 1937, p. 163. A formal account was published by the Accademia del Cimento: *Saggi di naturali Esperienze fatte nell'Accademia del Cimento*, Cocchini, Firenze, 1666. This is reprinted, with the above-mentioned letters, in *Neudrucke von Schriften und Karten über Meteorologie und Erdmagnetismus* (ed. G. HELLMANN) No. 7; Asher, Berlin, 1897.

TRAVERS, MORRIS WILLIAM: *See* RAMSAY.

VOLTA, ALESSANDRO. On the electricity excited by the mere contact of conducting substances of different kinds (in French). *Phil. Trans. Roy. Soc. London* 403 (1800). Reprinted in his *Opere* (Edizione Nazionale), Hoepli, Milano, 1918, Vol. I, p. 563.

WALLIS, JOHN. A summary account given by Dr. John Wallis of the general laws of motion. *Phil. Trans. Roy. Soc. London* 3, no. 43, p. 864 (1668); Lowthrop's abridged *Philosophical Transactions*, London, 1749, Vol. I, p. 457. See also his *Opera Mathematica*, E Theatro Sheldoniano, Oxoniae, 1699, Vol. I, p. 573.

WARBURG, EMIL: *See* KUNDT.

WATERSTON, JOHN JAMES. On the physics of media that are composed of free and perfectly elastic molecules in a state of motion. *Proceedings of the Royal Society (London)* 5, 604 (1846) (abstract); *Phil. Trans. Roy. Soc. London* **183A**, 5 (1893). Reprinted in *The Collected Scientific Papers of John James Waterston* (ed. J. S. HALDANE), Oliver and Boyd, Edinburgh, 1928, p. 207.

WATERSTON, JOHN JAMES. On a general theory of gases. British Association report, 21st meeting, Transactions of the Sections, p. 6 (1851); reprinted in *Phil. Trans. Roy. Soc. London* **183A**, 79 (1893) and *Papers*, p. 318.

WREN, CHRISTOPHER. Lex Naturae de Collisione Corporum. *Phil. Trans. Roy. Soc. London* 3, no. 43, p. 867 (1668); Lowthrop's abridged *Philosophical Transactions*, London, 1749, Vol. I, p. 459.

1

The Spring of the Air*

ROBERT BOYLE

SUMMARY†

A qualitative atomic theory of the " spring " of the air—i.e. its property of resisting compression by exerting pressure on a surface in contact with it—is proposed. Boyle conceives the air to be similar to " a heap of little bodies, lying one upon another, as may be resembled to a fleece of wool ". Each body is like a little spring, which can be easily bent or rolled up, but also tries to stretch itself out again. This tendency to expand is characteristic not only of air that has been compressed, but also of the ordinary air in the atmosphere, which has to support the weight of a column of air many miles in height above it. In support of this theory, the experiment of Pascal is mentioned, in which it was shown that the height of mercury in a barometer is less at the top of a mountain than at the bottom. (Boyle assumes that the mercury is supported by air pressure rather than by nature's abhorrence of a vacuum.)

For the more easy understanding of the experiments triable by our engine, I thought it not superfluous nor unreasonable in the recital of this first of them, to insinuate that notion, by which it seems likely, that most, if not all of them, will prove explicable. Your Lordship will easily suppose, that the notion I speak of is, that there is a spring, or elastical power in the air we live in. By which $\dot{\epsilon}\lambda\alpha\tau\eta\rho$ or

---

* The following selection is from the book *New Experiments Physico-Mechanical, touching the spring of the air, and its effects; made, for the most part in a new pneumatical engine. Written by way of letter to the right Honourable Charles Lord Viscount of Dungarvan, eldest son to the Earl of Corke.* Oxford, 1660. We have used the revised edition of Boyle's *Works*, published by Thomas Birch, London, 1772; the passage quoted is from pp. 11–15.

† All Summaries are by the Author.

spring of the air, that which I mean is this; that our air either consists of, or at least abounds with, parts of such a nature, that in case they be bent or compressed by the weight of the incumbent part of the atmosphere, or by any other body, they do endeavour, as much as in them lieth, to free themselves from that pressure, by bearing against the contiguous bodies that keep them bent; and, as soon as those bodies are removed, or reduced to give them way, by presently unbending and stretching out themselves, either quite, or so far forth as the contiguous bodies that resist them will permit, and thereby expanding the whole parcel of air, these elastical bodies compose.

This notion may perhaps be somewhat further explained, by conceiving the air near the earth to be such a heap of little bodies, lying one upon another, as may be resembled to a fleece of wool. For this (to omit other likenesses betwixt them) consists of many slender and flexible hairs; each of which may indeed, like a little spring, be easily bent or rolled up; but will also, like a spring, be still endeavouring to stretch itself out again. For though both these hairs, and the aëreal corpuscles to which we liken them, do easily yield to external pressures; yet, each of them (by virtue of its structure) is endowed with a power or principle of self-dilatation; by virtue whereof, though the hairs may by a man's hand be bent and crowded closer together, and into a narrower room than suits best with the nature of the body; yet, whilst the compression lasts, there is in the fleece they compose an endeavour outwards, whereby it continually thrusts against the hand than opposes its expansion. And upon the removal of the external pressure, by opening the hand more or less, the compressed wool doth, as it were, spontaneously expand or display itself towards the recovery of its former more loose and free condition, till the fleece hath either regained its former dimensions, or at least approached them as near as the compressing hand (perchance not quite opened) will permit. This power of self-dilatation is somewhat more conspicuous in a dry spunge compressed, than in a fleece of wool. But yet we rather chose to employ the latter on this occasion, because it is not, like a spunge, an entire body, but a number of slender and flexible bodies, loosely complicated, as the air itself seems to be.

There is yet another way to explicate the spring of the air:

namely, by supposing with that most ingenious gentleman, Monsieur *Des Cartes*,† that the air is nothing but a congeries or heap of small and (for the most part) of flexible particles, of several sizes, and of all kind of figures, which are raised by heat (especially that of the sun) into that fluid and subtle ethereal body that surrounds the earth; and by the restless agitation of that celestial matter, wherein those particles swim, are so whirled round, that each corpuscle endeavours to beat off all others from coming within the little sphere requisite to its motion about its own centre; and in case any, by intruding into that sphere, shall oppose its free rotation, to expel or drive it away: so that, according to this doctrine, it imports very little, whether the particles of the air have the structure requisite to springs, or be of any other form (how irregular soever) since their elastical power is not made to depend upon their shape or structure, but upon the vehement agitation, and (as it were) brandishing motion, which they receive from the fluid æther, that swiftly flows between them, and whirling about each of them (independently from the rest) not only keeps those slender aëreal bodies separated and stretched out (at least, as far as the neighbouring ones will permit) which otherwise, by reason of their flexibleness and weight, would flag or curl; but also makes them hit against, and knock away each other, and consequently require more room than that, which, if they were compressed, they would take up.

By these two differing ways, my Lord, may the springs of the air be explicated. But though the former of them be that, which by reason of its seeming somewhat more easy, I shall for the most part make use of in the following discourse; yet am I not willing to declare peremptorily for either of them against the other. And indeed, though I have in another treatise endeavoured to make it probable, that the returning of elastical bodies (if I may so call them) forcibly bent, to their former position, may be mechanically explicated; yet I must confess, that to determine whether the motion of restitution in bodies proceed from this, that the parts of a body of a peculiar structure are put into motion by the bending of the spring, or from the endeavour of some subtle ambient body, whose passage

---

† See Descartes' *Principia Philosophiae* (1644), Pars Quarta, §§ 29–47; *Oeuvres*, Vol. 8, pp. 218–31 (reprint of the original Latin) and Vol. 9, pp. 215–27 (French translation).

may be opposed or obstructed, or else its pressure unequally resisted by reason of the new shape or magnitude, which the bending of a spring may give the pores of it: to determine this, I say, seems to me a matter of more difficulty, than at first sight one would easily imagine it. Wherefore I shall decline meddling with a subject, which is much more hard to be explicated than necessary to be so by him, whose business it is not, in this letter, to assign the adequate cause of the spring of the air, but only to manifest, that the air hath a spring, and to relate some of its effects.

I know not whether I need annex, that though either of the abovementioned hypotheses, and perhaps some others, may afford us an account plausible enough of the air's spring; yet I doubt, whether any of them gives us a sufficient account of its nature. And of this doubt I might here mention some reasons, but that, peradventure, I may (God permitting) have a fitter occasion to say something of it elsewhere. And therefore I should now proceed to the next experiment, but that I think it requisite, first, to suggest to your Lordship what comes into my thoughts, by way of answer to a plausible objection, which I foresee you may make against our proposed doctrine, touching the spring of the air. For it may be alledged, that though the air were granted to consist of springy particles (if I may so speak) yet thereby we could only give an account of the dilatation of the air in wind-guns, and other pneumatical engines, wherein the air hath been compressed, and its springs violently bent by an apparent external force; upon the removal of which, it is no wonder, that the air should, by the motion of restitution, expand itself till it hath recovered its more natural dimensions: whereas, in our abovementioned first experiment, and in almost all others triable in our engine, it appears not, that any compression of the air preceded its spontaneous dilatation or expansion of itself. To remove this difficulty, I must desire your Lordship to take notice, that of whatever nature the air, very remote from the earth, may be, and whatever the schools may confidently teach to the contrary, yet we have divers experiments to evince, that the atmosphere we live in is not (otherwise than comparatively to more ponderous bodies) light, but heavy. And did not their gravity hinder them, it appears not why the streams of the terraqueous globe, of which our air in great part consists, should not rise much higher, than the refractions of the sun, and other stars, give men ground to think, that the atmosphere, (even in the judgment of those recent astronomers, who seem willing to enlarge its bounds as much as they dare,) doth reach.

But lest you should expect my seconding this reason by experience; and lest you should object, that most of the experiments, that have been proposed to prove the gravity of the air, have been either barely proposed, or perhaps not accurately tried; I am content, before I pass further, to mention here, that I found a dry lamb's bladder containing near about two thirds of a pint, and compressed by a packthread tied about it, to lose a grain and the eighth part of a grain of its former weight, by the recess of the air upon my having prickt it: and this with a pair of scales, which, when the full bladder and the correspondent weight were in it, would manifestly turn either way with the 32^d part of the grain. And if it be further objected, that the air in the bladder was violently compressed by the packthread and the sides of the bladder, we might probably (to wave prolix answers) be furnished with a reply, by setting down the differing weight of our receiver, when emptied, and when full of uncompressed air, if we could here procure scales fit for so nice an experiment; since we are informed, that in the German experiment, commended at the beginning of this letter, the ingenious triers of it found, that their glass vessel, of the capacity of 32 measures, was lighter when the air had been drawn out of it, than before, by no less than one ounce and $\frac{3}{10}$, that is, an ounce and very near a third. But of the gravity of the air we may elsewhere have occasion to make further mention.

Taking it then for granted, that the air is not devoid of weight, it will not be uneasy to conceive, that that part of the atmosphere, wherein we live, being the lower part of it, the corpuscles, that compose it, are very much compressed by the weight of all those of the like nature, that are directly over them; that is, of all the particles of air, that being piled up upon them, reach to the top of the atmosphere. And though the height of this atmosphere, according to the famous *Kepler*,† and some others, scarce exceeds eight common miles; yet other eminent and later astronomers would promote the confines of

---

† JOHANN KEPLER, *Ad Vitellionem Paralipomena, quibus Astronomiae pars Optica traditur*, etc., Claudium Marnium et Haeredes Joannis Aubrii, Francofurti, 1604, Caput IV. 6, Propositio XI, Problema IV, " Ex refractionibus altitudinem aëris a Terra investigare ". Reprinted in

the atmosphere to exceed six or seven times that number of miles. And the diligent and learned *Ricciolo* makes it probable, that the atmosphere may, at least in divers places, be at least fifty miles high.† So that, according to a moderate estimate of the thickness of the atmosphere, we may well suppose, that a column of air, of many miles in height, leaning upon some springy corpuscles of air here below, may have weight enough to bend their little springs, and keep them bent: as, (to resume our former comparison,) if there were fleeces of wool piled up to a mountainous height one upon another, the hairs, that compose the lowermost locks, which support the rest, would, by the weight of all the wool above them, be as well strongly compressed, as if a man should squeeze them together in his hands, or employ any such other moderate force to compress them. So that we need not wonder, that upon the taking off the incumbent air from any parcel of the atmosphere here below, the corpuscles, whereof that undermost air consists, should display themselves, and take up more room than before.

And if it be objected, that in water, the weight of the upper and of the lower part is the same; I answer, that, (besides that it may be well doubted whether the observation, by reason of the great difficulty, hath been exactly made,) there is a manifest disparity betwixt the air and water: for I have not found, upon an experiment purposely made, (and in another treatise recorded) that water will suffer any considerable compression; whereas we may observe in wind-guns, (to mention now no other engines) that the air will suffer itself to be crouded into a comparatively very little room; insomuch, that a very diligent examiner of the phænomena of wind-guns would have us believe, that in one of them, by condensation, he reduced the

---

his *Astronomi Opera Omnia* (ed. Ch. Frisch), Heyder & Zimmer, Francofurti a. M., Volumen II, pp. 207–8, 1859, and in his *Gesammelte Werke*, Beck, Munchen, Band II, pp. 120–1, 1939. (I am indebted to Stillman Drake for this reference.) German translation: *J. Keplers Grundlagen der geometrischen Optik* (im *Anschluss an die Optik des Witelo* (trans. F. PLEHN, ed. M. VON ROHR), Akademische Verlagsgesellschaft m. b. H., Leipzig, 1922, pp. 113–14.

† GIOVANNI RICCIOLI (1598–1671) was an Italian astronomer; see his *Almagestum novum Astronomiam novamque complectens observationibus aliorum*, etc., Benatij, Bononiae, 1651.

air into a space at least eight times narrower than it before possessed. And to this, if we add a noble phænomenon of the experiment *de vacuo*; these things put together may for the present suffice to countenance our doctrine. For that noble experimenter, Monsieur *Pascal* (the son) had the commendable curiosity to cause the Torricellian experiment to be tried at the foot, about the middle, and at the top of that high mountain (in *Auvergne,* if I mistake not) commonly called *Le Puy de Domme*; whereby it was found, that the mercury in the tube fell down lower, about three inches, at the top of the mountain, than at the bottom.† And a learned man awhile since informed me, that a great Virtuoso, friend to us both, hath, with not unlike success, tried the same experiment in the lower and upper parts of a mountain in the west of *England.* Of which the reason seems manifestly enough to be this, that upon the tops of high mountains, the air, which bears against the restagnant quicksilver, is less pressed by the less ponderous incumbent air; and consequently is not able totally to hinder the descent of so tall and heavy a cylinder of quicksilver, as at the bottom of such mountains did but maintain an æquilibrium with the incumbent atmosphere.

And if it be yet further objected against what hath been proposed touching the compactness and pressure of the inferiour air, that we find this very air to yield readily to the motion of little flies, and even to that of feathers, and such other light and weak bodies; which seems to argue, that the particles of our air are not so compressed as we have represented them, especially since, by our former experiment, it appears, that the air readily dilated itself downward, from the receiver into the pump, when it is plain, that it is not the incumbent atmosphere, but only the subjacent air in the brass cylinder that hath been removed: If this, I say, be objected, we may reply, that, when a man squeezeth a fleece of wool in his hand, he may feel, that the wool incessantly bears against his hand, as that which hinders the hairs it consists of, to recover their former and more natural extent. So each parcel of the air about the earth doth constantly endeavour to thrust away all those contiguous bodies, (whether aëreal or more gross,) that keep it bent, and hinder the expansion of its parts, which will dilate themselves, or fly abroad towards that part, (whether upwards

---

† B. PASCAL and F. PERIER, *Récit de la grande experience des liqueurs,* Savreux, Paris, 1648.

or downwards,) where they find their attempted dilatation of themselves less resisted by the neighbouring bodies. Thus the corpuscles of that air we have been all this while speaking of, being unable, by reason of their weight, to ascend above the convexity of the atmosphere, and by reason of the resistance of the surface of the earth and water, to fall down lower, they are forced, by their own gravity and this resistance, to expand and diffuse themselves about the terrestrial globe; whereby it comes to pass, that they must as well press the contiguous corpuscles of air, that on either side oppose their dilatation, as they must press upon the surface of the earth; and, as it were recoiling thence, endeavour to thrust away those upper particles of air that lean upon them.

And, as for the easy yielding of the air to the bodies that move in it, if we consider, that the corpuscles, whereof it consists, though of a springy nature, are yet so very small, as to make up (which it is manifest to do) a fluid body, it will not be difficult to conceive, that in the air, as in other bodies that are fluid, the little bodies it consists of, are in an almost restless motion, whereby they become (as we have more fully discoursed in another treatise†) very much disposed to yield to other bodies, or easy to be displaced by them; and that the same corpuscles are likewise so variously moved, as they are entire corpuscles, that if some strive to push a body placed among them towards the right hand (for instance,) others, whose motion hath an opposite determination, as strongly thrust the same body towards the left; whereby neither of them proves able to move it out of its place, the pressure on all hands being reduced as it were to an æquilibrium: so that the corpuscles of the air must be as well sometimes considered under the notion of little springs, which remaining bent, are in their entire bulk transported from place to place; as under the notion of springs displaying themselves, whose parts fly abroad, whilst, as to their entire bulk, they scarce change place: as the two ends of a bow, shot off, fly from one another; whereas the bow itself may be held fast in the archer's hand. And that it is the equal pressure of the air on all sides upon the bodies that are in it, which causeth the easy cession of its parts, may be argued from hence; that if by the help of

---

† In a discourse touching Fluidity and Firmness. (*The History of Fluidity and Firmness*. Physiological Essays (1661), No. 6 and 7; reprinted in Boyle's *Works*, Vol. I, pp. 377–442.)

our engine the air be but in great part, though not totally, drawn away from one side of a body without being drawn away from the other, he that shall think to move that body to and fro, as easily as before, will find himself much mistaken.

2
The Repulsion Theory *

ISAAC NEWTON

SUMMARY

The purpose of natural philosophy is to develop mathematical principles so that from the observed phenomena of motions one may investigate the forces of nature, and then from these forces demonstrate other phenomena. Newton suspects that all the phenomena of nature may depend on "... certain forces by which the particles of bodies ... are either mutually impelled towards one another, and cohere in regular figures, or are repelled and recede from one another." One example of this kind of investigation is a demonstration that "... particles fleeing from each other, with forces that are inversely proportional to the distances of their centres, compose an elastic fluid, whose density is as the compression." (In this way Boyle's experiments on the spring of the air may be explained by assuming that each atom exerts repulsive forces on its neighbors.)

Since the ancients (as we are told by *Pappus*†) esteemed the science

---

* These selections are from Newton's *Philosophiae Naturalis Principia Mathematica*, London, 1687; third edition, 1726. The first paragraph is taken from Newton's Preface to the First Edition; the rest is from Book II, Section V, pp. 292–4 of the Third Edition. We have used Florian Cajori's revision of Andrew Motte's translation (1729), *Sir Isaac Newton's Mathematical Principles of Natural Philosophy and his System of the World*, University of California Press, Berkeley, California, 1934 (reprinted by permission of the University of California Press), pp. xvii–xviii and 300–302. For a critical discussion of this edition, see I. B. COHEN, *Isis* **54**, 319 (1963).

† Pappus of Alexandria, a Greek geometer, lived *circa* A.D. 320. See: *Pappi Alexandrini Collectionis quae supersunt e libris manu scriptis edidit Latina interpretatione et commentariis instruxit Fridericus Hultsch*,

of mechanics of greatest importance in the investigation of natural things, and the moderns, rejecting substantial forms and occult qualities, have endeavoured to subject the phenomena of nature to the laws of mathematics, I have in this treatise cultivated mathematics as far as it relates to philosophy. The ancients considered mechanics in a twofold respect; as rational, which proceeds accurately by demonstration, and practical. To practical mechanics all the manual arts belong, from which mechanics took its name. . . . But I consider philosophy rather than arts and write not concerning manual but natural powers, and consider chiefly those things which relate to gravity, levity, elastic force, the resistance of fluids, and the like forces, whether attractive or impulsive; and therefore I offer this work as the mathematical principles of philosophy, for the whole burden of philosophy seems to consist in this—from the phenomena of motions to investigate the forces of nature, and then from these forces to demonstrate the other phenomena; and to this end the general propositions in the first and second Books are directed. In the third Book I give an example of this in the explication of the System of the World; for by the propositions mathematically demonstrated in the former Books, in the third I derive from the celestial phenomena the forces of gravity with which bodies tend to the sun and the several planets. Then from these forces, by other propositions which are also mathematical I deduce the motions of the planets, the comets, the moon, and the sea. I wish we could derive the rest of the phenomena of Nature by the same kind of reasoning from mechanical principles, for I am induced by many reasons to suspect that they may all depend upon certain forces by which the particles of bodies, by some causes hitherto unknown, are either mutually impelled towards one another, and cohere in regular figures, or are repelled and recede from one another. These forces being unknown, philosophers have hitherto attempted to search of Nature in vain; but I hope the principles here laid down will afford some light either to this or some truer method of philosophy.

Berolini, Weidmannos, 1878 (esp. Libri VIII in Vol. III, Tom. I). French translation: Pappus d'Alexandrie, *La Collection Mathematique* (translated with Introduction and Notes by P. VER ECKE). Brouer, Paris and Bruges, 1933.

PROPOSITION XXIII. THEOREM XVIII

If a fluid be composed of particles fleeing from each other, and the density be as the compression, the centrifugal forces of the particles will be inversely proportional to the distances of their centres. And, conversely, particles fleeing from each other, with forces that are inversely proportional to the distances of their centres, compose an elastic fluid, whose density is as the compression.

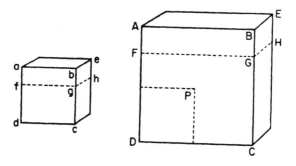

Let the fluid be supposed to be included in a cubic space ACE, and then to be reduced by compression into a lesser cubic space *ace*; and the distances of the particles retaining a like situation with respect to each other in both the spaces, will be as the sides AB, *ab* of the cubes; and the densities of the mediums will be inversely as the containing spaces AB^3, ab^3. In the plane side of the greater cube ABCD take the square DP equal to the plane side *db* of the lesser cube; and, by the supposition, the pressure with which the square DP urges the inclosed fluid will be to the pressure with which that square *db* urges the inclosed fluid as the densities of the mediums are to each other, that is, as ab^3 to AB^3. But the pressure with which the square DB urges the included fluid is to the pressure with which the square DP urges the same fluid as the square DB to the square DP, that is, as AB^2 to ab^2. Therefore, multiplying together corresponding terms of the proportions, the pressure with which the square DB urges the fluid is to the pressure with which the square *db* urges the fluid as *ab* to AB. Let the planes FGH, *fgh* be drawn through the interior of the two cubes, and divide the fluid into two parts. These parts will press each other with the same forces with which they are themselves pressed by the planes AC, ac, that is, in the proportion of ab to AB: and therefore the centrifugal forces by which these pressures are sustained are in the same ratio. The number of the particles being equal, and the situation alike, in both cubes, the forces which all the particles exert, according to the planes FGH, fgh, upon all, are as the forces which each exerts on each. Therefore the forces which each exerts on each, according to the plane FGH in the greater cube, are to the forces which each exerts on each, according to the plane fgh in the lesser cube, as ab to AB, that is, inversely as the distances of the particles from each other. Q.E.D.

And, conversely, if the forces of the single particles are inversely as the distances, that is, inversely as the sides of the cubes AB, ab; the sums of the forces will be in the same ratio, and the pressures of the sides DB, db as the sums of the forces; and the pressure of the square DP to the pressure of the side DB as ab^2 to AB^2. And, multiplying together corresponding terms of the proportions, one obtains the pressure of the square DP to the pressure of the side db as ab^3 to AB^3; that is, the force of compression in the one is to the force of compression in the other as the density in the former to the density in the latter. Q.E.D.

SCHOLIUM

By a like reasoning, if the centrifugal forces of the particles are inversely as the square of the distances between the centres, the cubes of the compressing forces will be as the fourth power of the densities. If the centrifugal forces be inversely as the third or fourth power of the distances, the cubes of the compressing forces will be as the fifth or sixth power of the densities. And universally, if D be put for the distance, and E for the density of the compressed fluid, and the centrifugal forces be inversely as any power D^n of the distance, whose index is the number n, the compressing forces will be as the cube roots of the power E^{n+2}, whose index is the number $n + 2$; and conversely. All these things are to be understood of particles whose centrifugal forces terminate in those particles that are next them, or are diffused not much farther. We have an example of this in magnetic bodies. Their attractive force is terminated nearly in bodies of their own kind that are next them. The force of the magnet is reduced by the interposition of an iron plate, and is almost terminated at it: for bodies farther off are not attracted by the magnet so much as by the iron plate. If in this manner particles repel others of their own kind that lie next them, but do not exert their force on the more remote, particles of this kind will compose such fluids as are treated of in this Proposition. If the force of any particle diffuse itself every way *in infinitum*, there will be required a greater force to produce an equal condensation of a greater quantity of the fluid. But whether elastic fluids do really consist of particles so repelling each other, is a physical question. We have here demonstrated mathematically the property of fluids consisting of particles of this kind, that hence philosophers may take occasion to discuss that question.

3
On the Properties and Motions of Elastic Fluids, Especially Air*

DANIEL BERNOULLI

SUMMARY

The properties of elastic fluids (i.e. gases) depend especially on the facts that: (1) they possess weight; (2) they expand in all directions unless restrained; and (3) they allow themselves to be more and more compressed as the force of compression increases. These properties can be explained if we assume a fluid to consist of a very large number of small particles in rapid motion. If such a fluid is placed in a cylindrical container with a movable piston on which there rests a weight P, then the particles will strike against the piston and hold it up by their impacts; the fluid will expand as the weight P is moved or reduced, and will become denser if the weight is increased. We can find the relation between P and the density of the fluid by computing the effect of changing the volume underneath the piston. If the volume is decreased, the fluid will exert a greater force on the piston, first, because the number of particles is now greater in proportion to the smaller space in which they are confined, and secondly because any given particle makes more frequent impacts. If in addition we assume that the particles themselves are of negligible size, then it follows that the force of compression is approximately inversely proportional to the volume occupied by the air, if the temperature is fixed; and, if the temperature changes, the force of compression will also change, so that it will be proportional to the square of the velocity of the particles. This result enables us to measure the temperature of a gas by the pressure it exerts; the standard temperature from which the rest are measured should be obtained from boiling rainwater, because this no doubt

---

* From *Hydrodynamica, sive de vivibus et motibus fluidorum commentarii.* Sectio Decima, " De affectionibus atque motibus fluidorum elasticorum, praecipue autem aëris." (pp. 200–204) Argentorati, Sumptibus Johannes Reinholdi Dulseckeri, 1738. Translated from Latin by J. P. BERRYMAN. Portions of this work are similar to the translation by WILLIAM FRANCIS MAGIE in his *Source Book in Physics*, published by McGraw-Hill in 1935; permission to reprint this translation has been granted by the present copyright holder, The President and Fellows of Harvard College (Harvard University Press, Copyright 1935, 1963).

has very nearly the same temperature all over the earth. Experiments to measure temperature and pressure are described. The theory is applied to atmospheric phenomena and to the firing of projectiles from cannons.

§1. Proceeding now to consider elastic fluids, we may assume that they have the same characteristics as everything else we have so far examined; thus we shall have a method for dealing with other phenomena not yet adequately studied. However, the properties of elastic fluids depend especially on the facts that: 1. they possess weight; 2. they expand in all directions unless restrained; and 3. they allow themselves to be more and more compressed as the force of compression increases. Such are the properties of air to which in particular we now direct our attention.

§2. Imagine therefore a cylindrical container $ACDB$ placed in a vertical position (Fig. 56) and in it a movable piston EF on which there rests a weight P. Let the space $ECDF$ contain very small particles in very rapid motion; as they strike against the piston EF and hold it up by their impact, they constitute an elastic fluid which expands as the weight P is removed or reduced; but if P is increased it becomes denser and presses on the horizontal case CD just as if it were endowed with no elastic property. For whether the particles are at rest or are in motion, they do not change their weight, so that the base bears the weight and elasticity of the fluid simultaneously. Let us therefore substitute for the air a fluid that conforms to the principal properties of elastic fluids, and thus we shall explain some of its properties that have already been noted and throw light on others that have not yet been considered.

§3. We shall consider the particles enclosed in our hollow cylinder as infinite in number, and when they fill the space $ECDF$ we shall then say that they constitute ordinary air, which will be our standard of reference. The weight P holding down the piston in the position EF is the same as the weight of the overlying atmosphere, which we shall designate P in what follows.

Let it be noted, however, that this pressure is by no means equal to the mere weight of the vertical cylinder of atmospheric air that rests on top of the piston EF, as some previous authors have carelessly asserted; the pressure is, in fact, equal to the fourth proportional to the area of the Earth's surface, the size of the piston *EF*, and the weight of the whole atmosphere on the Earth's surface.

§4. Now let us find a weight π sufficient to compress the air *ECDF* into the space *eCDf* assuming that the particles of the air have the

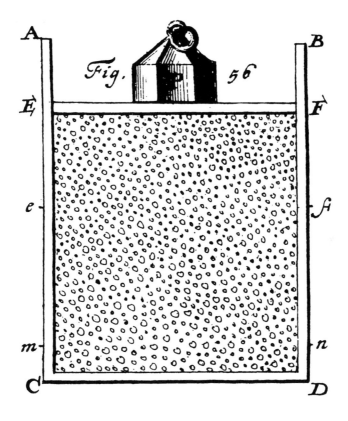

same velocity before and after compression. Let $EC = 1$ and $eC = s$. Now when the piston *EF* is moved to *ef*, it is subjected to a greater force by the fluid in two ways: first because the number of particles is now greater in proportion to the smaller space in which they are confined, and secondly because any given particle makes more frequent impacts. In order correctly to calculate the increment due to the first cause, we shall consider the particles as being at rest, and put equal to n the number of particles adjacent to the piston in its position *EF*; this number will become $n/(eC/EC)^{2/3}$ or $n/s^{2/3}$ when the piston is at *ef*.

Let it be noted, however, that we are regarding the fluid as no more compressed in its lower part than in its upper; such is the case when the weight P is infinitely greater than the fluid's own weight; hence it is clear that the pressure of the fluid is changed in the ratio $n : n/s^{2/3}$, i.e. $s^{2/3} : 1$. Now as for the increment arising from the second cause, this is found by considering the motion of the particles; thus it is evident that the impacts occur more frequently in proportion to the reduction in the mutual distances of the particles; evidently the number of impacts will be inversely proportional to the mean distance between the surfaces of the particles; these mean distances will be determined as follows.

Let us assume the particles to be spherical, and that the mean distance between their centres is D when the piston is at EF. Let the diameter of a sphere be d; then the mean distance between the surfaces of the spheres will be $D - d$; it is evident that in the position ef of the piston the mean distance between centres is $D\sqrt[3]{s}$ and hence the mean distance between surfaces is $D\sqrt[3]{s} - d$. Therefore, in respect of the second cause, the ratio of the pressure of the ordinary gas $ECDF$ to that of the compressed gas $eCDf$ is

$$\frac{1}{D-d} : \frac{1}{D\sqrt[3]{s} - d} \quad \text{or} \quad D\sqrt[3]{s} - d : D - d.$$

Taking the two causes together, we find that the pressure will be in the ratio

$$s^{2/3} \times (D\sqrt[3]{s} - d) : D - d.$$

For the ratio $D : d$ we can substitute another one that is easier to grasp: if we suppose that the piston EF descends under an infinitely great weight to the position mn at which all the particles touch each other, and if we denote by m the length mC, then $D : d = 1 : \sqrt[3]{m}$; finally, on substituting this ratio, the pressures of the air before and after compression will be in the ratio

$$s^{2/3} (\sqrt[3]{s} - \sqrt[3]{m}) : 1 - \sqrt[3]{m}$$

i.e.

$$s - \sqrt[3]{ms^2} : 1 - \sqrt[3]{m}$$

so that

$$\pi = \frac{(1 - \sqrt[3]{m}) P}{s - \sqrt[3]{ms^2}}.$$

§5. From all observations we may conclude that ordinary air can be considerably compressed, and reduced to an almost infinitesimal volume. Putting $m = 0$, we have $\pi = P/s$, so that the force of compression is approximately inversely proportional to the volume occupied by the air. This is confirmed by a variety of experiments. This law can certainly be safely applied to air less dense than normal; though I have not adequately examined whether it also applies to air very much more dense; experiments of the necessary degree of accuracy have not yet been put in hand. An experiment is needed to fix the value of the quantity m but it must be made with very great accuracy, even when the air is strongly compressed; the temperature of the air should be carefully kept constant while it is being compressed.

§6. The pressure of the air is increased not only by reduction in volume but also by rise in temperature. As it is well known that heat is intensified as the internal motion of the particles increases, it follows that any increase in the pressure of air that has not changed its volume indicates more intense motion of its particles, which is in agreement with our hypothesis; for it is clear that the weight P needs to be the greater to contain the air in the volume $ECDF$, according as the particles are in more violent motion. Moreover it is not difficult to see that the weight P will vary as the square of the particle velocity, on account of the fact that through this increased velocity the number of impacts and their intensity will increase equally at the same time; in fact, each increase is individually proportional to the weight P.

Therefore if the velocity of the particles is called v, there will be a weight which the air if able to hold in the position EF of the piston, equal to v^2P, and in the position ef, equal to

$$\frac{1 - \sqrt[3]{m}}{s - \sqrt[3]{ms^2}} \times v^2 P$$

or approximately v^2P/s, because, as we have seen, the number m is quite small compared with unity or s.

§7. The theorem set out in the preceding paragraph—namely, that in any air of whatever density, but at a given temperature, the pressure varies as the density, and furthermore, that increases of pressure arising from equal increases of temperature are proportional to the density—this theorem was discovered by experiment by Dr. Amontons; he has given an account of it in the Memoires of the Royal Academy of Sciences of Paris for 1702.† The implication of the theorem is that if, for example, ordinary air at normal temperature can hold a weight of 100 lb distributed over a given surface area, and then its temperature is increased until it can hold 120 lb over the same area and at the same volume, then when the same air is compressed to half its volume and held at the same temperatures, it can support 200 and 240 lb respectively; thus the increases of 20 and 40 lb produced by the increase of temperature are proportional to the density. He asserts, moreover, that air which he calls temperate has a pressure approximately 3/4 that of air at the temperature of boiling water, or more exactly 55/73. But in experiments that I have undertaken I found that the warmest air here at the height of summer has not the elasticity that Dr. Amontons attributed to " temperate " air; in fact I cannot believe that even at the equator such warm air ever exists. I think my experiments are more reliable than his, because in his the air did not keep its volume and no account of the variation of volume was taken, in his calculations. I have found that the elasticity of air that was at its coldest here at St. Petersburg on 25 December, 1731 (old style) was 523/1000 of the elasticity of like air at the temperature of boiling water. On 21st January, 1733 the cold was much more intense and I noticed that the corresponding pressure of the air was less than half what it was at the temperature of boiling water.

---

† G. AMONTONS, *Memoirs de l'Academie Royale des Sciences, Paris*, p. 155 (1702). See the extract in Magie's *Source Book in Physics*, McGraw-Hill, New York, 1935, pp. 128–31 (reprinted by Harvard University Press, 1963).

It is tempting to credit Bernoulli with the view that the heat content of a gas *is* just the kinetic energy of molecular motion, on the basis of this derivation. This was not the interpretation that an 18th-century reader would necessarily make; for example, Jean Trembley, while accepting Bernoulli's explanation of the pressure of a gas, complained that he still had not explained how heat increases the motion of the particles. See *Mém. Acad. Roy. Sci., Berlin*, pp. 77–8 (1796).

But when the air temperature in the shade was at its highest in 1731, it had a pressure about 4/3 (more precisely 100/76) times that of the very cold air and 2/3 of that at boiling water temperature. And so the greatest variations of air temperature in these parts are confined within limits proportional to the numbers 3 and 4, and in England (so I read) the corresponding figures are 7 and 8. But I consider that air hot enough to have a pressure equal to 3/4 of that of air at boiling water temperature would be practically intolerable to a living creature.

§8. From the known ratio between the different pressures of a given amount of air enclosed in a given volume, it is easy to deduce a measure of the temperature of the air, if only we agree in defining what we mean by " twice the temperature, 3 times the temperature," and so on; such a definition is arbitrary and not inherent in nature; in fact, it seems reasonable to me to determine the temperature of air by reference to its pressure, provided it is of standard density. The standard temperature from which the rest are measured should be obtained from boiling rainwater, because this no doubt has very nearly the same temperature all over the earth.

Accepting this, the temperatures of boiling water, of the warmest air in summer, and of the coldest air in winter in this country are as 6 : 4 : 3. I will now state how I have found these numbers, so that judgment may be passed on the accuracy of my experiments, the outcome of which is quite different from those of Amontons . . .

[There follows a description of the experimental procedure for measuring pressure and temperature. Bernoulli then considers the pressure of air at different places in the atmosphere. He concludes that if two regions of air at the same height are subjected to the same pressure, the densities will be in approximately inverse ratio to the square of the velocities of the particles in those regions; and that " . . . the pressure of air must be the same in all places at equal heights above sea-level, if the atmosphere is regarded as in a steady state of equilibrium and not stirred by any winds, regardless of the difference in temperature between the various parts of the atmosphere " (§§ 13, 14). Further consideration of the equilibrium of columns of air with different densities and temperatures leads to the conclusion that when a barometer is carried from a low place to a higher one, and the mercury level descends, ". . . it does not follow that the reduction in weight of the mercury column is equal to the weight of an air column of the same diameter and height, as inferred by some authors." On this basis Bernoulli criticizes the method employed in England for investigating the ratio of the specific gravities of air and mercury, as reported by Duhamel in the records of the Paris Academy of Sciences. This method leads to a determination, not of the specific gravity of the air at the site of the experiment, but rather of the mean specific gravity of the air in the whole surrounding region of the earth. Taking the specific gravity of mercury as 1, the mean specific gravity of air comes out to be 0·000105, which seems to be rather higher than expected (§ 16). It is pointed out that large variations in the pressure recorded by a barometer may be caused by rather small changes in the local temperature of the air. Barometric variations may also be caused by rapid changes of temperature in cavities under ground (§ 20). Most of the other possible causes appear to be too slow, and to be dispersed over the whole mass of air, so that they cannot account for observed rapid variations. The variation of mean barometric level with altitude above sea-level is also investigated; some experiments are reported, and a theory is tentatively proposed, involving the variation of particle velocity with height. The result of the theory is that the ratio of pressure at height x to pressure at sea-level is $22000/(22000 + x)$ (x in feet) and this seems to fit the data better than the usual logarithmic law. The formula is applied to find the astronomical refraction of a star for any apparent altitude (§ 30). Various problems involving the flow of air from a container through a small hole into a vacuum are discussed, and the amount of energy released in such processes is calculated. Air occupying a volume a cannot be compressed into a volume $a - x$ without the expenditure of the amount of energy generated by the descent of a weight P through a distance $a \log[a/(a - x)]$, however that compression may have been achieved (§ 39). For example, if one has a cubic foot of air at twice normal density, the energy that it loses in acquiring the density of the normal surrounding air is found to be 865 ft lb. "It is clear from this agreement between the conservation of energy in compressed air and in a body falling from a given height that no special advantage is to be hoped for from the idea of compressing air when designing machines, and that the principles displayed in the previous part of this book are valid everywhere. But because it happens in many ways that air can be compressed naturally without applied force, or can acquire abnormal elasticity, there is certainly hope that powerful devices can be invented for operating machines, in the way Dr. Amontons has already described using the energy of fire. I believe that if all the energy which is latent in a cubic foot of coal and is drawn out of it by burning were usefully expended in operating machines, more could be achieved than by a day's labour by 8 or 10 men. Besides, the combustion of coal not only enormously increases the elasticity of air but also generates a huge quantity of new air." (§ 43.) The power of compressed air produced by gunpowder in firing projectiles from cannons is discussed; the optimum length of the barrel of the gun is computed. A method is given for deducing the power of the propellant by observing the altitude attained by the projectile. Application of the theory to observations of the flight of cannon-balls leads to the conclusion that either the compressed gas obtained from burning gunpowder is not ordinary air, or that its pressure is disproportionately greater than its density.]

4
The Existence of Fire*

GEORGE GREGORY

SUMMARY

The arguments advanced by theorists who believe that heat is nothing more than molecular motion are summarized. However, it is asserted that " . . . the hypothesis which assigns existence to the principle of fire (heat), as a distinct elementary principle, is supported by more numerous facts, and by more decisive reasons; it accounts better for all the phenomena of nature, and even for those very phenomena which are adduced in support of the contrary opinion." After stating the various arguments in favor of the materiality of heat, Gregory gives a brief account of the properties of this substance and its effects on matter.

The element of fire is only known by its effects; so subtile and evasive indeed is this wonderful fluid, so various are the forms which it assumes in the different departments of nature which it occupies, that its very existence, we have seen, has been questioned by some philosophers.

Heat, say these theorists, is nothing more than an intestine motion of the most subtile particles of bodies. Fire is no other than this motion increased to a certain degree, in other words, a body heated very hot; and flame is no more than ignited vapour, that is, vapour, the particles of which are agitated in an extraordinary degree.

In support of this theory it is alleged, 1. That motion in all cases is known to generate heat; and if continued to a certain degree, actual ignition will be produced, as the friction of two pieces of wood will first produce heat, and afterwards fire; and the motion of a glass

---

* From *The Economy of Nature, explained and illustrated on the principles of modern philosophy*, London, second edition, 1798, Vol. I, pp. 93–96.

globe upon an elastic cushion will cause a stream of fire to be copiously emitted. 2ndly, Bodies which are most susceptible of intestine motion, are most readily heated. 3rdly, Motion always accompanies fire or heat, as is evident on mixing oil of turpentine and vitriolic acid; and the heat seems in most cases to bear a proportion to the degree of motion or agitation. In the boiling of water, and in the hissing of heated iron when applied to a fluid, this motion is evidently manifested. 4thly, If the particles of any body are excited to a violent degree of intestine motion, by attrition, fermentation, etc., if they do not actually emit flames, they will yet be disposed to catch fire with the utmost facility; as in the distillation of spirits, if the head of the still is removed, the vapours will instantly be converted into flame if brought into contact with a lighted candle, or any other ignited body. Lastly, Heated bodies receive no accession of weight, which they apparently ought to do, on another body being introduced into their pores.

Plausible as this reasoning appears at first sight, the hypothesis which assigns existence to the principle of fire, as a distinct elementary principle, is supported by more numerous facts, and by more decisive reasons; it accounts better for all the phenomena of nature, and even for those very phenomena which are adduced in support of the contrary opinion.

1st. If it is admitted, as I apprehend it must, by the advocates for the contrary opinion, that the internal motion or agitation, which they say constitutes heat, is not equally felt by all the component particles of bodies, but only by the minuter and more subtile particles; and that these particles being afterwards thrown into a projectile state produce the effect of light; these concessions will almost amount to the establishment of the principle of fire as an elementary principle.

2ndly. That fire is really a substance, and not a quality, appears from its acting upon other substances, the reality of which has never been doubted . . .

3rdly. All the evidence of our senses, and many indubitable experiments, provide that light, which many suppose to be fire in a projectile state, is a substance. Boerhaave† concentrated the rays of

---

† Herman Boerhaave (1668–1738) was a Dutch physician and chemist who wrote a well-known textbook, *Elementa Chemiae* (1732).

the sun in a very strong burning-glass, and by throwing them upon the needle of a compass, the needle was put into motion by the force of the rays, as it would have been by a blast of air, or a stroke from some other body. . . . This experiment I think a sufficient demonstration, if any demonstration was wanted, that light at least is a substance. Of the identity of light, heat, and fire, I shall have occasion afterwards to treat.

4thly. The electric fire affects bodies with a true corporeal percussion;[†] and that this effect is not owing to the vibration of the air, or any medium but that of fire itself, is proved by many experiments in vacuo, etc. Now, if one species of fire is allowed to be material, there seems to be no reason why we should deny the same attribute to the rest.

5thly. It is not easy to conceive how a body can be expanded by motion alone; and it is much more natural to suppose, that bodies are expanded by the interposition of an extremely active and elastic substance between their component particles.

6thly. It is well known that there can be no ignition or combustion, that is, there can be no very high degree of heat, without a supply of air; a candle, for instance, will cease to burn in vacuo, or in air, the pure part of which is destroyed by burning or respiration. This is a fact which cannot be accounted for on the principle that all heat is no other than intestine motion; but is easily explained if we suppose fire a distinct elementary substance, which is contained in pure air, and is yielded by the air to the force of a superior attraction.

7thly. That heat is generally accompanied by motion, is no proof that heat and motion are the same; on the contrary, nothing is more natural than that the entrance of an exceedingly elastic substance into the pores of another body should excite some degreee of intestine motion, as well as the emission of the same substance, which must occasion some degree of contraction in the particles of the body. . . . There is no increase of gravity in heated bodies, because of the great elasticity of caloric or the matter of fire, which expands

---

[†] (WILLIAM) JONES, *Physiological Disquisitions* (*or Discourses on the Natural Philosophy of the Elements*, Rivington, London, 1781).

the bodies into which it enters, and consequently rather diminishes their specific gravities.

8thly. All the other phenomena of nature are more satisfactorily accounted for, on the principle that fire is a distinct substance, than on that which supposes it a mere quality, depending on the tremor or intestine motion of bodies . . .

I conceive fire therefore, or caloric, as termed by the French chemists, to be the elementary principle or cause of heat and light. Caloric in a disengaged state, or in the act of passing from one body to another, impresses our organs with the sense of heat; and in a rarefied and projectile state, it probably constitutes the matter of light. . . . cold is universally allowed to be a mere negative quality, and to mean nothing more than the absence of heat or fire . . .

The particles of *fire* appear to be more *minute* than those of any other substance whatever. It penetrates all bodies with the utmost ease. . . . The matter of fire is attracted more or less by all bodies. . . . It does not appear, however, that all bodies have an equal attraction for the matter of fire. . . . The power of conducting the matter of fire seems to depend upon the texture of bodies, that is, upon the contact of their parts; hence the excessive slowness with which heat is communicated to bodies of a rare and spongy texture . . .

The matter of fire will exist in a state of combination. I do not contend for the term chemical combination, in the strict and literal sense of the word; it is sufficient if it can be proved, that caloric may exist in bodies in a *latent* state, or in a state not perceptible to our senses . . .

Fluids, from their very nature and constitution, contain a greater quantity of caloric in a latent state than solid bodies: indeed it is now universally admitted, and may be easily proved, that the fluidity of all bodies is altogether owing to the quantity of fire which they retain in this latent or combined state, the elasticity of which keeps their particles remote from each other, and prevents their fixing into a solid mass . . .

The matter of fire is elastic, as is proved evidently from all its effects. There is indeed reason to believe that caloric is the only fluid in nature which is permanently elastic, and that it is the cause of the elasticity of all fluids which are esteemed so.

From the elasticity of this element it results that all natural bodies can only retain a certain quantity of it, without undergoing an alteration in their state and form. Thus a moderate quantity of fire admitted into a solid body expands it; a still larger quantity renders it fluid; and if the quantity is still increased, it will be converted into vapour.

5
The Forces of Inorganic Nature*
ROBERT MAYER

SUMMARY

Various properties of forces are stated and illustrated. (By a " force " Mayer means a particular form of what we would now call energy.) Forces are causes, and must therefore be equal to their effects; in other words, they are indestructible; when one force disappears, it must be replaced by another that is equal to it. In contrast to matter, forces have no weight. One should not confuse a force with a property, since a property lacks the essential attribute of a force, the union of indestructibility and convertibility. (Thus gravity is a property, not a force; but the separation in space of ponderable objects *is* a force, which Mayer calls " falling force ".)

We often see motion cease without having caused another motion or the lifting of a weight. Since a force once in existence cannot be annihilated but can only change its form, we are led to ask whether the motion has been converted into heat. If so, then the heat produced must be equivalent to the motion lost, and we have to determine, for example, how much heat corresponds to a given quantity of motion or " falling force ". It is found that the heat needed to warm a given weight of water from $0°$ to $1°$ C corresponds to the fall of an equal weight from the height of about 365 metres, the latter phrase being the measure of a quantity of falling force.

---

* Mayer's ideas were first conceived in 1840 and his first paper on this subject was submitted to the *Annalen der Physik* in 1841. It was rejected by Poggendorff, the editor. The first paper to be published was written half a year later, in the beginning of 1842, and was submitted to Liebig who accepted it for the *Annalen der Chemie und Pharmacie*. It was published under the title " Bemerkungen über die Kräfte der unbelebten Natur ", Vol. 42, pp. 233–40 (1842). We reprint here the English translation by G. C. Foster, first published in the *Philosophical Magazine*, Vol. 24, pp. 371–7 (1862).

The following pages are designed as an attempt to answer the questions, What are we to understand by " Forces "? and how are different forces related to each other? Whereas the term *matter* implies the possession, by the object to which it is applied, of very definite properties, such as weight and extension, the term *force* conveys for the most part the idea of something unknown, unsearchable, and hypothetical. An attempt to render the notion of force equally exact with that of matter, and so to denote by it only objects of actual investigation, is one which, with the consequences that flow from it, ought not to be unwelcome to those who desire that their views of nature may be clear and unencumbered by hypotheses.

Forces are causes: accordingly, we may in relation to them make full application of the principle—*causa aequat effectum*. If the cause c has the effect e, then $c = e$; if, in its turn, e is the cause of a second effect f, we have $e = f$, and so on: $c = e = f \ldots = c$.—In a chain of causes and effects, a term or a part of a term can never, as plainly appears from the nature of an equation, become equal to nothing. This first property of all causes we call their *indestructibility*.

If the given cause c has produced an effect e equal to itself, it has in that very act ceased to be: c has become e; if, after the production of e, c still remained in whole or in part, there must be still further effects corresponding to this remaining cause: the total effect of c would thus be $> e$, which would be contrary to the supposition $c = e$. Accordingly, since c becomes e, and e becomes f, etc., we must regard these various magnitudes as different forms under which one and the same object makes its appearance. This capability of assuming various forms is the second essential property of all causes. Taking both properties together, we may say, causes are (quantitatively) *indestructible* and (qualitatively) *convertible* objects.

Two classes of causes occur in nature, which, so far as experience goes, never pass one into another. The first class consists of such causes as possess the properties of weight and impenetrability; these are kinds of Matter: the other class is made up of causes which are wanting in the properties just mentioned, namely Forces, called also Imponderables, from the negative property that has been indicated. Forces are therefore *indestructible, convertible, imponderable objects*.

We will in the first instance take matter, to afford us an example of causes and effects. Explosive gas, H + O, and water, HO, are related to each other as cause and effect, therefore H + O = HO. But if H + O becomes HO, heat, *cal.*, make is appearance as well as water; this heat must likewise have a cause, x, and we have therefore, H + O + x = HO + *cal.* It might, however, be asked whether H + O is really = HO, and x = *cal.*, and not perhaps H + O = *cal.*, and x = HO, whence the above equation could equally be deduced; and so in many other cases. The phlogistic chemists recognized the equation between *cal.* and x, or Phlogiston as they called it, and in so doing made a great step in advance; but they involved themselves again in a system of mistakes by putting $-x$ in place of O; thus, for instance, they obtained H = HO + x.

Chemistry, whose problem it is to set forth in equations the causal connexion existing between the different kinds of matter, teaches us that matter, as a cause, has matter for its effect; but we are equally justified in saying that to force as cause, corresponds force as effect. Since $c = e$, and $e = c$, it is unnatural to call one term of an equation a force, and the other an effect of force or phenomenon, and to attach different notions to the expressions Force and Phenomenon. In brief, then, if the cause is matter, the effect is matter; if the cause is a force, the effect is also a force.

A cause which brings about the raising of a weight is a force; its effect (*the raised weight*) is, accordingly, equally a force; or, expressing this relation in a more general form, *separation in space of ponderable objects is a force*; since this force causes the fall of bodies, we call it *falling force*. Falling force and fall, or, more generally still, falling force and motion, are forces which are related to each other as cause and effect—forces which are convertible one into the other— two different forms of one and the same object. For example, a weight resting on the ground is not a force: it is neither the cause of motion, nor of the lifting of another weight; it becomes so, however, in proportion as it is raised above the ground: the cause—the distance between a weight and the earth—and the effect—the quantity of motion produced—bear to each other, as we learn from mechanics, a constant relation.

Gravity being regarded as the cause of the falling of dodies, a gravitating force is spoken of, and so the notions of *property* and of *force* are confounded with each other: precisely that which is the essential attribute of every force—the *union* of indestructibility with convertibility—is wanting in every property: between a property and a force, between gravity and motion, it is therefore impossible to establish the equation required for a rightly conceived causal relation. If gravity be called a force, a cause is supposed which produces effects without itself diminishing, and incorrect conceptions of the causal connexion of things are thereby fostered. In order that a body may fall, it is no less necessary that it should be lifted up, than that it should be heavy or possess gravity; the fall of bodies ought not therefore to be ascribed to their gravity alone.

It is the problem of Mechanics to develope the equations which subsist between falling force and motion, motion and falling force, and between different motions: here we will call to mind only one point. The magnitude of the falling force v is directly proportional (the earth's radius being assumed $= \infty$) to the magnitude of the mass m, and the height d to which it is raised; that is, $v = md$. If the height $d = 1$, to which the mass m is raised, is transformed into the final velocity $c = 1$ of this mass, we have also $v = mc$; but from the known relations existing between d and c, it results that, for the other values of d or of c, the measure of the force v is mc^2; accordingly $v = md = mc^2$: the law of the conservation of *vis viva* is thus found to be based on the general law of the indestructibility of causes.

In numberless cases we see motion cease without having caused another motion or the lifting of a weight; but a force once in existence cannot be annihilated, it can only change its form: and the question therefore arises. What other forms is force, which we have become acquainted with as falling force and motion, capable of assuming? Experience alone can lead us to a conclusion on this point. In order to experiment with advantage, we must select implements which, besides causing a real cessation of motion, are as little as possible altered by the objects to be examined. If, for example, we rub together two metal plates, we see motion disappear, and heat, on the other hand, make its appearance, and we have now only to ask whether *motion* is the cause of heat. In order to come to a decision on this point, we must discuss the question whether, in the numberless cases in which the expenditure of motion is accompanied by the appearance of heat, the motion has not some other effect than the production of heat, and the heat some other cause than the motion.

An attempt to ascertain the effects of ceasing motion has never yet been seriously made; without, therefore, wishing to exclude *à priori* the hypotheses which it may be possible to set up, we observe only that, as a rule, this effect cannot be supposed to be an alteration in the state of aggregation of the moved (that is, rubbing, etc.) bodies. If we assume that a certain quantity of motion v is expended in the conversion of a rubbing substance m into n, we must then have $m + v = n$, and $n = m + v$; and when n is reconverted into m, v must appear again in some form or other. By the friction of two metallic plates continued for a very long time, we can gradually cause the cessation of an immense quantity of movement; but would it ever occur to us to look for even the smallest trace of the force which has disappeared in the metallic dust that we could collect, and to try to regain it thence? We repeat, the motion cannot have been annihilated; and contrary, or positive and negative, motions cannot be regarded as $= 0$, any more than contrary motions can come out of nothing, or a weight can raise itself.

Without the recognition of a causal connexion between motion and heat, it is just as difficult to explain the production of heat as it is to give any account of the motion that disappears. The heat cannot be derived from the diminution of the volume of the rubbing substances. It is well known that two pieces of ice may be melted by rubbing them together *in vacuo*; but let any one try to convert ice into water by pressure, however enormous. Water undergoes, as was found by the author, a rise of temperature when violently shaken. The water so heated (from 12° to 13° C) has a greater bulk after being shaken than it had before; whence now comes this quantity of heat, which by repeated shaking may be called into existence in the same apparatus as often as we please? The vibratory hypothesis of heat is an approach towards the doctrine of heat being the effect of motion, but it does not favour the admission of this causal relation in its full generality; it rather lays the chief stress on uneasy oscillations (*unbehagliche Schwingungen*).

If it be now considered as established that in many cases (*exceptio confirmat regulam*) no other effect of motion can be traced except heat, and that no other cause than motion can be found for the heat that is produced, we prefer the assumption that heat proceeds from motion, to the assumption of a cause without effect and of an effect without a cause,—just as the chemist, instead of allowing oxygen and hydrogen to disappear without further investigation, and water to be produced in some inexplicable manner, establishes a connexion between oxygen and hydrogen on the one hand and water on the other.

The natural connexion existing between falling force, motion, and heat may be conceived of as follows. We know that heat makes its appearance when the separate particles of a body approach nearer to each other: condensation produces heat. And what applies to the smallest particles of matter, and the smallest intervals between them, must also apply to large masses and to measurable distances. The falling of a weight is a real diminution of the bulk of the earth, and must therefore without doubt be related to the quantity of heat thereby developed; this quantity of heat must be proportional to the greatness of the weight and its distance from the ground. From this point of view we are very easily led to the equations between falling force, motion, and heat, that have already been discussed.

But just as little as the connexion between falling force and motion authorizes the conclusion that the essence of falling force is motion, can such a conclusion be adopted in the case of heat. We are, on the contrary, rather inclined to infer that, before it can become heat, motion—whether simple, or vibratory as in the case of light and radiant heat, etc.—must cease to exist as motion.

If falling force and motion are equivalent to heat, heat must also naturally be equivalent to motion and falling force. Just as heat appears as an *effect* of the diminution of bulk and of the cessation of motion, so also does heat disappear as a *cause* when its effects are produced in the shape of motion, expansion, or raising of weight.

In water-mills, the continual diminution in bulk which the earth undergoes, owing to the fall of the water, gives rise to motion, which afterwards disappears again, calling forth unceasingly a great quantity of heat; and inversely, the steam-engine serves to decompose heat again into motion or the raising of weights. A locomotive engine with its train may be compared to a distilling apparatus; the heat applied under the boiler passes off as motion, and this is deposited again as heat at the axles of the wheels.

We will close our disquisition, the propositions of which have resulted as necessary consequences from the principle "*causa aequat effectum,*" and which are in accordance with all the phenomena of Nature, with a practical deduction. The solution of the equations subsisting between falling force and motion requires that the space fallen through in a given time, *e.g.* the first second, should be experimentally determined; in like manner, the solution of the equations subsisting between falling force and motion on the one hand and heat on the other, requires an answer to the question, How great is the quantity of heat which corresponds to a given quantity of motion or falling force? For instance, we must ascertain how high a given weight requires to be raised above the ground in order that its falling force may be equivalent to the raising of the temperature of an equal weight of water from 0° to 1° C. The attempt to show that such an equation is the expression of a physical truth may be regarded as the substance of the foregoing remarks.

By applying the principles that have been set forth to the relations subsisting between the temperature and the volume of gases, we find that the sinking of a mercury column by which a gas is compressed is equivalent to the quantity of heat set free by the compression; and hence it follows, the ratio between the capacity for heat of air under constant pressure and its capacity under constant volume being taken as $= 1\cdot 421$, that the warming of a given weight of water from 0° to 1° C corresponds to the fall of an equal weight from the height of about 365 metres. If we compare with the result the working of our best steam-engines, we see how small a part only of the heat applied under the boiler is really transformed into motion or the raising of weights; and this may serve as justification for the attempts at the profitable production of motion by some other method than the expenditure of the chemical difference between carbon and oxygen—more particularly by the transformation into motion of electricity obtained by chemical means.

6
On Matter, Living Force, and Heat*

JAMES JOULE

SUMMARY

Matter may be said to be defined by the attributes of *impenetrability* and *extension*; it also has a number of properties, such as gravity or weight, repulsion, inertia, and living force (kinetic energy). The last property is simply the force of bodies in motion, and is proportional to their weight and to the square of their velocity. Living force may be transferred from one body to another in collisions, or by attraction through a certain distance. It would seem absurd that this living force can be destroyed, or even lessened, without producing the equivalent of attraction through a given distance or heat, despite the fact that until very recently a contrary opinion has been generally held. We have reason to believe that the manifestations of living force on our globe are, at the present time, as extensive as those which have ever existed, and that the motions of air and water are not annihilated by friction but only converted to heat. The quantity of heat produced is always in proportion to the amount of living force absorbed; the living force possessed by a body of 817 lb when moving with the velocity of eight feet per second is convertible into the quantity of heat which can raise the temperature of 1 lb of water by one degree Fahrenheit.

Until recently it was thought that heat is a substance; but we have shown that it can be converted into living force and into attraction through space, so that it must therefore consist of either living force or of attraction through space. It seems to me that both hypotheses are probably true. There is reason to suppose that the particles of all bodies are in a state of rapid motion; the velocity of the atoms of water, for instance, is at least a mile per second. Similarly, the heat required to melt ice must be that needed to overcome the attraction of the particles for each other.

* A lecture at St. Ann's Church Reading-Room, Manchester, 28 April, 1847; reported in the Manchester Courier, 5 and 12 May, 1847; reprinted in *The Scientific Papers of James Prescott Joule*, The Physical Society, London, 1884, pp. 265–76.

In our notion of matter two ideas are generally included, namely those of *impenetrability* and *extension*. By the extension of matter we mean the space which it occupies; by its impenetrability we mean that two bodies cannot exist at the same time in the same place. Impenetrability and extension cannot with much propriety be reckoned among the *properties* of matter, but deserve rather to be called its *definitions*, because nothing that does not possess the two qualities bears the name of matter. If we conceive of impenetrability and extension we have the idea of matter, and of matter only.

Matter is endowed with an exceedingly great variety of wonderful properties, some of which are common to all matter, while others are present variously, so as to constitute a difference between one body and another. Of the first of these classes, the attraction of gravitation is one of the most important. We observe its presence readily in all solid bodies, the component parts of which are, in the opinion of Majocci, held together by this force. If we break the body in pieces, and remove the separate pieces to a distance from each other, they will still be found to attract each other, though in a very slight degree, owing to the force being one which diminishes very rapidly as the bodies are removed farther from one another. The larger the bodies are the more powerful is the force of attraction subsisting between them. Hence, although the force of attraction between small bodies can only be appreciated by the most delicate apparatus except in the case of contact, that which is occasioned by a body of immense magnitude, such as the earth, becomes very considerable. This attraction of bodies towards the earth constitutes what is called their *weight* or *gravity*, and is always exactly proportional to the quantity of matter. Hence, if any body be found to weigh 2 lb., while another only weighs 1 lb., the former will contain exactly twice as much matter as the latter; and this is the case, whatever the bulk of the bodies may be: 2-lb. weight of air contains exactly twice the quantity of matter that 1 lb. of lead does.

Matter is sometimes endowed with other kinds of attraction besides the attraction of gravitation; sometimes also it possesses the faculty of *repulsion*, by which force the particles tend to separate farther from each other. Wherever these forces exist, they do not supersede the attraction of gravitation. Thus the weight of a piece of iron or steel is in no way affected by imparting to it the magnetic virtue.

Besides the force of gravitation, there is another very remarkable property displayed in an equal degree by every kind of matter—its perseverance in any condition, whether of rest or motion, in which it may have been placed. This faculty has received the name of *inertia*, signifying passiveness, or the inability of any thing to change its own state. It is in consequence of this property that a body at rest cannot be set in motion without the application of a certain amount of force to it, and also that when once the body has been set in motion it will never stop of itself, but continue to move straight forwards with a uniform velocity until acted upon by another force, which, if applied contrary to the direction of motion, will retard it, if in the same direction will accelerate it, and if sideways will cause it to move in a curved direction. In the case in which the force is applied contrary in direction, but equal in degree to that which set the body first in motion, it will be entirely deprived of motion whatever time may have elapsed since the first impulse, and to whatever distance the body may have travelled.

From these facts it is obvious that the force expended in setting a body in motion is carried by the body itself, and exists with it and in it, throughout the whole course of its motion. This force possessed by moving bodies is termed by mechanical philosophers *vis viva*, or *living force*. The term may be deemed by some inappropriate, inasmuch as there is no life, properly speaking, in question; but it is *useful*, in order to distinguish the moving force from that which is stationary in its character, as the force of gravity. When, therefore, in the subsequent parts of this lecture I employ the term *living force*, you will understand that I simply mean the force of bodies in motion. The living force of bodies is regulated by their weight and by the velocity of their motion. You will readily understand that if a body of a certain weight possesses a certain quantity of living force, twice as much living force will be possessed by a body of twice the weight, provided both bodies move with equal velocity. But the law by which the *velocity* of a body regulates its living force is not so obvious. At first sight one would imagine that the living force would be simply proportional to the velocity, so that if a body moved twice as fast as another, it would have twice the impetus or living force.

Such, however, is not the case; for if three bodies of equal weight move with the respective velocities of 1, 2, and 3 miles per hour, their living forces will be found to be proportional to those numbers multiplied by themselves, viz. to $1 \times 1, 2 \times 2, 3 \times 3$, or 1, 4, and 9, the squares of 1, 2, and 3. This remarkable law may be proved in several ways. A bullet fired from a gun at a certain velocity will pierce a block of wood to only one quarter of the depth it would if propelled at twice the velocity. Again, if a cannon-ball were found to fly at a certain velocity when propelled by a given charge of gunpowder, and it were required to load the cannon so as to propel the ball with twice that velocity, it would be found necessary to employ four times the weight of powder previously used. Thus, also, it will be found that a railway-train going at 70 miles per hour possesses 100 times the impetus, or living force, as it does when travelling at 7 miles per hour.

A body may be endowed with living force in several ways. It may receive it by the impact of another body. Thus, if a perfectly elastic ball be made to strike another similar ball of equal weight at rest, the striking ball will communicate the whole of its living force to the ball struck, and, remaining at rest itself, will cause the other ball to move in the same direction and with the same velocity that it did itself before the collision. Here we see an instance of the facility with which living force may be transferred from one body to another. A body may also be endowed with living force by means of the action of gravitation upon it through a certain distance. If I hold a ball at a certain height and drop it, it will have acquired when it arrives at the ground a degree of living force proportional to its weight and the height from which it has fallen. We see, then, that living force may be produced by the action of gravity through a given distance or space. We may therefore say that the former is of equal value, or *equivalent*, to the latter. Hence, if I raise a weight of 1 lb. to the height of one foot, so that gravity may act on it through that distance, I shall communicate to it that which is of equal value or equivalent to a certain amount of living force; if I raise the weight to twice the height, I shall communicate to it the equivalent of twice the quantity of living force. Hence, also, when we compress a spring, we communicate to it the equivalent to a certain amount of living force; for in that case we produce molecular attraction between the particles of the spring through the distance they are forced asunder, which is strictly analogous to the production of the attraction of gravitation through a certain distance.

You will at once perceive that the living force of which we have been speaking is one of the most important qualities with which matter can be endowed, and, as such, that it would be absurd to suppose that it can be destroyed, or even lessened, without producing the equivalent of attraction through a given distance of which we have been speaking. You will therefore be surprised to hear that until very recently the universal opinion has been that living force could be absolutely and irrevocably destroyed at any one's option. Thus, when a weight falls to the ground, it has been generally supposed that its living force is absolutely annihilated, and that the labour which may have been expended in raising it to the elevation from which it fell has been entirely thrown away and wasted, without the production of any permanent effect whatever. We might reason, *à priori*, that such absolute destruction of living force cannot possibly take place, because it is manifestly absurd to suppose that the powers with which God has endowed matter can be destroyed any more than that they can be created by man's agency; but we are not left with this argument alone, decisive as it must be to every unprejudiced mind. The common experience of every one teaches him that living force is not *destroyed* by the friction or collision of bodies. We have reason to believe that the manifestations of living force on our globe are, at the present time, as extensive as those which have existed at any time since its creation, or, at any rate, since the deluge—that the winds blow as strongly, and the torrents flow with equal impetuosity now, as at the remote period of 4000 or even 6000 years ago; and yet we are certain that, through that vast interval of time, the motions of the air and of the water have been incessantly obstructed and hindered by friction. We may conclude, then, with certainty, that these motions of air and water, constituting living force, are not *annihilated* by friction. We lose sight of them, indeed, for a time; but we find them again reproduced. Were it not so, it is perfectly obvious that long ere this all nature would have come to a dead standstill. What, then, may we inquire, is the cause of this apparent anomaly? How comes it to pass that, though in almost all natural phenomena we witness the arrest of motion and the apparent destruction of living force, we find that no waste or loss of living force has actually occurred? Experiment has enabled us to answer these questions in a satisfactory manner; for it has shown that, wherever living force is *apparently* destroyed, an equivalent is produced which in process of time may be reconverted into living force. This equivalent is *heat*. Experiment has shown that wherever living force is apparently destroyed or absorbed, heat is produced. The most frequent way in which living force is thus converted into heat is by means of friction. Wood rubbed against wood or against any hard body, metal rubbed against metal or against any other body—in short, all bodies, solid or even liquid, rubbed against each other are invariably heated, sometimes even so far as to through space. All, three, therefore—namely, heat, living force, and attraction through space (to which I might also add *light*, were it consistent with the scope of the present lecture)—are mutually convertible into one another. In these conversions nothing is ever lost. The same quantity of heat will always be converted into the same quantity of living force. We can therefore express the equivalency in definite language applicable at all times and under all circumstances. Thus the attraction of 817 lb. through the space of one foot is equivalent to, and convertible into, the living force possessed by a body of the same weight of 817 lb. when moving with the velocity of eight feet per second, and this living force is again convertible into the quantity of heat which can increase the temperature of one pound of water by one degree Fahrenheit. The knowledge of the equivalency of heat to mechanical power is of great value in solving a great number of interesting and important questions. In the case of the steam-engine, by ascertaining the quantity of heat produced by the combustion of coal, we can find out how much of it is converted into mechanical power, and thus come to a conclusion how far the steam-engine is susceptible of further improvements. Calculations made upon this principle have shown that at least ten times as much power might be produced as is now obtained by the combustion of coal. Another interesting conclusion is, that the animal frame, though destined to fulfil so many other ends, is as a machine more perfect than the best contrived steam-engine—that is, is capable of more work with the same expenditure of fuel.

Behold, then, the wonderful arrangements of creation. The earth in its rapid motion round the sun possesses a degree of living force so vast that, if turned into the equivalent of heat, its temperature would be rendered at least 1000 times greater than that of red-hot iron, and the globe on which we tread would in all probability be rendered equal in brightness to the sun itself. And it cannot be doubted that if the course of the earth were changed so that it might fall into the sun, that body, so far from being cooled down by the contact of a comparatively cold body, would actually blaze more brightly than before in consequence of the living force with which the earth struck the sun being converted into its equivalent of heat. Here we see that our existence depends upon the *maintenance* of the living force of the earth. On the other hand, our safety equally depends in some instances upon the *conversion* of living force into heat. You have, no doubt, frequently observed what are called *shooting-stars*, as they appear to emerge from the dark sky of night, pursue a short and rapid course, burst, and are dissipated in shining fragments. From the velocity with which these bodies travel, there can be little doubt that they are small planets which, in the course of their revolution round the sun, are attracted and drawn to the earth. Reflect for a moment on the consequences which would ensue, if a hard meteoric stone were to strike the room in which we are assembled with a velocity sixty times as great as that of a cannon-ball. The dire effects of such a collision are effectually prevented by the atmosphere surrounding our globe, by which the velocity of the meteoric stone is checked and its living force converted into heat, which at last becomes so intense as to melt the body and dissipate it into fragments too small probably to be noticed in their fall to the ground. Hence it is that, although multitudes of shooting-stars appear every night, few meteoric stones have been found, those few corroborating the truth of our hypothesis by the marks of intense heat which they bear on their surfaces.

Descending from the planetary space and firmament to the surface of our earth, we find a vast variety of phenomena connected with the conversion of living force and heat into one another, which speak in language which cannot be misunderstood of the wisdom and beneficence of the Great Architect of nature. The motion of air which we call *wind* arises chiefly from the intense heat of the torrid zone compared with the temperature of the temperate and frigid zones. Here we have an instance of heat being converted into the living force of currents of air. These currents of air, in their progress across the sea, lift up its waves and propel the ships; whilst in passing across the land they shake the trees and disturb every blade of grass. The waves by their violent motion, the ships by their passage through a resisting medium, and the trees by the rubbing of their branches together and the friction of their leaves against themselves and the air, each and all of them generate heat equivalent to the diminution of the living force of the air which they occasion. The heat thus restored may again contribute to raise fresh currents of air; and thus the phenomena may be repeated in endless succession and variety.

When we consider our own animal frames, " fearfully and wonderfully made," we observe in the motion of our limbs a continual conversion of heat into living force, which may be either converted back again into heat or employed in producing an attraction through space, as when a man ascends a mountain. Indeed the phenomena of nature, whether mechanical, chemical, or vital, consist almost entirely in a continual conversion of attraction through space, living force, and heat into one another. Thus it is that order is maintained in the universe—nothing is deranged, nothing ever lost, but the entire machinery, complicated as it is, works smoothly and harmoniously. And though, as in the awful vision of Ezekiel, " wheel may be in the middle of wheel," and every thing may appear complicated and involved in the apparent confusion and intricacy of an almost endless variety of causes, effects, conversions, and arrangements, yet is the most perfect regularity preserved—the whole being governed by the sovereign will of God.

A few words may be said, in conclusion, with respect to the real nature of heat. The most prevalent opinion, until of late, has been that it is a *substance* possessing, like all other matter, impenetrability and extension. We have, however, shown that heat can be converted into living force and into attraction through space. It is perfectly clear, therefore, that unless matter can be converted into attraction through space, which is too absurd an idea to be entertained for a moment, the hypothesis of heat being a substance must fall to the ground. Heat must therefore consist of either living force or of attraction through space. In the former case we can conceive the constituent particles of heated bodies to be, either in whole or in part, in a state of motion. In the latter we may suppose the particles to be removed by the process of heating, so as to exert attraction through greater space. I am inclined to believe that both of these hypotheses will be found to hold good,—that in some instances, particularly in the case of *sensible* heat, or such as is indicated by the thermometer, heat will be found to consist in the living force of the particles of the bodies in which it is induced; whilst in others, particularly in the case of *latent* heat, the phenomena are produced by the separation of particle from particle, so as to cause them to attract one another through a greater space. We may conceive, then, that the communication of heat to a body consists, in fact, in the communication of impetus, or living force, to its particles. It will perhaps appear to some of you something strange that a body apparently quiescent should in reality be the seat of motions of great rapidity; but you will observe that the bodies themselves, considered as wholes, are not supposed to be in motion. The constituent particles, or atoms of the bodies, are supposed to be in motion, without producing a gross motion of the whole mass. These particles, or atoms, being far too small to be seen even by the help of the most powerful microscopes, it is no wonder that we cannot observe their motion. There is therefore reason to suppose that the particles of all bodies, their constituent atoms, are in a state of motion almost too rapid for us to conceive, for the phenomena cannot be otherwise explained. The velocity of the atoms of water, for instance, is at least equal to a mile per second of time. If, as there is reason to think, some particles are at rest while others are in motion, the velocity of the latter will be proportionally greater. An increase of the velocity of revolution of the particles will constitute an increase of temperature, which may be distributed among the neighbouring bodies by what is called *conduction*—that is, on the present hypothesis, by the communication of the increased motion from the particles of one body to those of another. The velocity of the particles being further increased, they will tend to fly from each other in consequence of the centrifugal force overcoming the attraction subsisting between them. This removal of the particles from each other will constitute a new condition of the body—it will enter into the state of fusion, or become melted. But, from what we have already stated, you will perceive that, in order to remove the particles violently attracting one another asunder, the expenditure of a certain amount of living force or heat will be required. Hence it is that heat is always absorbed when the state of a body is changed from solid to liquid, or from liquid to gas. Take, for example, a block of ice cooled down to zero; apply heat to it, and it will gradually arrive at 32°, which is the number conventionally employed to represent the temperature at which ice begins to melt. If, when the ice has arrived at this temperature, you continue to apply heat to it, it will become melted; but its temperature will not increase beyond 32° until the whole has been converted into water. The explanation of these facts is clear on our hypothesis. Until the ice has arrived at the temperature of 32° the application of heat increases the velocity of rotation of its constituent particles; but the instant it arrives at that point, the velocity produces such an increase of the centrifugal force of the particles that they are compelled to separate from each other. It is in effecting this separation of particles strongly attracting one another that the heat applied is *then* spent; not in increasing the velocity of the particles. As soon, however, as the separation has been effected, and the fluid water produced, a further application of heat will cause a further increase of the velocity of the particles, constituting an increase of temperature, on which the thermometer will immediately rise above 32°. When the water has been raised to the temperature of 212°, or the boiling point, a similar phenomenon will be repeated; for it will be found impossible to increase the temperature beyond that point, because the heat then applied is employed in separating the particles of water so as to form steam, and not in increasing their velocity and living force. When, again, by the application of cold we condense the steam into water, and by a further abstraction of heat we bring the water to the solid condition of ice, we witness the repetition of similar phenomena in the reverse order. The particles of steam, in assuming the condition of water, fall together through a certain space. The living force thus produced becomes converted into heat, which must be removed before any more steam can be converted into water. Hence it is always necessary to abstract a great quantity of heat in order to convert steam into water, although the temperature will all the while remain exactly at 212°; but the instant that all the steam has been condensed, the further abstraction of heat will cause a diminution of temperature, since it can only be employed in diminishing the velocity of revolution of the atoms of water. What has been said with regard to the condensation of steam will apply equally well to the congelation of water.

I might proceed to apply the theory to the phenomena of combustion, the heat of which consists in the living force occasioned by the powerful attraction through space of the combustile for the oxygen, and to a variety of other thermochemical phenomena; but you will doubtless be able to pursue the subject further at your leisure.

I do assure you that the principles which I have very imperfectly advocated this evening may be applied very extensively in elucidating many of the abstruse as well as the simple points of science, and that patient inquiry on these grounds can hardly fail to be amply rewarded.

7
The Conservation of Force*

HERMANN VON HELMHOLTZ

SUMMARY

The basic principles on which are based the propositions in this memoir may be taken to be either of the following: (1) it is not possible by any means to produce an unlimited amount of mechanical forces; (2) all actions in nature can be explained in terms of attractive or repulsive forces between points, depending on the distances between those points. The two principles are essentially equivalent.

The objective of theoretical physics is to explain all natural phenomena in terms of motions of material particles exerting forces on each other. The first principle to be used is that of the conservation of *vis viva* (living force). For example, the work required to raise a body of weight m to a height h against the force of gravity g is mgh; a body falling through the same height attains the velocity $v = \sqrt{2gh}$, hence its *vis viva* is $\frac{1}{2}mv^2 = mgh$. The same is valid in general when the particles act on each other with any kind of forces, depending only on their distance. In general, the sum of the " tensions " (potential energies) and *vires vivae* (kinetic energies) of any system of particles remains constant, and this we may call the *principle of conservation of force*. The application of the principle in mechanical theorems is discussed; for example, the motions of perfectly elastic bodies and the transmission of waves are considered.

Can we consider heat to be an equivalent of force? The experiments of Rumford, Joule and others indicate that heat is produced by motion, and we conclude that heat *is* really some kind of molecular motion, so that the same principles may be applied to it.

---

* Read before the Physical Society of Berlin, 23 July, 1847; published under the title *Ueber die Erhaltung der Kraft. Eine physikalische Abhandlung*. G. Reimer, Berlin, 1847. Translated and edited by JOHN TYNDALL, in Taylor's *Scientific Memoirs* (1853) p. 114.

INTRODUCTION

The principal contents of the present memoir show it to be addressed to physicists chiefly, and I have therefore thought it judicious to lay down its fundamental principles in the form of a physical premise, and independent of metaphysical considerations—to develop the consequences of these principles, and to submit them to a comparison with what experience has established in the various branches of physics. The deduction of the propositions contained in the memoir may be based on either of two maxims; either on the maxim that it is not possible by any combination whatever of natural bodies to derive an unlimited amount of mechanical forces, or on the assumption that all actions in nature can be ultimately referred to attractive or repulsive forces, the intensity of which depends solely upon the distances between the points by which the forces are exerted. That both these propositions are identical is shown at the commencement of the memoir itself. Meanwhile the important bearing which they have upon the final aim of the physical sciences may with propriety be made the subject of a special introduction.

The problem of the sciences just alluded to is, in the first place, to seek the laws by which the particular processes of nature may be referred to, and deduced from, general rules. These rules,—for example, the law of the reflexion and refraction of light, the law of Mariotte and Gay-Lussac regarding the volumes of gases,—are evidently nothing more than general ideas by which the various phænomena which belong to them are connected together. The finding out of these is the office of the experimental portion of our science. The theoretic portion seeks, on the contrary, to evolve the unknown causes of the processes from the visible actions which they present; it seeks to comprehend these processes according to the laws of causality. We are justified, and indeed impelled in this proceeding, by the conviction that every change in nature *must* have a sufficient cause. The proximate causes to which we refer phænomena may, in themselves, be either variable or invariable; in the former case the above conviction impels us to seek for causes to account for the change, and thus we proceed until we at length arrive at final causes which are unchangeable, and which therefore must, in all cases where the exterior conditions are the same, produce the same invariable effects. The final aim of the theoretic natural sciences is therefore to discover the ultimate and unchangeable causes of natural phænomena. Whether all the processes of nature be actually referrible to such,—whether nature is capable of being completely comprehended, or whether changes occur which are not subject to the laws of necessary causation, but spring from spontaneity or freedom, this is not the place to decide; it is at all events clear that the science whose subject it is to comprehend nature must proceed from the assumption that it is comprehensible, and in accordance with this assumption investigate and conclude until, perhaps, she is at length admonished by irrefragable facts that there are limits beyond which she cannot proceed.

Science regards the phænomena of the exterior world according to two processes of abstraction: in the first place it looks upon them as simple existences, without regard to their action upon our organs of sense or upon each other; in this aspect they are named *matter*. The existence of matter in itself is to us something tranquil and devoid of action: in it we distinguish merely the relations of space and of quantity (mass), which is assumed to be eternally unchangeable. To matter, thus regarded, we must not ascribe qualitative differences, for when we speak of different kinds of matter we refer to differences of action, that is, to differences in the forces of matter. Matter in itself can therefore partake of one change only,—a change which has reference to space, that is, motion. Natural objects are not, however, thus passive; in fact we come to a knowledge of their existence solely from their actions upon our organs of sense, and infer from these actions a something which acts. When, therefore, we wish to make actual application of our idea of matter, we can only do it by means of a second abstraction, and ascribe to it properties which in the first case were excluded from our idea, namely the capability of producing effects, or, in other words, of exerting force. It is evident that in the application of the ideas of matter and force to nature the two former should never be separated: a mass of pure matter would, as far as we and nature are concerned, be a nullity, inasmuch as no action could be wrought by it either upon our organs of sense or upon the remaining portion of nature. A pure force would be something which must have a basis, and yet which has no basis, for the basis we name matter. It would be just as erroneous to define matter as something which has an actual existence, and force as an idea which has no corresponding reality. Both, on the contrary, are abstractions from the actual, formed in precisely similar ways. Matter is only discernible by its forces, and not by itself.

We have seen above that the problem before us is to refer back the phænomena of nature to unchangeable final causes. This requirement may now be expressed by saying that for final causes unchangeable forces must be found. Bodies with unchangeable forces have been named in science (chemistry) elements. Let us suppose the universe decomposed into elements possessing unchangeable qualities, the only alteration possible to such a system is an alteration of position, that is, motion; hence, the forces can be only moving forces dependent in their action upon conditions of space.

To speak more particularly: the phænomena of nature are to be referred back to motions of material particles possessing unchangeable moving forces, which are dependent upon conditions of space alone.

Motion is the alteration of the conditions of space. Motion, as a matter of experience, can only appear as a change in the relative position of at least two material bodies. Force, which originates motion, can only be conceived of as referring to the relation of at least two material bodies towards each other; it is therefore to be defined as the endeavour of two masses to alter their relative position. But the force which two masses exert upon each other must be resolved into those exerted by all their particles upon each other; hence in mechanics we go back to forces exerted by material points. The relation of one point to another, as regards space, has reference solely to their distance apart: a moving force, therefore, exerted by each upon the other, can only act so as to cause an alteration of their distance, that is, it must be either attractive or repulsive.

Finally, therefore, we discover the problem of physical natural science to be, to refer natural phænomena back to unchangeable attractive and repulsive forces, whose intensity depends solely upon distance. The solvability of this problem is the condition of the complete comprehensibility of nature. In mechanical calculations this limitation of the idea of moving force has not yet been assumed: a great number, however, of general principles referring to the motion of compound systems of bodies are only valid for the case that these bodies operate upon each other by unchangeable attractive or repulsive forces; for example, the principle of virtual velocities; the conservation of the motion of the centre of gravity; the conservation of the principal plane of rotation; of the moment of rotation of free systems, and the conservation of *vis viva*. In terrestrial matters application is made chiefly of the first and last of these principles, inasmuch as the others refer to systems which are supposed to be completely free; we shall however show that the first is only a special case of the last, which therefore must be regarded as the most general and important consequence of the deduction which we have made.

Theoretical natural science therefore, if she does not rest contented with half views of things, must bring her notions into harmony with the expressed requirements as to the nature of simple forces, and with the consequences which flow from them. Her vocation will be ended as soon as the reduction of natural phænomena to simple forces is complete, and the proof given that this is the only reduction of which the phænomena are capable.

I. *The principle of the Conservation of* vis viva.

We will set out with the assumption that it is impossible, by any combination whatever of natural bodies, to produce force continually from nothing. By this proposition Carnot and Clapeyron have deduced theoretically a series of laws, part of which are proved by experiment and part not yet submitted to this test, regarding the latent and specific heats of various natural bodies.† The object of the present memoir is to carry the same principle, in the same manner, through all branches of physics; partly for the purpose of showing its applicability in all those cases where the laws of the phænomena have been sufficiently investigated, partly, supported by the manifold analogies of the known cases, to draw further conclusions regarding

---

† S. CARNOT, *Reflexions sur la puissance motrice du feu*, Paris, 1824.
E. CLAPEYRON, *J. Ecole Polyt.* **14**, 153 (1834).

laws which are as yet but imperfectly known, and thus to indicate the course which the experimenter must pursue.

The principle mentioned can be represented in the following manner:—Let us imagine a system of natural bodies occupying certain relative positions towards each other, operated upon by forces mutually exerted among themselves, and caused to move until another definite position is attained; we can regard the velocities thus acquired as a certain mechanical work and translate them into such. If now we wish the same forces to act a second time, so as to produce again the same quantity of work, we must, in some way, by means of other forces placed at our disposal, bring the bodies back to their original position, and in effecting this a certain quantity of the latter forces will be consumed. In this case our principle requires that the quantity of work gained by the passage of the system from the first position to the second, and the quantity lost by the passage of the system from the second position back again to the first, are always equal, it matters not in what way or at what velocity the change has been effected. For were the quantity of work greater in one way than another, we might use the former for the production of work and the latter to carry the bodies back to their primitive positions, and in this way procure an indefinite amount of mechanical force. We should thus have built a *perpetuum mobile* which could not only impart motion to itself, but also to exterior bodies.

If we inquire after the mathematical expression of this principle, we shall find it in the known law of the conservation of *vis viva*. The quantity of work which is produced and consumed may, as is known, be expressed by a weight m, which is raised to a certain height h; it is then mgh, where g represents the force of gravity. To rise perpendicularly to the height h, the body m requires the velocity $v + \sqrt{2gh}$, and attains the same by falling through the same height. Hence we have $\frac{1}{2}mv^2 = mgh$; and hence we can set the half of the product mv^2, which is known in mechanics under the name of the *vis viva* of the body m, in the place of the quantity of work. For the sake of better agreement with the customary manner of measuring the intensity of forces, I propose calling the quantity $\frac{1}{2}mv^2$ the quantity of *vis viva*, by which it is rendered identical with the quantity of work. For the applications of the doctrine of *vis viva* which have been hitherto made this alteration is of no importance, but we shall derive much advantage from it in the following. The principle of the conservation of *vis viva*, as is known, declares that when any number whatever of material points are set in motion, solely by such forces as they exert upon each other, or as are directed against fixed centres, the total sum of the *vires vivæ*, at all times when the points occupy the same relative position, is the same, whatever may have been their paths or their velocities during the intervening times. Let us suppose the *vires vivæ* applied to raise the parts of the system or their equivalent masses to a certain height, it follows from what has just been shown, that the quantities of work, which are represented in a similar manner, must also be equal under the conditions mentioned. This principle however is not applicable to all possible kinds of forces; in mechanics it is generally derived from the principle of virtual velocities, and the latter can only be proved in the case of material points endowed with attractive or repulsive forces. We will now show that the principle of the conservation of *vis viva* is alone valid where the forces in action may be resolved into those of material points which act in the direction of the lines which unite them, and the intensity of which depends only upon the distance. In mechanics such forces are generally named central forces. Hence, conversely, it follows that in all actions of natural bodies upon each other, where the above principle is capable of general application, even to the ultimate particles of these bodies, such central forces must be regarded as the simplest fundamental ones.

Let us consider the case of a material point with the mass m, which moves under the influence of several forces which are united together in a fixed system A; by mechanics we are enabled to determine the velocity and position of this point at any given time. We should therefore regard the time t as primitive variable, and render dependent upon it,—the ordinates x, y, z of m in a system of co-ordinates, definite as regards A, the tangential velocity q, the components of the latter parallel to the axes, $u = dx/dt$, $v = dy/dt$, $w = dz/dt$, and finally the components of the acting forces

$$X = m\frac{du}{dt}, \quad Y = m\frac{dv}{dt}, \quad Z = m\frac{dw}{dt}.$$

Now according to our principle $\frac{1}{2}mq^2$, and hence also q^2, must be always the same when m occupies the same position relative to A;

it is not therefore to be regarded merely as a function of the primitive variable t, but also as a function of the coordinates x, y, z only; so that

$$d(q^2) = \frac{d(q^2)}{dx} dx + \frac{d(q^2)}{dy} dy + \frac{d(q^2)}{dz} dz. \tag{1}$$

As $q^2 = u^2 + v^2 + w^2$, we have $d(q^2) = 2u\,du + 2v\,dv + 2w\,dw$. Instead of u let us substitute its value dx/dt, and instead of du its value $X dt/m$, the corresponding values of v and w being also used, we have

$$d(q^2) = \frac{2X}{m} dx + \frac{2Y}{m} dy + \frac{2Z}{m} dz. \tag{2}$$

As the equations (1) and (2) must hold good together for all values whatever of dx, dy, dz, it follows that

$$\frac{d(q^2)}{dx} = \frac{2X}{m}, \quad \frac{d(q^2)}{dy} = \frac{2Y}{m} \quad \text{and} \quad \frac{d(q^2)}{dz} = \frac{2Z}{m}.$$

But if q^2 is a function of x, y, and z merely, it follows that X, Y, and Z, that is, the direction and magnitude of the acting forces, are purely functions of the position of m in respect to A.

Let us now imagine, instead of the system A, a single material point a, it follows from what has been just proved, that the direction and magnitude of the force exerted by a upon m is only affected by the position which m occupies with regard to a. But the only circumstance, as regards position, that can affect the action between the two points is the distance ma; the law, therefore, in this case would require to be so modified, that the direction and magnitude of the force must be functions of the said distance, which we shall name r. Let us suppose the coordinates referred to any system of axes whatever whose origin lies in a, we have then

$$m\,d(q^2) = 2X\,dx + 2Y\,dy + 2Z\,dz = 0 \tag{3}$$

as often as

$$d(r^2) = 2x\,dx + 2y\,dy + 2z\,dz = 0$$

that is, as often as

$$dz = -\frac{xdx + ydy}{z};$$

setting this value in equation (3), we obtain

$$\left(X - \frac{x}{z}Z\right)dx + \left(Y - \frac{y}{z}Z\right)dy = 0$$

for any values whatever of dx and dy; hence also singly

$$X = \frac{x}{z}Z \quad \text{and} \quad Y = \frac{y}{z}Z,$$

that is to say, the resultant must be directed towards the origin of co-ordinates, or towards the point a.

Hence in systems to which the principle of the conservation of force can be applied, in all its generality, the elementary forces of the material points must be central forces.

II. *The principle of the Conservation of Force.*

We will now give the law for the cases where the central forces act, a still more general expression.

Let ϕ be the intensity of the force which acts in the direction of r, which is to be regarded as positive when it attracts, and as negative when it repels, then we have

$$X = -\frac{x}{r}\phi; \quad Y = -\frac{y}{r}\phi; \quad Z = -\frac{z}{r}\phi; \tag{1}$$

and from equation (2) of the foregoing section, we have

$$md(q^2) = -2\frac{\phi}{r}(xdx + ydy + zdz); \quad \text{hence}$$

$$\tfrac{1}{2}md(q^2) = -\phi dr;$$

or when Q and R, q and r represent corresponding tangential velocities and distances,

$$\tfrac{1}{2}mQ^2 - \tfrac{1}{2}mq^2 = -\int_r^R \phi dr. \tag{2}$$

Let us regard this equation more closely; we find at the left-hand side the difference of the *vires vivæ* possessed by m at two different distances. To understand the import of the quantity $\int_r^R \phi dr$, let us suppose the intensities of ϕ which belong to different points of the connecting line ma erected as ordinates at these points, then the above quantity would denote the superficial content of the space enclosed between the two ordinates r and R. As this surface may be regarded as the sum of the infinite number of ordinates which lie between r and R, it therefore represents the sum of the intensities of the forces which act at all distances between R and r. Calling the forces which tend to move the point m, before the motion has actually taken place, *tensions*, in opposition to that which in mechanics is named *vis viva*, then the quantity $\int_r^R \phi dr$ would be *the sum of the tensions* between the distances R and r, and the above law would be thus expressed:— The increase of *vis viva* of a material point during its motion under the influence of a central force is equal to the sum of the tensions which correspond to the alteration of its distance.

Let us suppose the case of two points operated upon by an attractive force, at the distance R; by the action of the force they will be drawn to less distances r, their velocity, and consequently *vis viva*, will be increased; but if they should be driven to greater distances r, their *vis viva* must diminish and must finally be quite consumed. We can therefore distinguish, in the case of attractive forces, the sum of the tensions for the distances between $r = 0$ and $r = R.\int_0^R \phi dr$, as those which yet remain, but those between $r = R$ and $r = \infty$ as those already consumed; the former can immediately act, the latter can only be called into action by an equivalent loss of *vis viva*. It is the reverse with repulsive forces. If the points are situated at the distance R, as the distance becomes greater *vis viva* will be gained, and the still existing tensions are those between $r = R$ and $r = \infty$, those lost are between $r = 0$ and $r = R$.

To carry our law through in quite a general manner, let us suppose any number whatever of material points with the masses m_1, m_2, m_3, &c. denoted generally by m_a; let the components of the forces which act upon these parallel to the axes be X_a, Y_a, Z_a, the components of the velocities along the same axes u_a, v_a, w_a, the tangential velocity q_a; let the distance between m_a and m_b be r_{ab}, the central force between both being ϕ_{ab}. For the single point m_n we have, analogous to equation (1),

$$X_n = \Sigma\left[(x_a - x_n)\frac{\phi_{an}}{r_{an}}\right] = m_n \frac{du_n}{dt},$$

$$Y_n = \Sigma\left[(y_a - y_n)\frac{\phi_{an}}{r_{an}}\right] = m_n \frac{dv_n}{dt},$$

$$Z_n = \Sigma\left[(z_a - z_n)\frac{\phi_{an}}{r_{an}}\right] = m_n \frac{dw_n}{dt},$$

where the sign of summation Σ includes all members which are obtained by putting in the place of the index a the separate indices 1, 2, 3, &c., with the exception of n.

Multiplying the first equation by $dx_n = u_n dt$, the second by $dy_n = v_n dt$, the third by $dz_n = w_n dt$, and supposing the three equations thus obtained to be formed for every single point of m_b, as it is already done for m_n; adding all together, we obtain

$$\Sigma\left[(x_a - x_b)dx_b \frac{\phi^{ab}}{r_{ab}}\right] = \Sigma[\tfrac{1}{2}m_a d(u_a^2)]$$

$$\Sigma\left[(y_a - y_b)dy_b \frac{\phi^{ab}}{r_{ab}}\right] = \Sigma[\tfrac{1}{2}m_a d(v_a^2)]$$

$$\Sigma\left[(z_a - z_b)dz_b \frac{\phi^{ab}}{r_{ab}}\right] = \Sigma[\tfrac{1}{2}m_a d(cw_a^2)].$$

The members of the left-hand series will be obtained by placing instead of a all the single indices 1, 2, 3, &c., and in each case for b also all the values of b, which are greater or smaller than a already possesses. The sums divide themselves therefore into two portions, in one of which a is always greater than b, and in the other always smaller, and it is clear that for every member of the one portion

$$(x_p - x_q)dx_q \frac{\phi_{pq}}{r_{pq}},$$

a member

$$(x_q - x_p)dx_p \frac{\phi_{pq}}{r_{pq}}$$

must appear in the other portion: adding both together, we obtain

$$- (x_p - x_q)(dx_p - dx_q) \frac{\phi_{pq}}{r_{pq}}:$$

drawing the sums thus together, adding all three and setting

$$\tfrac{1}{2}d[(x_a - x_b)^2 + (y_a - y_b)^2 + (z_a - z_b)^2] = r_{ab}dr_{ab},$$

we obtain

$$- \Sigma[\phi_{ab}dr_{ab}] = \Sigma[\tfrac{1}{2}m_a d(q_a^2)], \tag{3}$$

or

$$- \Sigma\left[\int_{r_{ab}}^{R_{ab}} \phi_{ab}dr_{ab}\right] = \Sigma[\tfrac{1}{2}m_a Q_a^2] - \Sigma[\tfrac{1}{2}m_a q_a^2], \tag{4}$$

where R and Q, as well as r and q, denote contemporaneous values.

We have here at the left-hand side again the sum of the tensions consumed, on the right the *vis viva* of the entire system, and we can now express the law as follows:—In all cases of the motion of free material points under the influence of their attractive and repulsive forces, whose intensity depends solely upon distance, the loss in tension is always equal to the gain in *vis viva*, and the gain in the former equal to the loss in the latter. Hence *the sum of the existing tensions and vires vivæ is always constant.* In this most general form we can distinguish our law as *the principle of the conservation of force.*

In the deduction of the law as given above, nothing is changed if a number of the points, which we will denote generally by the letter d, are supposed to be fixed, so that q_d is constantly $= 0$; the form of the law will then be

$$\Sigma[\phi_{ab}dr_{ab}] + \Sigma[\phi_{ad}dr_{ad}] = - \Sigma[\tfrac{1}{2}m_b d(q_b^2)]. \tag{5}$$

It remains to be shown in what relation the principle of the conservation of force stands to the most general law of statics, the so-called principle of virtual velocities. This follows immediately from equations (3) and (5). If equilibrium is to set in when a certain arrangement of the points ma takes place, that is, if in case these points come to rest, hence $q_a = 0$, they remain at rest, hence $dq_a = 0$, it follows from equation (3),

$$\Sigma[\phi_{ab}dr_{ab}] = 0; \tag{6}$$

or in case that forces act upon them from points m_b without the systems, by equation (5),

$$\Sigma[\phi_{ab}dr_{ab}] + \Sigma[\phi_{ad}dr_{ad}] = 0 . \tag{7}$$

In these equations under dr are understood alterations of distance consequent on the small displacements of the point m_a, which are permitted by the conditions of the system. We have seen, in the former deductions, that an increase of *vis viva*, hence a transition from rest to motion, can only be effected by an expenditure of tension; in correspondence with this, the last equations declare that in cases where in no single one of the possible directions of motion tension in the first moment is consumed, the system once at rest must remain so.

It is known that all the laws of statics may be deduced from the above equations. The most important consequence as regards the nature of the acting forces is this: instead of the arbitrary small displacements of the points m, let us suppose such introduced as might take place were the system in itself firmly united, so that in equation (7) every $dr_{ab} = 0$, it follows singly,

$$\Sigma[\phi_{ab}dr_{ab}] = 0, \text{ and}$$
$$\Sigma[\phi_{ab}dr_{ab}] = 0.$$

Then the exterior, as well as the interior forces, must satisfy among themselves the conditions of equilibrium. Hence, if any system whatever of natural bodies be brought by the action of exterior forces into a certain position of equilibrium, the equilibrium will not be destroyed—1, if we imagine the single points of the system in their present position to be rigidly united to each other; and 2, if we then remove the forces which the points exert upon each other. From this however it follows further: If the forces which two material points exert upon each other be brought into equilibrium by the action of exterior forces, the equilibrium must continue, when, instead of the mutual forces of the points, a rigid connexion between them is substituted. Forces, however, which are applied to two points of a rigid right line can only be in equilibrium when they lie in this line and are equal and opposite. It follows therefore for the forces of the points themselves, which are equal and opposed to the exterior ones, that they must act in the direction of the line of connexion, and hence must be either attractive or repulsive.

The preceding propositions may be collected together as follows:—

1. Whenever natural bodies act upon each other by attractive or repulsive forces, which are independent of time and velocity, the sum of their *vires vivæ* and tensions must be constant; the maximum quantity of work which can be obtained is therefore a limited quantity.

2. If, on the contrary, natural bodies are possessed of forces which depend upon time and velocity, or which act in other directions than the lines which unite each two separate material points, for example, rotatory forces, then combinations of such bodies would be possible in which force might be either lost or gained *ad infinitum*.

3. In the case of the equilibrium of a system of bodies under the operation of central forces, the exterior and the interior forces must, each system for itself, be in equilibrium, if we suppose that the bodies of the system cannot be displaced, the whole system only being moveable in regard to bodies which lie without it. A rigid system of such bodies can therefore never be set in motion by the action of its interior forces, but only by the operation of exterior forces. If, however, other than central forces had an existence, rigid combinations of natural bodies might be formed which could move of themselves without needing any relation whatever to other bodies.

III. *The application of the principle in Mechanical Theorems.*

We will now turn to the special application of the law of the constancy of force. In the first place, we will briefly notice those cases in which the principle of the conservation of *vis viva* has been heretofore recognized and made use of.

1. *All motions which proceed under the influence of the general force of gravitation*; hence those of the heavenly and the ponderable terrestrial bodies. In the former case the law pronounces itself in the increase of velocity which takes place when the paths of the planets approach the central body, in the unchangeableness of the greater axes of their orbits, their time of rotation and orbital revolution. In the latter case, by the known law that the terminal velocity depends only upon the perpendicular height fallen through, and that this velocity, when it is not destroyed by friction or by unelastic concussion, is exactly sufficient to carry the body to the same height as

that from which it has fallen; that the height of ascent of a certain weight is used as the unit of measure in our machines has been already mentioned.

2. *The transmission of motion through the incompressibly solid and fluid bodies*, where neither friction nor concussion takes place. Our general principle finds for this case expression in the known fact, that a motion transmitted and altered by mechanical powers, diminishes in force as it increases in velocity. Let us suppose that by means of any machine whatever, to which mechanical force is uniformly applied, the weight m is raised with the velocity c, by another mechanical arrangement the weight nm may be raised, but only with the velocity c/n, so that in both cases the quantity of tension developed by the machine in the unit of time is expressed by mgc, where g represents the intensity of gravity.

3. *The motions of perfectly elastic, solid, and fluid bodies.* As condition of complete elasticity, we must to the ordinary definition, that the body which has been changed in form or volume completely regains its primitive condition, add, that no friction takes place between the particles in the interior. In the laws of these motions our principle was first recognized and most frequently made use of. Among the most common cases of its application to solids may be mentioned the collision of elastic bodies, the laws of which may be readily deduced from our principle; the conservation of the centre of gravity, and the manifold elastic vibrations which continue without fresh excitement, until, through the friction of the interior parts and the yielding up of motion to exterior bodies, they are destroyed. In fluid bodies, liquid (evidently also elastic, but endowed with a high modulus of elasticity and a position of equilibrium of the particles) as well as gaseous, (with low modulus of elasticity and without position of equilibrium) motions are in general propagated by undulations. To these belong the waves on the surfaces of liquids, the motion of sound, and probably also those of light and radiant heat.

The *vis viva* of a single particle Δm in a medium which is traversed by a train of waves, is evidently to be determined from the velocity which it possesses as its position of equilibrium. The general equation of waves determines, as is known, the velocity u, when a^2 is the intensity, λ the length of the wave, α the velocity of propagation, x the abscissa, and t the time, as follows:—

$$u = a \cdot \cos\left[\frac{2\pi}{\lambda}(x - \alpha t)\right].$$

For the position of equilibrium u is $= a$, hence the *vis viva* of the particle Δm during the undulatory motion $\frac{1}{2}\Delta m a^2$ is proportional to the intensity. If the waves expand spherically from a centre, masses continually increasing in bulk are set in motion, and hence the intensity must diminish, if the *vis viva* is to remain the same. Now as the masses embraced by the waves increase as the square of the distance, the law follows as a consequence, that the intensities diminish in the reciprocal ratio.

The laws of reflexion, refraction and polarization of light at the limit of two media of different wave-velocity, are known to have been deduced by Fresnel from the assumption that the motion of the limiting particles in both media is the same, and from the conservation of *vis viva*. By the interference of two trains of waves we have no destruction of *vis viva*, but merely another distribution. Two trains of waves of the intensities a^2 and b^2 which do not interfere, give to all points on which they strike the intensity $a^2 + b^2$; if they interfere, the maxima possess the intensity $(a + b)^2$, that is, $2ab$ more, and the minima $(a - b)^2$, just as much less than $a^2 + b^2$.

The *vis viva* of elastic waves is only destroyed by such processes as we denominate absorption. The absorption of sonorous waves we find to be chiefly effected by concussion against yielding unelastic bodies, for example, curtains and coverlets; they may therefore be regarded as a communication of motion to the bodies in question, in which the motion is destroyed by friction. Whether motion can be destroyed by the friction of the air-particles against each other is a question which cannot yet perhaps be decided. The absorption of rays of heat is accompanied by a proportional development of heat; how far the latter corresponds to a certain equivalent of force, we will consider in the next section. The conservation of force would take place if the quantity of heat radiated from one body appeared again in the body into which it was radiated, provided that none was lost by conduction, and no portion of the rays escaped elsewhere. The theorem is certainly assumed in investigations upon radiant heat, but

I am aware of no experiments which furnish the proof of it. As regards the absorption of light by imperfectly transparent or totally opake bodies, we are acquainted with three peculiarities. In the first place, phosphorescent bodies absorb the light in such a manner that they yield it up again afterwards. Secondly, most luminous rays, perhaps all of them, appear to excite heat. The obstacles to the belief in the identity of the luminous, calorific, and chemical rays of the spectrum have been lately disappearing more and more; the heat-equivalent of the chemical and luminous rays appears to be very inconsiderable in comparison to their intense actions upon the eye. If, however, the similarity of these differently acting rays does not permit of being established, then the end of the motion of light must undoubtedly be declared to be unknown. Thirdly, in many cases the light absorbed developes chemical action. We must here distinguish two species of action; first, where the mere incitement to chemical activity is communicated, as in the case of those bodies which induce catalytic action, for example, the action upon a mixture of chlorine and hydrogen; and secondly, those in which it is opposed to chemical action, as in the decomposition of salts of silver and the action upon the green portions of plants. In most of these processes however, the effect of light is so little known, that we are able to form no judgement regarding the magnitude of the forces developed. The latter appear to be considerable in quantity and intensity only in the actions on the green portions of plants.

IV. *The Force-equivalent of Heat.*

The mechanical processes in which an absolute loss of force has been heretofore assumed, are—

1. *The collision of unelastic bodies.*—The loss is mostly connected with the change of form and the compression of the body struck; hence with the increase of the tensions, we also find, by the repetition of such shocks, a considerable development of heat; for example, by hammering a piece of metal; finally, a portion of the motion will be communicated as sound to the contiguous solid and gaseous bodies.

2. *Friction*, both at the surface of two bodies which move over each other, and also that arising in the interior from the displacement

of the particles. In the case of friction certain small changes in the molecular constitution of the bodies take place, especially when they commence to rub against each other; afterwards the surfaces generally accommodate themselves to each other, so that this change becomes a vanishing quantity in the further course of the motion. In many cases such changes do not at all appear, for example, when fluids rub against solid bodies or against each other. Besides those already mentioned, thermic and electric changes always take place.

It is customary in mechanics to represent friction as a force which acts against the existing motion, and the intensity of which is a function of the velocity. This mode of representing the subject is only made use of for the sake of calculation, and is evidently an extremely incomplete expression of the complicated process of action and reaction of the molecular forces. From this customary manner of regarding the subject, it was inferred that by friction *vis viva* was absolutely lost; and the same was assumed in the case of unelastic collision. It is not however here taken into account that, disregarding the increase of the tensions caused by the compression of the body rubbed or struck, the heat developed is also the representant of a force by which we can develop mechanical actions; the electricity developed, whose attractions and repulsions are direct mechanical actions, and the heat it excites, an indirect one, has also been neglected. It remains therefore to be asked whether the sum of these forces always corresponds to the mechanical force which has been lost. In those cases where the molecular changes and the development of electricity are to a great extent avoided, the question would be, whether for a certain loss of mechanical force a definite quantity of heat is always developed, and how far can a quantity of heat correspond to a mechanical force. For the solution of the first question but few experiments have yet been made. Joule[†] has measured the heat developed by the friction of water in narrow tubes, and that developed in vessels in which the water was set in motion by a paddle-wheel; in the first case he found that the heat which raises 1 kilogramme of water 1°, was sufficient to raise 452 kilogrammes through the height of 1 metre; in the second case he found the weight to be 521 kilogrammes. His method of measurement however meets the

---

† J. P. JOULE, *Phil. Mag.* [3] **27**, 205 (1845).

difficulty of the investigation so imperfectly, that the above results can lay little claim to accuracy. Probably the above numbers are too high, inasmuch as in his proceeding a quantity of heat might have readily escaped unobserved, while the necessary loss of mechanical force in other portions of the machine is not taken into account.

Let us now turn to the further question, how far heat can correspond to an equivalent of force. The material theory of heat must necessarily assume the quantity of caloric to be constant; it can therefore develope mechanical forces only by its effort to expand itself. In this theory the force-equivalent of heat can only consist in the work produced by the heat in its passage from a warmer to a colder body; in this sense the problem has been treated by Carnot and Clapeyron, and all the consequences of the assumption, at least with gases and vapours, have been found corroborated.

To explain the heat developed by friction, the material theory must either assume that it is communicated by conduction as supposed by Henry,† or that it is developed by the compression of the surfaces and of the particles rubbed away, as supposed by Berthollet.‡ The first of these assumptions lacks all experimental proof; if it were true, then in the neighbourhood of the rubbed portions a cold proportionate to the intense heat often developed must be observed. The second assumption, without dwelling upon the altogether improbable magnitude of action, which according to it must be ascribed to the almost imperceptible compression of the hydrostatic balance, breaks down completely when it is applied to the friction of fluids, or to the experiments where wedges of iron have been rendered red-hot and soft by hammering and pieces of ice melted by friction††; for here the softened iron and the water of the melted ice could not remain in a compressed condition. Besides this, the development of heat by the motion of electricity proves that the quantity of heat can be actually increased. Passing by frictional and voltaic electricity—because it might here be suspected that, by some hidden relation of caloric to electricity, the former was transferred from the place where it was originated and deposited in the heated wire—two other ways of

---

† W. HENRY, *Mem. Lit. Phil. Soc. Manchester* 5, 603 (1802).

‡ C. L. BERTHOLLET, *Statique Chimique*, Vol. 1, p. 247. (Paris, 1803.)

†† H. DAVY, " Essay on Heat, Light, and the Combinations of Light." (*Contributions to physical and medical knowledge*, Bristol, 1799.)

producing electric tensions by purely mechanic agencies in which heat does not at all appear, are still open to us, namely, by induction and by the motion of magnets. Suppose we possess a completely insulated body positively electric and which cannot part with its electricity; an insulated conductor brought near to it will show free $+ E$, we can discharge this upon the interior coating of a battery and remove the conductor, which will then show $- E$; this latter can of course be discharged upon the exterior surface of the first or upon a second battery. By repeating this process, it is evident that we can charge a battery of any magnitude whatever as often as we please, and by means of its discharge can develope heat, which nowhere disappears. We shall, on the contrary, have consumed a certain amount of force, for at each removal of the negatively-charged conductor from the inducing body the attraction between both is to be overcome. This process is essentially carried out when the electrophorus is used to charge a Leyden jar. The same takes place in magneto-electric machines; as long as magnet and keeper are moved opposite to each other, electric currents are excited which develope heat in the connecting wire; and inasmuch as they constantly act in a sense contrary to the motion of the keeper, they destroy a certain amount of mechanical force. Here evidently heat *ad infinitum* may be developed by the bodies constituting the machine, while it nowhere disappears. That the magneto-electric current developes heat instead of cold, in the portion of the spiral directly under the influence of the magnet, Joule has endeavoured to prove experimentally.† From these facts, it follows that the quantity of heat can be absolutely increased by mechanical forces, that therefore calorific phænomena cannot be deduced from the hypothesis of a species of matter, the mere presence of which produces the phænomena, but that they are to be referred to changes, to motions, either of a peculiar species of matter, or of the ponderable or imponderable bodies already known, for example of electricity or the luminiferous æther. That which has been heretofore named the quantity of heat, would, according to this, be the expression, first, of the quantity of *vis viva* of the calorific motion, and, secondly, of the quantity of those tensions between the atoms, which, by changing the arrangement of the latter, such a motion can

---

† J. P. JOULE, *Phil. Mag.* [3] (1844). (Page reference not given.)

develope. The first portion would correspond to that which has been heretofore called free heat, the second with that which has been named latent heat. If it be permitted to make an attempt at rendering the idea of this motion still clearer, the view derived from the hypothesis of Ampère seems best suited to the present state of science. Let us imagine the bodies formed of atoms which themselves are composed of subordinate particles (chemical elements, electricity, &c.), in such an atom three species of motion may be distinguished,— 1, displacement of the centre of gravity; 2, rotation round the centre of gravity; 3, displacement of the particles of the atom among themselves. The two first would be compensated by the forces of the neighbouring atoms, and hence transmitted to these in the form of undulations, a species of propagation which corresponds to the radiation of heat, but not to its conduction. Motions of the single particles of the atoms among themselves, would be compensated by the forces existing within the atom, and would communicate motion but slowly to the surrounding atoms, as a vibrating string sets a second in motion and thereby loses an equal quantity of motion itself; this description of motion seems to be similar to the conduction of heat. It is also clear, that such motions in the atoms may cause changes in the molecular forces, and consequently give rise to expansion or an alteration of the state of aggregation. Of what nature the motion is, we have no means whatever of ascertaining; the possibility of conceiving the phænomena of heat as being due to motion is, however, sufficient for our present object. The conservation of force in the case of these motions will hold good in all cases where hitherto the conservation of caloric has been assumed; for example, in all phænomena of radiation and conduction of heat from one body to another, and in the case of the appearance and disappearance of heat during changes of aggregation.

Of the different modes in which heat manifests itself, we have considered the cases where one body radiates into another, and where it is produced by mechanical force; further on we will examine the heat generated by electricity. It remains to consider the development of heat in chemical processes. It has been heretofore referred to the setting free of caloric which was previously latent in the combining bodies. According to this, we must ascribe to every simple body, and every chemical combination which is capable of entering into still further combinations of a higher order, a definite quantity of latent heat which is necessary to its chemical constitution. From this we derive the law, which has been also partially verified by experience, that when several bodies unite together to a chemical compound, the same quantity of heat is developed, no matter in what order the combination may have been effected.† According to our way of viewing the subject, the quantity of heat developed by chemical processes would be the quantity of *vis viva* produced by the chemical attractions, and the above law would be the expression for the principle of the conservation of force in this case.

As little as the conditions and laws of the generation of heat have been investigated, although such a generation undoubtedly occurs, this has been done with reference to the disappearance of heat. Hitherto we are only acquainted with cases in which chemical combinations have been decomposed, or less dense states of aggregation brought about, and thus heat rendered latent. Whether by the development of mechanical force heat disappears, which would be a necessary postulate of the conservation of force, nobody has troubled himself to inquire. I can only in respect to this cite an experiment by Joule,‡ which seems to have been carefully made. He found that air while streaming from a reservoir with a capacity of 136·5 cubic inches, in which it was subjected to a pressure of 22 atmospheres, cooled the surrounding water 4°·085 Fahr. when the air issued into the atmosphere, and therefore had to overcome the resistance of the latter. When, on the contrary, the air rushed into a vessel of equal size which had been exhausted of air, thus finding no resistance and exerting no mechanical force, no change of temperature took place.

[In the remainder of this section, Helmholtz compares the previous work of Clapeyron and Holtzmann with his own views. Section V is devoted to a discussion of the force-equivalent of electric processes, including galvanism and thermo-electricity. In Section VI, he discusses magnetism and electromagnetism, and concludes that the general principle of conservation of force can be applied to all physical processes.]

---

† H. Hess, *Ann. Phys.* [2] **50**, 392 (1840); **56**, 598 (1842).
‡ J. P. Joule, *Phil. Mag.* [3] **26**, 369 (1845).

8
The Nature of the Motion which we call Heat*

RUDOLF CLAUSIUS

SUMMARY

Krönig has recently proposed that the molecules of a gas do not oscillate about definite positions of equilibrium, but that they move with constant velocity in straight lines until they strike against other molecules, or against the surface of the container. I share this view, but believe that rotary as well as rectilineal motions are present, since the *vis viva* of the translatory motion alone is too small to represent the whole heat present in the gas. I assert that the translatory motion of the molecules will also be in a constant ratio to the motions of the constituents of the molecules among themselves, because the transfer of *vis viva* from translatory to internal motions, and conversely, by collisions, must eventually reach an equilibrium.

The molecular conditions that must be satisfied in order that a gas may behave as an *ideal* gas are indicated. Types of molecular motion occurring in the solid, liquid, and gaseous states are described, and a qualitative theory of evaporation is given.

In order to explain the relations of volumes of gases combining by chemical reactions, it is assumed that in simple gases two or more atoms are combined to form one molecule.

By considering collisions of molecules of mass m against the wall of a container of volume v, assuming that all n molecules move with velocity u, Clausius shows that the pressure of the gas is equal to $(mnu^2/3v)$. The absolute temperature is proportional to $(\frac{1}{2}nmu^2)$. The actual velocities of gas molecules can be calculated in this way; for example, at the temperature of melting ice, the velocity of an oxygen molecule is, on the average, 461 metres per second.

---

* Originally published under the title " Ueber die Art der Bewegung, welche wir Wärme nennen ", *Annalen der Physik*, Vol. 100, pp. 353–80 (1857); English translation in *Philosophical Magazine*, Vol. 14, pp. 108–27 (1857).

The ratio of the *vis viva* of translatory motion to the total *vis viva* is found to be equal to $3(\gamma' - \gamma)/2\gamma$, where γ is the specific heat of the gas at constant volume (for unit volume) and γ' is specific heat at constant pressure. For air, $\gamma'/\gamma = 1\cdot421$, and hence this ratio is $0\cdot6315$.

1. Before writing my first memoir on heat, which was published in 1850,† and in which heat is assumed to be a motion, I had already formed for myself a distinct conception of the nature of this motion, and had even employed the same in several investigations and calculations. In my former memoirs I intentionally avoided mentioning this conception, because I wished to separate the conclusions which are deducible from certain general principles from those which presuppose a particular kind of motion, and because I hoped to be able at some future time to devote a separate memoir to my notion of this motion and to the special conclusions which flow therefrom. The execution of this project, however, has been retarded longer than I at first expected, inasmuch as the difficulties of the subject, as well as other occupations, have hitherto prevented me from giving to its development that degree of completeness which I deemed necessary for publication.

A memoir has lately been published by Krönig, under the title *Grundzüge einer Theorie der Gase*,‡ in which I have recognized some of my own views. Seeing that Krönig has arrived at these views just as independently as I have, and has published them before me, all claim to priority on my part is of course out of the question; nevertheless, the subject having once been mooted in this memoir, I feel myself induced to publish those parts of my own views which I have not yet found in it. For the present, I shall confine myself to a brief indication of a few principal points, and reserve a more complete analysis for another time.††

---

† *Ann. Phys.* **79**, 368, 500 (1850).

‡ This was first printed separately by A. W. Hayn in Berlin, and afterwards appeared in Poggendorff's *Annalen*, vol. xcix, p. 315.

†† I must not omit to mention here, that some time ago Mr. William Siemens of London, when on a visit to Berlin, informed me that Joule had also expressed similar ideas in the *Memoirs of the Literary and Philosophical Society of Manchester*. My views being consequently no longer completely new, this was an additional reason why I should hasten their publication less than I otherwise probably should have done. Hitherto I

2. Krönig assumes that the molecules of gas do not oscillate about definite positions of equilibrium, but that they move with constant velocity in straight lines until they strike against other molecules, or against some surface which is to them impermeable. I share this view completely, and I also believe that the expansive force of the gas arises from this motion. On the other hand, I am of opinion that this is not the only motion present.

In the first place, the hypothesis of a rotatory as well as a progressive motion of the molecules at once suggests itself; for at every impact of two bodies, unless the same happens to be central and rectilineal, a rotatory as well as a translatory motion ensues.

I am also of opinion that vibrations take place within the several masses in a state of progressive motion. Such vibrations are conceivable in several ways. Even if we limit ourselves to the consideration of the atomic masses solely, and regard these as absolutely rigid, it is still possible that a molecule, which consists of several atoms, may not also constitute an absolutely rigid mass, but that within it the several atoms are to a certain extent moveable, and thus capable of oscillating with respect to each other.

I may also remark, that by thus ascribing a movement to the atomic masses themselves, we do not exclude the hypothesis that each atomic mass may be provided with a quantity of finer matter, which, without separating from the atom, may still be moveable in its vicinity.

By means of a mathematical investigation given at the end of the present memoir, it may be proved that the *vis viva* of the translatory motion alone is too small to represent the whole heat present in the gas; so that without entering into the probability of the same, we are thus compelled to assume one or more motions of another kind.

---

have not been able to procure the memoir of Joule in question, and therefore I am ignorant how far he has pursued the subject, and whether his views coincide with mine in all points. It is to be regretted that Joule did not publish his memoir in a more widely circulated periodical. (See the Bibliography for Joule's paper; this note by Clausius induced Joule to send his paper to *Philosophical Magazine*, requesting that it be reprinted, but observing that " The *Memoirs of the Literary and Philosophical Society of Manchester* are at present regularly forwarded to the principal scientific societies of Europe and America.")

According to this calculation, the excess of the whole *vis viva* over that of the translatory motion alone is particularly important in gases of a complicated chemical constitution, in which each molecule consists of a great number of atoms.

3. In one and the same gas the translatory motion of the whole molecules will always have a constant relation to the several motions which, in addition to the above, the constituents of the molecules likewise possess. For brevity I will call the latter the *motions of the constituents*.

Conceive a number of molecules whose constituents are in active motion, but which have no translatory motion. It is evident the latter will commence as soon as two molecules in contact strike against each other in consequence of the motion of their constituents. The translatory motion thus originated will of course occasion a corresponding loss of *vis viva* in the motion of the constituents. On the other hand, if the constituents of a number of molecules in a state of translatory motion were motionless, they could not long remain so, in consequence of the collisions between the molecules themselves, and between them and fixed sides or walls. It is only when all possible motions have reached a certain relation towards one another, which relation will depend upon the constitution of the molecules, that they will cease mutually to increase or diminish each other.

When two molecules whose constituents are in motion come into collision they will not rebound, like two elastic balls, according to the ordinary laws of elasticity; for their velocities and directions after collision will depend, not only upon the motion which the whole molecules had before impact, but also upon the motion of those constituents which are nearest each other at the moment of collision. After the equalization of the several motions, however, when the translatory motion is, on the whole, neither increased nor diminished by the motions of the constituents, we may, in our investigation of the total action of a great number of molecules, neglect the irregularities occurring at the several collisions, and assume that, in reference to the translatory motion, the molecules follow the common laws of elasticity.

4. The explanation of the expansive force of gases and its dependence upon volume and temperature, as given by Krönig, suffers no essential modification through the introduction of other motions.

The pressure of the gas against a fixed surface is caused by the molecules in great number continually striking against and rebounding from the same. The force which must thence arise is, in the first place, by equal velocity of motion inversely proportional to the volume of the given quantity of gas; and secondly, by equal volume proportional to the *vis viva* of the translatory motion: the other motions do not here immediately come into consideration.

On the other hand, from Gay-Lussac's law we know that, under constant volume, the pressure of a perfect gas increases in the same ratio as the temperature calculated from $-273°$ C., which we call the absolute temperature. Hence, according to the above, it follows that the absolute temperature is proportional to the *vis viva* of the translatory motion of the molecules. But as, according to a former remark, the several motions in one and the same gas bear a constant relation to each other, it is evident that the *vis viva* of the translatory motion forms an aliquot part of the total *vis viva*, so that the absolute temperature is also proportional to the whole *vis viva* in the gas.

These considerations, together with others connected therewith to be given hereafter, induced me, in my memoir " On the Moving Force of Heat," to express the opinion that the specific heat of gases was constant; which opinion was in opposition to the experiments then known.† The quantity of heat which must be imparted to a gas, under constant volume, in order to raise its temperature is to be considered as the increase of the *vis viva* in the gas, inasmuch as in this case no work is done whereby heat could be consumed. The specific heat *under constant volume*, therefore, is in a perfect gas the magnitude which Rankine calls the *true* specific heat. Now the assertion that the true specific heat of a gas is constant, is simply equivalent to the assertion that *the total* vis viva *in the gas has a constant ratio to the* vis viva *of the translatory motion which serves us as a measure of the temperature*. With respect to the specific heat under constant pressure, I have proved in the memoir before cited, and by means of a hypothesis proceeding from the same considerations, that it differs only by a constant magnitude from the true specific heat.

---

† Poggendorff's *Annalen*, vol. lxxix, p. 393. *Phil. Mag.*, vol. ii, pp. 1, 102.

5. The foregoing is true for permanent gases only, and even for these only approximatively. In general, the small deviations which present themselves can be easily accounted for.

In order that Mariotte's and Gay-Lussac's laws, as well as others in connexion with the same, may be strictly fulfilled, the gas must satisfy the following conditions with respect to its molecular condition:—

(1) The space actually filled by the molecules of the gas must be infinitesimal in comparison to the whole space occupied by the gas itself.

(2) The duration of an impact, that is to say, the time required to produce the actually occurring change in the motion of a molecule when it strikes another molecule or a fixed surface, must be infinitesimal in comparison to the interval of time between two successive collisions.

(3) The influence of the molecular forces must be infinitesimal. Two conditions are herein involved. In the first place, it is requisite that the force with which all the molecules at their mean distances attract each other, vanish when compared with the expansive force due to the motion. But the molecules are not always at their mean distances asunder; on the contrary, during their motion a molecule is often brought into close proximity to another, or to a fixed surface consisting of active molecules, and in such moments the molecular forces will of course commence their activity. The second condition requires, therefore, that those parts of the path described by a molecule under the influence of the molecular forces, when the latter are capable of altering appreciably the direction or velocity of the molecule's motion, should vanish when compared with those parts of its path with respect to which the influence of these forces may be regarded as zero.

If these conditions are not fulfilled, deviations in several ways from the simple laws of gases necessarily arise; and these deviations become more important the less the molecular condition of the gas fulfils the conditions in question.

On becoming acquainted with the celebrated investigations of Regnault on the deviations of gases from Mariotte's and Gay-Lussac s laws, I attempted, by means of the principles above intimated, to deduce some conclusions with respect to the molecular condition of several gases from the nature of the deviations which Regnault detected in the same. A description of this method, however, would be too prolix; and even the results, in consequence of the many difficulties encountered in actual calculation, are too uncertain to merit being here adduced.

Whenever, therefore, in the sequel a gas is spoken of, we shall, as before, conceive it to be one which *perfectly* fulfils the above conditions, and which Regnault calls an *ideal* gas, inasmuch as all known gases present but an approximation to this condition.

6. After these considerations on the *gaseous* condition, the question at once arises in what manner the *solid* and *liquid* conditions differ from the gaseous. Although a definition of these states of aggregation, in order to be satisfactory in all its details, would require a more complete knowledge than we at present possess of the condition of the individual molecules, yet it appears to me that several fundamental distinctions may be advanced with tolerable probability.

A motion of the molecules takes place in all three states of aggregation.

In the *solid* state, the motion is such that the molecules move about certain positions of equilibrium without ever forsaking the same, unless acted upon by foreign forces. In solid bodies, therefore, the motion may be characterized as a vibrating one, which may, however, be of a very complicated kind. In the first place, the constituents of a molecule may vibrate among themselves; and secondly, the molecule may vibrate as a whole: again, the latter vibrations may consist in oscillations to and fro of the centre of gravity, as well as in rotatory oscillations around this centre of gravity. In cases where external forces act on the body, as in concussions, the molecules may also be permanently displaced.

In the *liquid* state the molecules have no longer any definite position of equilibrium. They can turn completely around their centres of gravity; and the latter, too, may be moved completely out of its place. The separating action of the motion is not, however, sufficiently strong, in comparison to the mutual attraction between the molecules, to be able to separate the latter entirely. Although a molecule no longer adheres to definite neighbouring molecules, still it does not spontaneously forsake the latter, but only under the united actions of forces proceeding from other molecules, with respect to which it then occupies the same position as it formerly did with respect to its neighbouring molecules. In liquids, therefore, an oscillatory, a rotatory, and a translatory motion of the molecules takes place, but in such a manner that these molecules are not thereby separated from each other, but, even in the absence of external forces, remain within a certain volume.

Lastly, in the *gaseous* state the motion of the molecules entirely transports them beyond the spheres of their mutual attraction, causing them to recede in right lines according to the ordinary laws of motion. If two such molecules come into collision during their motion, they will in general fly asunder again with the same vehemence with which they moved towards each other; and this will the more readily occur, since a molecule will be attracted with much less force by another single molecule than by all the molecules which in the liquid or solid state surround it.

7. The phænomenon of *evaporation* appearing peculiarly interesting to me, I have attempted to account for the same in the following manner.

It has been stated above, that in liquids a molecule, during its motion, either remains within the sphere of attraction of its neighbouring molecules, or only leaves the same in order to take up a corresponding position with respect to other neighbouring molecules. This applies only to the mean value of the motions, however; and as the latter are quite irregular, we must assume that the velocities of the several molecules deviate within wide limits on both sides of this mean value.

Taking next the surface of a liquid into consideration, I assume that, amongst the varied motions to and fro, it happens that under the influence of a favourable cooperation of the translatory, oscillatory, and rotatory motions, a molecule separates itself with such violence from its neighbouring molecules that it has already receded from the sphere of their action before losing all its velocity under the influence of their attracting forces, and thus that it continues its flight into the space above the liquid.

Conceive this space to be enclosed, and at the commencement empty; it will gradually become more and more filled with these expelled molecules, which will now deport themselves in the space exactly as a gas, and consequently in their motion strike against the enclosing surfaces. The liquid itself, however, will form one of these surfaces; and when a molecule strikes against the same, it will not in general be driven back, but rather retained, and, as it were, absorbed in consequence of the renewed attraction of the other molecules into whose vicinity it has been driven. A state of equilibrium will ensue when the number of molecules in the superincumbent space is such, that on the average as many molecules strike against, and are retained by the surface of the liquid in a given time, as there are molecules expelled from it in the same time. The resulting state of equilibrium, therefore, is not a state of rest or a cessation of evaporation, but a state in which evaporation and condensation continually take place and compensate each other in consequence of their equal intensity.

The density of the vapour necessary for this compensation depends upon the number of molecules expelled from the surface of the liquid in the unit of time; and this number is again evidently dependent upon the activity of the motion within the liquid, that is to say, upon its temperature. I have not yet succeeded in deducing from these considerations the law according to which the pressure of vapour must increase with the temperature.

The preceding remarks on the deportment of the surface of the liquid towards the superincumbent vapour, apply in a similar manner to the other surfaces which enclose the space filled with vapour. The vapour is in the first place condensed on these surfaces, and the liquid thus produced then suffers evaporation, so that here also a state must be attained in which condensation and evaporation become equal. The requisite quantity of condensed vapour on these surfaces depends upon the density of the vapour in the enclosed space, upon the temperature of the vapour and of the enclosing surfaces, and upon the force with which the molecules of vapour are attracted towards these surfaces. In this respect a maximum will occur when the enclosing surfaces are completely moistened with the condensed liquid; and as soon as this takes place, these surfaces deport themselves exactly like a single surface of the same liquid.

8. The reason why the presence of another gas above the liquid cannot impede the evaporation of the same may now be immediately explained.

The pressure of the gas on the liquid arises solely from the fact, that here and there single molecules of gas strike against the surface of the liquid. In other respects, however, inasmuch as the molecules of gas themselves actually fill but a very small part of the superincumbent space, the latter must be considered as empty, and as offering a free passage to the molecules of the liquid. In general these molecules will only come into collision with those of the gas at comparatively great distances from the surface, and the former will then deport themselves towards the latter as would the molecules of any other admixed gas. We must conclude, therefore, that the liquid also expels its molecules into the space filled with gas; and that in this case also the quantity of vapour thus mixed with the gas continues to increase until, on the whole, as many molecules of vapour strike against and are absorbed by the surface of the liquid as the latter itself expels; and the number of molecules of vapour to the unit of volume requisite hereto, is the same whether the space does or does not contain additional molecules of gas.

The pressure of the gas, however, exercises a different influence on the interior of the liquid. Here also, or at places where the mass of liquid is bounded by a side of the vessel, it may happen that the molecules separate from each other with such force that for a moment the continuity of the mass is broken. The small vacuum thus produced, however, is surrounded on all sides by masses which do not admit of the passage of the moved molecules; and hence this vacuum will only then become magnified into a bubble of vapour, and be able to continue as such, when the number of molecules expelled from its enclosing liquid walls is sufficient to produce an internal vapour-pressure capable of holding in equilibrium the pressure which acts externally and tends to compress the bubble again. Hence the expansive force of the enclosed vapour must be greater, the greater the pressure to which the liquid is exposed, and thus is explained the relation which exists between the pressure and the temperature of the boiling-point.

The relations will be more complicated when the gas above the liquid is itself condensable, and forms a liquid which mixes with the given one, for then of course the tendency of the two kinds of matter to mix enters as a new force. I shall not here enter into these phænomena.

As in liquids, so also in solids the possibility of an evaporation may be comprehended; nevertheless it does not follow from this that, on the contrary, an evaporation *must* take place on the surface of all bodies. It is, in fact, readily conceivable that the mutual cohension of the molecules of a body may be so great, that, so long as the temperature does not exceed a certain limit, even the most favourable combination of the several molecular motions is not able to overcome this cohesion.

9. The explanation of the *generation* and *consumption* of heat accompanying changes of the state of aggregation and of the volume of a body, follows immediately, according to the above principles, from the consideration of the positive and negative work done by the active forces.

In the first place, let us consider the *interior* work.

When the molecules of a body change their relative positions, the change may be either in the same direction as that according to which the forces resident in the molecules tend to move the latter, or in a direction opposite to this. In the first case, a certain velocity is imparted by the forces to the molecules during their passage from one position to the other, and the *vis viva* of this velocity is immediately transformed into heat; in the second case, and disregarding for the present the action of extraordinary foreign forces, it is in virtue of heat that the molecules move in directions opposed to the interior forces, and the retardation which these molecules thereby suffer through the action of the opposing forces appears as a diminution of the motion of heat.

In the passage from the solid to the liquid state the molecules do not, indeed, recede beyond the spheres of their mutual action; but, according to the above hypothesis, they pass from a definite and, with respect to the molecular forces, suitable position, to other irregular positions, in doing which the forces which tend to retain the molecules in the former position have to be overcome.

In evaporation, the complete separation which takes place between the several molecules and the remaining mass evidently again necessitates the overcoming of opposing forces.

With respect to gaseous bodies, it is evident from what has been above remarked concerning the requisites of a perfect gaseous state, that as soon as a gas has once attained this state, molecular attractions have no longer to be overcome during its further expansion, so that during the changes of volume of an ideal gas no interior work has to be performed.

10. Besides the interior, we have in the next place to consider the *exterior* work, and the corresponding change in the quantity of heat.

In the first place, with respect to the changes in the heat of a permanent gas subjected to pressure or expanding by overcoming a counter-pressure, Krönig has already remarked at the conclusion of his memoir, that when one of the sides against which the molecules of gas strike is itself in motion, these molecules will not rebound with velocities equal to those they possessed when moving towards the side. If the side approaches the molecules, the velocity of the molecules on leaving the same is in general greater than before, if it recedes less. By means of special mathematical considerations, it may without difficulty be proved that the increase or decrease of the *vis viva* thereby produced must exactly correspond to the external work done by the expansive force of the gas; nevertheless it is not even necessary to give a special proof of this theorem, since the same is an immediate consequence of the general law of equivalence between *vis viva* and work.

If the side moves so slowly that the pressure of the gas against the moving side is just as great as against a stationary one, then, in determining the work, the velocity of the side no longer enters into consideration, but merely the whole path described by the same. On the contrary, if the velocity of the side is so great that in the vicinity of the same a sensible compression or rarefaction of the gas thereby ensues, then the pressure actually exercised by the gas during the motion must always be brought into calculation.

When an overflow takes place between two vessels filled with gases of different densities, or between a full and an empty vessel, on the whole no work will be performed, and therefore no change in the total quantity of heat can occur. It is not here asserted that no change in the quantity of heat takes place in either of the two vessels considered separately, for a mass of gas whose molecules move principally in a definite direction deports itself towards adjoining gaseous masses in the same manner as a moved side; and when the moved gaseous mass strikes against stationary walls, just as much motion of heat makes its appearance as *vis viva* is lost by the common translatory motion of the whole mass.

Just as in the changes of volume of gaseous bodies, so also in other cases the external work must be taken into consideration; as, for instance, the work which during the evaporation is employed to overcome external resistance, and thus to make room for the vapour as it is generated. In solid and liquid bodies which only suffer small changes of volume, the external work is also for the most part only small; nevertheless here also cases occur in which its influence becomes considerable.

11. Lastly, I must mention a phænomenon the explanation of which appears to me to be of great importance, viz. *when two gases combine with each other, or when a gas combines with another body, and the combination is also gaseous, the volume of the compound gas bears a simple ratio to the volumes of the single constituents, at least when the latter are gaseous.*

Krönig has already proved that the pressure exerted by a gas on the unit of its enclosing surface must be proportional to the number of molecules contained in the unit of volume, and to the *vis viva* of the several molecules arising from their translatory motion, the only one which Krönig considers.

If we apply this to simple gases, and assume that, when pressure and temperature are the same, equal volumes contain the same number of atoms—a hypothesis which for other reasons is very probable,—it follows that, in reference to their translatory motion, the atoms of different gases must have the same *vis viva*.

We will next examine in what manner this theorem remains true when applied to the molecules of compound gases.

12. In the first place, let us compare compound gases amongst themselves, *e.g.* two gases to form which the constituents have combined in ratios of volume respectively equal to 1 : 1 and 1 : 2. Nitric acid and nitrous acid may serve as examples.

With respect to these two gases, we know that quantities containing the same amount of oxygen occupy the same volume. Hence here, too, equal volumes contain the same number of molecules, although in the one gas each molecule consists of two, and in the other of three atoms; and we must further conclude, that even these differently constituted molecules have the same *vis viva* with respect to their translatory motion.

In most other compound gases we are led to the same conclusion; and in cases which do not submit themselves to this rule, it does not appear to me impossible that the discrepancy may be accounted for in one or both of two ways: either the gas was not sufficiently removed from its points of condensation when its volume was determined, or the chemical formula hitherto employed does not properly represent the manner in which the atoms are combined to form molecules.

On comparing compound and simple gases, however, an unmistakeable deviation from the foregoing rule shows itself, inasmuch as the space corresponding to an atom of the simple gas does not correspond to a molecule of the compound one. When two simple gases combine in equal volumes, it is well known that no change of volume takes place, whilst according to the above rule the volume ought to be diminished in the ratio of 2 : 1. Again, when a volume of one gas combines with two or three volumes of another, the combination is found to occupy two volumes, whereas according to rule it ought only to occupy one volume, and so on.

13. On seeking to explain these curious anomalies, and especially to find a common law governing the relations of volume in gases, I was led to adopt the following view as being most plausible. I beg to offer the same to the scientific public as a hypothesis which is at least worthy of further examination.

I assume that the force which determines chemical combination, and which probably consists in a kind of polarity of the atoms, is already active in simple substances, and that *in these likewise two or more atoms are combined to form one molecule.*

For instance, let equal volumes of oxygen and nitrogen be given. A mixture of these gases contains a certain number of molecules, which consist either of two atoms of oxygen or of two atoms of nitrogen. Conceive the mixture to pass into a chemical compound, and the latter then contains just as many molecules, which are merely constituted in a different manner, inasmuch as each consists of an atom of oxygen and an atom of nitrogen. Hence there is no reason why a change of volume should take place. If, on the other hand, one volume of oxygen and two of nitrogen are given, then in the mixture each molecule consists of two, and in the compound of three atoms. The chemical combination, therefore, has caused the number of molecules to diminish in the ratio of 3 : 2, and consequently the volume ought to diminish in the same ratio.

It is well known that some simple substances do not, in the gaseous form, occupy the volume which their atomic weights and the volumes of their combinations would lead us to anticipate, but another, and in most cases a smaller volume, which bears to the former a simple ratio. A special investigation of these substances would here be out of place, more especially as two of them, sulphur and phosphorus, deport themselves in other respects in so remarkable a manner, in consequence of the variety of conditions they are capable of assuming, that we may reasonably expect further discoveries from chemistry with respect to these bodies; and then, perhaps, besides other irregularities, those of the volumes of their vapours will be explained. Nevertheless I may here recall one circumstance which in some cases may possibly facilitate this explanation. I refer to the fact, that the above hypothesis, according to which the molecules of simple substances each consist of *two* atoms, may not be the only possible one.

On comparing with each other all cases of simple and compound gases, we must not expect to find immediately a perfect agreement throughout. I am of opinion, however, that, under the present uncertainty with respect to the inner constitution of several bodies, and particularly of those which possess a complicated chemical composition, too great weight out not to be laid upon individual anomalies; and I deem it probable, that, by means of the above hypothesis respecting the molecules of simple substances, all relations of volume in gases may be referred back to the theorem, *that the several molecules of all gases possess equal* vis viva *in reference to their translatory motion.*

14. Proceeding to treat the subject mathematically, we will first deduce the expression which shows in what manner the pressure of the gas on the sides of the vessel depends upon the motion of its molecules.

As the shape of the vessel is indifferent, we will select that which is most convenient for our purpose. We will assume the vessel to be very flat, and that two of its sides consist of parallel planes so close to one another that their distance asunder is infinitesimal when compared with the other dimensions of the vessel. Hence we need not consider the cases where the molecules strike against one of the narrow strips of sides, and we may assume that each moves in a right line until it either strikes against another molecule or against one of the large parallel sides. In fact, to take the small sides into consideration would change nothing in the final result, and would only make the development more prolix.

Let us consider one only of the two large sides; during the unit of time it is struck a certain number of times by molecules moving in all possible directions compatible with an approach towards the surface. We must first determine the number of such shocks, and how many correspond on the average to each direction.

15. Hereafter we shall always assume the gas to be an *ideal* one; in other words, we shall disregard the irregularities proceeding from an imperfect gaseous state, so that in determining the pressure we may, with Krönig, introduce certain simplifications in place of considering the motion exactly as it takes place.

The whole number of shocks received by the side remains unchanged when we assume that the molecules do not disturb each other in their motion, but that each pursues its rectilineal path until it arrives at the side.

Further, although it is not actually necessary that a molecule should obey the ordinary laws of elasticity with respect to elastic spheres and a perfectly plane side, in other words, that when striking the side, the angle and velocity of incidence should equal those of reflexion, yet, according to the laws of probability, we may assume that there are as many molecules whose angles of reflexion fall within a certain interval, *e.g.* between 60° and 61°, as there are molecules whose angles of incidence have the same limits, and that, on the whole, the velocities of the molecules are not changed by the side. No difference will be produced in the final result, therefore, if we assume that for each molecule the angle and velocity of reflexion are equal to those of incidence. According to this, each molecule would move to and fro between the large parallel sides, in the same directions as those chosen by a ray of light between two plane mirrors, until at length it would come in contact with one of the small sides; from this it would be reflected, and then commence a similar series of journeys to and fro, and so forth.

Lastly, there is no doubt that actually the greatest possible variety exists amongst the velocities of the several molecules. In our considerations, however, we may ascribe a certain mean velocity to all molecules. It will be evident from the following formulæ, that, in order to maintain an equal pressure, this mean velocity must be so chosen that with it the total *vis viva* of all the molecules may be the same as that corresponding to their actual velocities.

16. According to these assumptions, it is evident, that, during the unit of time, each molecule will strike the side under consideration just as often as during that time it can, by following its peculiar direction, travel from the side in question to the other and back again. Let h be the distance between the large parallel sides, and ϑ the acute angle between the normal and the direction of motion; then $h/\cos\vartheta$ is the length of the path from one side to the other, and

$$\frac{u \cdot \cos\vartheta}{2h} \tag{1}$$

the number of impulses given to the side, u being the velocity of the molecule.

With respect to the directions of the several molecules, we must assume that on the average each direction is equally represented. From this it follows, that the number of molecules moving in directions which form with the normal angles included between ϑ and $\vartheta + d\vartheta$, has to the whole number of molecules the same ratio that the surface of the spherical zone, whose limiting circles correspond to the angles ϑ and $d\vartheta$, has to the surface of the hemisphere, that it, the ratio

$$2\pi \sin\vartheta d\vartheta : 2\pi.$$

Hence if n represents the whole number of molecules, the number which corresponds to the angular interval between ϑ and $\vartheta + d\vartheta$ will be

$$n \sin\vartheta d\vartheta,$$

and the number of shocks imparted by them will be

$$\frac{nu}{2h} \cos\vartheta \sin\vartheta d\vartheta. \tag{2}$$

17. In order to determine the intensity of a shock, the whole velocity must be resolved into two components, one parallel and the other perpendicular to the side. Of these components, the first will not be affected by the shock, and will not enter into consideration in determining its intensity; the second, however, whose magnitude is represented by $u \cos \vartheta$, will be changed by the shock into an equal velocity in the opposite direction. The action of the side upon the molecule, therefore, consists in depriving it in one direction of the velocity $u \cos \vartheta$, calculated according to the normal, and of imparting to it an equal velocity in an opposite direction; in other words, of imparting to it a velocity of $2u \cdot \cos \vartheta$ in the latter direction. Hence the quantity of motion imparted to the molecule will be

$$2mu \cdot \cos \vartheta, \tag{3}$$

where m is the mass of the molecule.

Applying this to all molecules which correspond to the interval between ϑ and $\vartheta + d\vartheta$, we obtain during the unit of time,

$$\frac{nu}{2h} \cos \vartheta \sin \vartheta d\vartheta$$

times the same action, hence the quantity of motion imparted to these molecules during the unit of time is

$$\frac{nmu^2}{h} \cos^2 \vartheta \cdot \sin \vartheta \cdot d\vartheta. \tag{4}$$

Integrating this expression between the limits $\vartheta = 0$ and $\vartheta = \pi/2$, we find the motion imparted by the side to all the molecules which strike against it during the unit of time to be

$$\frac{nmu^2}{3h}. \tag{5}$$

Let us now conceive the side to be capable of moving freely; then in order that it may not recede before the shocks of the molecules, it must be acted upon on the other side by a counter force, which latter may in fact be regarded as continuous, in consequence of the great number of shocks and the feebleness of each. The intensity of this force must be such as to enable it, during the unit of time, to generate the quantity of motion represented by the above expression. Since all forces, however, are measured by the quantity of motion they can produce in the unit of time, the above expression at once represents this force as well as the pressure exerted by the gas, the latter being equilibrated by the former.

If α be the superficial area of the side and p the pressure on the unit of surface, then

$$p = \frac{mnu^2}{3\alpha h}.$$

The product αh here involved gives the volume of the vessel or gas; hence representing the same by v, we have

$$p = \frac{mnu^2}{3v}. \tag{6}$$

The same formula would have been obtained if, with Krönig, we had, for the sake of simplification, assumed that one third of the whole molecules move perpendicularly to the side under consideration, and the two remaining thirds in two other directions parallel to the side. Nevertheless I preferred deducing the formula for the pressure without using this simplifying hypothesis.

If we write the last equation in the form

$$\tfrac{3}{2} pv = \frac{nmu^2}{2}, \tag{6a}$$

the right-hand side then denotes *the* vis viva *of the translatory motion of the molecules.*† But, according to Mariotte's and Gay-Lussac's laws,

$$pv = T \cdot \text{const.},$$

---

† In accordance with a practice lately become general, and with what I have myself done in former memoirs, I call the *semi*-product of the mass into the square of the velocity the *vis viva*, because it is only with this definition of the notion that we can, without the addition of a coefficient, equate the expressions representing a quantity of work and the increase or decrease of *vis viva* which corresponds to the same.

where T is the absolute temperature; hence

$$\frac{nmu^2}{2} = T \cdot \text{const.};$$

and, as before stated, the *vis viva* of the translatory motion is proportional to the absolute temperature.

18. We may now make an interesting application of the above equations by determining the velocity u with which the several molecules of gas move.

The product nm represents the mass of the whole given quantity of gas, whose weight we will call q. Then g being the force of gravity,

$$nm = \frac{q}{g};$$

and from equation (6) we deduce

$$u^2 = \frac{3gpv}{q}. \tag{7}$$

Adopting the metre as unit of length, and the kilogramme as unit of weight, let us suppose a kilogramme of gas under the pressure of 1 atmosphere—10333 kilogrammes on the square metre—to be given. Then

$$g = 9^m \cdot 80896,$$
$$p = 10333,$$
$$q = 1.$$

To determine v, we know that, according to Regnault, a kilogramme of atmospheric air under the pressure of 1 atmosphere, and at the temperature of melting ice, occupies 0·7733 cubic metre. Hence ρ being the specific gravity of the gas under consideration, its volume at the temperature of melting ice will be

$$\frac{0 \cdot 7733}{\rho};$$

and at the absolute temperature T, assuming 273° to be the absolute temperature of melting ice, it will be

$$\frac{0 \cdot 7733}{\rho} \cdot \frac{T}{273}.$$

Hence we deduce

$$u^2 = 3 \cdot 9{\cdot}80896 \cdot 10333 \cdot 0{\cdot}7733 \cdot \frac{T}{273 \cdot \rho}$$

$$= 235130 \frac{T}{273 \cdot \rho},$$

and consequently

$$u = 485^m \sqrt{\frac{T}{273 \cdot \rho}}. \tag{8}$$

As particular cases, we obtain the following numbers corresponding to the temperature of melting ice:—

for oxygen .... 461^m,

for nitrogen.... 492^m,

for hydrogen .. 1844^m.

These numbers are the mean velocities which, for the totality of molecules, give the same *vis viva* as would their actual velocities. At the same time, however, it is possible that the actual velocities of the several molecules differ materially from their mean value.

19. By means of the equations above established, we will lastly examine *what relation exists between the* vis viva *of the translatory motion and the whole* vis viva, *or heat, in the gas.*

In doing so we will conceive the quantity of heat to be measured, not by the ordinary unit of heat, but by the mechanical unit of *vis viva*, or what is equivalent, by the unit of work. To this end we have only to divide the quantity of heat measured in the ordinary manner by the thermal equivalent of the unit of work, which as before I will denote by A. Let H be the quantity of heat thus determined.

Further, let c be the specific heat of the gas under constant volume, in other words, the true specific heat; then the increase of the quantity of heat in the quantity q of gas corresponding to an elevation of temperature dT is

$$dH = \frac{qc}{A} dT.$$

Integrating this equation, we have

$$H = \frac{qc}{A} T. \tag{9}$$

No constant need be added, since, as before remarked, the heat in the gas is proportional to the *vis viva* of the translatory motion, and hence also to the absolute temperature.

The expression on the right of this equation may be replaced by another which is very convenient for our present investigation.

The quantity of heat which must be imparted to the quantity of gas q in order to elevate its temperature by dT and its volume by dv is expressed thus,

$$\frac{qc}{A} dT + pdv,$$

wherein the first term represents the increase of the heat contained in the gas, and the second the quantity of heat consumed by work. If we assume the gas to be heated under constant pressure, the relation between dT and dv is thereby defined. For we have generally

$$pv = T \cdot \text{const.};$$

and differentiating, under the supposition that p is constant, we obtain

$$pdv = dT \cdot \text{const.};$$

whence the undetermined constant may be eliminated by means of the foregoing equation, and we have

$$dv = \frac{v}{T} dT.$$

Let us substitute this value of dv in the above equation, and at the same time note that, c' being the specific heat under constant pressure, the whole quantity of heat imparted to the gas in the case under consideration may be represented by $(qc'/A)dT$. In this manner we arrive at the equation

$$\frac{qc'}{A} dT = \frac{qc}{A} dT + \frac{pv}{T} dT,$$

whence we conclude that

$$\frac{q(c' - c)}{A} \cdot T = pv. \tag{10}$$

By means of this equation (9) becomes

$$H = \frac{c}{c' - c} \cdot pv. \tag{11}$$

20. Let us now return to the equation (6a) before established, and for brevity let us denote the *vis viva* of the translatory motion by K, then

$$K = \frac{3}{2}pv.$$

By combining this with the foregoing equation we obtain

$$\frac{K}{H} = \frac{3}{2}\left(\frac{c'}{c} - 1\right). \tag{12}$$

The ratio of the *vis viva* of the translatory motion to the whole *vis viva* is thus reduced to the ratio between the two specific heats.

In order to compare with each other the values of the ratio K/H corresponding to different gases, it will be found convenient to introduce in the above formula, in place of the specific heats calculated with reference to the unit of weight, those calculated according to the unit of volume, which for distinction may be represented by γ and γ'. The equation then becomes

$$\frac{K}{H} = \frac{3}{2} \cdot \frac{\gamma' - \gamma}{\gamma}. \tag{13}$$

If we neglect deviations which arise from an imperfect gaseous condition, and conceive all gases to be in the ideal state, then, as I have shown in my memoir " On the Moving Force of Heat†," the difference $\gamma' - \gamma$ is the same for all gases. Hence the ratio K/H is inversely proportional to the true specific heat of the gas calculated according to the unit of volume.

---

† Poggendorff's *Annalen*, vol. lxxix, p. 394. *Phil. Mag.*, vol. ii, p. 1.

For those simple gases which manifest no irregularities with respect to their volume, and for those compound ones which suffered no diminution of volume during the act of combination, γ, and therefore K/H also, has the same value. For these gases we have approximately

$$\frac{\gamma'}{\gamma} = 1\cdot 421,$$

hence

$$\frac{K}{H} = 0\cdot 6315.$$

On the contrary, for those compound gases whose volumes during combination have been diminished, γ is greater, and consequently K/H less; and, in fact, it is less the smaller the volume of the combination compared with the sum of the volumes of its constituents, the latter being all considered gaseous.

Thus it is corroborated what was before stated, that the *vis viva* of the translatory motion does not alone represent the whole quantity of heat in the gas, and that the difference is greater the greater the number of atoms of which the several molecules of the combination consist. We must conclude, therefore, that besides the translatory motion of the molecules as such, the constituents of these molecules perform other motions, whose *vis viva* also forms a part of the contained quantity of heat.

Zurich, January 5, 1857.

9

On the Mean Lengths of the Paths Described by the Separate Molecules of Gaseous Bodies*

RUDOLF CLAUSIUS

SUMMARY

The theory of gases proposed by Joule, Krönig and Clausius (see preceding selection) was criticized on the ground that if the molecules really move great distances in straight lines, then two gases in contact with each other would rapidly mix, in contrast to experience. The objection may be answered by taking account of the fact that real gases are not " ideal " and therefore the portion of the time during which molecular forces act is not vanishingly small compared with the entire path of a molecule. A quantitative theory may be constructed if one assumes the molecules to be characterized by a certain distance ρ, such that if the centres of gravity of two molecules pass by each other at a distance greater than ρ the intermolecular repulsive forces do not come into play and there is only a slight deflection resulting from attractive forces, while if the distance is less than ρ the repulsive forces cause the molecules to rebound from each other. The spherical volume of radius ρ around the molecule is called the *sphere of action* of the molecule. We ask: how far on an average can a

---

* Originally published under the title " Ueber die mittlere Länge der Wege, welche bei Molecularbewegung gasförmigen Körper von den einzelnen Moleculen zurückgelegt werden, nebst einigen anderen Bemerkungen über die mechanischen Wärmetheorie ", *Annalen der Physik*, Vol. 105, pp. 239–58 (1858); English translation by DR. F. GUTHRIE, in *Phil. Mag.*, Vol. 17, pp. 81–91 (1859).

molecule move before its centre of gravity comes into the sphere of action of another molecule? This average distance is called the mean free path l.

It is shown that the mean length of path of a molecule is in the same proportion to the radius of the sphere of action as the entire space occupied by the gas, to that portion of the space which is actually filled up by the spheres of action of the molecules: $l/\rho = V/(\frac{4}{3}\pi\rho^3)$. (It is assumed that all the molecules move with the same velocity.) For example, if this ratio is taken to be 1000:1, and the average distance between molecules is $\lambda = V^{1/3}$, we find that $l = 62\lambda$. Although the number of molecules in a given volume of gas is as yet unknown, we must assume that it is quite large, and that λ is very small compared to our usual units of length. Thus it is plausible that the mean path is really quite small, and the objection cited is not valid.

(1.) The February number of Poggendorff's *Annalen* contains a paper by Buijs-Ballot " On the Nature of the Motion which we call Heat and Electricity."† Amongst the objections which the author there makes against the views advanced by Joule, Krönig and myself concerning the molecular motion of gaseous bodies, the following deserves especial consideration. Attention is drawn to the circumstance that, if the molecules moved in straight lines, volumes of gases in contact would necessarily speedily mix with one another,—a result which does not actually take place. To prove that such mixture does not occur, the following facts are adduced (p. 250):— " How then does it happen that tobacco-smoke, in rooms, remains so long extended in immoveable layers?" Mention is also made of the same appearance with clouds of smoke in the open air. Further, " If sulphuretted hydrogen or chlorine be evolved in one corner of a room, entire minutes elapse before they are smelt in another corner, although the particles of gas must have had to traverse the room hundreds of times in a second." Further, " How could carbonic acid gas remain so long in an open vessel?"

These objections may, at first glance, appear to have very great weight; and I consider it therefore necessary to prove, by special considerations, that the facts adduced are perfectly reconcileable with the theory of the rectilineal motion of the molecules. Indeed, I rejoice at the discussion of this point by M. Buijs-Ballot, inasmuch

---

† C. H. D. Buijs-Ballot (Buys-Ballot), *Ann. Phys.* **103**, 240 (1858).

as it affords me a desired opportunity of completing this part of my theory (which was perhaps discussed too briefly in my paper), and to prevent thereby further misunderstandings.

(2.) It is assumed in the objections, that the molecules traverse considerable spaces in straight lines; this appears prominently in the second objection, in which it is said that a molecule must have had to traverse the room many times in a second. This assumption can, however, in nowise be considered as a necessary consequence of the views advanced by me concerning the conditions of gases. Amongst the conditions which must be satisfied if Mariotte and Gay-Lussac's law for a gas is true with perfect strictness, I have adduced the following,—" that those portions of the path of a molecule throughout which the molecular forces are of influence in sensibly altering the motion of the molecule, either in direction or velocity, must be of vanishing value compared with those portions of the path throughout which such forces may be considered as inactive." Now in actually existing gases, Mariotte and Gay-Lussac's law is not strictly, but only approximately true; and it hence follows that in them such first portions of the paths of the molecules must be small, but not vanishingly small, compared with the entire paths. Inasmuch, now, as one of the fundamental conditions upon which the whole theory rests is that the molecular forces are only effective at small distances from the molecules, a path which is very great in comparison to the sphere of action of a molecule may yet, considered absolutely, be very small.

By a few simple considerations, an approximate idea may be formed of the mean magnitude of the paths traversed by the separate molecules: I purpose endeavouring to elucidate this in what immediately follows.

(3.) For this purpose it will be advisable to prefix some remarks concerning the manner in which it is possible to view the molecular forces, and what has accordingly to be understood by the sphere of action. These remarks are not to be considered as an essential part of the subsequent development, but are merely intended to fix our ideas.

If we do not take into account the forces of chemical affinity, and only consider such molecules as are chemically indifferent to one another, I imagine that there are still two forces which are to be distinguished. I believe, namely, that when two molecules approach one another an attraction is at first exerted, which begins to be of sensible effect even at some distance, and which increases as the distance diminishes; but that, when the molecules have arrived into the immediate neighbourhood of one another, a force comes into play which seeks to drive them asunder. For the view which it is here intended to take, it is indifferent what kind of force this repulsive one is supposed to be, that is, whether, as in the case of solid elastic bodies, it only strives to separate the molecules when they are in actual contact with a force equal to that with which they are pressed together, or whether it is one which begins to act before the actual contact of the molecules. In the same manner, we need not here discuss the question as to the source of these forces, whether they are both to be ascribed to the particles of ponderable matter themselves, or whether one of them is to be referred to a more subtle substance, with which the ponderable particles of matter are furnished.

Let us now imagine two molecules moving in directions such that, if they preserved them unchanged, they would not strike one another, but pass by at some distance. Two cases may here occur. If the distance is very small, the molecules which were drawn towards one another, even from some distance, by the force of attraction, approach so closely that the repulsive force comes into play, and a rebounding of the molecules results. If the distance be somewhat greater, the paths of the molecules only suffer a certain change of direction through the attractive force, without the repulsive force being able to act. Finally, at still greater distances, the effect of the molecules upon one another may be altogether neglected.

How great the distances must be in order that the one or other case might occur, could not be determined universally, even if we possessed exact knowledge of the molecular forces; for the velocity of the molecules and the reciprocal inclination of their paths are of influence. Nevertheless, *mean* values of these distances may be obtained. We will therefore suppose that the distance ρ is given for such a mean value, which forms the boundary between the first and second case, and the meaning of which we will define with greater precision in the following manner:—If the centres of gravity of two molecules have such directions of motion that if they were to proceed in those directions in straight lines they would pass by one another at a distance greater than ρ, then the molecules only change their courses to some extent through reciprocal attraction, without the repulsive force coming into action between them. If, on the other hand, this distance is less than ρ, the latter force also comes into play, and a rebounding of the molecules takes place.

If, now, the latter case alone be considered as one of *impact*, and we do not concern ourselves with the changes of direction which the force of attraction effects at greater distances, we may, for what we have here to consider, represent a sphere of radius ρ, described around a molecule and having its centre of gravity for a centre, by the term *sphere of action* of the molecule.

I again call attention to the fact that the special hypotheses here made, concerning the nature of the molecular forces, are not to be viewed as a necessary condition for the developments which follow; their only purpose is to facilitate the comprehension by giving something definite to the imagination. It is of no import how we consider the forces by reason of which the molecules change the directions of their motions; if we but admit that their effects are only sensible at very small distances, we may assume some distances as limiting value for the purpose of being able to neglect the actions from greater distances, and only regard those for smaller ones. A sphere described at this distance may be called a sphere of action.

(4.) If, now, in a given space, we imagine a great number of molecules moving irregularly about amongst one another, and if we select one of them to watch, such a one would ever and anon impinge upon one of the other molecules, and bound off from it. We have now, therefore, to solve the question as to how great is the mean length of the path between two such impacts; or more exactly expressed, *how far on an average can the molecule move, before its centre of gravity comes into the sphere of action of another molecule.*

We will not discuss this question, however, immediately in the form just given: we will propose instead a somewhat simpler one, which is related to the other in such a manner that the solution of the one may be derived from that of the other.

If we assume that not all the molecules present in the space are in motion, but that the one chosen for observation is the only one which moves, and all the rest remain fixed in position, the moving molecule in these circumstances also would strike here and there upon one of the others, and the number of blows which it suffers in this case during one unit of time may be compared with the numbers which it would experience in event of universal movement. On considering the matter more attentively, we are soon convinced that the number of blows amongst moving molecules must be greater than amongst stationary ones, or, which comes to the same thing, that the mean length of the paths which the molecule watched passes over between two consecutive impacts, must be less in the first case than in the second. The relation between the lengths of the two paths may be definitely found as soon as the velocity of the remaining molecules, in comparison with that of the one watched, is known. For our investigations, that case only is of special interest where the velocities of all the molecules are *on an average* equally great. In this case, if we only consider the *mean* velocities, we may more simply assume that all molecules move at the same rate; and for this case we obtain the following result:—*The mean lengths of path for the two cases* (1) *where the remaining molecules move with the same velocity as the one watched*; and (2) *where they are at rest, bear the proportion to one another of* $\frac{3}{4}$ *to* 1.

It would not be difficult to prove the correctness of this relation: it is, however, unnecessary for us to devote our time to it; for, in our consideration of the mean path, it is not the question to determine exactly its numerical value, but merely to obtain an approximate notion of its magnitude; and hence the exact knowledge of this relation is not necessary. It is even sufficient for our purpose if we may assume as certain that the mean path among moving molecules *cannot be greater* than among stationary ones; this will certainly be at once admitted. Under this hypothesis, we will confine the discussion of the question to that case *where the molecule watched alone moves, while all the others remain at rest*.

Moreover, without affecting the question in anything, we may suppose a mere moving point in place of the moving molecule; for it is in fact only the centre of gravity of the molecule which has to be considered.

(5.) Suppose, then, there is a space containing a great number of molecules, and that these are not regularly arranged, the only condition being that the density is the same throughout, *i.e.* in equal parts of the space there are the same numbers of molecules. The determination of the density may be performed conveniently for our investigation by knowing how far apart two neighbouring molecules would be separated from one another if the molecules were arranged cubically, that is, so arranged that the whole space might be supposed divided into a number of equal very small cubic spaces, in whose corners the centres of the molecules were situated. We shall denote this distance, that is, the side of one of these little cubes, by λ, and shall call it *the mean distance of the neighbouring molecules*.

If, now, a point moves through this space in a straight line, let us suppose the space to be divided into parallel layers perpendicular to the motion of the point, and let us determine *how great is the probability that the point will pass freely through a layer of the thickness* x *without encountering the sphere of action of a molecule*.

Let us first take a layer of the thickness 1, and let us denote by the fraction of unity a the probability of the point passing through this layer without meeting with any sphere of action: then the corresponding probability for a thickness 2 is a^2; for if such a layer be supposed divided into two layers of the thickness 1, the probability of the points passing free through the first layer, and thereby arriving at the second, must be multiplied by the probability of its passing through the latter one. Similarly, for a layer of the thickness 3, we have a^3, &c., and for a layer of any thickness x we may accordingly write a^x. Let us transform this expression by putting $e^{-\alpha}$ for a, in which e is the base of the natural logarithms, and $-\alpha = \log_e a$, which logarithm must be negative, because a is less than 1. If now we denote the probability of the free passage through a layer of the thickness x by W, we have the equation

$$W = e^{-\alpha x}, \qquad (1)$$

and we have only to determine here the constant α.

Again, let us consider a layer of such thinness that the higher powers of the thickness may be neglected in comparison with the first. Calling this thickness δ and the corresponding probability W_δ, the former equation becomes

$$W_\delta = e^{-\alpha\delta} = 1 - \alpha\delta. \qquad (2)$$

The probability in this case may also be determined from special considerations. Let us direct our attention to any plane in the layer parallel to one of the bounding planes of the layer, and let us suppose all the molecules whose centres lie in the layer to be so moved perpendicular to the layer that their centres all fall upon this plane; we have now only to inquire how great the probability is that the point, in its passage through this plane, meets with no sphere of action; such probability may be simply represented by the proportion of two superficial areas. Of the entire part of the plane which falls within the given space, a certain portion is covered by the great circles of the spheres of action whose centres fall upon it, while the remaining portion is free for the passage; and the probability of the uninterrupted passage is therefore expressed *by the relation of the free portion of the plane to the whole plane.*

From the manner in which the density was determined at the beginning of this article, it follows that in a layer of thickness λ, so many molecules must be contained, that, if they be supposed brought into one and the same plane parallel to the bounding plane, and to be arranged still quadratically in this plane, then the side of the small square in whose corners would be situated the centres of the molecules would be equal to λ. Hence it follows, that the part of the plane which would be covered by the great circles of the spheres of action, would be related to the remainder of the plane as a great circle would to a square of side λ, so that, accordingly, the covered superficial area would be expressed by the fraction

$$\frac{\pi \rho^2}{\lambda^2}$$

of the entire superficial area. In order to ascertain the corresponding magnitude for a layer of the thickness δ, we have only to multiply the previous fraction by δ/λ, that is,

$$\frac{\pi \rho^2}{\lambda^3} \delta;$$

and if this magnitude be subtracted from 1, the difference represents the free portion of the plane as a fraction of the whole plane.

Hence the probability that the point will pass through our plane, or, which comes to the same thing, through a layer of thickness δ, without obstruction, is determined by the equation

$$W_\delta = 1 - \frac{\pi\rho^2}{\lambda^3}\delta; \tag{3}$$

and on comparing this expression for W_δ with that given in equation (2), we find that

$$\alpha = \frac{\pi\rho^2}{\lambda^3}, \tag{4}$$

and hence the general equation (1) is transformed into

$$W = e^{-(\pi\rho^2/\lambda^3)\,x}. \tag{5}$$

(6.) By means of this equation we can now determine the mean value of the path which the point has to traverse before it meets with a sphere of action.

Let us suppose that a great number (N) of points are thrown through space in one direction, and let us suppose the space to be divided into very thin layers perpendicular to the direction of motion; then a small number of the points would be detained in the first layer by the spheres of action, another lot in the second, another in the third, and so on. If, now, each of these small numbers be multiplied by the length of path, the products added, and the sum obtained divided by the whole number N, the quotient will be the mean length of the path which we seek.

According to equation (5), the number of points which either reach or pass the distance x from the commencement of the motion is represented by

$$Ne^{-(\pi\rho^2/\lambda^3)x};$$

and accordingly the number which reach or pass the distance $x + dx$ is expressed by

$$Ne^{-(\pi\rho^2/\lambda^3)(x+dx)} = Ne^{-(\pi\rho^2/\lambda^3)\,x}\left(1 - \frac{\pi\rho^2}{\lambda^3}dx\right).$$

The difference of these two expressions, namely,

$$Ne^{-(\pi\rho^2/\lambda^3)\,x} \cdot \frac{\pi\rho^2}{\lambda^3}dx,$$

represents the number of those points which are detained between x and $x + dx$. The path traversed by these points may be considered as x if we neglect infinitely small differences; and hence the above expression must be multiplied by this length in order to obtain one of the products mentioned before, namely,

$$Ne^{-(\pi\rho^2/\lambda^3)x} \cdot \frac{\pi\rho^2}{\lambda^3} \, xdx.$$

If, now, it be desired to obtain the sum of all products of this kind which correspond to the several layers of the thickness dx, this must of course, in the case in point where the layers are infinitely thin, be effected by integration. Hence the above formula has to be integrated from $x = 0$ to $x = \infty$, whence the following expression is obtained,

$$N\frac{\lambda^3}{\pi\rho^2}.$$

This expression has now only to be divided by N in order to arrive at the mean length of path required. If this be called l', the equation is

$$l' = \frac{\lambda^3}{\pi\rho^2}. \tag{6}$$

In the case where not one molecule only is in motion while all the others are at rest, but where all molecules move with equal velocity, the mean length of way, as mentioned before, is less than that above considered in the proportion of $\frac{3}{4}$ to 1. Hence if we put the simple letter l for this case, we have

$$l = \tfrac{3}{4}\frac{\lambda^3}{\pi\rho^2}. \tag{7}$$

Writing this equation in the form

$$\frac{l}{\rho} = \frac{\lambda^3}{\tfrac{4}{3}\pi\rho^3}, \tag{7a}$$

a simple law results. It follows from the manner in which we determined the density, that the part of the given space filled by the spheres of action of the molecules is related to the whole given space as a sphere of action to a cube of the side λ, that is, as

$$\tfrac{4}{3}\pi\rho^3 : \lambda^3.$$

Accordingly the meaning of the previous equation may be so put:—
The mean length of path of a molecule is in the same proportion to the radius of the sphere of action as the entire space occupied by the gas, to that portion of the space which is actually filled up by the spheres of action of the molecules.

(7.) In order to have a definite numerical example, let us assume, in round numbers, that the spheres of action of the molecules are so small that only a thousandth of the space occupied by the gas is really filled out by the spheres of action, and that the whole remaining space be free for motion.

For this case we have

$$\frac{\lambda^3}{\frac{4}{3}\pi\rho^3} = 1000,$$

whence it follows that

$$\frac{\lambda}{\rho} = 16\cdot12. \tag{8}$$

On applying these values we obtain from equations (6) and (7),

$$l' = 1333\rho = 83\lambda, \tag{9}$$

$$l = 1000\rho = 62\lambda. \tag{10}$$

The first expressions in both equations show that, under the assumption made, the mean path has a considerable length in comparison to the radius of the spheres of action, and that therefore, as far as the effect of this circumstance is concerned, Mariotte and Gay-Lussac's law may be very nearly true for the gas. By a simple calculation it may be shown that the relation of 1000 to 1 completely suffices, even for those approximations found by Regnault with permanent gases. It follows that the magnitude of the spheres of action which was taken for illustration, although arbitrarily chosen, may yet be regarded as one within the bounds of possibility.

But if we now regard this same mean value of the length of path in such a manner as to compare it, not with the sizes of molecules, but with our usual units of length, we obtain totally different relations. In all physical and chemical investigations in which opportunity presents itself for drawing conclusions concerning the weight and size of the separate molecules, we are invariably led to the conclusion that, compared with all measurable magnitudes, molecules must be of extraordinarily small size. As yet, no one has been able to establish a bounding line on the other side (for smallness). Accordingly, when an ordinary unit of measure, *e.g.* a litre, is filled with gas at the ordinary atmospheric pressure, we must assume that the number of molecules present is very great, and that consequently the distances between the molecules is very small. Accordingly the values previously found for l' and l, namely, 83λ and 62λ, must only be regarded as small magnitudes.

(8.) After the above determination of the length of the *mean* path, we still have to consider how the separate paths which really occur are related to the mean path.

The first question which presents itself is, in what proportion is the number of cases in which the real path is less than the mean path, to that of the cases in which it is greater. For answering this question, use is made of (5), in which we have only to substitute the mean value l' for x in order to find what probability there is that the true path is equal to or greater than the mean one. If for l' we here make use of the expression in (6), and denote the corresponding value of W by W_1, then

$$W_1 = e^{-1} = 0.3679. \tag{11}$$

From the above equation it follows, that out of N cases only $0.3679\ N$ occur in which the real path is equal to, or greater than the mean one, while in the

$$0.6321\ N$$

cases the true path is the smaller one.

If, further, it be required to know the number of cases in which the true path is equal to or above the double, treble, &c. of the mean one, the same process may be adopted as before. Calling the probabilities in question W_2, W_3, &c., we have

$$\left. \begin{array}{l} W_2 = e^{-2} \\ W_3 = e^{-3} \\ \&c. \end{array} \right\} \tag{12}$$

These numbers evidently diminish very rapidly, since, for instance, $e^{-10} = 0.000045$; and we gather from this that, although in isolated cases a molecule may traverse a path considerably longer than the mean one, such cases are comparatively rare, and that in the majority of cases the actual path is smaller or very little larger than the small mean value found above.

(9.) If, now, these results be applied to the externally recognizable behaviours of a gas, in which it is presumed that no other motion common to the whole mass besides the molecular one is present, it is easy to convince oneself that the theory which explains the expansive force of gases does not lead to the conclusion that two quantities of gas bounding one another must mix with one another quickly and violently, but that only a comparatively small number of atoms can arrive quickly at a great distance, while the chief quantities only gradually mix at the surface of their contact.

From this it is clear why clouds of smoke only slowly lose their form on quiet days. Even when the air is in motion, provided such motion consists of a uniform one of the entire current, a cloud of smoke may be carried off without quickly losing its form. But the other facts adduced by Buijs-Ballot also admit of simple explanation. The remark made by him attached to one of his objections, that the molecules of gas in a room must traverse the room many hundred times in one second, is completely foreign to the theory. Perhaps it might be said of a remark which occurs in the mathematical development attached to my previous paper, that it afforded reason for such an idea. I assumed there, namely, that the gas was in a very flat vessel, and I then assumed that the molecules of gas without disturbing one another, sped backwards and forwards between the two great parallel sides. Nevertheless, this assumption was there introduced with the following words: " In estimating the pressure, *instead of regarding the motion as it really occurs*, we may introduce certain simplifications." I believe I thereby called sufficient attention to the fact that this assumption should not serve to furnish an image of the real process, but only to simplify the calculation there intended, the result of which could not be thereby changed.

10
Illustrations of the Dynamical Theory of Gases *

JAMES CLERK MAXWELL

SUMMARY

In view of the current interest in the theory of gases proposed by Bernoulli (Selection 3), Joule, Krönig, Clausius (Selections 8 and 9) and others, a mathematical investigation of the laws of motion of a large number of small, hard, and perfectly elastic spheres acting on one another only during impact seems desirable.

It is shown that the number of spheres whose velocity lies between v and $v + dv$ is

$$N \frac{4}{\alpha^3 \sqrt{\pi}} v^2 e^{-v^2/\alpha^2} dv,$$

where N is the total number of spheres, and α is a constant related to the average velocity:

$$\text{mean value of } v^2 = \tfrac{3}{2} \alpha^2.$$

If two systems of particles move in the same vessel, it is proved that the mean kinetic energy of each particle will be the same in the two systems.

Known results pertaining to the mean free path and pressure on the surface of the container are rederived, taking account of the fact that the velocities are distributed according to the above law.

The internal friction (viscosity) of a system of particles is predicted to be independent of density, and proportional to the square root of the

---

* Originally published in *Phil. Mag.*, Vol. 19, pp. 19–32; Vol. 20, pp. 21–37 (1860); reprinted in *The Scientific Papers of James Clerk Maxwell* (ed. W. D. NIVEN), Cambridge University Press, 1890 Vol. I, pp. 377–409.

absolute temperature; there is apparently no experimental evidence to confirm this prediction for real gases.

A discussion of collisions between perfectly elastic bodies of any form leads to the conclusion that the final equilibrium state of any number of systems of moving particles of any form is that in which the average kinetic energy of translation along each of the three axes is the same in all the systems, and equal to the average kinetic energy of rotation about each of the three principal axes of each particle (equipartition theorem). This mathematical result appears to be in conflict with known experimental values for the specific heats of gases.

PART I

On the Motions and Collisions of Perfectly Elastic Spheres.

So many of the properties of matter, especially when in the gaseous form, can be deduced from the hypothesis that their minute parts are in rapid motion, the velocity increasing with the temperature, that the precise nature of this motion becomes a subject of rational curiosity. Daniel Bernoulli, Herapath, Joule, Krönig, Clausius, etc.† have shewn that the relations between pressure, temperature, and density in a perfect gas can be explained by supposing the particles to move with uniform velocity in straight lines, striking against the sides of the containing vessel and thus producing pressure. It is not necessary to suppose each particle to travel to any great distance in the same straight line; for the effect in producing pressure will be the same if the particles strike against each other; so that the straight line described may be very short. M. Clausius‡ has determined the mean length of path in terms of the average distance of the particles, and the distance between the centres of two particles when collision takes place. We have at present no means of ascertaining either of these distances; but certain phenomena, such as the internal friction of gases, the conduction of heat through a gas, and the diffusion of one gas through another, seem to indicate the possibility of determining accurately the mean length of path which a particle describes between two successive collisions. In order to

---

† See the Bibliography and Selections 3, 8 and 9 in this volume.

‡ See Selection 9.

lay the foundation of such investigations on strict mechanical principles, I shall demonstrate the laws of motion of an indefinite number of small, hard, and perfectly elastic spheres acting on one another only during impact.

If the properties of such a system of bodies are found to correspond to those of gases, an important physical analogy will be established, which may lead to more accurate knowledge of the properties of matter. If experiments on gases are inconsistent with the hypothesis of these propositions, then our theory, though consistent with itself, is proved to be incapable of explaining the phenomena of gases. In either case it is necessary to follow out the consequences of the hypothesis.

Instead of saying that the particles are hard, spherical, and elastic, we may if we please say that the particles are centres of force, of which the action is insensible except at a certain small distance, when it suddenly appears as a repulsive force of very great intensity. It is evident that either assumption will lead to the same results. For the sake of avoiding the repetition of a long phrase about these repulsive forces, I shall proceed upon the assumption of perfectly elastic spherical bodies. If we suppose those aggregate molecules which move together to have a bounding surface which is not spherical, then the rotatory motion of the system will store up a certain proportion of the whole *vis viva*, as has been shewn by Clausius, and in this way we may account for the value of the specific heat being greater than on the more simple hypothesis.

On the Motion and Collision of Perfectly Elastic Spheres.

Prop. I. Two spheres moving in opposite directions with velocities inversely as their masses strike one another; to determine their motions after impact.

Let P and Q be the position of the centres at impact; AP, BQ the directions and magnitudes of the velocities before impact; Pa, Qb the same after impact; then, resolving the velocities parallel and perpendicular to PQ the line of centres, we find that the velocities parallel to the line of centres are exactly reversed, while those perpendicular to that line are unchanged. Compounding these velocities again, we find that the velocity of each ball is the same before and after impact, and that the directions before and after impact lie in the same plane with the line of centres, and make equal angles with it.

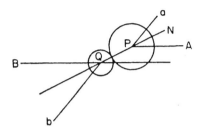

Prop. II. To find the probability of the direction of the velocity after impact lying between given limits.

In order that a collision may take place, the line of motion of one of the balls must pass the centre of the other at a distance less than the sum of their radii; that is, it must pass through a circle whose centre is that of the other ball, and radius (s) the sum of the radii of the balls. Within this circle every position is equally probable, and therefore the probability of the distance from the centre being between r and $r + dr$ is

$$\frac{2r dr}{s^2}.$$

Now let ϕ be the angle APa between the original direction and the direction after impact, than $APN = \frac{1}{2}\phi$, and $r = s \sin \frac{1}{2}\phi$, and the probability becomes

$$\tfrac{1}{2} \sin \phi d\phi.$$

The area of a spherical zone between the angles of polar distance ϕ and $\phi + d\phi$ is

$$2\pi \sin \phi d\phi;$$

therefore if ω be any small area on the surface of a sphere, radius unity, the probability of the direction of rebound passing through this area is

$$\frac{\omega}{4\pi};$$

so that the probability is independent of ϕ, that is, all directions of rebound are equally likely.

Prop. III. Given the direction and magnitude of the velocities of two spheres before impact, and the line of centres at impact; to find the velocities after impact.

Let OA, OB represent the velocities before impact, so that if there had been no action between the bodies they would have been at A and B at the end of a second. Join AB, and let G be their centre of

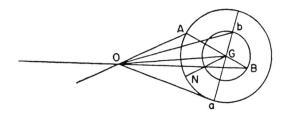

gravity, the position of which is not affected by their mutual action. Draw GN parallel to the line of centres at impact (not necessarily in the plane AOB). Draw aGb in the plane AGN, making $NGa = NGA$, and $Ga = GA$ and $Gb = GB$; then by Prop. I. Ga and Gb will be the velocities relative to G; and compounding these with OG, we have Oa and Ob for the true velocities after impact.

By Prop. II. all directions of the line aGb are equally probable. It appears therefore that the velocity after impact is compounded of the velocity of the centre of gravity, and of a velocity equal to the velocity of the sphere relative to the centre of gravity, which may with equal probability be in any direction whatever.

If a great many equal spherical particles were in motion in a perfectly elastic vessel, collisions would take place among the particles, and their velocities would be altered at every collision; so that after a certain time the *vis viva* will be divided among the particles according to some regular law, the average number of particles whose velocity lies between certain limits being ascertainable, though the velocity of each particle changes at every collision.

Prop. IV. To find the average number of particles whose velocities lie between given limits, after a great number of collisions among a great number of equal particles.

Let N be the whole number of particles. Let x, y, z be the components of the velocity of each particle in three rectangular directions, and let the number of particles for which x lies between x and $x + dx$, be $Nf(x)dx$, where $f(x)$ is a function of x to be determined.

The number of particles for which y lies between y and $y + dy$ will be $Nf(y)dy$; and the number for which z lies between z and $z + dz$ will be $Nf(z)dz$, where f always stands for the same function.

Now the existence of the velocity x does not in any way affect that of the velocities y or z, since these are all at right angles to each other and independent, so that the number of particles whose velocity lies between x and $x + dx$, and also between y and $y + dy$, and also between z and $z + dz$, is

$$Nf(x)f(y)f(z)dx\,dy\,dz.$$

If we suppose the N particles to start from the origin at the same instant, then this will be the number in the element of volume $(dx\,dy\,dz)$ after unit of time, and the number referred to unit of volume will be

$$Nf(x)f(y)f(z).$$

But the directions of the coordinates are perfectly arbitrary, and therefore this number must depend on the distance from the origin alone, that is

$$f(x)f(y)f(z) = \phi(x^2 + y^2 + z^2).$$

Solving this functional equation, we find

$$f(x) = Ce^{Ax^2}, \quad \phi(r^2) = C^3 e^{Ar^2}.$$

If we make A positive, the number of particles will increase with the velocity, and we should find the whole number of particles infinite. We therefore make A negative and equal to $-1/\alpha^2$, so that the number between x and $x + dx$ is

$$NCe^{-(x^2/\alpha^2)}\,dx.$$

Integrating from $x = -\infty$ to $x = +\infty$, we find the whole number of particles,

$$NC\sqrt{\pi}\alpha = N, \quad \therefore C = \frac{1}{\alpha\sqrt{\pi}},$$

$f(x)$ is therefore
$$\frac{1}{\alpha\sqrt{\pi}} e^{-(x^2/\alpha^2)}.$$

Whence we may draw the following conclusions:—

1st. The number of particles whose velocity, resolved in a certain direction, lies between x and $x + dx$ is

$$N \frac{1}{\alpha\sqrt{\pi}} e^{-(x^2/\alpha^2)} dx. \tag{1}$$

2nd. The number whose actual velocity lies between v and $v + dv$ is

$$N \frac{4}{\alpha^3\sqrt{\pi}} v^2 e^{-(v^2/\alpha^2)} dv. \tag{2}$$

3rd. To find the mean value of v, add the velocities of all the particles together and divide by the number of particles; the result is

$$\text{mean velocity} = \frac{2\alpha}{\sqrt{\pi}}. \tag{3}$$

4th. To find the mean value of v^2, add all the values together and divide by N,

$$\text{mean value of } v^2 = \tfrac{3}{2}\alpha^2. \tag{4}$$

This is greater than the square of the mean velocity, as it ought to be.

It appears from this proposition that the velocities are distributed among the particles according to the same law as the errors are distributed among the observations in the theory of the " method of least squares." The velocities range from 0 to ∞, but the number of those having great velocities is comparatively small. In addition to these velocities, which are in all directions equally, there may be a general motion of translation of the entire system of particles which must be compounded with the motion of the particles relatively to one another. We may call the one the motion of translation, and the other the motion of agitation.

Prop. V. Two systems of particles move each according to the law stated in Prop. IV.; to find the number of pairs of particles, one of each system, whose relative velocity lies between given limits.

Let there be N particles of the first system, and N' of the second, then NN' is the whole number of such pairs. Let us consider the velocities in the direction of x only; then by Prop. IV. the number of the first kind, whose velocities are between x and $x + dx$, is

$$N \frac{1}{\alpha \sqrt{\pi}} e^{-(x^2/\alpha^2)} dx.$$

The number of the second kind, whose velocity is between $x + y$ and $x + y + dy$, is

$$N' \frac{1}{\beta \sqrt{\pi}} e^{-((x + y)^2/\beta^2)} dy,$$

where β is the value of α for the second system.

The number of pairs which fulfil both conditions is

$$NN' \frac{1}{\alpha \beta \pi} e^{-((x^2/\alpha^2) + (x + y)^2/\beta^2)} dx\, dy.$$

Now x may have any value from $-\infty$ to $+\infty$ consistently with the difference of velocities being between y and $y + dy$; therefore integrating between these limits, we find

$$NN' \frac{1}{\sqrt{\alpha^2 + \beta^2} \sqrt{\pi}} e^{-(y^2/\alpha^2 + \beta^2)} dy \qquad (5)$$

for the whole number of pairs whose difference of velocity lies between y and $y + dy$.

This expression, which is of the same form with (1) if we put NN' for N, $\alpha^2 + \beta^2$ for α^2, and y for x, shews that the distribution of relative velocities is regulated by the same law as that of the velocities themselves, and that the mean relative velocity is the square root of the sum of the squares of the mean velocities of the two systems.

Since the direction of motion of every particle in one of the systems may be reversed without changing the distribution of velocities, it follows that the velocities compounded of the velocities of two particles, one in each system, are distributed according to the same formula (5) as the relative velocities.

Prop. VI. Two systems of particles move in the same vessel; to prove that the mean *vis viva* of each particle will become the same in the two systems.

Let P be the mass of each particle of the first system, Q that of each particle of the second. Let p, q be the mean velocities in the two systems before impact, and let p', q' be the mean velocities after one impact. Let $OA = p$ and $OB = q$, and let AOB be a right angle; then, by Prop. V., AB will be the mean relative velocity, OG will be

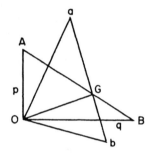

the mean velocity of the centre of gravity; and drawing aGb at right angles to OG, and making $aG = AG$ and $bG = BG$, then Oa will be the mean velocity of P after impact, compounded of OG and Ga, and Ob will be that of Q after impact.

Now $\quad AB = \sqrt{p^2 + q^2}, \quad AG = \dfrac{Q}{P + Q} \sqrt{p^2 + q^2},$

$\quad\quad BG = \dfrac{P}{P + Q} \sqrt{p^2 + q^2}, \quad OG = \dfrac{\sqrt{P^2 p^2 + Q^2 q^2}}{P + Q},$

therefore $\quad p' = Oa = \dfrac{\sqrt{Q^2 (p^2 + q^2) + P^2 p^2 + Q^2 q^2}}{P + Q},$

and $\quad p' = Ob = \dfrac{\sqrt{P^2 (p^2 + q^2) + P^2 p^2 + Q^2 q^2}}{P + Q},$

and $\quad Pp'^2 - Qq'^2 = \left(\dfrac{P - Q}{P + Q}\right)^2 (Pp^2 - Qq^2). \quad\quad (6)$

It appears therefore that the quantity $Pp^2 - Qq^2$ is diminished at every impact in the same ratio, so that after many impacts it will vanish, and then

$$Pp^2 = Qq^2.$$

Now the mean *vis viva* is $\frac{3}{2}P\alpha^2 = (3\pi/8)Pp^2$ for P, and $(3\pi/8)Qq^2$ for Q; and it is manifest that these quantities will be equal when $Pp^2 = Qq^2$.

If any number of different kinds of particles, having masses P, Q, R and velocities p, q, r respectively, move in the same vessel, then after many impacts

$$Pp^2 = Qq^2 = Rr^2, \&c. \tag{7}$$

Prop. VII. A particle moves with velocity r relatively to a number of particles of which there are N in unit of volume; to find the number of these which it approaches within a distance s in unit of time.

If we describe a tubular surface of which the axis is the path of the particle, and the radius the distance s, the content of this surface generated in unit of time will be πrs^2, and the number of particles included in it will be

$$N\pi rs^2, \tag{8}$$

which is the number of particles to which the moving particle approaches within a distance s.

Prop. VIII. A particle moves with velocity v in a system moving according to the law of Prop. IV.; to find the number of particles which have a velocity relative to the moving particle between r and $r + dr$.

Let u be the actual velocity of a particle of the system, v that of the original particle, and r their relative velocity, and θ the angle between v and r, then

$$u^2 = v^2 + r^2 - 2vr \cos \theta.$$

If we suppose, as in Prop. IV., all the particles to start from the origin at once, then after unit of time the "density" or number of particles to unit of volume at distance u will be

$$N \frac{1}{\alpha^3 \pi^{\frac{3}{2}}} e^{-(u^2/\alpha^2)}.$$

From this we have to deduce the number of particles in a shell whose centre is at distance v, radius $= r$, and thickness $= dr$,

$$N \frac{1}{\alpha \sqrt{\pi}} \frac{r}{v} \{e^{-((r-v)^2/\alpha^2)} - e^{-((r+v)^2/\alpha^2)}\} dr, \qquad (9)$$

which is the number required.

Cor. It is evident that if we integrate this expression from $r = 0$ to $r = \infty$, we ought to get the whole number of particles $= N$, whence the following mathematical result,

$$\int_0^\infty dx \cdot x \, (e^{-((x-a)^2/\alpha^2)} - e^{-((x+a)^2/\alpha^2)}) = \sqrt{\pi} a \alpha. \qquad (10)$$

Prop. IX. Two sets of particles move as in Prop. V.; to find the number of pairs which approach within a distance s in unit of time.

The number of the second kind which have a velocity between v and $v + dv$ is

$$N' \frac{4}{\beta^3 \sqrt{\pi}} v^2 e^{-(v^2/\beta^2)} \, dv = n'.$$

The number of the first kind whose velocity relative to these is between r and $r + dr$ is

$$N \frac{1}{\alpha \sqrt{\pi}} \frac{r}{v} (e^{-((r-v)^2/\alpha^2)} - e^{-((r+v)^2/\alpha^2)}) \, dr = n,$$

and the number of pairs which approach within distance s in unit of time is

$$nn' \pi r s^2,$$

$$= NN' \frac{4}{\alpha \beta^3} s^2 r^2 v e^{-(v^2/\beta)} \{e^{-((v-r)^2/\alpha^2)} - e^{-((v+r)^2/\alpha^2)}\} \, dr \, dv.$$

By the last proposition we are able to integrate with respect to v, and get

$$NN' \frac{4\sqrt{\pi}}{(\alpha^2 + \beta^2)^{\frac{3}{2}}} s^2 r^3 e^{-(r^2/\alpha^2 + \beta^2)} dr.$$

Integrating this again from $r = 0$ to $r = \infty$,

$$2NN' \sqrt{\pi} \sqrt{\alpha^2 + \beta^2} \, s^2 \tag{11}$$

is the number of collisions in unit of time which take place in unit of volume between particles of different kinds, s being the distance of centres at collision.

The number of collisions between two particles of the first kind, s_1 being the striking distance, is

$$2N^2 \sqrt{\pi} \sqrt{2\alpha^2} \, s_1^2 \, ;$$

and for the second system it is

$$2N'^2 \sqrt{\pi} \sqrt{2\beta^2} \, s_2^2 \, .$$

The mean velocities in the two systems are $2\alpha/\sqrt{\pi}$ and $2\beta/\sqrt{\pi}$; so that if l_1 and l_2 be the mean distances travelled by particles of the first and second systems between each collision, then

$$\frac{1}{l_1} = \pi N_1 \sqrt{2} \, s_1^2 + \pi N_2 \frac{\sqrt{\alpha^2 + \beta^2}}{\alpha} s^2,$$

$$\frac{1}{l_2} = \pi N_1 \frac{\sqrt{\alpha^2 + \beta^2}}{\beta} s^2 + \pi N_2 \sqrt{2} \, s_2^2.$$

Prop. X. To find the probability of a particle reaching a given distance before striking any other.

Let us suppose that the probability of a particle being stopped while passing through a distance dx, is αdx; that is, if N particles arrived at a distance x, $N\alpha dx$ of them would be stopped before getting to a distance $x + dx$. Putting this mathematically,

$$\frac{dN}{dx} = -N\alpha, \quad \text{or} \quad N = Ce^{-\alpha x}.$$

Putting $N = 1$ when $x = 0$, we find $e^{-\alpha x}$ for the probability of a particle not striking another before it reaches a distance x.

The *mean distance* travelled by each particle before striking is $1/\alpha = l$. The probability of a particle reaching a distance $= nl$ without being struck is e^{-n}. (See a paper by M. Clausius, *Philosophical Magazine*, February 1859.)†

If all the particles are at rest but one, then the value of α is

$$\alpha = \pi s^2 N,$$

where s is the distance between the centres at collision, and N is the number of particles in unit of volume. If v be the velocity of the moving particle relatively to the rest, then the number of collisions in unit of time will be

$$v\pi s^2 N;$$

and if v_1 be the actual velocity, then the number will be $v_1 \alpha$; therefore

$$\alpha = \frac{v}{v_1} \pi s^2 N,$$

where v_1 is the actual velocity of the striking particle, and v its velocity relatively to those it strikes. If v_2 be the actual velocity of the other particles, then $v = \sqrt{v_1^2 + v_2^2}$. If $v_1 = v_2$, then $v = \sqrt{2} v_1$, and

$$\alpha = \sqrt{2}\, \pi s^2 N.$$

Note‡.†† M. Clausius makes $\alpha = \frac{4}{3} \pi s^2 N$.

---

† See Selection 9.

‡ [In the *Philosophical Magazine* of 1860, Vol. I, pp. 434–6, Clausius explains the method by which he found his value of the mean relative velocity. It is briefly as follows: If u, v be the velocities of two particles their relative velocity is $\sqrt{u^2 + v^2 - 2uv \cos \theta}$ and the mean of this as regards direction only, all directions of v being equally probable, is shewn to be

$$v + \frac{1}{3}\frac{u^2}{v} \text{ when } u < v, \text{ and } u + \frac{1}{3}\frac{v^2}{u} \text{ when } u > v.$$

If $v = u$ these expressions coincide. Clausius in applying this result and putting u, v for the mean velocities assumes that the mean relative velocity is given by expressions of the same form, so that when the mean velocities

Prop. XI. In a mixture of particles of two different kinds, to find the mean path of each particle.

Let there be N_1 of the first, and N_2 of the second in unit of volume. Let s_1 be the distance of centres for a collision between two particles of the first set, s_2 for the second set, and s' for collision between one of each kind. Let v_1 and v_2 be the coefficients of velocity, M_1, M_2 the mass of each particle.

The probability of a particle M_1 not being struck till after reaching a distance x_1 by another particle of the same kind is

$$e^{-\sqrt{2}\,\pi s_1^2 N_1 x}.$$

---

are each equal to u the mean relative velocity would be $\tfrac{4}{3}u$. This step is, however, open to objection, and in fact if we take the expressions given above for the mean velocity, treating u and v as the velocities of two particles which may have any values between 0 and ∞, to calculate the mean relative velocity we should proceed as follows: Since the number of particles with velocities between u and $u + du$ is

$$N \frac{4}{\alpha^3 \sqrt{\pi}} u^2 e^{-(u^2/\alpha^2)} \, du,$$

the mean relative velocity is

$$\frac{16}{\alpha^3 \beta^2 \pi} \int_0^\infty \int_v^\infty u^2 v^2 e^{-(u^2/\alpha^2 + v^2/\beta^2)} \left(u + \frac{1}{3}\frac{v^2}{u}\right) du\, dv +$$

$$\frac{16}{\alpha^3 \beta^2 \pi} \int_0^\infty \int_0^v u^2 v^2 e^{-(u^2/\alpha^2 + v^2/\beta^2)} \left(v + \frac{1}{3}\frac{u^2}{v}\right) du\, dv.$$

This expression, when reduced, leads to

$$\frac{2}{\sqrt{\pi}} \sqrt{\alpha^2 + \beta^2},$$

which is the result in the text. *Ed.* (W. D. Niven).]

†† In a letter to William Thomson in 1871, Maxwell makes the following remark on this discrepancy in numerical factors: " Clausius made objection No. 1 to an integration founded on his theory of uniform velocity of molecules. (This is the first commitment of Clausius to such a theory.) As he was sure to be converted & I was lazy, I said 0. Objection No 2 &c. to theory of diffusion and conduction were well founded . . . " (see H. T. BERNSTEIN, *Isis* **54**, 212, 214 (1963)). As it turned out, Clausius was indeed converted (*Lumière Electrique* **17**, 241 (1885)) without any further effort on Maxwell's part. The above note by W. D. NIVEN in Maxwell's *Scientific Papers*, Vol. I, p. 387, gives a concise explanation.

The probability of not being struck by a particle of the other kind in the same distance is

$$e^{-\sqrt{1+(v_2^2/v_1^2)}\pi s'^2 N_2 x}.$$

Therefore the probability of not being struck by any particle before reaching a distance x is

$$e^{-\pi(\sqrt{2}s_1^2 N_1 + \sqrt{1+(v_2^2/v_1^2)}s'^2 N_2)x};$$

and if l_1 be the *mean distance* for a particle of the first kind,

$$\frac{1}{l_1} = \sqrt{2}\,\pi s_1^2 N_1 + \pi\sqrt{1+\frac{v_2^2}{v_1^2}}\,s'^2 N_2. \tag{12}$$

Similarly, if l_2 be the mean distance for a particle of the second kind,

$$\frac{1}{l_2} = \sqrt{2}\,\pi s_2^2 N_2 + \pi\sqrt{1+\frac{v_1^2}{v_2^2}}\,s'^2 N_1. \tag{13}$$

The mean density of the particles of the first kind is $N_1 M_1 = \rho_1$, and that of the second $N_2 M_2 = \rho_2$. If we put

$$A = \sqrt{2}\,\frac{\pi s_1^2}{M_1},\quad B = \pi\sqrt{1+\frac{v_2^2}{v_1^2}}\,\frac{s'^2}{M_2},\quad C = \pi\sqrt{1+\frac{v_1^2}{v_2^2}}\,\frac{s'^2}{M_1},$$

$$D = \sqrt{2}\,\frac{\pi s_2^2}{M_2}, \tag{14}$$

$$\frac{1}{l_1} = A\rho_1 + B\rho_2,\quad \frac{1}{l_2} = C\rho_1 + D\rho_2 \tag{15}$$

and

$$\frac{B}{C} = \frac{M_1 v_2}{M_2 v_1} = \frac{v_2^3}{v_1^3}. \tag{16}$$

Prop. XII. To find the pressure on unit of area of the side of the vessel due to the impact of the particles upon it.

Let N = number of particles in unit of volume;
M = mass of each particle;
v = velocity of each particle;
l = mean path of each particle;

then the number of particles in unit of area of a stratum dz thick is

$$Ndz. \tag{17}$$

The number of collisions of these particles in unit of time is

$$Ndz \frac{v}{l}. \qquad (18)$$

The number of particles which after collision reach a distance between nl and $(n + dn)\, l$ is

$$N \frac{v}{l} e^{-n}\, dz\, dn. \qquad (19)$$

The proportion of these which strike on unit of area at distance z is

$$\frac{nl - z}{2nl}; \qquad (20)$$

the mean velocity of these in the direction of z is

$$v \frac{nl + z}{2nl}. \qquad (21)$$

Multiplying together (19), (20), and (21), and M, we find the momentum at impact

$$MN \frac{v^2}{4n^2 l^3} (n^2 l^2 - z^2)\, e^{-n}\, dz\, dn. \qquad (22)$$

Integrating with respect to z from 0 to nl, we get

$$\tfrac{1}{6} MNv^2\, ne^{-n}\, dn.$$

Integrating with respect to n from 0 to ∞, we get

$$\tfrac{1}{6} MNv^2$$

for the momentum in the direction of z of the striking particles; for the momentum of the particles after impact is the same, but in the opposite direction; so that the whole pressure on unit of area is twice this quantity, or

$$p = \tfrac{1}{3} MNv^2.$$

This value of p is independent of l the length of path. In applying this result to the theory of gases, we put $MN = \rho$, and $v^2 = 3k$, and then

$$p = k\rho,$$

which is Boyle and Mariotte's law. By (4) we have

$$v^2 = \tfrac{3}{2}\alpha^2, \quad \therefore \quad \alpha^2 = 2k. \tag{23}$$

We have seen that, on the hypothesis of elastic particles moving in straight lines, the pressure of a gas can be explained by the assumption that the square of the velocity is proportional directly to the absolute temperature, and inversely to the specific gravity of the gas at constant temperature, so that at the same pressure and temperature the value of NMv^2 is the same for all gases. But we found in Prop. VI. that when two sets of particles communicate agitation to one another, the value of Mv^2 is the same in each. From this it appears that N, the number of particles in unit of volume, is the same for all gases at the same pressure and temperature. This result agrees with the chemical law, that equal volumes of gases are chemically equivalent.

We have next to determine the value of l, the mean length of the path of a particle between consecutive collisions. The most direct method of doing this depends upon the fact, that when different strata of a gas slide upon one another with different velocities, they act upon one another with a tangential force tending to prevent this sliding, and similar in its results to the friction between two solid surfaces over each other in the same way. The explanation of gaseous friction, according to our hypothesis, is, that particles having the mean velocity of translation belonging to one layer of the gas, pass out of it into another layer having a different velocity of translation; and by striking against the particles of the second layer, exert upon it a tangential force which constitutes the internal friction of the gas. The whole friction between two portions of gas separated by a plane surface, depends upon the total action between all the layers on the one side of that surface upon all the layers on the other side.

Prop. XIII. To find the internal friction in a system of moving particles.

Let the system be divided into layers parallel to the plane of xy, and let the motion of translation of each layer be u in the direction of x, and let $u = A + Bz$. We have to consider the mutual action between the layers on the positive and negative sides of the plane xy. Let us first determine the action between two layers dz and dz', at distances z and $-z'$ on opposite sides of this plane, each unit of area. The number of particles which, starting from dz in unit of time, reach a distance between nl and $(n + dn)\, l$ is by (19),

$$N \frac{v}{l} e^{-n}\, dz\, dn.$$

The number of these which have the ends of their paths in the layer dz' is

$$N \frac{v}{2nl^2} e^{-n}\, dz\, dz'\, dn.$$

The mean velocity in the direction of x which each of these has before impact is $A + Bz$, and after impact $A + Bz'$; and its mass is M, so that a mean momentum $= MB(z - z')$ is communicated by each particle. The whole action due to these collisions is therefore

$$NMB \frac{v}{2nl^2} (z - z')\, e^{-n}\, dz\, dz'\, dn.$$

We must first integrate with respect to z' between $z' = 0$ and $z' = z - nl$; this gives

$$\tfrac{1}{2} NMB \frac{v}{2nl^2} (n^2 l^2 - z^2)\, e^{-n}\, dz\, dn$$

for the action between the layer dz and all the layers below the plane xy. Then integrate from $z = 0$ to $z = nl$,

$$\tfrac{1}{6} NMB lv n^2 e^{-n}\, dn.$$

Integrate from $n = 0$ to $n = \infty$, and we find the whole friction between unit of area above and below the plane to be

$$F = \tfrac{1}{3} MNlvB = \tfrac{1}{3}\rho lv \frac{du}{dz} = \mu \frac{du}{dz},$$

where μ is the ordinary coefficient of internal friction,

$$\mu = \tfrac{1}{3}\rho lv = \frac{1}{3\sqrt{2}} \frac{Mv}{\pi s^2}. \qquad (24)$$

where ρ is the density, l the mean length of path of a particle, and v the mean velocity $v = 2\alpha/\sqrt{\pi} = 2\sqrt{2k/\pi}$,

$$l = \tfrac{3}{2} \frac{\mu}{\rho} \sqrt{\frac{\pi}{2k}}. \qquad (25)$$

Now Professor Stokes finds by experiments on air,

$$\sqrt{\frac{\mu}{\rho}} = \cdot 116.$$

If we suppose $\sqrt{k} = 930$ feet per second for air at $60°$, and therefore the mean velocity $v = 1505$ feet per second, then the value of l, the mean distance travelled over by a particle between consecutive collisions, $= \frac{1}{447000}$th of an inch, and each particle makes 8,077,200,000 collisions per second.

A remarkable result here presented to us in equation (24), is that if this explanation of gaseous friction be true, the coefficient of friction is independent of the density. Such a consequence of a mathematical theory is very startling, and the only experiment I have met with on the subject does not seem to confirm it. We must next compare our theory with what is known of the diffusion of gases, and the conduction of heat through a gas.

PART II

On the process of diffusion of two or more kinds of moving particles among one another.

[This Part has been omitted because its methods and conclusions were later found to be incorrect by Clausius, and the errors were admitted by Maxwell. Maxwell's improved theory of viscosity, diffusion, and heat conduction in gases, first published in 1866, will be reprinted in the next *Kinetic Theory* volume of this series of *Selected Readings in Physics*. It is recommended that the student who wishes to learn how the mean-free-path theory accounts for transport processes in gases should first consult a modern textbook on kinetic theory, and then read the omitted section of this paper in Maxwell's *Scientific Papers*, Vol. 1, pp. 392–405. There is a useful note by W. D. NIVEN on p. 392 which cites Clausius' objection.]

PART III

On the collision of perfectly elastic bodies of any form.

When two perfectly smooth spheres strike each other, the force which acts between them always passes through their centres of gravity; and therefore their motions of rotation, if they have any, are not affected by the collision, and do not enter into our calculations. But, when the bodies are not spherical, the force of impact will not, in general, be in the line joining their centres of gravity; and therefore the force of impact will depend both on the motion of the centres and the motions of rotation before impact, and it will affect both these motions after impact.

In this way the velocities of the centres and the velocities of rotation will act and react on each other, so that finally there will be some relation established between them; and since the rotations of the particles about their three axes are quantities related to each other in the same way as the three velocities of their centres, the reasoning of Prop. IV. will apply to rotation as well as velocity, and both will be distributed according to the law

$$\frac{dN}{dx} = N \frac{1}{\alpha\sqrt{\pi}} e^{-x/\alpha^2}.$$

Also, by Prop. V., if x be the average velocity of one set of particles, and y that of another, then the average value of the sum or difference of the velocities is

$$\overline{\sqrt{x^2 + y^2}};$$

from which it is easy to see that, if in each individual case

$$u = ax + by + cz,$$

where x, y, z are independent quantities distributed according to the law above stated, then the *average values* of these quantities will be connected by the equation

$$u^2 = a^2 x^2 + b^2 y^2 + c^2 z.$$

Prop. XXII. *Two perfectly elastic bodies of any form strike each other: given their motions before impact, and the line of impact, to find their motions after impact.*

Let M_1 and M_2 be the centres of gravity of the two bodies. $M_1 X_1$, $M_1 Y_1$, and $M_1 Z_1$ the principal axes of the first; and $M_2 X_2$, $M_2 Y_2$, and $M_2 Z_2$ those of the second. Let I be the point of impact, and $R_1 I R_2$ the line of impact.

Let the co-ordinates of I with respect to M_1 be $x_1 y_1 z_1$, and with respect to M_2 let them be $x_2 y_2 z_2$.

Let the direction-cosines of the line of impact $R_1 I R_2$ be $l_1 m_1 n_1$ with respect to M_1, and $l_2 m_2 n_2$ with respect to M_2.

Let M_1 and M_2 be the masses, and $A_1 B_1 C_1$ and $A_2 B_2 C_2$ the moments of inertia of the bodies about their principal axes.

Let the velocities of the centres of gravity, resolved in the direction of the principal axes of each body, be

$$U_1, V_1, W_1, \quad \text{and} \quad U_2, V_2, W_2, \quad \text{before impact,}$$

and $\quad U'_1, V'_1, W'_1, \quad$ and $\quad U'_2, V'_2, W'_2, \quad$ after impact.

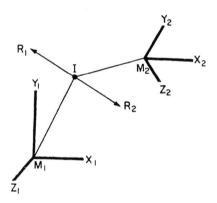

Let the angular velocities round the same axes be

$$p_1, q_1, r_1, \quad \text{and} \quad p_2, q_2, r_2, \quad \text{before impact,}$$

and $\quad p'_1, q'_1, r'_1, \quad$ and $\quad p'_2, q'_2, r'_2, \quad$ after impact.

Let R be the impulsive force between the bodies, measured by the momentum it produces in each.

Then, for the velocities of the centres of gravity, we have the following equations:

$$U'_1 = U_1 + \frac{R l_1}{M_1}, \quad U'_2 = U_2 - \frac{R l_2}{M_2}, \tag{62}$$

with two other pairs of equations in V and W.

The equations for the angular velocities are

$$p'_1 = p_1 + \frac{R}{A_1}(y_1 n_1 - z_1 m_1), \quad p'_2 = p_2 - \frac{R}{A_2}(y_2 n_2 - z_2 m_2), \tag{63}$$

with two other pairs of equations for q and r.

The condition of perfect elasticity is that the whole *vis viva* shall be the same after impact as before, which gives the equation

$$M_1 (U'^2_1 - U^2_1) + M_2 (U'^2_2 - U^2_2) + A_1 (p'^2_1 - p^2_1)$$
$$+ A_2 (p'^2_2 - p^2_2) + \&c. = 0. \quad (64)$$

The terms relating to the axis of x are here given; those relating to y and z may be easily written down.

Substituting the values of these terms, as given by equations (62) and (63), and dividing by R, we find

$$l_1 (U'_1 + U_1) - l_2 (U'_2 + U_2) + (y_1 n_1 - z_1 m_1)(p'_1 + p_1)$$
$$- (y_2 n_2 - z_2 m_2)(p'_2 + p_2) + \&c. = 0. \quad (65)$$

Now if v_1 be the velocity of the striking-point of the first body before impact, resolved along the line of impact,

$$v_1 = l_1 U_1 + (y_1 n_1 - z_1 m_1) p_1 + \&c.;$$

and if we put v_2 for the velocity of the other striking-point resolved along the same line, and v'_1 and v'_2 the same quantities after impact, we may write, equation (65),

$$v_1 + v'_1 - v_2 - v'_2 = 0, \quad (66)$$

or
$$v_1 - v_2 = v'_2 - v'_1, \quad (67)$$

which shows that the velocity of separation of the striking-points resolved in the line of impact is equal to that of approach.

Substituting the values of the accented quantities in equation (65) by means of equations (63) and (64), and transposing terms in R, we find

$$2\{U_1 l_1 - U_2 l_2 + p_1 (y_1 n_1 - z_1 m_1) - p_2 (y_2 n_2 - z_2 m_2)\} + \&c.$$
$$= - R \left\{ \frac{l_1^2}{M_1} + \frac{l_2^2}{M_2} + \frac{(y_1 n_1 - z_1 m_1)^2}{A_1} + \frac{(y_2 n_2 - z_2 m_2)^2}{A_2} + \&c., \right. \quad (68)$$

the other terms being related to y and z as these are to x. From this equation we may find the value of R; and by substituting this in equations (63), (64), we may obtain the values of all the velocities after impact.

We may, for example, find the value of U'_1 from the equation

$$\begin{aligned}U'_1 &\left\{\frac{l_1^2}{M_1} + \frac{l_2^2}{M_2} + \frac{(y_1n_1 - z_1m_1)^2}{A_1} + \frac{(y_2n_2 - z_2m_2)^2}{A_2} + \&c.\right\}\frac{M_1}{l_1} \\ = U_1 &\left\{-\frac{l_1^2}{M_1} + \frac{l_2^2}{M_2} + \frac{(y_1n_1 - z_1m_1)^2}{A_1} + \frac{(y_2n_2 - z_2m_2)^2}{A_2} + \&c.\right\}\frac{M_1}{l_1} \\ &+ 2U_2l_2 - 2p_1(y_1n_1 - z_1m_1) + 2p_2(y_2n_2 - z_2m_2) - \&c.\end{aligned}$$
(69)

Prop. XXIII. *To find the relations between the average velocities of translation and rotation after many collisions among many bodies.*

Taking equation (69), which applies to an individual collision, we see that U'_1 is expressed as a linear function of U_1, U_2, p_1, p_2, &c., all of which are quantities of which the values are distributed among the different particles according to the law of Prop. IV. It follows from Prop. V., that if we square every term of the equation, we shall have a new equation between the *average values* of the different quantities. It is plain that, as soon as the required relations have been established, they will remain the same after collision, so that we may put $U'^2_1 = U_1^2$ in the equation of averages. The equation between the verage values may then be written

$$(M_1U_1^2 - M_2U_2^2)\frac{l_2^2}{M_2} + (M_1U_1^2 - A_1p_1^2)\frac{(y_1n_1 - z_1m_1)^2}{A_1}$$
$$+ (M_1U_1^2 - A_2p_2^2)\frac{(y_2n_2 - z_2m_2)^2}{A_2} + \&c. = 0.$$

Now since there are collisions in every possible way, so that the values of l, m, n, &c. and x, y, z, &c. are infinitely varied, this equation cannot subsist unless

$$M_1U_1^2 = M_2U_2^2 = A_1p_1^2 = A_2p_2^2 = \&c.$$

The final state, therefore, of any number of systems of moving particles of any form is that in which the average *vis viva* of translation along each of the three axes is the same in all the systems, and equal to the average *vis viva* of rotation about each of the three principal axes of each particle.

Adding the *vires vivæ* with respect to the other axes, we find that the whole *vis viva* of translation is equal to that of rotation in each system of particles, and is also the same for different systems, as was proved in Prop. VI.

This result (which is true, however nearly the bodies approach the spherical form, provided the motion of rotation is at all affected by the collisions) seems decisive against the unqualified acceptation of the hypothesis that gases are such systems of hard elastic particles. For the ascertained fact that γ, the ratio of the specific heat at constant pressure to that at constant volume, is equal to 1·408, requires that the ratio of the whole *vis viva* to the *vis viva* of translation should be

$$\beta = \frac{2}{3(\gamma - 1)} = 1\cdot 634 ;$$

whereas, according to our hypothesis, $\beta = 2$.

We have now followed the mathematical theory of the collisions of hard elastic particles through various cases, in which there seems to be an analogy with the phenomena of gases. We have deduced, as others have done already, the relations of pressure, temperature, and density of a single gas. We have also proved that when two different gases act freely on each other (that is, when at the same temperature), the mass of the single particles of each is inversely proportional to the square of the molecular velocity; and therefore, at equal temperature and pressure, *the number of particles in unit of volume is the same*.

We then offered an explanation of the internal friction of gases, and deduced from experiments a value of the mean length of path of a particle between successive collisions.

We have applied the theory to the law of diffusion of gases, and, from an experiment of olefiant gas, we have deduced a value of the length of path not very different from that deduced from experiments on friction.

Using this value of the length of path between collisions, we found that the resistance of air to the conduction of heat is 10,000,000 times that of copper, a result in accordance with experience.

Finally, by establishing a necessary relation between the motions of translation and rotation of all particles not spherical, we proved that a system of such particles could not possibly satisfy the known relation between the two specific heats of all gases.

11

On a Mechanical Theorem Applicable to Heat*

RUDOLF CLAUSIUS

SUMMARY

The following theorem (virial theorem) is proved: let there be any system of material points in stationary motion. (Stationary motion means that the points are constrained to move in a limited region of space, and the velocities only fluctuate within certain limits, with no preferential direction.) Denote by X, Y, Z the components of the force acting on a particle at position x, y, z in space, and form the quantity

$$-\tfrac{1}{2}\Sigma\,(Xx + Yy + Zz)$$

where the sum is extended over all points in the system. The mean value of this quantity is called the *virial* of the system. Then: *the mean kinetic energy of the system is equal to its virial.*

In the particular case when the points are confined to a volume v by an external pressure p, and the force $\phi(r)$ between two points depends only on their distance, r, the theorem indicates that the kinetic energy of the internal motions (which we call heat) is given by the equation

$$h = \tfrac{1}{2}\Sigma\,r\phi(r) + \tfrac{3}{2}p\,v.$$

(This equation permits one to calculate the relation between pressure, volume, and temperature—the " equation of state "—if the force law $\phi(r)$ is given.)

---

* Originally published under the title " Ueber einen auf die Wärme anwendbaren mechanischen Satz ", *Sitzungsberichte der Niedderrheinischen Gesellschaft, Bonn*, pp. 114–19 (1870); English translation in *Philosophical Magazine*, Vol. 40, pp. 122–7 (1870).

In a treatise which appeared in 1862, on the mechanical theory of heat,† I advanced a theorem which, in its simplest form, may be thus expressed:—*The effective force of heat is proportional to the absolute temperature.* From this theorem, in conjunction with that of the equivalence of heat and work, I have, in the subsequent portion of that treatise, deduced various conclusions concerning the deportment of bodies towards heat. As the theorem of the equivalence of heat and work may be reduced to a simple mechanical one, namely that of the equivalence of *vis viva* and mechanical work, I was convinced *à priori* that there must be a mechanical theorem which would explain that of the increase of the effective force of heat with the temperature. This theorem I think I shall be able to communicate in what follows.

Let there be any system whatever of material points in stationary motion. By stationary motion I mean one in which the points do not continually remove further and further from their original position, and the velocities do not alter continuously in the same direction, but the points move within a limited space, and the velocities only fluctuate within certain limits. Of this nature are all periodic motions—such as those of the planets about the sun, and the vibrations of elastic bodies,—further, such irregular motions as are attributed to the atoms and molecules of a body in order to explain its heat.

Now let m, m', m'', &c. be the given material points, $x, y, z, x', y', z', x'', y'', z''$, &c. their rectangular coordinates at the time t, and $X, Y, Z, X', Y', Z', X'', Y'', Z''$, &c. the components, taken in the directions of the coordinates, of the forces acting upon them. Then we form first the sum

$$\sum \frac{m}{2}\left[\left(\frac{dx}{dt}\right)^2 + \left(\frac{dy}{dt}\right)^2 + \left(\frac{dz}{dt}\right)^2\right],$$

for which, v, v', v'', &c. being the velocities of the points, we may write, more briefly,

$$\sum \frac{m}{2} v^2,$$

---

† *Phil. Mag.* [4], vol. xxiv, pp. 81, 201; *The Mechanical Theory of Heat*, p. 215.

which sum is known under the name of the *vis viva* of the system. Further, we will form the following expression:—

$$- \tfrac{1}{2}\Sigma(Xx + Yy + Zz).$$

The magnitude represented by this expression depends, as is evident, essentially upon the forces acting in the system, and, if with given coordinates all the forces varied in equal ratio, would be proportional to the forces. We will therefore give to the mean value which this magnitude has during the stationary motion of the system the name of *Virial* of the system, from the Latin word *vis* (force).

In relation to these two magnitudes the following theorem may now be advanced:—

The mean vis viva *of the system is equal to its virial.*

Distinguishing the mean value of a magnitude from its variable value by drawing a horizontal line over the formula which represents the latter, we can express our theorem by the following equation:—

$$\sum \frac{m}{2}\overline{v^2} = -\tfrac{1}{2}\Sigma\overline{(Xx + Yy + Zz)}.$$

As regards the value of the virial, in the most important of the cases occurring in nature it takes a very simple form. For example, the forces which act upon the points of the mass may be attractions or repulsions which those points exert upon one another, and which are governed by some law of the distance. Let us denote, then, the reciprocal force between two points of the mass, m and m', at the distance r from each other, by $\phi(r)$, in which an attraction will reckon as a positive, and repulsion as a negative force; we thus have, for the reciprocal action:—

$$Xx + X'x' = \phi(r)\frac{x'-x}{r}x + \phi(r)\frac{x-x'}{r}x' = -\phi(r)\frac{(x'-x)^2}{r}.$$

And since for the two other coordinates corresponding equations may be formed, there results

$$-\tfrac{1}{2}(Xx + Yy + Zz + X'x' + Y'y' + Z'z') = \tfrac{1}{2}r\phi(r).$$

Extending this result to the whole system of points, we obtain
$$-\tfrac{1}{2}\Sigma(Xx + Yy + Zz) = \tfrac{1}{2}\Sigma r\phi(r),$$
in which the sign of summation on the right-hand side of the equation relates to all combinations of the points of the mass in pairs. Thence comes for the virial the expression
$$\tfrac{1}{2}\overline{\Sigma r\phi(r)} \; ;$$
and we immediately recognize the analogy between this expression and that which serves to determine the work accomplished in the motion. Introducing the function $\Phi(r)$ with the signification
$$\Phi(r) = \int \phi(r)dr,$$
we obtain the familiar equation
$$-\Sigma(Xdx + Ydy + Zdz) = d\Sigma\Phi(r).$$

The sum $\Sigma\Phi(r)$ is that which, in the case of attractions and repulsions, which act inversely as the square of the distance, is named, irrespective of the sign, the reciprocal *potential* of the system of points. As it is advisable to have a convenient name† for the case in which the attractions and repulsions are governed by any law whatever, or, more generally still, for every case in which the work accomplished in an infinitely small motion of the system may be represented by the differential of any magnitude dependent only on the space-coordinates of the points, I propose to name the magnitude whose differential represents the negative value of the work, from the Greek word εργον (work), the *ergal* of the system. The theorem of the equivalence of *vis viva* and work can then be expressed very simply; and in order to exhibit distinctly the analogy between this theorem and that respecting the virial, I will place the two in juxtaposition:—

(1) The sum of the *vis viva* and the ergal is constant.

(2) The mean *vis viva* is equal to the virial.

In order to apply our theorem to heat, let us consider a body as a system of material points in motion. With respect to the forces which act upon these points we have a distinction to make: in the

---

† The term *force-function*, besides some inconvenience, has the disadvantage of having been already used for another magnitude, which stands to the one in question in a relation similar to that in which the potential-function stands to the potential.

first place, the elements of the body exert upon one another attractive or repulsive forces; and, secondly, forces may act upon the body from without. Accordingly we can divide the virial into two parts, which refer respectively to the internal and the external forces, and which we will call the *internal* and the *external virial*.

Provided that the whole of the internal forces can be reduced to central forces, the internal virial is represented by the formula above given for a system of points acting by way of attraction or repulsion upon one another. It is further to be remarked that, with a body in which innumerable atoms move irregularly but in essentially like circumstances, so that all possible phases of motion occur simultaneously, it is not necessary to take the mean value of $r\phi(r)$ for each pair of atoms, but the values of $r\phi(r)$ may be taken for the precise position of the atoms at a certain moment, as the sum formed therefrom does not importantly differ from their total value throughout the course of the individual motions. Consequently we have for the internal virial the expression

$$\tfrac{1}{2}\Sigma r\phi(r).$$

As to the external forces, the case most frequently to be considered is where the body is acted upon by a uniform pressure normal to the surface. The virial relative to this can be expressed very simply; for, p signifying the pressure, and v the volume of the body, it is represented by

$$\tfrac{3}{2}pv.$$

Denoting, further, by h the *vis viva* of the internal motions (which we call heat), we can form the following equation:—

$$h = \tfrac{1}{2}\Sigma r\phi(r) + \tfrac{3}{2}pv.$$

We have still to adduce the proof of our theorem of the relation between the *vis viva* and the virial, which can be done very easily.

The equations of the motion of a material point are:—

$$m\frac{d^2x}{dt^2} = X; \quad m\frac{d^2y}{dt^2} = Y; \quad m\frac{d^2z}{dt^2} = Z.$$

But we have

$$\frac{d^2(x^2)}{dt^2} = 2\frac{d}{dt}\left(x\frac{dx}{dt}\right) = 2\left(\frac{dx}{dt}\right)^2 = 2x\frac{d^2x}{dt^2},$$

or, differently arranged,

$$2\left(\frac{dx}{dt}\right)^2 = -2x\frac{d^2x}{dt^2} + \frac{(d^2x^2)}{dt^2}.$$

Multiplying this equation by $m/4$, and putting the magnitude X for $m(d^2x/dt^2)$, we obtain

$$\frac{m}{2}\left(\frac{dx}{dt}\right)^2 = -\tfrac{1}{2}Xx + \frac{m}{4}\cdot\frac{d^2(x^2)}{dt^2}.$$

The terms of this equation may now be integrated for the time from 0 to t, and the integral divided by t; we thereby obtain

$$\frac{m}{2t}\int_0^t\left(\frac{dx}{dt}\right)^2 dt = -\frac{1}{2t}\int_0^t Xx\,dt + \frac{m}{4t}\left[\frac{d(x^2)}{dt} - \left(\frac{d(x^2)}{dt}\right)_0\right],$$

in which $(d(x^2)/dt)_0$ denotes the initial value of $d(x^2)/dt$.

The formulæ

$$\frac{1}{t}\int_0^t\left(\frac{dx}{dt}\right)^2 dt \quad \text{and} \quad \frac{1}{t}\int_0^t Xx\,dt,$$

occurring in the above equation, represent, if the duration of time t is properly chosen, the mean values of $(dx/dt)^2$ and Xx, which were denoted above by $\overline{(dx/dt)^2}$ and $\overline{Xx}$. For a periodic motion the duration of a period may be taken as the time t; but for irregular motions (and, if we please, also for periodic ones) we have only to consider that the time t, in proportion to the times during which the point moves in the same direction in respect of any one of the directions of coordinates is very great, so that in the course of the time t many changes of motion have taken place, and the above expressions of the mean values have become sufficiently constant.

The last term of the equation, which has its factor included in the square brackets, becomes, when the motion is periodic, $= 0$ at the end of each period, as at the end of the period $d(x^2)/dt$ resumes the initial value $(d(x^2)/dt)_0$. When the motion is not periodic, but irregularly varying, the factor in brackets does not so regularly become $= 0$; yet its value cannot continually increase with the time, but can only fluctuate within certain limits; and the divisor t, by which the term is affected, must accordingly cause the term to become vanishingly small with very great values of t. Hence, omitting it, we may write

$$\frac{m}{2}\overline{\left(\frac{dx}{dt}\right)^2} = -\tfrac{1}{2}\overline{Xx}.$$

As the same equation is valid also for the remaining coordinates, we have

$$\frac{m}{2}\left[\overline{\left(\frac{dx}{dt}\right)^2} + \overline{\left(\frac{dy}{dt}\right)^2} + \overline{\left(\frac{dz}{dt}\right)^2}\right] = -\tfrac{1}{2}\overline{(Xx + Yy + Zz)},$$

or, more briefly,

$$\frac{m}{2}\overline{v^2} = -\tfrac{1}{2}\overline{(Xx + Yy + Zz)},$$

and for a system of any number of points we have the perfectly corresponding one

$$\sum \frac{m}{2}\overline{v^2} = -\tfrac{1}{2}\Sigma\overline{(Xx + Yy + Zz)}.$$

Hence our theorem is demonstrated; and at the same time it is evident that it is not merely valid for the whole system of material points, and for the three directions of coordinates together, but also for each material point and for each direction separately.

Part II

Irreversible Processes

Introduction

THE works reprinted in the first section of this book were concerned with establishing the fundamental nature of heat and of gases. By about 1865 this question seemed to have been definitely settled by the general adoption of thermodynamics and the kinetic theory of gases. Scientists could now turn their attention to working out the detailed consequences of these theories and comparing them with experiments.

Knowing the empirical laws relating pressure, volume, and temperature, many scientists had discovered for themselves the kinetic explanation based on the collisions of molecules with the walls of the container. The ideal gas laws could be deduced by assuming that the space occupied by the molecules themselves is negligible compared to that total volume of the gas; it was only a natural extension of the theory to try to explain deviations from the ideal gas laws by dropping this assumption, although it was not until 1873 that J. D. van der Waals found the most fruitful mathematical formulation of this extension. But accurate experimental data on viscosity, heat conductivity, and diffusion were not available until well after 1850, and the first kinetic theorists had to make their calculations of these properties without the advantage of knowing the answer beforehand. Moreover, the type of assumption that one makes about short-range intermolecular forces determines the first approximation to the transport coefficients, whereas it comes in only in the second approximation to the equilibrium properties. Of course, such circumstances made it all the more convincing when experiments confirmed the theoretical predictions; for example, Maxwell's discovery that the viscosity of a gas is independent of density (Selection 10, Vol. 1) was very important in establishing the kinetic theory.

One may distinguish two types of irreversible process: in the first, there is a continuous flow of mass, momentum, or energy resulting from variations in concentration, velocity, and temperature imposed externally; in the second, a gas that is initially in a non-equilibrium state spontaneously moves toward equilibrium in the absence of external interference. (The two types are, of course, intimately related, as shown by Maxwell's theory of relaxation processes (Selection 1) and by modern work on the " fluctuation-dissipation theorem "—transport coefficients can be expressed in terms of the relaxation of fluctuations, and conversely.)

Following the introduction of the " mean-free-path " concept by Clausius in 1858 (Vol. 1, Selection 9) and Maxwell's theory of transport processes based on it (Vol. 1, Selection 10) there was a large amount of theoretical work by such scientists as O. E. Meyer, J. Stefan, P. G. Tait, G. Jäger, and J. H. Jeans.† This work was devoted to refining the mean-free-path theory and applying it to various phenomena. Maxwell himself realized almost immediately that the mean-free-path method was inadequate as a foundation for kinetic theory, even though it might be useful for rough approximations, and while he revised his own original treatment of diffusion and heat conduction he did not publish any further works based on this method.‡ Instead he developed the much more accurate technique based on transfer equations, described in Selection 1.

Maxwell's transfer equations provide a method for computing the rate of transport of any quantity such as mass, momentum, and energy which can be defined in terms of molecular properties, consistently with the macroscopic equations such as the Navier-Stokes equation of hydrodynamics. However, the actual calculation of transport coefficients depends in general on a knowledge of

---

† See the works by Boltzmann (1896–8), Brush (1962), Jeans (1925) Kennard (1938) and Loeb (1934), cited in the Bibliography.

‡ For diffusion theory, see Furry (1948). It is hoped that Maxwell's manuscripts on heat conduction and other topics in kinetic theory will be published shortly.

the velocity–distribution function, which is no longer assumed to be the same as the equilibrium Maxwell distribution; yet there is no direct way to compute this function. It was not until 1916 that Sydney Chapman worked out a systematic method for calculating the velocity–distribution function and all the transport coefficients of gases by means of the transfer equations. However, Maxwell discovered that in one special case, when the force between the molecules varies inversely as the fifth power of the distance, the distribution function enters the equations in such a way that one can calculate the transport coefficients without knowing it. In this case it turns out that the viscosity coefficient is directly proportional to the temperature, whereas in the case of hard spheres it is proportional to the square root of the temperature. Maxwell's own measurements of the viscosity of air, which he had been carrying out at about the same time, led him to believe that the former result is in agreement with experiment.† The inverse fifth-power force, which is an exactly soluble case for purely mathematical reasons, therefore seemed to be at the same time a good model for real gases, and Maxwell adopted it provisionally as the basis for his later work.‡

Boltzmann

Ludwig Boltzmann (1844–1906), whose writings occupy the major part of this volume, was born and educated in Vienna, and taught there most of his life. He is best known for his work on the kinetic theory of gases and statistical mechanics. His equation for the velocity–distribution function of molecules (first derived in the work reprinted here as Selection 2) has been used extensively to study transport properties such as viscosity, thermal conductivity, and diffusion. Boltzmann's equation is actually equivalent to

---

† J. C. Maxwell, *Phil. Trans. Roy. Soc. London* **156**, 249 (1866).

‡ J. C. Maxwell, *Trans. Cambridge Phil. Soc.* **12**, 547 (1879); see also Boltzmann (1896), Chapter III. The later work of Chapman and Enskog showed that thermal diffusion, which takes place in systems of molecules with any general force law, happens to be absent in the special case of inverse fourth-power forces.

Maxwell's set of transfer equations, and its solution presents exactly the same difficulties. Just as Chapman solved Maxwell's equations, so did David Enskog (1917) solve Boltzmann's equation in the general case, and the Maxwell–Boltzmann–Chapman–Enskog theory constitutes the completion of the classical kinetic theory of low-density gases. The basic theory has remained essentially unchanged since 1917; with appropriate modifications, it forms the basis for many modern theories of liquids, solids, plasmas, and neutron transport.

Though Boltzmann himself did not obtain the complete solution to his equation for the distribution function, he did deduce from it one important consequence, which later† came to be known as the H-theorem (see the last part of the first section of Selection 2). The H-theorem attempts to explain the irreversibility of natural processes by showing how molecular collisions tend to increase entropy; any initial distribution of molecular positions and velocities will almost certainly evolve into an equilibrium state, in which the velocities are distributed according to Maxwell's law.

The Maxwell velocity distribution law was also generalized by Boltzmann so as to apply to physical systems in which interatomic forces and external fields must be taken into account.‡ The so-called Boltzmann factor, $e^{-E/kT}$, which gives the relative probability of a configuration of energy E at a temperature T, provides the basis for all calculations of the equilibrium properties of matter from the molecular viewpoint. Boltzmann further showed that the entropy of a system in any physical state may be calculated from its probability by counting the number of molecular configurations corresponding to that state.§ This result, and the

---

† Boltzmann himself originally used the letter E, and did not change to H until 1895; the first use of H for this quantity was apparently by S. H. Burbury, *Phil. Mag.* **30**, 301 (1890). A rumour has circulated among modern physicists that H was intended to be a capital eta, the letter eta having been used for entropy by Gibbs and other writers [see S. Chapman, *Nature* **139**, 931 (1937)]. This can hardly be true if in fact the usage was originated by Burbury, since he makes no such suggestion in his paper.

‡ L. Boltzmann, *Wien. Ber.* **58**, 517 (1868).

§ L. Boltzmann, *Wien. Ber.* **76**, 373 (1877).

Boltzmann–Stefan formula for the energy of black-body radiation,† aided Max Planck in his development of quantum theory in 1900.‡

Boltzmann's work was followed with considerable interest by British scientists, and conversely Boltzmann helped to convert German scientists to Maxwell's electromagnetic theory.§ However, he did not always find support for his own theories in Germany. In the 1890's he had to defend the kinetic theory, and even the existence of atoms, against the attacks of Wilhelm Ostwald, Pierre Duhem, Ernst Mach and others. Boltzmann thought himself to be on the losing side in this battle, and feared that " Energetics " and the " reaction against materialism " would obliterate his life-work.‖ However, within a few years of his suicide in 1906, the existence of atoms had been definitely established by experiments such as those on Brownian motion, and even Ostwald had to admit that he had been wrong.¶

Dissipation of Energy

The modern idea of irreversibility in physical processes is based on the second law of thermodynamics in its generalized form. As is well known, the second law was first enunciated by Sadi Carnot in 1824 in his memoir on the efficiency of steam engines,†† and the

---

† Stated as an empirical generalization by J. Stefan, *Wien. Ber.* **79,** 391 (1879), and derived from thermodynamics and Maxwell's electromagnetic theory by L. Boltzmann, *Ann. Phys.* **22,** 291 (1884).

‡ See M. J. Klein, *Arch. Hist. Exact Sci.* **1,** 459 (1962); *The Natural Philosopher* **1,** 83 (1963).

§ L. Boltzmann, *Vorlesungen über Maxwells Theorie der Elektricität und des Lichts*, J. A. Barth, Leipzig, 1891 and 1893.

‖ See Boltzmann (1896–8), pp. 13–17, 23–8, 215–16 of the English translation.

¶ Compare W. Ostwald, *Nature* **70,** 15 (1904) (before) and *Grundriss der allgemeinen Chemie*, Engelmann, Leipzig, 4th ed., 1909, Vorbericht (after). [English translation of the latter published as *Outlines of General Chemistry*, Macmillan, London, 1912.]

†† See Carnot (1824), cited in Bibliography.

problem of obtaining the maximum amount of work from a given amount of fuel seems to have provided the motivation for many of the nineteenth-century researches on heat and its transformations.

Although one can find scattered statements in the technical literature before 1850 to the effect that something is always lost or dissipated when heat is used to produce mechanical work, it was not until 1852 that William Thomson (later Lord Kelvin) asserted the existence of "A universal tendency in nature to the dissipation of mechanical energy ".† At the same time he mentioned his conclusion that " within a finite period of time past, the earth must have been, and within a finite period of time to come the earth must again be, unfit for the habitation of man as at present constituted, unless operations going on at present in the material world are subject ".

The consequences of Thomson's Dissipation Principle were elaborated further by Hermann von Helmholtz in a lecture 2 years later, in which he described the final state of the universe: all energy will eventually be transformed into heat at uniform temperature, and all natural processes must cease; "the universe from that time forward would be condemned to a state of eternal rest ". Thus was made explicit the concept of the "heat death" of the universe.‡

The modern statement of the Dissipation Principle involves the notion of " entropy ", introduced by Clausius in 1865.§ It should be emphasized that no new physical content was being added to the second law in this formulation of Clausius, yet the mere act of giving a new short name to a physical quantity that had previously been represented only by mathematical formulae and awkward circumlocutions had an undeniable influence on the subsequent history of the subject. Entropy is taken from the Greek $\tau\rho o\pi\acute{\eta}$, meaning transformation, and was intentionally chosen to have a resemblance to the word energy. The two laws of thermodynamics are then stated as follows:

---

† See Kelvin (1852).

‡ H. von Helmholtz, *Popular Scientific Lectures* (translated from German and edited by M. Kline), Dover Publications, New York, 1962, p. 59.

§ R. Clausius, *Ann. Phys.* **125**, 353 (1865).

1. The energy of the universe is constant.
2. The entropy of the universe tends toward a maximum.

Before returning to the role of energy dissipation in the history of the kinetic theory, we should mention the primary use which Thomson made of his Dissipation Principle. One of the major goals of mathematical physics in the eighteenth century had been to deduce the stability of the solar system from Newtonian mechanics, and by the beginning of the nineteenth century this goal had apparently been reached by the work of such scientists as Lagrange, Laplace and Poisson. Consequently it was generally believed that the earth had remained at the same average distance from the sun for an indefinitely long time in the past, and that physical conditions on the surface of the earth had been roughly the same as they are now for countless millions of years. This assumption was the basis for the " Uniformitarian " theory in geology, advocated by Hutton and Lyell, which gradually replaced the earlier " Catastrophist " doctrine. According to the Uniformitarians, the present appearance of the earth's surface was to be explained as a result of physical causes, like erosion, whose operation can still be observed, rather than by postulating catastrophic upheavals in the past such as the Flood.† Likewise, the Darwin–Wallace theory of biological evolution by natural selection assumes that the origin and development of species could have taken place by gradual changes over a very long period of time, during which physical conditions did not change very much, and indeed Darwin relied heavily on Uniformitarian geology in establishing his theory.‡ It was just this assumption of constant physical conditions that Thomson attacked, using the Dissipation Principle in conjunction with Fourier's theory of heat conduction. He showed that if one assumed that the earth had once been

---

† C. C. Gillispie, *Genesis and Geology*, Harvard University Press, 1951; A Wolf, *A History of Science, Technology, and Philosophy in the 18th Century* Macmillan, New York, 2nd edition 1952, Vol. I, Chap. XV; S. J. Gould, *Amer. J. Sci.* **263**, 223 (1965).

‡ See " On the Imperfection of the Geological Record " which is Chapter IX in the first edition of the *Origin of Species* (1859) and Chapter X in later editions.

very hot and was slowly losing its heat by conduction outward from the centre, and that no other sources of heat are present, then it follows from experimental data on the present rate of heat loss and certain plausible assumptions about the conductivity of the interior that present conditions can have lasted no more than about 20 million years. Before that time the temperature of the entire earth must have been so high that the whole globe was liquid. Therefore Uniformitarian geology, whose advocates had assumed constant conditions over periods of hundreds of millions of years, must be wrong.†

There followed a bitter controversy between physicists (led by Thomson and P. G. Tait) and geologists (led by T. H. Huxley) about the validity of various methods for estimating the age of the earth.‡ Religious views played some role in this dispute, for Thomson believed that he had discovered a mathematical proof that there must have been a creation by solving Fourier's heat equation.§ If he could discredit Uniformitarian geology, he would also have removed one of the major supporting arguments from the theory of evolution by natural selection, a theory which he considered " did not sufficiently take into account a continually guiding and controlling intelligence ".‖ Darwin himself was well aware of the fact that Thomson's attack weakened his own theory, and one historian of evolution has noted that the physicists " forced Darwin, before his death, into an awkward retreat which mars in some degree the final edition of the *Origin* ".¶

---

† W. Thomson, *Trans. Roy. Soc. Edinburgh* **13**, 157 (1864); *Proc. Roy. Soc. Edinburgh* **5**, 512 (1866); *Trans. Glasgow. Geol. Soc.* **3**, 1, 215 (1871); *A Source Book in Astronomy* (ed. H. Shapley and H. E. Howarth), McGraw-Hill, New York, 1929.

‡ T. H. Huxley, *Q. J. Geol. Soc. London*, **25**, xxviii (1869); anonymous article in the *North British Review*, **50**, 406 (1869), attributed to P. G. Tait by Darwin, in a letter to J. D. Hooker, in *More Letters of Charles Darwin*, Appleton, New York, 1903, **1**, 313–14.

§ S. P. Thompson, *The Life of William Thomson, Baron Kelvin of Largs*, Macmillan, London, 1910, **1**, 111.

‖ W. Thomson, *British Assn. Rept.* (1871) p. lxxxiv; *Nature* **4**, 262 (1871).

¶ L. Eiseley, *Darwin's Century*, Doubleday, Garden City, 1958, p. 245.

The dispute was finally settled in favour of the geologists by the discovery of radioactivity, which not only provided a source of heat which Thomson had not taken into account in his calculations, but also made it possible to determine the age of rocks fairly accurately. Thomson himself lived long enough to see the publication of these studies which showed that the earth must be at least 2 thousand million years old.†

Cultural Influence of the Dissipation of Energy

The English philosopher Herbert Spencer was one of the first writers to take up the idea of dissipation of energy and incorporate it into a general system of philosophy.‡ Despite the apparently pessimistic consequences of the heat death, Spencer manages to draw from the second law " a warrant for the belief, that Evolution can end only in the establishment of the greatest perfection and the most complete happiness ". He also predicts that after the motion of all the stars has become equilibrated and degraded to heat energy, there will be a process of concentration under the action of gravity, followed by dispersion, and an infinite sequence of alternate eras of Evolution and Dissolution.

The most remarkable application of the Dissipation Principle was made by the American historian Henry Adams. In three essays, later published under the title *The Degradation of the Democratic Dogma*, he argued that a science of history could be based on the general properties of energy as discovered by the physical sciences.§ He saw a general process of degradation and deterioration in human history, and noted that the cheerful

---

† R. J. Strutt, *Proc. Roy. Soc.* London **A76**, 88 (1905); B. B. Boltwood, *Amer. J. Sci.* **23**, 77 (1907); see also E. N. da C. Andrade, *Rutherford and the Nature of the Atom*, Doubleday, Garden City, 1964, p. 80.

‡ H. Spencer, *First Principles*, London, 1870, pp. 514–17.

§ H. Adams, " The Tendency of History " (1894); "A letter to American Teachers of History " (1910); " The Rule of Phase applied to History " (1909), reprinted with an Introduction by Brooks Adams in *The Degradation of the Democratic Dogma*, Macmillan, New York, 1919.

optimism inspired by Darwin had given way to *fin-de-siècle* pessimism toward the end of the nineteenth century. To the historian, the concept of entropy " meant only that the ash-heap was constantly increasing in size ".

Aside from these and a few other examples, the direct influence of the Dissipation Principle on European thought in the nineteenth century was remarkably small. It is only at the end of the century, and in the first part of the twentieth century, that one finds an increasing number of references to the second law of thermodynamics, and attempts to connect it with general historical tendencies. The writings of Sir James Jeans and Sir Arthur Eddington in the 1920's and 1930's have now made the " heat death " an integral part of the modern educated layman's knowledge of cosmology.†

The Statistical Nature of the Second Law

Naturally not everyone was satisfied with the pessimistic consequences that William Thomson and others believed to follow from the second law, and there were many attempts to disprove or circumvent them. The best-known scheme for this purpose is based on the concept of " Maxwell's Demon ", originally invented in order to illustrate the statistical nature of irreversibility. The Demon was conceived in discussions among Maxwell, Tait and Thomson; he was first described in a letter from Maxwell to Tait in 1867, and publicly introduced in Maxwell's *Theory of Heat* published in 1871.‡ He is " a being whose faculties are so sharpened that he can follow every molecule in its course ". A vessel filled with gas is divided into two parts by a partition with a small hole, and the demon opens and closes the hole so as to allow

---

† A. S. Eddington, *The Nature of the Physical World*, Cambridge University Press, 1928; J. H. Jeans, *The Universe Around Us*, Cambridge University Press, 1933.

‡ See C. G. Knott, *Life and Scientific Work of Peter Guthrie Tait*, Cambridge University Press, 1911, pp. 213–14; J. C. Maxwell, *Theory of Heat*, Longmans, London, 1871, Chapter XXII.

only the swifter molecules to pass through in one direction, and only the slower ones in the other. He thus produces a temperature difference without any expenditure of work, violating the second law. The point of this imaginary construction is that the second law does not necessarily apply to events on a molecular level, if a being with sufficient knowledge about the details of molecular configurations is present to manipulate things.

Boltzmann's H-theorem shows the statistical nature of the second law in a more quantitative though less picturesque manner. In his proof of the theorem (Selection 2), Boltzmann first introduced a definition of entropy in terms of the molecular velocity distribution, then showed that if certain assumptions are granted, the entropy as thus defined will always increase as a result of collisions among molecules. The assumptions are of two types: first, statistical assumptions about the random distribution of the velocities of two molecules *before* they collide; and second, mechanical assumptions about the existence of " inverse " collisions. The first assumption is needed in order the calculate the number of collisions of various types; the second is needed in order to balance the effects of some collisions against others in which the initial and final configurations are interchanged, in order to obtain a final result having the desired mathematical properties. The limitations on the validity of these assumptions were not explicitly recognized by Boltzmann until later, and in fact the H-theorem inspired a large amount of controversy during the late nineteenth century and afterwards.† Roughly speaking, what the H-theorem proves is not so much a property of real physical systems as a property of our information about those systems. The equilibrium Maxwell distribution is the most random one possible, in the sense that it represents the minimum amount of information. If we start with partial information about molecular velocities (so that the distribution is not Maxwellian) then this information will be gradually lost as collisions occur. On the other hand, if we had started with complete information and

---

† See P. and T. Ehrenfest (1911); ter Haar (1954); Dugas (1959); Bryan (1891–4).

could follow the motions as determined by the laws of (classical) mechanics, no information need ever be lost in principle, and the entropy would remain constant.†

Thomson's 1874 paper on the dissipation of energy (Selection 3) contains a substantial part of the modern interpretation of irreversibility, though it is seldom cited. Thomson notes the contrast between " abstract dynamics ", which is reversible, and " physical dynamics ", which is not, and shows how this " reversibility paradox " may be explained by taking account of the large number of molecules involved in physical processes.

The reversibility paradox is usually attributed to Josef Loschmidt, who mentioned it very briefly in the first of four articles on the thermal equilibrium of a system of bodies subject to gravitational forces.‡ Loschmidt was attempting to show in these papers that equilibrium was possible without equality of temperature; in this way he hoped to demonstrate that the heat death of the universe is not inevitable. He claimed that the second law could be correctly formulated as a mechanical principle without reference to the sequence of events in time; he thought that he could thus " destroy the terroristic nimbus of the second law, which has made it appear to be an annihilating principle for all living beings of the universe; and at the same time open up the comforting prospect that mankind is not dependent on mineral coal or the sun for transforming heat into work, but rather may have available forever an inexhaustible supply of transformable heat ". After attacking Maxwell's conclusion that the temperature of the gas in a column should be independent of height, and proposing a model which supposedly violates this law, he says that in any system " the entire course of events will be retraced if at some instant the velocities of all its parts are reversed ". Loschmidt's application of this reversibility principle to the validity of the second law is somewhat obscurely stated, but Boltzmann quickly got the point and immediately published a reply (Selection 4) in which he gave a thorough discussion of the reversibility

---

† See Tolman (1938) Chapter VI.
‡ J. Loschmidt, *Wien. Ber.* **73**, 128, 366 (1876); **75**, 287; **76**, 209 (1877).

paradox, ending up with a conclusion very similar to that of Thomson.

The Eternal Return and the Recurrence Paradox

The notion that history repeats itself—that there is no progress or decay in the long run, but only a cycle of development that always returns to its starting point—has been inherited from ancient philosophy and primitive religion. It has been noted by some scholars that belief in recurrence, as opposed to unending progress, is intimately connected with man's view of his place in the universe, as well as with his concept of history. Starting, in most cases, from a pessimistic view of the present and immediate future, it denies the reality or validity of human actions and historical events by themselves; actions and events are real only insofar as they can be understood as the working out of timeless archetypal patterns of behaviour in the mythology of the society. This attitude is said to be illustrated in classical Greek and Roman art and literature, where there is no consciousness of past or future, but only of eternal principles and values. By contrast the modern Western view, as a result of the influence of Christianity, is deeply conscious of history as progress toward a goal.† Nevertheless, the cyclical view has by no means died out, and can easily be recognized in the persistent tendency to draw historical analogies and comparisons.

The suggestion that eternal recurrence might be proved as a theorem of physics, rather than as a religious or philosophical doctrine, seems to have occurred at about the same time to the German philosopher Friedrich Nietzsche and the French mathematician Henri Poincaré. Nietzsche encountered the idea of recurrence in his studies of classical philology, and again in a book

---

† M. Eliade, *The Myth of the Eternal Return*, Pantheon Books, New York, 1954; A. Rey, *Le Retour Eternel et la Philosophie de la Physique*, Flammarion, Paris, 1927; J. Baillie, *The Belief in Progress*, Oxford University Press, 1950, §10; P. Sorokin, *Social and Cultural Dynamics*, American Book Co., New York, 1937, Vol. II, Chap. 10.

by Heine.† It was not until 1881 that he began to take it seriously, however, and then he devoted several years to studying physics in order to find a scientific foundation for it.‡ Poincaré, on the other hand, was led to the subject by his attempts to complete Poisson's proof of the stability of the solar system. Both Nietzsche and Poincaré were trying, though in very different ways, to attack the " materialist " or " mechanist " view of the universe.

Nietzsche's " proof " of the necessity of eternal recurrence (written during the period 1884–88 but not published until after his death in 1900) is as follows: " If the universe has a goal, that goal would have been reached by now " since the universe, he thinks, has always existed; the concept of a world " created " at some finite time in the past is considered a meaningless relic of the superstitious ages. He absolutely rejects the idea of a " final state " of the universe, and further remarks that " if, for instance, materialism cannot consistently escape the conclusion of a finite state, which William Thomson has traced out for it, then materialism is thereby refuted ". He continues:

> If the universe may be conceived as a definite quantity of energy, as a definite number of centres of energy—and every other concept remains indefinite and therefore useless—it follows therefrom that the universe must go through a calculable number of combinations in the great game of chance which constitutes its existence. In infinity, at some moment or other, every possible combination must once have been realized; not only this, but it must have been realized an infinite number of times. And inasmuch as between every one of these combinations and its next recurrence every other possible combination would necessarily have been undergone, and since every one of these combinations would determine the whole series in the same order, a circular movement of absolutely identical series is thus demonstrated: the universe is thus shown to be a circular movement which has already repeated itself an infinite number of times, and which plays its game for all eternity.§.

---

† See W. A. Kaufman, *Nietzsche: Philosopher, Psychologist, Antichrist*, Princeton University Press, 1950, Chapter 11.

‡ See C. Andler, *Nietzsche, sa Vie et sa Pensée*, Gallimard, Paris, 1958, Vol. 4, Livre 2, Chap. I and Livre 3, Chap. I.

§ F. Nietzsche, *Der Wille zur Macht*, in his *Gesammelte Werke*, Musarion Verlag, Munich, 1926, Vol. 19, Book 4, Part 3; English translation by O. Manthey-Zorn in Nietzsche, *An Anthology of his works*, Washington Square Press, New York, 1964, p. 90.

Nietzsche thought that his doctrine was not materialistic because materialism entailed the irreversible dissipation of energy and the ultimate heat death of the universe. In fact, the discussion of Poincaré's theorem (Selections 6–10) showed that on the contrary it is precisely the mechanistic view of the universe that has recurrence as its inevitable consequence. Since Zermelo and some other scientists believed that the second law must have absolute rather than merely statistical validity, they thought that the mechanistic theory was refuted by the " recurrence paradox ".†
The effect of Nietzsche's argument is actually just the opposite of what he thought it should be: if there *is* eternal recurrence, so that the second law of thermodynamics cannot always be valid, then the materialist view (as represented by Boltzmann's interpretation) would be substantiated.

While Poincaré's version of the theorem, which was used by Zermelo to attack the kinetic theory (Selections 7 and 9), had an important influence on the history of theoretical physics whereas Nietzsche's had none, it should not be thought that the former was a rigorous mathematical proof whereas the latter was merely another aphorism of the mad philosopher. Nietzsche's version presents the essential point clearly and plausibly enough, while Poincaré and Zermelo, with all their concern for mathematical rigor, still fail to come up to the standards of modern mathematics. Their proof is inadequate because they lack the precise concept of the " measure " of a set of points, introduced by Lebesgue in 1902 and used to prove the recurrence theorem by Carathéodory in 1918.‡ Poincaré's " exceptional " non-recurrent trajectories, which can be proved to exist and are in fact infinite in

---

† See also F. Wald, *Die Energie und ihre Entwerthung*, Englemann, Leipzig, 1889, p. 104; E. Mach, *Die Prinzipien der Wärmelehre*, Barth, Leipzig, 1896, p. 362; G. Helm, *Die Lehre von der Energie historisch-kritisch entwickelt*, Felix, Leipzig, 1887; *Grundzüge der mathematischen Chemie*, Veit, Leipzig, 1898; P. Duhem, *Traité d'energetique*, Gauthier-Villars, Paris, 1911; H. Poincaré, *Thermodynamique*, Gauthier-Villars, Paris, 1892; *Nature* **45**, 414, 485 (1892).

‡ H. Lebesgue, *Annali Mat. pura e appl.* **7**, 231 (1902); C. Carathéodory, *Sitzber. Preuss. Akad. Wiss.* 579 (1919).

number, yet have zero " probability ", are a " set of measure zero " with respect to the others, in modern terminology.

Who won the debate between Zermelo and Boltzmann? The reader may of course decide for himself, but modern physicists almost unanimously follow Boltzmann's views. To some extent this is because the alternative of rejecting all atomic theories is no longer open to us, as it was in the 1890's, and we are therefore forced to rely on something like an H-theorem in order to understand irreversibility. While physicists do not generally go so far as Reichenbach in accepting Boltzmann's notion of alternating time-directions in the universe,† they do accept the statistical interpretation of the second law of thermodynamics.‡

---

† H. Reichenbach, *The Direction of Time*. University of California Press, Berkeley, 1956.

‡ A concrete example of the behaviour of the H-curve postulated by Boltzmann was provided by the Ehrenfest urn-model: see P. and T. Ehrenfest, *Phys. z.* **8,** 311 (1907), and the book by ter Haar (1954).

Bibliography

BOLTZMANN, LUDWIG. *Vorlesungen über Gastheorie.* J. A. Barth, Leipzig, Part I, 1896, Part II, 1898. English translation by S. G. Brush, *Lectures on Gas Theory*, University of California Press, Berkeley, 1964.

BRUSH, STEPHEN G. Development of the Kinetic Theory of Gases, VI. Viscosity. *Amer. J. Phys.* **30**, 269 (1962).

BRUSH, STEPHEN G. Thermodynamics and History, *The Graduate Journal* (in press).

BRYAN, GEORGE H. Researches related to the connection of the second law with dynamical principles, *Brit. Assoc. Rept.* **61**, 85 (1891).

BRYAN, GEORGE H. The laws of distribution of energy and their limitations, *Brit. Assoc. Rept.* **64**, 64 (1894).

CARNOT, SADI. *Reflexions sur la puissance motrice de feu et sur les machines propres à developper cette puissance.* Bachelier, Paris, 1824. Reprinted with additional material by Blanchard, Paris, 1878. English translation by R. H. Thurston, together with other papers by Clapeyron and Clausius, in *Reflections on the Motive Power of Fire*, etc., Dover, New York, 1960.

CHAPMAN, SYDNEY. On the law of distribution of molecular velocities, and on the theory of viscosity and thermal conduction, in a non-uniform simple monatomic gas, *Phil. Trans. Roy. Soc. London*, **A216**, 279 (1916).

CHAPMAN, SYDNEY. On the kinetic theory of a gas, Part II, a composite monatomic gas, diffusion, viscosity, and thermal conduction, *Phil. Trans. Roy. Soc. London*, **A217**, 115 (1917).

CHAPMAN, SYDNEY and COWLING, T. G. *The Mathematical Theory of Non-Uniform Gases*, Cambridge University Press, 1939, 2nd edition, 1952.

DUGAS, RENÉ. *La Théorie Physique au sens de Boltzmann et ses prolongements modernes.* Éditions du Griffon, Neuchâtel-Suisse, 1959.

EHRENFEST, PAUL and TATIANA. Begriffliche Grundlagen der statistischen Auffassung in der Mechanik. *Encyklopädie der mathematischen Wissenschaften*, Vol. 4, Part 32, Teubner, Leipzig, 1911. English translation by M. J. Moravcsik, *The Conceptual Foundations of the Statistical Approach in Mechanics*, Cornell University Press, 1959.

ENSKOG, DAVID. *Kinetische Theorie der Vorgänge in mässig verdünnten Gasen*, Almqvist and Wiksell, Uppsala, 1917.

FURRY, WENDELL H. On the elementary explanation of diffusion phenomena in Gases, *Amer. J. Phys.* **16**, 63 (1948).

HAAR, D. TER. *Elements of Statistical Mechanics*, Rinehart, New York, 1954.

JEANS, JAMES H. *The Dynamical Theory of Gases*, Cambridge University Press, 1904; 4th edition, 1925, reprinted by Dover, New York, 1954.

KELVIN, WILLIAM THOMSON. On a universal tendency in nature to the dissipation of mechanical energy, *Phil Mag.* [4] **4**, 304 (1852); *Proc. Roy. Soc. Edinburgh* **3**, 139 (1857); reprinted in his *Mathematical and Physical Papers*, Cambridge University Press, 2nd edition, 1882–1911, Vol. 1. p. 511.

KENNARD, EARLE H. *Kinetic Theory of Gases with an Introduction to Statistical Mechanics*, McGraw-Hill, New York, 1938.

LEOB, LEONARD B. *Kinetic Theory of Gases*, McGraw-Hill, New York, 1922, 2nd edition, 1934.

TOLMAN, RICHARD C. *The Principles of Statistical Mechanics*, Oxford University Press, 1938.

1

On the Dynamical Theory of Gases*

JAMES CLERK MAXWELL

SUMMARY†

The theory of transport processes in gases—such as diffusion, heat conduction, and viscosity—is developed on the basis of the assumption that the molecules behave like point-centres of force. The method of investigation consists in calculating mean values of various functions of the velocity of all the molecules of a given kind within an element of volume, and the variations of these mean values due, first, to the encounters of the molecules with others of the same or a different kind; second, to the action of external forces such as gravity; and third, to the passage of molecules through the boundary of the element of volume.

The encounters are analysed of molecules repelling each other with forces inversely as the nth power of the distance. In general the variation of mean values of functions of the velocity due to encounters depends on the relative velocity of the two colliding molecules, and unless the gas is in thermal equilibrium the velocity distribution is unknown so that these variations cannot be calculated directly. However, in the case of inverse fifth-power forces the relative velocity drops out, and the calculations can be carried out. It is found that in this special case the viscosity coefficient is proportional to the absolute temperature, in agreement with experimental results of the author. An expression for the diffusion coefficient is also derived, and compared with experimental results published by Graham.

A new derivation is given of the velocity–distribution law for a gas in thermal equilibrium. The theory is also applied to give an explanation of the

---

* Originally published in the *Philosophical Transactions of the Royal Society of London*, **157**, 49–88 (1867), and in *Philosophical Magazine*, **32**, 390–3 (1866), **35**, 129–45, 185–217 (1868); reprinted in *The Scientific Papers of James Clerk Maxwell*, Cambridge University Press, 1890, **2**, 26–78.

† All Summaries by S. G. B.

Law of Equivalent Volumes, the conduction of heat through gases, the hydrodynamic equations of motion corrected for viscosity (Navier–Stokes equation), the relaxation of inequalities of pressure, and the final equilibrium of temperature in a column of gas under the influence of gravity.

Theories of the constitution of bodies suppose them either to be continuous and homogeneous, or to be composed of a finite number of distinct particles or molecules.

In certain applications of mathematics to physical questions, it is convenient to suppose bodies homogeneous in order to make the quantity of matter in each differential element a function of the co-ordinates, but I am not aware that any theory of this kind has been proposed to account for the different properties of bodies. Indeed the properties of a body supposed to be a uniform *plenum* may be affirmed dogmatically, but cannot be explained mathematically.

Molecular theories suppose that all bodies, even when they appear to our senses homogeneous, consist of a multitude of particles, or small parts the mechanical relations of which constitute the properties of the bodies. Those theories which suppose that the molecules are at rest relative to the body may be called statical theories, and those which suppose the molecules to be in motion, even while the body is apparently at rest, may be called dynamical theories.

If we adopt a statical theory, and suppose the molecules of a body kept at rest in their positions of equilibrium by the action of forces in the directions of the lines joining their centres, we may determine the mechanical properties of a body so constructed, if distorted so that the displacement of each molecule is a function of its co-ordinates when in equilibrium. It appears from the mathematical theory of bodies of this kind, that the forces called into play by a small change of form must always bear a fixed proportion to those excited by a small change of volume.

Now we know that in fluids the elasticity of form is evanescent, while that of volume is considerable. Hence such theories will not apply to fluids. In solid bodies the elasticity of form appears in many cases to be smaller in proportion to that of volume than the theory gives,† so that we are forced to give up the theory of molecules whose displacements are functions of their co-ordinates when at rest, even in the case of solid bodies.

The theory of moving molecules, on the other hand, is not open to these objections. The mathematical difficulties in applying the theory are considerable, and till they are surmounted we cannot fully decide on the applicability of the theory. We are able, however, to explain a great variety of phenomena by the dynamical theory which have not been hitherto explained otherwise.

The dynamical theory supposes that the molecules of solid bodies oscillate about their positions of equilibrium, but do not travel from one position to another in the body. In fluids the molecules are supposed to be constantly moving into new relative positions, so that the same molecule may travel from one part of the fluid to any other part. In liquids the molecules are supposed to be always under the action of the forces due to neighbouring molecules throughout their course, but in gases the greater part of the path of each molecule is supposed to be sensibly rectilinear and beyond the sphere of sensible action of the neighbouring molecules.

I propose in this paper to apply this theory to the explanation of various properties of gases, and to shew that, besides accounting for the relations of pressure, density, and temperature in a single gas, it affords a mechanical explanation of the known chemical relation between the density of a gas and its equivalent weight, commonly called the Law of Equivalent Volumes. It also explains the diffusion of one gas through another, the internal friction of a gas, and the conduction of heat through gases.

The opinion that the observed properties of visible bodies apparently at rest are due to the action of invisible molecules in rapid motion is to be found in Lucretius. In the exposition which he gives of the theories of Democritus as modified by Epicurus, he

---

† In glass, according to Dr. Everett's second series of experiments (1866), the ratio of the elasticity of form to that of volume is greater than that given by the theory. In brass and steel it is less. March 7, 1867. [J. D. Everett, *Phil. Trans.* **156**, 185 (1866).]

describes the invisible atoms as all moving downwards with equal velocities, which, at quite uncertain times and places, suffer an imperceptible change, just enough to allow of occasional collisions taking place between the atoms. These atoms he supposes to set small bodies in motion by an action of which we may form some conception by looking at the motes in a sunbeam. The language of Lucretius must of course be interpreted according to the ideas of his age, but we need not wonder that it suggested to Le Sage the fundamental conception of his theory of gases, as well as his doctrine of ultramundane corpuscles.†

Professor Clausius, to whom we owe the most extensive developments of dynamical theory of gases, has given‡ a list of authors who have adopted or given countenance to any theory of invisible particles in motion. Of these, Daniel Bernoulli, in the tenth section of his *Hydrodynamics*, distinctly explains the pressure of air by the impact of its particles on the sides of the vessel containing it.§

Clausius also mentions a book entitled *Deux Traités de Physique Mécanique*, publiés par Pierre Prevost, comme simple Éditeur du premier et comme Auteur du second, Genève et Paris, 1818. The first memoir is by G. Le Sage, who explains gravity by the impact of " ultramundane corpuscles " on bodies. These corpuscles also set in motion the particles of light and various ethereal media, which in their turn act on the molecules of gases and keep up their motions. His theory of impact is faulty, but his explanation of the expansive force of gases is essentially the same as in the dynamical theory as it now stands. The second memoir, by Prevost, contains new applications of the principles of Le Sage to gases and to light. A more extensive application of the theory of

---

† [G. L. LeSage, *Physique Mecanique* (1746), published by P. Prevost under the title *Deux Traites de Physique Mecanique*, Geneva, 1818; see also LeSage's article " Lucrèce Newtonien " in *Nouveaux Mémoires de l'Academie Royale des Sciences et Belles-Lettres*, Berlin, 1782, p. 404.]

‡ Poggendorff's *Annalen*, January 1862. Translated by G. C. Foster, B.A., *Phil. Mag.* June 1862. [*Ann. Phys.* 115, 1; *Phil. Mag.* 23, 417, 512.]

[See Vol. 1 of this series, Selection 3.]

moving molecules was made by Herapath.† His theory of the collisions of perfectly hard bodies, such as he supposes the molecules to be, is faulty, inasmuch as it makes the result of impact depend on the absolute motion of the bodies, so that by experiments on such hard bodies (if we could get them) we might determine the absolute direction and velocity of the motion of the earth.‡ This author, however, has applied his theory to the numerical results of experiment in many cases, and his speculations are always ingenious, and often throw much real light on the questions treated. In particular, the theory of temperature and pressure in gases and the theory of diffusion are clearly pointed out.

Dr Joule§ has also explained the pressure of gases by the impact of their molecules, and has calculated the velocity which they must have in order to produce the pressure observed in particular gases.

It is to Professor Clausius, of Zurich, that we owe the most complete dynamical theory of gases. His other researches on the general dynamical theory of heat are well known, and his memoirs *On the kind of Motion which we call Heat*, are a complete exposition of the molecular theory adopted in this paper. After reading his investigation‖ of the distance described by each molecule between successive collisions, I published some propositions¶ on the motions and collisions of perfectly elastic spheres, and deduced several properties of gases, especially the law of equivalent volumes, and the nature of gaseous friction. I also gave a theory of diffusion of gases, which I now know to be erroneous, and there were several errors in my theory of the conduction of heat in gases

---

† *Mathematical Physics*, etc., by John Herapath, Esq. 2 vols. London: Whittaker and Co., and Herapath's *Railway Journal* Office, 1847. [See also *Annals of Philosophy* [2] **1**, 273, 340, 401; **2**, 50, 89, 201, 257, 363, 435; **3**, 16 (1821).]

‡ *Mathematical Physics*, etc., p. 134.

§ *Some Remarks on Heat and the Constitution of Elastic Fluids*, October 3, 1848. *Memoirs of the Manchester Literary and Philosophical Society* **9**, 107 (1851).

‖ *Phil. Mag.* February 1859. [Part A, Selection 9.]

¶ " Illustrations of the Dynamical Theory of Gases," *Phil. Mag.* January and July 1860. [Vol. 1, Selection 10.]

which M. Clausius has pointed out in an elaborate memoir on that subject.†

M. O. E. Meyer‡ has also investigated the theory of internal friction on the hypothesis of hard elastic molecules.

In the present paper I propose to consider the molecules of a gas, not as elastic spheres of definite radius, but as small bodies or groups of smaller molecules repelling one another with a force whose direction always passes very nearly through the centres of gravity of the molecules, and whose magnitude is represented very nearly by some function of the distance of the centres of gravity. I have made this modification of the theory in consequence of the results of my experiments on the viscosity of air at different temperatures, and I have deduced from these experiments that the repulsion is inversely as the *fifth* power of the distance.

If we suppose an imaginary plane drawn through a vessel containing a great number of such molecules in motion, then a great many molecules will cross the plane in either direction. The excess of the mass of those which traverse the plane in the positive direction over that of those which traverse it in the negative direction, gives a measure of the flow of gas through the plane in in the positive direction.

If the plane be made to move with such a velocity that there is no excess of flow of molecules in one direction through it, then the velocity of the plane is the mean velocity of the gas resolved normal to the plane.

There will still be molecules moving in both directions through the plane, and carrying with them a certain amount of momentum into the portion of gas which lies on the other side of the plane.

The quantity of momentum thus communicated to the gas on the other side of the plane during a unit of time is a measure of the force exerted on this gas by the rest. This force is called the pressure of the gas.

---

† Poggendorff, January 1862; *Phil Mag.* June 1862. Translated by G. C. Foster, B.A., *Phil. Mag.* June 1862. [*Ann. Phys.* 115, 1; *Phil. Mag.* 23, 417, 512.]

‡ " Ueber die innere Reibung der Gase " (Poggendorff, Vol. CXXV. 1865).

If the velocities of the molecules moving in different directions were independent of one another, then the pressure at any point of the gas need not be the same in all directions, and the pressure between two portions of gas separated by a plane need not be perpendicular to that plane. Hence, to account for the observed equality of pressure in all directions, we must suppose some cause equalizing the motion in all directions. This we find in the deflection of the path of one particle by another when they come near one another. Since, however, this equalization of motion is not instantaneous, the pressures in all directions are perfectly equalized only in the case of a gas at rest, but when the gas is in a state of motion, the want of perfect equality in the pressures gives rise to the phenomena of viscosity or internal friction. The phenomena of viscosity in all bodies may be described, independently of hypothesis, as follows:

A distortion or strain of some kind, which we may call S, is produced in the body by displacement. A state of stress or elastic force which we may call F is thus excited. The relation between the stress and the strain may be written $F = ES$, where E is the coefficient of elasticity for that particular kind of strain. In a solid body free from viscosity, F will remain $= ES$, and

$$\frac{dF}{dt} = E\frac{dS}{dt}$$

If, however, the body is viscous, F will not remain constant, but will tend to disappear at a rate depending on the value of F, and on the nature of the body. If we suppose this rate proportional to F, the equation may be written

$$\frac{dF}{dt} = E\frac{dS}{dt} - \frac{F}{T}$$

which will indicate the actual phenomena in an empirical manner. For if S be constant,

$$F = ESe^{-t/T}$$

showing that F gradually disappears, so that if the body is left to itself it gradually loses any internal stress, and the pressures are finally distributed as in a fluid at rest.

If dS/dl is constant, that is, if there is a steady motion of the body which continually increases the displacement,

$$F = ET\frac{dS}{dt} + Ce^{-t/T}$$

showing that F tends to a constant value depending on the rate of displacement. The quantity ET, by which the rate of displacement must be multiplied to get the force, may be called the coefficient of viscosity. It is the product of a coefficient of elasticity, E, and a time T, which may be called the " time of relaxation " of the elastic force. In mobile fluids T is a very small fraction of a second, and E is not easily determined experimentally. In viscous solids T may be several hours or days, and then E is easily measured. It is possible that in some bodies T may be a function of F, and this would account for the gradual untwisting of wires after being twisted beyond the limit of perfect elasticity. For if T diminishes as F increases, the parts of the wire furthest from the axis will yield more rapidly than the parts near the axis during the twisting process, and when the twisting force is removed, the wire will at first untwist till there is equilibrium between the stresses in the inner and outer portions. These stresses will then undergo a gradual relaxation; but since the actual value of the stress is greater in the outer layers, it will have a more rapid rate of relaxation, so that the wire will go on gradually untwisting for some hours or days, owing to the stress on the interior portions maintaining itself longer than that of the outer parts. This phenomenon was observed by Weber in silk fibres, by Kohlrausch in glass fibres, and by myself in steel wires.

In the case of a collection of moving molecules such as we suppose a gas to be, there is also a resistance to change of form, constituting what may be called the linear elasticity, or " rigidity " of the gas, but this resistance gives way and diminishes at a rate depending on the amount of the force and on the nature of the gas.

Suppose the molecules to be confined in a rectangular vessel with perfectly elastic sides, and that they have no action on one another, so that they never strike one another, or cause each other to deviate from their rectilinear paths. Then it can easily be shewn that the pressures on the sides of the vessel due to the impacts of the molecules are perfectly independent of each other, so that the mass of moving molecules will behave, not like a fluid, but like an elastic solid. Now suppose the pressures at first equal in the three directions perpendicular to the sides, and let the dimensions a, b, c of the vessel be altered by small quantities, δa, δb, δc.

Then if the original pressure in the direction of a was p, it will become

$$p\left(1 - 3\frac{\delta a}{a} - \frac{\delta b}{b} - \frac{\delta c}{c}\right);$$

or if there is no change of volume,

$$\frac{\delta p}{p} = -2\frac{\delta a}{a},$$

shewing that in this case there is a "longitudinal" elasticity of form of which the coefficient is $2p$. The coefficient of "Rigidity" is therefore $= p$.

This rigidity, however, cannot be directly observed, because the molecules continually deflect each other from their rectilinear courses, and so equalize the pressure in all directions. The rate at which this equalization takes place is great, but not infinite; and therefore there remains a certain inequality of pressure which constitutes the phenomenon of viscosity.

I have found by experiment that the coefficient of viscosity in a given gas is independent of the density, and proportional to the absolute temperature, so that if ET be the viscosity, $ET \propto p/\rho$.

But $E = p$, therefore T, the time of relaxation, varies inversely as the density and is independent of the temperature. Hence, the number of collisions producing a given deflection which take place in unit of time is independent of the temperature, that is, of the velocity of the molecules, and is proportional to the number of molecules in unit of volume. If we suppose the molecules hard elastic bodies, the number of collisions of a given kind will be proportional to the velocity, but if we suppose them centres of force, the angle of deflection will be smaller when the velocity is greater; and if the force is inversely as the fifth power of the distance, the number of deflections of a given kind will be independent of the velocity. Hence I have adopted this law in making my calculations.

The effect of the mutual action of the molecules is not only to equalize the pressure in all directions, but, when molecules of different kinds are present, to communicate motion from the one kind to the other. I formerly shewed that the final result in the case of hard elastic bodies is to cause the average *vis viva* of a molecule to be the same for all the different kinds of molecules. Now the pressure due to each molecule is proportional to its *vis viva*, hence the whole pressure due to a given number of molecules in a given volume will be the same whatever the mass of the molecules, provided the molecules of different kinds are permitted freely to communicate motion to each other.

When the flow of *vis viva* from the one kind of molecules to the other is zero, the temperature is said to be the same. Hence, equal volumes of different gases at equal pressures and temperatures contain equal numbers of molecules.

This result of the dynamical theory affords the explanation of the " law of equivalent volumes " in gases.

We shall see that this result is true in the case of molecules acting as centres of force. A law of the same general character is probably to be found connecting the temperatures of liquid and solid bodies with the energy possessed by their molecules, although our ignorance of the nature of the connexions between the molecules renders it difficult to enunciate the precise form of the law.

The molecules of a gas in this theory are those portions of it which move about as a single body. These molecules may be mere points, or pure centres of force endowed with inertia, or the capacity of performing work while losing velocity. They may be systems of several such centres of force, bound together by their mutual actions, and in this case the different centres may either be separated, so as to form a group of points, or they may be actually coincident, so as to form one point.

Finally, if necessary, we may suppose them to be small solid bodies of a determinate form; but in this case we must assume a new set of forces binding the parts of these small bodies together, and so introduce a molecular theory of the second order. The doctrines that all matter is extended, and that no two portions of matter can coincide in the same place, being deductions from our experiments with bodies sensible to us, have no application to the theory of molecules.

The actual energy of a moving body consists of two parts, one due to the motion of its centre of gravity, and the other due to the motions of its parts relative to the centre of gravity. If the body is of invariable form, the motions of its parts relative to the centre of gravity consist entirely of rotation, but if the parts of the body are not rigidly connected, their motions may consist of oscillations of various kinds, as well as rotation of the whole body.

The mutual interference of the molecules in their courses will cause their energy of motion to be distributed in a certain ratio between that due to the motion of the centre of gravity and that due to the rotation, or other internal motion. If the molecules are pure centres of force, there can be no energy of rotation, and the whole energy is reduced to that of translation; but in all other cases the whole energy of the molecule may be represented by $\frac{1}{2}Mv^2\beta$, where β is the ratio of the total energy to the energy of translation. The ratio β will be different for every molecule, and will be different for the same molecule after every encounter with another molecule, but it will have an average value depending on the nature of the molecules, as has been shown by Clausius.† The value of β can be determined if we know either of the specific heats of the gas, or the ratio between them.

The method of investigation which I shall adopt in the following paper, is to determine the mean values of the following functions of the velocity of all the molecules of a given kind within an element of volume:

---

† [Vol. 1, Selection 8.]

(α) the mean velocity resolved parallel to each of the coordinate axes;

(β) the mean values of functions of two dimensions of these component velocities;

(γ) The mean values of functions of three dimensions of these velocities.

The rate of translation of the gas, whether by itself, or by diffusion through another gas, is given by (α), the pressure of the gas on any plane, whether normal or tangential to the plane, is given by (β), and the rate of conduction of heat through the gas is given by (γ).

I propose to determine the variations of these quantities, due, 1st, to the encounters of the molecules with others of the same system or of a different system; 2nd, to the action of external forces such as gravity; and 3rd, to the passage of molecules through the boundary of the element of volume.

I shall then apply these calculations to the determination of the statical cases of the final distribution of two gases under the action of gravity, the equilibrium of temperature between two gases, and the distribution of temperature in a vertical column. These results are independent of the law of force between the molecules. I shall also consider the dynamical cases of diffusion, viscosity, and conduction of heat, which involve the law of force between the molecules.

On the Mutual Action of Two Molecules

Let the masses of these molecules be M_1, M_2, and let their velocities resolved in three directions at right angles to each other be ξ_1, η_1, ζ_1 and ξ_2, η_2, ζ_2. The components of the velocity of the centre of gravity of the two molecules will be

$$\frac{\xi_1 M_1 + \xi_2 M_2}{M_1 + M_2}, \quad \frac{\eta_1 M_1 + \eta_2 M_2}{M_1 + M_2}, \quad \frac{\zeta_1 M_1 + \zeta_2 M_2}{M_1 + M_2}$$

The motion of the centre of gravity will not be altered by the mutual action of the molecules, of whatever nature that action may be. We may therefore take the centre of gravity as the origin of a system of coordinates moving parallel to itself with uniform velocity, and consider the alteration of the motion of each particle with reference to this point as origin.

If we regard the molecules as simple centres of force, then each molecule will describe a plane curve about this centre of gravity, and the two curves will be similar to each other and symmetrical with respect to the line of apses. If the molecules move with sufficient velocity to carry them out of the sphere of their mutual action, their orbits will each have a pair of asymptotes inclined at an angle $\pi/2 - \theta$ to the line of apses. The asymptotes of the orbit of M_1 will be at a distance b_1 from the centre of gravity, and those of M_2 at a distance b_2, where

$$M_1 b_1 = M_2 b_2$$

The distance between two parallel asymptotes, one in each orbit, will be

$$b = b_1 + b_2$$

If, while the two molecules are still beyond each other's action, we draw a straight line through M_1 in the direction of the relative velocity of M_1 to M_2, and draw from M_2 a perpendicular to this line, the length of this perpendicular will be b, and the plane including b and the direction of relative motion will be the plane of the orbits about the centre of gravity.

When, after their mutual action and deflection, the molecules have again reached a distance such that there is no sensible action between them, each will be moving with the same velocity relative to the centre of gravity that it had before the mutual action, but the direction of this relative velocity will be turned through an angle 2θ in the plane of the orbit.

The angle θ is a function of the relative velocity of the molecules and of b, the form of the function depending on the nature of the action between the molecules.

If we suppose the molecules to be bodies, or systems of bodies, capable of rotation, internal vibration, or any form of energy other than simple motion of translation, these results will be modified. The value of θ and the final velocities of the molecules will depend on the amount of internal energy in each molecule before the encounter, and on the particular form of that energy at every instant during the mutual action. We have no means of determining such intricate actions in the present state of our knowledge of molecules, so that we must content ourselves with the assumption that the value of θ is, on an average, the same as for pure centres of force, and that the final velocities differ from the initial velocities only by quantities which may in each collision be neglected, although in a great many encounters the energy of translation and the internal energy of the molecules arrive, by repeated small exchanges, at a final ratio, which we shall suppose to be that of 1 to $\beta-1$.

We may now determine the final velocity of M_1 after it has passed beyond the sphere of mutual action between itself and M_2.

Let V be the velocity of M_1 relative to M_2, then the components of V are

$$\xi_1-\xi_2, \quad \eta_1-\eta_2, \quad \zeta_1-\zeta_2$$

The plane of the orbit is that containing V and b. Let this plane be inclined ϕ to a plane containing V and parallel to the axis of x; then, since the direction of V is turned round an angle 2θ in the plane of the orbit, while its magnitude remains the same, we may find the value of ξ_1 after the encounter. Calling it ξ_1',

$$\left. \xi_1' = \xi_1 + \frac{M_2}{M_1+M_2}\{(\xi_2-\xi_1)2\sin^2\theta \right. \\ \left. +\sqrt{(\eta_2-\eta_1)^2+(\zeta_2-\zeta_1)^2}\sin 2\theta \cos\phi\} \right\} \quad (1)$$

There will be similar expressions for the components of the final velocity of M_1 in the other coordinate directions.

If we know the initial positions and velocities of M_1 and M_2, we can determine V, the velocity of M_1 relative to M_2; b the shortest distance between M_1 and M_2 if they had continued to move with uniform velocity in straight lines; and ϕ the angle which determines the plane in which V and b lie. From V and b we can determine θ, if we know the law of force, so that the problem is solved in the case of two molecules.

When we pass from this case to that of two systems of moving molecules, we shall suppose that the time during which a molecule is beyond the action of other molecules is so great compared with the time during which it is deflected by that action, that we may neglect both the time and distance described by the molecules during the encounter, as compared with the time and the distance described while the molecules are free from disturbing force. We may also neglect those cases in which three or more molecules are within each other's spheres of action at the same instant.

On the Mutual Action of Two Systems of Moving Molecules

Let the number of molecules of the first kind in unit of volume be N_1, the mass of each being M_1. The velocities of these molecules will in general be different both in magnitude and direction. Let us select those molecules the components of whose velocities lie between

$$\xi_1 \text{ and } \xi_1 + d\xi_1, \quad \eta_1 \text{ and } \eta_1 + d\eta_1, \quad \zeta_1 \text{ and } \zeta_1 + d\zeta_1$$

and let the number of these molecules be dN_1. The velocities of these molecules will be very nearly equal and parallel.

On account of the mutual actions of the molecules, the number of molecules which at a given instant have velocities within given limits will be definite, so that

$$dN_1 = f_1(\xi_1 \eta_1 \zeta_1) d\xi_1 d\eta_1 d\zeta_1 \qquad (2)$$

We shall consider the form of this function afterwards.

Let the number of molecules of the second kind in unit of volume be N_2, and let dN_2 of these have velocities between ξ_2 and $\xi_2 + d\xi_2$, η_2 and $\eta_2 + d\eta_2$, ζ_2 and $\zeta_2 + d\zeta_2$, where

$$dN_2 = f_2(\xi_2 \eta_2 \zeta_2) d\xi_2 d\eta_2 d\zeta_2$$

The velocity of any of the dN_1 molecules of the first system relative to the dN_2 molecules of the second system is V, and each molecule M_1 will in the time δt describe a relative path $V\delta t$ among the molecules of the second system. Conceive a space bounded by the following surfaces. Let two cylindrical surfaces have the common axis $V\delta t$ and radii b and $b+db$. Let two planes be drawn through the extremities of the line $V\delta t$ perpendicular to it. Finally, let two planes be drawn through $V\delta t$ making angles ϕ and $\phi+d\phi$ with a plane through V parallel to the axis of x. Then the volume included between the four planes and the two cylindric surfaces will be $Vbdbd\phi\delta t$.

If this volume includes one of the molecules M_2, then during the time δt there will be an encounter between M_1 and M_2, in which b is between b and $b+db$, and ϕ between ϕ and $\phi+d\phi$.

Since there are dN_1 molecules similar to M_1 and dN_2 similar to M_2 in unit of volume, the whole number of encounters of the given kind between the two systems will be

$$Vbdbd\phi\delta t dN_1 dN_2$$

Now let Q be any property of the molecule M_1, such as its velocity in a given direction, the square or cube of that velocity or any other property of the molecule which is altered in a known manner by an encounter of the given kind, so that Q becomes Q' after the encounter, then during the time δt a certain number of the molecules of the first kind have Q changed to Q', while the remainder retain the original value of Q, so that

$$\delta Q dN_1 = (Q'-Q)Vbdbd\phi\delta t dN_1 dN_2$$

or

$$\frac{\delta Q dN_1}{\delta t} = (Q'-Q)Vbdbd\phi dN_1 dN_2 \qquad (3)$$

Here $\delta Q dN_1/\delta t$ refers to the alteration in the sum of the values of Q for the dN_1 molecules, due to their encounters of the given kind with the dN_2 molecules of the second sort. In order to determine the value of $\delta QN_1/\delta t$, the rate of alteration of Q among all the molecules of the first kind, we must perform the following integrations:

1st, with respect to ϕ from $\phi = 0$ to $\phi = 2\pi$.

2nd, with respect to b from $b = 0$ to $b = \infty$. These operations will give the results of the encounters of every kind between the dN_1 and dN_2 molecules.

3rd, with respect to dN_2, or $f_2(\xi_2\eta_2\zeta_2)d\xi_2 d\eta_2 d\zeta_2$.

4th, with respect to dN_1, or $f_1(\xi_1\eta_1\zeta_1)d\xi_1 d\eta_1 d\zeta_1$.

These operations require in general a knowledge of the forms of f_1 and f_2.

1st. *Integration with respect to ϕ*

Since the action between the molecules is the same in whatever plane it takes place, we shall first determine the value of $\int_0^{2\pi}(Q'-Q)d\phi$ in several cases, making Q some function of ξ, η, and ζ.

(α) Let $Q = \xi_1$ and $Q' = \xi_1'$, then

$$\int_0^{2\pi}(\xi_1'-\xi_1)d\phi = \frac{M_2}{M_1+M_2}(\xi_2-\xi_1)4\pi \sin^2\theta \qquad (4)$$

(β) Let $Q = \xi_1^2$ and $Q' = \xi_1'^2$

$$\left.\begin{aligned}\int_0^{2\pi}(\xi_1'^2-\xi_1^2)d\phi &= \frac{M_2}{(M_1+M_2)^2}[(\xi_2-\xi_1)(M_1\xi_1+M_2\xi_2)\\ &\quad \times 8\pi\sin^2\theta + M_2\{(\eta_2-\eta_1)^2+(\zeta_2-\zeta_1)^2\\ &\quad -2(\xi_2-\xi_1)^2\}\pi\sin^2 2\theta]\end{aligned}\right\} \qquad (5)$$

By transformation of coordinates we may derive from this

$$\int_0^\pi (\xi_1'\eta_1' - \xi_1\eta_1)d\phi = \frac{M_2}{(M_1+M_2)^2}[\{M_2\xi_2\eta_2 - M_1\xi_1\eta_1 \\ + \tfrac{1}{2}(M_1-M_2)(\xi_1\eta_2+\xi_2\eta_1)\}\,8\pi\sin^2\theta \\ - 3M_2(\xi_2-\xi_1)(\eta_2-\eta_1)] \qquad (6)$$

with similar expressions for the other quadratic functions of ξ, η, ζ.

(γ)   Let $Q = \xi_1(\xi_1^2+\eta_1^2+\zeta_1^2)$, and $Q' = \xi_1'(\xi_1'^2+\eta_1'^2+\zeta_1'^2)$;

then putting

$$\xi_1^2+\eta_1^2+\zeta_1^2 = V_1^2, \quad \xi_1\xi_2+\eta_1\eta_2+\zeta_1\zeta_2 = U, \quad \xi_2^2+\eta_2^2+\zeta_2^2 = V_2^2,$$

and

$$(\xi_2-\xi_1)^2+(\eta_2-\eta_1)^2+(\zeta_2-\zeta_1)^2 = V^2$$

we find

$$\int_0^\pi (\xi_1'V_1'^2 - \xi_1 V_1^2)d\phi = \frac{M_2}{M_1+M_2}\,4\pi\sin^2\theta\{(\xi_2-\xi_1)V_1^2 \\ +2\xi_1(U-V_1^2)\} \\ + \left(\frac{M_2}{M_1+M_2}\right)^2 (8\pi\sin^2\theta - 3\pi\sin^2 2\theta)\times 2(\xi_2-\xi_1)(U-V_1^2) \\ + \left(\frac{M_2}{M_1+M_2}\right)^2 (8\pi\sin^2\theta + 2\pi\sin^2 2\theta)\times \xi_1 V^2 \\ + \left(\frac{M_2}{M_1+M_2}\right)^2 (8\pi\sin^2\theta - 2\pi\sin^2 2\theta)\times 2(\xi_2-\xi_1)V^2 \qquad (7)$$

These are the principal functions of ξ, η, ζ whose changes we shall have to consider; we shall indicate them by the symbols α, β, or γ, according as the function of the velocity is of one, two, or three dimensions.

2nd. *Integration with respect to b*

We have next to multiply these expressions by bdb, and to integrate with respect to b from $b = 0$ to $b = \infty$. We must bear in mind that θ is a function of b and V, and can only be determined when the law of force is known. In the expressions which we have to deal with, θ occurs under two forms only, namely, $\sin^2\theta$ and $\sin^2 2\theta$. If, therefore, we can find the two values of

$$B_1 = \int_0^\infty 4\pi bdb \sin^2\theta, \quad \text{and} \quad B_2 = \int_0^\infty \pi bdb \sin^2 2\theta \qquad (8)$$

we can integrate all the expressions with respect to b.

B_1 and B_2 will be functions of V only, the form of which we can determine only in particular cases, after we have found θ as a function of b and V.

Determination of θ for Certain Laws of Force

Let us assume that the force between the molecules M_1 and M_2 is repulsive and varies inversely as the nth power of the distance between them, the value of the moving force at distance unity being K, then we find by the equation of central orbits,

$$\frac{\pi}{2} - \theta = \int_0^{x'} \frac{dx}{\sqrt{1 - x^2 - \frac{2}{n-1}\left(\frac{x}{a}\right)^{n-1}}} \qquad (9)$$

where $x = b/r$, or the ratio of b to the distance of the molecules at a given time: x is therefore a numerical quantity; α is also a numerical quantity and is given by the equation

$$\alpha = b\left\{\frac{V^2 M_1 M_2}{K(M_1 + M_2)}\right\}^{1/(n-1)} \qquad (10)$$

The limits of integration are $x = 0$ and $x = x'$, where x' is the least positive root of the equation

$$1 - x^2 - \frac{2}{n-1}\left(\frac{x}{\alpha}\right)^{n-1} = 0 \tag{11}$$

It is evident that θ is a function of α and n, and when n is known θ may be expressed as a function of α only.

Also

$$b\,db = \left\{\frac{K(M_1+M_2)}{V^2 M_1 M_2}\right\}^{2/(n-1)} \alpha\,d\alpha \tag{12}$$

so that if we put

$$A_1 = \int_0^\infty 4\pi\alpha\,d\alpha \sin^2\theta, \quad A_2 = \int_0^\infty \pi\alpha\,d\alpha \sin^2 2\theta \tag{13}$$

A_1 and A_2 will be definite numerical quantities which may be ascertained when n is given, and B_1 and B_2 may be found by multiplying A_1 and A_2 by

$$\left\{\frac{K(M_1+M_2)}{M_1 M_2}\right\}^{2/(n-1)} V^{-4/(n-1)}$$

Before integrating further we have to multiply by V, so that the form in which V will enter into the expressions which have to be integrated with respect to dN_1 and dN_2 will be

$$V^{(n-5)/(n-1)}$$

It will be shewn that we have reason from experiments on the viscosity of gases to believe that $n = 5$. In this case V will disappear from the expressions of the form (3), and they will be capable of immediate integration with respect to dN_1 and dN_2.

If we assume $n = 5$ and put $\alpha^4 = 2\cot^2 2\phi$ and

$$x = \sqrt{1-\tan^2\phi}\,\cos\psi,$$

$$\left.\begin{array}{l}\dfrac{\pi}{2} - \theta = \sqrt{\cos 2\phi}\displaystyle\int_0^{\pi/2} \dfrac{d\psi}{\sqrt{1-\sin^2\phi\sin^2\psi}} \\ = \sqrt{\cos 2\phi}\,F_{\sin\phi}\end{array}\right\} \tag{14}$$

where $F_{\sin \phi}$ is the complete elliptic function of the first kind and is given in Legendre's Tables. I have computed the following Table of the distance of the asymptotes, the distance of the apse, the value of θ, and of the quantities whose summation leads to A_1 and A_2.

ϕ	b	Distance of apse	θ	$\dfrac{\sin^2 \theta}{\sin^2 2\phi}$	$\dfrac{\sin^2 2\theta}{\sin^2 2\phi}$
° ′			° ′		
0 0	infinite	infinite	0 0	0	0
5 0	2381	2391	0 31	·00270	·01079
10 0	1658	1684	1 53	·01464	·03689
15 0	1316	1366	4 47	·02781	·11048
20 0	1092	1172	8 45	·05601	·21885
25 0	916	1036	14 15	·10325	·38799
30 0	760	931	21 42	·18228	·62942
35 0	603	845	31 59	·31772	·71433
40 0	420	772	47 20	·55749	1·02427
41 0	374	758	51 32	·62515	·96763
42 0	324	745	56 26	·70197	·85838
43 0	264	732	62 22	·78872	·67868
44 0	187	719	70 18	·88745	·40338
44 30	132	713	76 1	·94190	·21999
45 0	0	707	90 0	1·00000	·00000

$$A_1 = \int 4\pi \alpha d\alpha \sin^2 \theta = 2\cdot 6595 \qquad (15)$$

$$A_2 = \int \pi \alpha d\alpha \sin^2 2\theta = 1\cdot 3682 \qquad (16)$$

The paths described by molecules about a centre of force S, repelling inversely as the fifth power of the distance, are given in Fig. 1 overleaf.

The molecules are supposed to be originally moving with equal velocities in parallel paths, and the way in which their deflections depend on the distance of the path from S is shewn by the different curves in the figure.

3rd. *Integration with respect to* dN_2

We have now to integrate expressions involving various functions of ξ, η, ζ, and V with respect to all the molecules of the second sort. We may write the expression to be integrated

$$\iiint QV^{(n-5)/(n-1)} f_2(\xi_2 \eta_2 \zeta_2) d\xi_2 d\eta_2 d\zeta_2$$

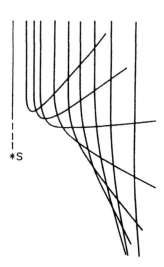

Fig. 1.

where Q is some function of ξ, η, ζ, etc., already determined, and f_2 is the function which indicates the distribution of velocity among the molecules of the second kind.

In the case in which $n = 5$, V disappears, and we may write the result of integration

$$\bar{Q} N_2,$$

where $\bar{Q}$ is the mean value of Q for all the molecules of the second kind, and N_1 is the number of those molecules.

If, however, n is not equal to 5, so that V does not disappear, we should require to know the form of the function f_2 before we could proceed further with the integration.

The only case in which I have determined the form of this function is that of one or more kinds of molecules which have by their continual encounters brought about a distribution of velocity such that the number of molecules whose velocity lies within given limits remains constant. In the *Philosophical Magazine* for January 1860, I have given an investigation of this case, founded on the assumption that the probability of a molecule having a velocity resolved parallel to x lying between given limits is not in any way affected by the knowledge that the molecule has a given velocity resolved parallel to y.† As this assumption may appear precarious, I shall now determine the form of the function in a different manner.

On the Final Distribution of Velocity among the Molecules of Two Systems acting on one another according to any Law of Force

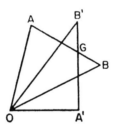

Fig. 2.

From a given point O let lines be drawn representing in direction and magnitude the velocities of every molecule of either kind in unit of volume. The extremities of these lines will be distributed over space in such a way that if an element of volume dV be taken anywhere, the number of such lines which will terminate within dV will be $f(r)dV$, where r is the distance of dV from O.

Let $OA = a$ be the velocity of a molecule of the first kind, and $OB = b$ that of a molecule of the second kind before they encounter one another, then BA will be the velocity of A relative

---

† [Vol. 1, Selection 10.]

to B; and if we divide AB in G inversely as the masses of the molecules, and join OG, OG will be the velocity of the centre of gravity of the two molecules.

Now let $OA' = a'$ and $OB' = b'$ be the velocities of the two molecules after the encounter, $GA = GA'$ and $GB = GB'$, and $A'GB'$ is a straight line not necessarily in the plane of OAB. Also $AGA' = 2\theta$ is the angle through which the relative velocity is turned in the encounter in question. The relative motion of the molecules is completely defined if we know BA the relative velocity before the encounter, 2θ the angle through which BA is turned during the encounter, and ϕ the angle which defines the direction of the plane in which BA and $B'A'$ lie. All encounters in which the magnitude and direction of BA, and also θ and ϕ, lie within certain almost contiguous limits, we shall class as encounters of the given kind. The number of such encounters in unit of time will be

$$n_1 n_2 F de \tag{17}$$

where n_1 and n_2 are the numbers of molecules of each kind under consideration, and F is a function of the relative velocity and of the angle θ, and de depends on the limits of variation within which we class encounters as of the same kind.

Now let A describe the boundary of an element of volume dV while AB and $A'B'$ move parallel to themselves, then B, A', and B' will also describe equal and similar elements of volume.

The number of molecules of the first kind, the lines representing the velocities of which terminate in the element dV at A, will be

$$n_1 = f_1(a) dV \tag{18}$$

The number of molecules of the second kind which have velocities corresponding to OB will be

$$n_2 = f_2(b) dV \tag{19}$$

and the number of encounters of the given kind between these two sets of molecules will be

$$f_1(a) f_2(b) (dV)^2 F de \tag{20}$$

The lines representing the velocities of these molecules after encounters of the given kind will terminate within elements of volume at A' and B', each equal to dV.

In like manner we should find for the number of encounters between molecules whose original velocities correspond to elements equal to dV described about A' and B', and whose subsequent velocities correspond to elements equal to dV described about A and B,

$$f_1(a')f_2(b')(dV)^2 F'de \tag{21}$$

where F' is the same function of $B'A'$ and $A'GA$ that F is of BA and AGA'. F is therefore equal to F'.

When the number of pairs of molecules which change their velocities from OA, OB to OA', OB' is equal to the number which change from OA', OB' to OA, OB, then the final distribution of velocity will be obtained, which will not be altered by subsequent exchanges. This will be the case when

$$f_1(a)f_2(b) = f_1(a')f_2(b') \tag{22}$$

Now the only relation between a, b and a', b' is

$$M_1 a^2 + M_2 b^2 = M_1 a'^2 + M_2 b'^2 \tag{23}$$

whence we obtain

$$f_1(a) = C_1 e^{-(a^2/\alpha^2)}, \quad f_2(b) = C_2 e^{-(b^2/\beta^2)} \tag{24}$$

where

$$M_1 \alpha^2 = M_2 \beta^2 \tag{25}$$

By integrating $\iiint C_1 e^{-\frac{\xi^2 + \eta^2 + \zeta^2}{\alpha^2}} d\xi d\eta d\zeta$, and equating the result to N_1, we obtain the value of C_1. If, therefore, the distribution of velocities among N_1 molecules is such that the number of molecules whose component velocities are between ξ and $\xi + d\xi$, η and $\eta + d\eta$, and ζ and $\zeta + d\zeta$ is

$$dN_1 = \frac{N_1}{\alpha^3 \pi^{3/2}} e^{-\frac{\xi^2 + \eta^2 + \zeta^2}{\alpha^2}} d\xi d\eta d\zeta \tag{26}$$

then this distribution of velocities will not be altered by the exchange of velocities among the molecules by their mutual action.

This is therefore a possible form of the final distribution of velocities. It is also the only form; for if there were any other, the exchange between velocities represented by OA and OA' would not be equal. Suppose that the number of molecules having velocity OA' increases at the expense of OA. Then since the total number of molecules corresponding to OA' remains constant, OA' must communicate as many to OA'', and so on till they return to OA.

Hence if OA, OA', OA'', etc. be a series of velocities, there will be a tendency of each molecule to assume the velocities OA, OA', OA'', etc. in order, returning to OA. Now it is impossible to assign a reason why the successive velocities of a molecule should be arranged in this cycle, rather than in the reverse order. If, therefore, the direct exchange between OA and OA' is not equal, the equality cannot be preserved by exchange in a cycle. Hence the direct exchange between OA and OA' is equal, and the distribution we have determined is the only one possible.

This final distribution of velocity is attained only when the molecules have had a great number of encounters, but the great rapidity with which the encounters succeed each other is such that in all motions and changes of the gaseous system except the most violent, the form of the distribution of velocity is only slightly changed.

When the gas moves in mass, the velocities now determined are compounded with the motion of translation of the gas.

When the differential elements of the gas are changing their figure, being compressed or extended along certain axes, the values of the mean square of the velocity will be different in different directions. It is probable that the form of the function will then be

$$f_1(\xi\eta\zeta) = \frac{N_1}{\alpha\beta\gamma\pi^{3/2}} e^{-\left(\frac{\xi^2}{\alpha^2}+\frac{\eta^2}{\beta^2}+\frac{\zeta^2}{\gamma^2}\right)} \tag{27}$$

where α, β, γ are slightly different. I have not, however attempted to investigate the exact distribution of velocities in this case, as the theory of motion of gases does not require it.

When one gas is diffusing through another, or when heat is being conducted through a gas, the distribution of velocities will be different in the positive and negative directions, instead of being symmetrical, as in the case we have considered. The want of symmetry, however, may be treated as very small in most actual cases.

The principal conclusions which we may draw from this investigation are as follows. Calling α the modulus of velocity,

1st. The mean velocity is

$$\bar{v} = \frac{2}{\sqrt{\pi}} \alpha \tag{28}$$

2nd. The mean square of the velocity is

$$\overline{v^2} = \frac{3}{2} \alpha^2 \tag{29}$$

3rd. The mean value of ξ^2 is

$$\overline{\xi^2} = \frac{1}{2} \alpha^2 \tag{30}$$

4th. The mean value of ξ^4 is

$$\overline{\xi^4} = \frac{3}{4} \alpha^4 \tag{31}$$

5th. The mean value of $\xi^2 \eta^2$ is

$$\overline{\xi^2 \eta^2} = \frac{1}{4} \alpha^4 \tag{32}$$

6th. When there are two systems of molecules

$$M_1 \alpha^2 = M_2 \beta^2 \tag{33}$$

whence

$$M_1 v_1^2 = M_2 v_2^2 \tag{34}$$

or the mean *vis viva* of a molecule will be the same in each system. This is a very important result in the theory of gases, and it is independent of the nature of the action between the molecules, as are all the other results relating to the final distribution of velocities. We shall find that it leads to the law of gases known as that of Equivalent Volumes.

Variation of Functions of the Velocity due to encounters between the Molecules

We may now proceed to write down the values of $\delta \overline{Q}/\delta t$ in the different cases. We shall indicate the mean value of any quantity for all the molecules of one kind by placing a bar over the symbol which represents that quantity for any particular molecule, but in expressions where all such quantities are to be taken at their mean values, we shall, for convenience, omit the bar. We shall use the symbols δ_1 and δ_2 to indicate the effect produced by molecules of the first kind and second kind respectively, and δ_3 to indicate the effect of external forces. We shall also confine ourselves to the case in which $n = 5$, since it is not only free from mathematical difficulty, but is the only case which is consistent with the laws of viscosity of gases.

In this case V disappears, and we have for the effect of the second system on the first,

$$\frac{\delta Q}{\delta t} = N_2 \left\{ \frac{K(M_1+M_2)}{M_1 M_2} \right\}^{\frac{1}{2}} A \int_0^\pi (Q'-Q) d\phi \qquad (35)$$

where the functions of ξ, η, ζ in $(Q'-Q)d\phi$ must be put equal to their mean values for all the molecules, and A_1 or A_2 must be put for A according as $\sin^2\theta$ or $\sin^2 2\theta$ occurs in the expressions in equations (4), (5), (6), (7). We thus obtain

$$(\alpha) \quad \frac{\delta_2 \xi_1}{\delta t} = \left\{ \frac{K}{M_1 M_2 (M_1+M_2)} \right\}^{\frac{1}{2}} N_2 M_2 A_1 (\xi_2 - \xi_1) \qquad (36)$$

(β) $\quad \dfrac{\delta_2 \xi_1^2}{\delta t} = \left\{\dfrac{K}{M_1 M_2(M_1+M_2)}\right\}^{\frac{1}{2}} \dfrac{N_2 M_2}{M_1+M_2}$

$\qquad \times \{2A_1(\xi_2-\xi_1)(M_1\xi_1+M_2\xi_2)$

$\qquad + A_2 M_2 \overline{(\eta_2-\eta_1^2 + \overline{\zeta_2-\zeta_1}^2 - 2\overline{\xi_2-\xi_1}^2)}\}$ $\qquad$ (37)

$\dfrac{\delta_2 \xi_1 \eta_1}{\delta t} = \left\{\dfrac{K}{M_1 M_2(M_1+M_2)}\right\}^{\frac{1}{2}} \dfrac{N_2 M_2}{M_1+M_2}$

$\qquad [A_1\{2M_2\xi_2\eta_2 - 2M_1\xi_1\eta_1$

$\qquad + (M_1-M_2)(\xi_1\eta_2+\xi_2\eta_1)\}$

$\qquad - 3A_2 M_2(\xi_2-\xi_1)(\eta_2-\eta_1)]$ $\qquad$ (38)

(γ) $\quad \dfrac{\delta_2 \xi_1 V_1^2}{\delta t} = \left\{\dfrac{K}{M_1 M_2(M_1+M_2)}\right\}^{\frac{1}{2}} N_2 M_2 \left[A_1\{\overline{\xi_2-\xi_1}V_1^2\right.$

$\qquad + 2\xi_1(U-V_1^2)\}$

$\qquad + \dfrac{M_2}{M_1+M_2}(2A_1-3A_2) 2(\xi_2-\xi_1) \times (U-V_1^2)$

$\qquad + \dfrac{M_2}{M_1+M_2}(2A_1+2A_2)\xi_1 V^2$ $\qquad$ (39)

$\qquad + \left(\dfrac{M_2}{M_1+M_2}\right)^2 (2A_1-2A_2) \times 2(\xi_2-\xi_1)V^2 \Big]$

using the symbol δ_2 to indicate variations arising from the action of molecules of the second system.

These are the values of the rate of variation of the mean values of ξ_1, ξ_1^2, $\xi_1\eta_1$, and $\xi_1 V_1^2$, for the molecules of the first kind due to their encounters with molecules of the second kind. In all of them we must multiply up all functions of ξ, η, ζ, and take the mean values of the products so found. As this has to be done for all such functions, I have omitted the bar over each function in these expressions.

To find the rate of variation due to the encounters among the particles of the same system, we have only to alter the suffix $_{(2)}$ into $_{(1)}$ throughout, and to change K, the coefficient of the force between M_1 and M_2 into K_1, that of the force between two molecules of the first system. We thus find

(α) $$\frac{\delta_1 \overline{\xi_1}}{\delta t} = 0 \tag{40}$$

(β) $$\frac{\delta_1 \overline{\xi_1^2}}{dt} = \left(\frac{K_1}{2M_1^3}\right)^{\frac{1}{2}} M_1 N_1 A_2 \{\overline{\eta_1^2} + \overline{\zeta_1^2} - 2\overline{\xi_1^2} - (\overline{\eta_1 \cdot \eta_1}$$
$$+ \overline{\zeta_1 \cdot \zeta_1} - 2\overline{\xi_1 \cdot \xi_1})\} \tag{41}$$

$$\frac{\delta_1 \overline{\xi_1 \eta_1}}{\delta t} = \left(\frac{K_1}{2M_1^3}\right)^{\frac{1}{2}} M_1 N_1 A_2 3\{\overline{\xi_1 \cdot \eta_1} - \overline{\xi_1 \eta_1}\} \tag{42}$$

(γ) $$\frac{\delta_1 \overline{\xi_1 V_1^2}}{\delta t} = \left(\frac{K_1}{2M_1^3}\right) M_1 N_1 A_2 3(\overline{\xi_1 \cdot V_1^2} - \overline{\xi_1 V_1^2}) \tag{43}$$

These quantities must be added to those in equations (36) to (39) in order to get the rate of variation in the molecules of the first kind due to their encounters with molecules of both systems. When there is only one kind of molecules, the latter equations give the rates of variation at once.

On the Action of External Forces on a System of Moving Molecules

We shall suppose the external force to be like the force of gravity, producing equal acceleration on all the molecules. Let the components of the force in the three coordinate directions be X, Y,

Z. Then we have by dynamics for the variations of ξ, ξ^2, and ξV^2 due to this cause,

$$(\alpha) \quad \frac{\delta_3 \xi}{\delta t} = X \tag{44}$$

$$(\beta) \quad \frac{\delta_3 \cdot \xi^2}{\delta t} = 2\xi X \tag{45}$$

$$\frac{\delta_3 \cdot \xi \eta}{\delta t} = \eta X + \xi Y \tag{46}$$

$$(\gamma) \quad \frac{\delta_3 \cdot \xi V^2}{\delta t} = 2\xi(\xi X + \eta Y + \zeta Z) + X V^2 \tag{47}$$

where δ_3 refers to variations due to the action of external forces.

On the total rate of change of the different functions of the velocity of the molecules of the first system arising from their encounters with molecules of both systems and from the action of external forces

To find the total rate of change arising from these causes, we must add

$$\frac{\delta_1 Q}{\delta t}, \quad \frac{\delta_2 Q}{\delta t}, \quad \text{and} \quad \frac{\delta_3 Q}{\delta t}$$

the quantities already found. We shall find it, however, most convenient in the remainder of this investigation to introduce a change in the notation, and to substitute for

$$\xi, \eta, \text{ and } \zeta, \quad u+\xi, \quad v+\eta, \text{ and } w+\zeta \tag{48}$$

where u, v, and w are so chosen that they are the mean values of the components of the velocity of all molecules of the same system in the immediate neighbourhood of a given point. We shall also write

$$M_1 N_1 = \rho_1, \quad M_2 N_2 = \rho_2 \tag{49}$$

where ρ_1 and ρ_2 are the densities of the two systems of molecules, that is, the mass in unit of volume. We shall also write

$$\left(\frac{K_1}{2M_1^2}\right)^{\frac{1}{2}} = k_1, \quad \left(\frac{K}{M_1M_2(M_1+M_2)}\right)^{\frac{1}{2}} = k, \quad \text{and} \quad \left(\frac{K_2}{2M_2^3}\right)^{\frac{1}{2}} = k_2 \tag{50}$$

$\rho_1, \rho_2, k_1, k_2,$ and k are quantities the absolute values of which can be deduced from experiment. We have not as yet experimental data for determining M, N, or K.

We thus find for the rate of change of the various functions of velocity,

$$(\alpha) \quad \frac{\delta u_1}{\delta t} = kA_1\rho_2(u_2-u_1)+X \tag{51}$$

$$(\beta) \quad \frac{\delta.\xi_1^2}{\delta t} = k_1A_2\rho_1\{\eta_1^2+\zeta_1^2-2\xi_1^2\} + k\rho_2\frac{M_2}{M_1+M_2}$$
$$\times \{2A_1(u_2-u_1)^2$$
$$+ A_2(\overline{v_2-v_1}^2 + \overline{w_2-w_1}^2 - \overline{2u_2-u_1}^2)\}$$
$$+ \frac{k\rho_2}{M_1+M_2}\{2A_1(M_2\xi_2^2-M_1\xi_1^2)$$
$$+ A_2M_2(\eta_1^2+\zeta_1^2-2\xi_1^2+\eta_2^2+\zeta_2^2-2\xi_2^2)\} \tag{52}$$

also

$$(\gamma) \quad \frac{\delta.\xi\eta}{\delta t} = -3k_1A_2\rho_1\xi_1\eta_1 + k\rho_2\frac{M_2}{M_1+M_2}$$
$$\times (2A_1-3A_2)(u_2-u_1)(v_2-v_1)$$
$$+ \frac{k\rho_2}{M_1+M_2}\{2A_1(M_2\xi_2\eta_2-M_1\xi_1\eta_1)$$
$$-3A_2M_2(\xi_1\eta_1+\xi_2\eta_2)\} \tag{53}$$

(γ) As the expressions for the variation of functions of three dimensions in mixed media are complicated, and as we shall not have occasion to use them, I shall give the case of a single medium,

$$\frac{\delta}{\delta t}(\xi_1^3 + \xi_1\eta_1^2 + \xi_1\zeta_1^2) = -3k_1\rho_1 A_2(\xi_1^3 + \xi_1\eta_1^2 + \xi_1\zeta_1^2)$$

$$+ X(3\xi_1^2 + \eta_1^2 + \zeta_1^2) + 2Y\xi_1\eta_1 + 2Z\xi_1\zeta_1 \quad (54)$$

Theory of a Medium composed of Moving Molecules

We shall suppose the position of every moving molecule referred to three rectangular axes, and that the component velocities of any one of them, resolved in the directions of x, y, z, are

$$u+\xi, \quad v+\eta, \quad w+\zeta$$

where u, v, w are the components of the mean velocity of all the molecules which are at a given instant in a given element of volume, and ξ, η, ζ are the components of the relative velocity of one of these molecules with respect to the mean velocity.

The quantities u, v, w may be treated as functions of x, y, z, and t, in which case differentiation will be expressed by the symbol d. The quantities ξ, η, ζ, being different for every molecule, must be regarded as functions of t for each molecule. Their variation with respect to t will be indicated by the symbol δ.

The mean values of ξ^2 and other functions of ξ, η, ζ for all the molecules in the element of volume may, however, be treated as functions of x, y, z, and t.

If we consider an element of volume which always moves with the velocities u, v, w, we shall find that it does not always consist of the same molecules, because molecules are continually passing through its boundary. We cannot therefore treat it as a mass moving with the velocity u, v, w, as is done in hydrodynamics, but we must consider separately the motion of each molecule. When we have occasion to consider the variation of the properties of this element during its motion as a function of the time we shall use the symbol ∂.

We shall call the velocities u, v, w the velocities of translation of the medium, and ξ, η, ζ the velocities of agitation of the molecules.

Let the number of molecules in the element $dx\,dy\,dz$ be $N\,dx\,dy\,dz$, then we may call N the number of molecules in unit of volume. If M is the mass of each molecule, and ρ the density of the element, then

$$MN = \rho \tag{55}$$

Transference of Quantities across a Plane Area

We must next consider the molecules which pass through a given plane of unit area in unit of time, and determine the quantity of matter, of momentum, of heat, etc. which is transferred from negative to the positive side of this plane in unit of time.

We shall first divide the N molecules in unit of volume into classes according to the value of ξ, η, and ζ for each, and we shall suppose that the number of molecules in unit of volume whose velocity in the direction of x lies between ξ and $\xi+d\xi$, η and $\eta+d\eta$, ζ and $\zeta+d\zeta$ is dN, dN will then be a function of the component velocities, the sum of which being taken for all the molecules will give N the total number of molecules. The most probable form of this function for a medium in its state of equilibrium is

$$dN = \frac{N}{\alpha^3 \pi^{\frac{3}{2}}} e^{-\frac{\xi^2+\eta^2+\zeta^2}{\alpha^2}} d\xi\,d\eta\,d\zeta \tag{56}$$

In the present investigation we do not require to know the form of this function.

Now let us consider a plane of unit area perpendicular to x moving with a velocity of which the part resolved parallel to x is u'. The velocity of the plane relative to the molecules we have been considering is $u'-(u+\xi)$, and since there are dN of these molecules in unit of volume it will overtake

$$\{u'-(u+\xi)\}dN$$

such molecules in unit of time, and the number of such molecules passing from the negative to the positive side of the plane, will be

$$(u+\xi-u')dN$$

Now let Q be any property belonging to the molecule, such as its mass, momentum, *vis viva*, etc., which it carries with it across the plane, Q being supposed a function of ξ or of ξ, η, and ζ, or to vary in any way from one molecule to another, provided it be the same for the selected molecules whose number is dN, then the quantity of Q transferred across the plane in the positive direction in unit of time is

$$\int(u-u'+\xi)QdN$$

or

$$(u-u')\int QdN + \int \xi QdN \tag{57}$$

If we put $\bar{Q}N$ for $\int QdN$, and $\overline{\xi Q}N$ for $\int \xi QdN$, then we may call $\bar{Q}$ the mean value of Q, and $\overline{\xi Q}$ the mean value of ξQ, for all the particles in the element of volume, and we may write the expression for the quantity of Q which crosses the plane in unit of time

$$(u-u')\bar{Q}N + \overline{\xi Q}N \tag{58}$$

(α) Transference of Matter across a Plane—Velocity of the Fluid

To determine the quantity of matter which crosses the plane, make Q equal to M the mass of each molecule; then, since M is the same for all molecules of the same kind, $\bar{M} = M$; and since the mean value of ξ is zero, the expression is reduced to

$$(u-u')MN = (u-u')\rho \tag{59}$$

If $u = u'$, or if the plane moves with velocity u, the whole excess of matter transferred across the plane is zero; the velocity of the fluid may therefore be defined as the velocity whose components are u, v, w.

(β) Transference of Momentum across a Plane—System of Pressures at any point of the Fluid

The momentum of any one molecule in the direction of x is $M(u+\xi)$. Substituting this for Q, we get for the quantity of momentum transferred across the plane in the positive direction

$$(u-u')u\rho + \overline{\xi^2}\rho \tag{60}$$

If the plane moves with the velocity u, this expression is reduced to $\overline{\xi^2}\rho$, where $\overline{\xi^2}$ represents the mean value of ξ^2.

This is the whole momentum in the direction of x of the molecules projected from the negative to the positive side of the the plane in unit of time. The mechanical action between the parts of the medium on opposite sides of the plane consists partly of the momentum thus transferred, and partly of the direct attractions or repulsions between molecules on opposite sides of the plane. The latter part of the action must be very small in gases, so that we may consider the pressure between the parts of the medium on opposite sides of the plane as entirely due to the constant bombardment kept up between them. There will also be a transference of momentum in the directions of y and z across the same plane,

$$(u-u')v\rho + \overline{\xi\eta}\rho \tag{61}$$

and

$$(u-u')w\rho + \overline{\xi\zeta}\rho \tag{62}$$

where $\overline{\xi\eta}$ and $\overline{\xi\zeta}$ represent the mean values of these products.

If the plane moves with the mean velocity u of the fluid, the total force exerted on the medium on the positive side by the projection of molecules into it from the negative side will be

a normal pressure $\overline{\xi^2}\rho$ in the direction of x,

a tangential pressure $\overline{\xi\eta}\rho$ in the direction of y,

and a tangential pressure $\overline{\xi\zeta}\rho$ in the direction of z.

If X, Y, Z are the components of the pressure on unit of area of a plane whose direction cosines are l, m, n,

$$\left. \begin{array}{l} X = l\overline{\xi^2}\rho + m\overline{\xi\eta}\rho + n\overline{\xi\zeta}\rho \\ Y = l\overline{\xi\eta}\rho + m\overline{\eta^2}\rho + n\overline{\eta\zeta}\rho \\ Z = l\overline{\xi\zeta}\rho + m\overline{\eta\zeta}\rho + n\overline{\zeta^2}\rho \end{array} \right\} \quad (63)$$

When a gas is not in a state of violent motion the pressures in all directions are nearly equal, in which case, if we put

$$\overline{\xi^2}\rho + \overline{\eta^2}\rho + \overline{\zeta^2}\rho = 3p \quad (64)$$

the quantity p will represent the mean pressure at a given point, and $\overline{\xi^2}\rho$, $\overline{\eta^2}\rho$, and $\overline{\zeta^2}\rho$ will differ from p only by small quantities; $\overline{\eta\zeta}\rho$, $\overline{\zeta\xi}\rho$, and $\overline{\xi\eta}\rho$ will then be also small quantities with respect to p.

Energy in the Medium—Actual Heat

The actual energy of any molecule depends partly on the velocity of its centre of gravity, and partly on its rotation or other internal motion with respect to the centre of gravity. It may be written

$$\tfrac{1}{2}M\{(u+\xi)^2 + (v+\eta)^2 + (w+\zeta)^2\} + \tfrac{1}{2}EM \quad (65)$$

where $\tfrac{1}{2}EM$ is the internal part of the energy of the molecule, the form of which is at present unknown. Summing for all the molecules in unit of volume, the energy is

$$\tfrac{1}{2}(u^2 + v^2 + w^2)\rho + \tfrac{1}{2}(\xi^2 + \eta^2 + \zeta^2)\rho + \tfrac{1}{2}E\rho \quad (66)$$

The first term gives the energy due to the motion of translation of the medium in mass, the second that due to the agitation of the centres of gravity of the molecules, and the third that due to the internal motion of the parts of each molecule.

If we assume with Clausius that the ratio of the mean energy of internal motion to that of agitation tends continually towards a definite value $(\beta-1)$,† we may conclude that, except in very violent disturbances, this ratio is always preserved, so that

$$\overline{E} = (\beta-1)(\xi^2+\eta^2+\zeta^2) \tag{67}$$

The total energy of the invisible agitation in unit of volume will then be

$$\tfrac{1}{2}\beta(\xi^2+\eta^2+\zeta^2) \tag{68}$$

or

$$\tfrac{3}{2}\beta p \tag{69}$$

This energy being in the form of invisible agitation, may be called the total heat in the unit of volume of the medium.

(γ) Transference of Energy across a Plane—Conduction of Heat

Putting

$$Q = \tfrac{1}{2}\beta(\xi^2+\eta^2+\zeta^2)M, \quad \text{and} \quad u = u' \tag{70}$$

we find for the quantity of heat carried over the unit of area by conduction in unit of time

$$\tfrac{1}{2}\beta(\overline{\xi^3}+\overline{\xi\eta^2}+\overline{\xi\zeta^2})\rho \tag{71}$$

where $\overline{\xi^3}$, etc. indicate the mean values of ξ^3, etc. They are always small quantities.

On the Rate of Variation of Q in an Element of Volume, Q being any property of the Molecules in that Element

Let Q be the value of the quantity for any particular molecule, and $\bar{Q}$ the mean value of Q for all the molecules of the same kind within the element.

The quantity $\bar{Q}$ may vary from two causes. The molecules within the element may by their mutual action or by the action of external forces produce an alteration of $\bar{Q}$, or molecules may pass

---

† [Vol. 1, Selection 8.]

into the element and out of it, and so cause an increase or diminution of the value of $\bar{Q}$ within it. If we employ the symbol δ to denote the variation of Q due to actions of the first kind on the individual molecules, and the symbol ∂ to denote the actual variation of Q in an element moving with the mean velocity of the system of molecules under consideration, then by the ordinary investigation of the increase or diminution of matter in an element of volume as contained in treatises on Hydrodynamics,

$$\frac{\partial \overline{QN}}{\partial t} = \frac{\delta \bar{Q}}{\delta t} N - \frac{d}{dx}\{(u-u')\bar{Q}N + \overline{\xi QN}\} \\ - \frac{d}{dy}\{(v-v')\bar{Q}N + \overline{\eta QN}\} - \frac{d}{dz}\{(w-w')\bar{Q}N + \overline{\zeta QN}\} \quad (72)$$

where the last three terms are derived from equation (58) and two similar equations, and denote the quantity of Q which flows out of an element of volume, that element moving with the velocities u', v', w'. If we perform the differentiations and then make $u' = u$, $v' = v$, and $w' = w$, then the variation will be that in an element which moves with the actual mean velocity of the system of molecules, and the equation becomes

$$\frac{\partial \bar{Q}N}{\delta t} + \bar{Q}N\left(\frac{du}{dx} + \frac{dv}{dy} + \frac{dw}{dz}\right) + \frac{d}{dx}(\overline{\xi Q}N) + \frac{d}{dy}(\overline{\eta Q}N) \\ + \frac{d}{dz}(\overline{\zeta Q}N) = \frac{\delta Q}{\delta t}N \quad (73)$$

Equation of Continuity

Put $Q = M$ the mass of a molecule; M is unalterable, and we have, putting $MN = \rho$,

$$\frac{\partial \rho}{\partial t} + \rho\left(\frac{du}{dx} + \frac{dv}{dy} + \frac{dw}{dz}\right) = 0 \quad (74)$$

which is the ordinary equation of continuity in hydrodynamics, the element being supposed to move with the velocity of the fluid. Combining this equation with that from which it was obtained, we find

$$N\frac{\partial \bar{Q}}{\partial t}+\frac{d}{dx}(\overline{\xi Q}N)+\frac{d}{dy}(\overline{\eta Q}N)+\frac{d}{dz}(\overline{\zeta Q}N) = N\frac{\delta Q}{\delta t} \qquad (75)$$

a more convenient form of the general equation.

Equations of Motion (α)

To obtain the Equation of Motion in the direction of x, put $Q = M_1(u_1+\xi_1)$, the momentum of a molecule in the direction of x.

We obtain the value of $\delta Q/\delta t$ from equation (51), and the equation may be written

$$\left.\begin{array}{c} \rho_1\dfrac{\partial u_1}{\partial t}+\dfrac{d}{dx}(\overline{\rho_1\xi_1^2})+\dfrac{d}{dy}(\overline{\rho_1\xi_1\eta_1})+\dfrac{d}{dz}(\overline{\rho_1\xi_1\zeta_1}) \\ \\ = kA_1\rho_1\rho_2(u_2-u_1)+X\rho_1 \end{array}\right\} \quad (76)$$

In this equation the first term denotes the efficient force per unit of volume, the second the variation of normal pressure, the third and fourth the variations of tangential pressure, the fifth the resistance due to the molecules of a different system, and the sixth the external force acting on the system.

The investigation of the values of the second, third, and fourth terms must be deferred till we consider the variations of the second degree.

Condition of Equilibrium of a Mixture of Gases

In a state of equilibrium u_1 and u_2 vanish, $\rho_1\xi_1^2$ becomes p_1, and the tangential pressures vanish, so that the equation becomes

$$\frac{dp_1}{dx} = X\rho_1 \qquad (77)$$

which is the equation of equilibrium in ordinary hydrostatics.

This equation, being true of the system of molecules forming the first medium independently of the presence of the molecules of the second system, shews that if several kinds of molecules are mixed together, placed in a vessel and acted on by gravity, the final distribution of the molecules of each kind will be the same as if none of the other kinds had been present. This is the same mode of distribution as that which Dalton considered to exist in a mixed atmosphere in equilibrium, the law of diminution of density of each constituent gas being the same as if no other gases were present.†

This result, however, can only take place after the gases have been left for a considerable time perfectly undisturbed. If currents arise so as to mix the strata, the composition of the gas will be made more uniform throughout.

The result at which we have arrived as to the final distribution of gases, when left to themselves, is independent of the law of force between the molecules.

Diffusion of Gases

If the motion of the gases is slow, we may still neglect the tangential pressures. The equation then becomes for the first system of molecules

$$\rho_1 \frac{\partial u_1}{\partial t} + \frac{dp_1}{dx} = kA_1 \rho_1 \rho_2 (u_2 - u_1) + X\rho_1 \qquad (78)$$

and for the second,

$$\rho_2 \frac{\partial u_2}{\partial t} + \frac{dp_2}{dx} = kA_1 \rho_1 \rho_2 (u_1 - u_2) + X\rho_2 \qquad (79)$$

In all cases of quiet diffusion we may neglect the first term of each equation. If we then put $p_1 + p_2 = p$, and $\rho_1 + \rho_2 = \rho$, we find by adding,

$$\frac{dp}{dx} = X\rho \qquad (80)$$

---

† [J. Dalton, *Memoirs of the Manchester Philosophical Society* **5**, 535 (1802).]

If we also put $p_1u_1+p_2u_2 = pu$, then the volumes transferred in opposite directions across a plane moving with velocity u will be equal, so that

$$p_1(u_1-u) = p_2(u-u_2) = \frac{p_1p_2}{p\rho_1\rho_2kA_1} \cdot \left(X\rho_1 - \frac{dp_1}{dx}\right) \quad (81)$$

Here $p_1(u_1-u)$ is the volume of the first gas transferred in unit of time across unit of area of the plane reduced to pressure unity, and at the actual temperature; and $p_2(u-u_2)$ is the equal volume of the second gas transferred across the same area in the opposite direction.

The external force X has very little effect on the quiet diffusion of gases in vessels of moderate size. We may therefore leave it out in our definition of the coefficient of diffusion of two gases.

When two gases not acted on by gravity are placed in different parts of a vessel at equal pressures and temperatures, there will be mechanical equilibrium from the first, and u will always be zero. This will also be approximately true of heavy gases, provided the denser gas is placed below the lighter. Mr Graham has described in his paper on the Mobility of Gases,† experiments which were made under these conditions. A vertical tube had its lower tenth part filled with a heavy gas, and the remaining nine-tenths with a lighter gas. After the lapse of a known time the upper tenth part of the tube was shut off, and the gas in it analyzed, so as to determine the quantity of the heavier gas which had ascended into the upper tenth of the tube during the given time.

In this case we have

$$u = 0 \quad (82)$$

$$p_1u_1 = -\frac{p_1p_2}{\rho_1\rho_2kA_1}\frac{1}{p}\frac{dp_1}{dx} \quad (83)$$

and by the equation of continuity,

$$\frac{dp_1}{dt} + \frac{d}{dx}(p_1u_1) = 0 \quad (84)$$

---

† *Philosophical Transactions*, 1863 [p. 385].

whence
$$\frac{dp_1}{dt} = \frac{p_1 p_2}{\rho_1 \rho_2 k A_1} \frac{1}{p} \frac{d^2 p_1}{dx^2} \qquad (85)$$
or if we put $D = \frac{p_1 p_2}{\rho_1 \rho_2 k A_1} \frac{1}{p}$,
$$\frac{dp_1}{dt} = D \frac{d^2 p_1}{dx^2} \qquad (86)$$

The solution of this equation is
$$p_1 = C_1 + C_2 e^{-n^2 Dt} \cos(nx + \alpha) + \&c. \qquad (87)$$

If the length of the tube is a, and if it is closed at both ends,
$$p_1 = C_1 + C_2 e^{-\frac{\pi^2 D}{a^2} t} \cos\frac{\pi x}{a} + C_3 e^{-4\frac{\pi^2 D}{a^2} t} \cos 2\frac{\pi x}{a} + \&c. \qquad (88)$$

where C_1, C_2, C_3 are to be determined by the condition that when $t = 0$, $p_1 = p$, from $x = 0$ to $x = \frac{1}{10}a$ and $p_1 = 0$ from $x = \frac{1}{10}a$ to $x = a$. The general expression for the case in which the first gas originally extends from $x = 0$ to $x = b$, and in which after a time t the gas from $x = 0$ to $x = c$ is collected, is

$$\frac{p_1}{p} = \frac{b}{a} + \frac{2a}{\pi^2 c} \left\{ e^{-\frac{\pi^2 D}{a^2} t} \sin\frac{\pi b}{a} \sin\frac{\pi c}{a} \right. \\ \left. + \frac{1}{2^2} e^{-4\frac{\pi^2 D}{a^2} t} \sin\frac{2\pi b}{a} \sin\frac{2\pi c}{a} + \&c. \right\} \qquad (89)$$

where p_1/p is the proportion of the first gas to the whole in the portion from $x = 0$ to $x = c$.

In Mr Graham's experiments, in which one-tenth of the tube was filled with the first gas, and the proportion of the first gas in the tenth of the tube at the other end ascertained after a time t, this proportion will be

$$\frac{p_1}{p} = \frac{1}{10} - \frac{20}{\pi^2} \left\{ e^{-\frac{\pi^2 D}{a^2} t} \sin^2\frac{\pi}{10} - e^{-2^2 \frac{\pi^2 D}{a^2} t} \sin^2 2\frac{\pi}{10} \right. \\ \left. + e^{-3^2 \frac{\pi^2 D}{a^2} t} \sin^2 3\frac{\pi}{10} - \&c. \right\} \qquad (90)$$

We find for a series of values of p_1/p taken at equal intervals of time T, where

$$T = \frac{\log_e 10}{10\pi^2} \frac{a^2}{D}$$

Time	$\dfrac{p_1}{p}$
0	0
T	·01193
$2T$	·02305
$3T$	·03376
$4T$	·04366
$5T$	·05267
$6T$	·06072
$8T$	·07321
$10T$	·08227
$12T$	·08845
∞	·10000

Mr Graham's experiments on carbonic acid and air, when compared with this Table give $T = 500$ seconds nearly for a tube 0·57 metre long. Now

$$D = \frac{\log_e 10}{10\pi^2} \frac{a^2}{T} \qquad (91)$$

whence

$$D_1 = \cdot 0235$$

for carbonic acid and air, in inch-grain-second measure.

Definition of the Coefficient of Diffusion

D is the volume of gas reduced to unit of pressure which passes in unit of time through unit of area when the total pressure is uniform and equal to p, and the pressure of either gas increases or diminishes by unity in unit of distance. D may be called the coefficient of diffusion. It varies directly as the square of the absolute temperature, and inversely as the total pressure p.

The dimensions of D are evidently L^2T^{-1}, where L and T are the standards of length and time.

In considering this experiment of the interdiffusion of carbonic acid and air, we have assumed that air is a simple gas. Now it is well known that the constituents of air can be separated by mechanical means, such as passing them through a porous diaphragm, as in Mr Graham's experiments on Atmolysis. The discussion of the interdiffusion of three or more gases leads to a much more complicated equation than that which we have found for two gases, and it is not easy to deduce the coefficients of interdiffusion of the separate gases. It is therefore to be desired that experiments should be made on the interdiffusion of every pair of the more important pure gases which do not act chemically on each other, the temperature and pressure of the mixture being noted at the time of experiment.

Mr Graham has also published in Brande's *Journal* for 1829, pt. 2, p. 74,† the results of experiments on the diffusion of various gases out of a vessel through a tube into air. The coefficients of diffusion deduced from these experiments are:

Air and Hydrogen	·026216
Air and Marsh-gas	·010240
Air and Ammonia	·00962
Air and Olefiant gas	·00771
Air and Carbonic acid	·00682
Air and Sulphurous acid	·00582
Air and Chlorine	·00486

The value for carbonic acid is only one-third of that deduced from the experiment with the vertical column. The inequality of composition of the mixed gas in different parts of the vessel is, however, neglected; and the diameter of the tube at the middle part, where it was bent, was probably less than that given.

---

† [*Quarterly Journal of Science.*]

Those experiments on diffusion which lasted ten hours, all give smaller values of D than those which lasted four hours, and this would also result from the mixture of the gases in the vessel being imperfect.

Interdiffusion through a small hole

When two vessels containing different gases are connected by a small hole, the mixture of gases in each vessel will be nearly uniform except near the hole; and the inequality of the pressure of each gas will extend to a distance from the hole depending on diameter of the hole, and nearly proportional to that diameter.

Hence in the equation

$$\rho_1 \frac{\partial u_1}{\partial t} + \frac{dp_1}{dx} = kA\rho_1\rho_2(u_2 - u_1) + X\rho \tag{92}$$

the term dp_1/dx will vary inversely as the diameter of the hole, while u_1 and u_2 will not vary considerably with the diameter.

Hence when the hole is very small the right-hand side of the equation may be neglected, and the flow of either gas through the hole will be independent of the flow of the other gas, as the term $kA\rho_1 p_2 (u_2 - u_1)$ becomes comparatively insignificant.

One gas therefore will escape through a very fine hole into another nearly as fast as into a vacuum; and if the pressures are equal on both sides, the volumes diffused will be as the square roots of the specific gravities inversely, which is the law of diffusion of gases established by Graham.[†]

Variation of the invisible agitation (β)

By putting for Q in equation (75)

$$Q = \frac{M}{2}\{(u_1+\xi_1)^2+(v_1+\eta_1)^2+(w_1+\zeta_1)^2+(\beta-1)(\xi_1^2+\eta_1^2+\zeta_1^2)\} \tag{93}$$

[†] *Trans. Royal Society of Edinburgh*, Vol. XII, p. 22 [(1834)].

and eliminating by means of equations (76) and (52), we find

$$\begin{aligned}
&\tfrac{1}{2}\rho_1\frac{\partial}{\partial t}\beta_1(\xi_1^2+\eta_1^2+\zeta_1^2)+\rho_1\xi_1^2\frac{du_1}{dx}+\rho_1\eta_1^2\frac{dv_1}{dy}+\rho_1\zeta_1^2\frac{dw_1}{dz}\\
&+\rho_1\eta_1\zeta_1\left(\frac{dv_1}{dz}+\frac{dw_1}{dy}\right)\\
&+\rho_1\zeta_1\xi_1\left(\frac{dw_1}{dx}+\frac{du_1}{dz}\right)+\rho_1\xi_1\eta_1\left(\frac{du_1}{dy}+\frac{dv_1}{dx}\right)\\
&+\beta_1\left\{\frac{d}{dx}(\rho_1\xi_1^3+\rho_1\xi_1\eta_1^2+\rho_1\xi_1\zeta_1^2)\right.\\
&+\frac{d}{dy}(\rho_1\eta_1\xi_1^2+\rho_1\eta_1^3+\rho_1\eta_1\zeta_1^2)\\
&\left.+\frac{d}{dz}(\rho_1\zeta_1\xi_1^2+\rho_1\zeta_1\eta_1^2+\rho_1\zeta_1^3)\right\}\\
&=\frac{k\rho_1\rho_2 A_1}{M_1+M_2}[M_2\{(u_2-u_1)^2+(v_2-v_1)^2+(w_2-w_1)^2\}\\
&+M_2(\xi_2^2+\eta_2^2+\zeta_2^2)-M_1(\xi_1^2+\eta_1^2+\zeta_1^2)]
\end{aligned} \quad (94)$$

In this equation the first term represents the variation of invisible agitation or heat; the second, third, and fourth represent the cooling by expansion; the fifth, sixth, and seventh the heating effect of fluid friction or viscosity; and the last the loss of heat by conduction. The quantities on the other side of the equation represent the thermal effects of diffusion, and the communication of heat from one gas to the other.

The equation may be simplified in various cases, which we shall take in order.

1st. Equilibrium of Temperature between two Gases—Law of Equivalent Volumes

We shall suppose that there is no motion of translation, and no transfer of heat by conduction through either gas. The equation (94) is then reduced to the following form,

$$\tfrac{1}{2}\rho_1 \frac{\partial}{\partial t}\beta_1(\xi_1^2+\eta_1^2+\zeta_1^2) = \frac{k\rho_1\rho_2 A_1}{M_1+M_2}\{M_2(\xi_2^2+\eta_2^2+\zeta_2^2) - M_1(\xi_1^2+\eta_1^2+\zeta_1^2)\} \quad (95)$$

If we put

$$\frac{M_1}{M_1+M_2}(\xi_1^2+\eta_1^2+\zeta_1^2) = Q_1, \quad \text{and} \quad \frac{M_2}{M_1+M_2}(\xi_2^2+\eta_2^2+\zeta_2^2) = Q_2 \quad (96)$$

we find

$$\frac{\partial}{\partial t}(Q_2-Q_1) = -\frac{2kA_1}{M_1+M_2}(M_2\rho_2\beta_1 + M_1\rho_1\beta_2)(Q_2-Q_1) \quad (97)$$

or

$$Q_2 - Q_1 = Ce^{-nt}, \text{ where } n = \frac{2kA_1}{M_1+M_2}(M_2\rho_2\beta_2 + M_1\rho_1\beta_1)\frac{1}{\beta_1\beta_2} \quad (98)$$

If, therefore, the gases are in contact and undisturbed, Q_1 and Q_2 will rapidly become equal. Now the state into which two bodies come by exchange of invisible agitation is called equilibrium of heat or equality of temperature. Hence when two gases are at the same temperature,

$$Q_1 = Q_2 \quad (99)$$

or

$$1 = \frac{Q_1}{Q_2} = \frac{M_1(\xi_1^2+\eta_1^2+\zeta_1^2)}{M_2(\xi_2^2+\eta_2^2+\zeta_2^2)}$$

$$= \frac{M_1\dfrac{p_1}{\rho_1}}{M_2\dfrac{p_2}{\rho_2}}$$

Hence if the pressures as well as the temperatures be the same in two gases,

$$\frac{M_1}{\rho_1} = \frac{M_2}{\rho_2} \tag{100}$$

or the masses of the individual molecules are proportional to the density of the gas.

This result, by which the relative masses of the molecules can be deduced from the relative densities of the gases, was first arrived at by Gay-Lussac from chemical considerations. It is here shewn to be a necessary result of the Dynamical Theory of Gases; and it is so, whatever theory we adopt as to the nature of the action between the individual molecules, as may be seen by equation (34), which is deduced from perfectly general assumptions as to the nature of the law of force.

We may therefore henceforth put s_1/s_2 for M_1/M_2, where s_1, s_2 are the specific gravities of the gases referred to a standard gas.

If we use θ to denote the temperature reckoned from absolute zero of a gas thermometer, M_0 the mass of a molecule of hydrogen, V_0^2 its mean square of velocity at temperature unity, s the specific gravity of any other gas referred to hydrogen, then the mass of a molecule of the other gas is

$$M = M_0 s \tag{101}$$

Its mean square of velocity,

$$V^2 = \frac{1}{s} V_0^2 \theta \tag{102}$$

Pressure of the gas,

$$p = \tfrac{1}{3} \frac{\rho}{s} \theta V_0^2 \tag{103}$$

We may next determine the amount of cooling by expansion.

Cooling by Expansion

Let the expansion be equal in all directions, then

$$\frac{du}{dx} = \frac{dv}{dy} = \frac{dw}{dz} = -\frac{1}{3\rho}\frac{\partial \rho}{\partial t} \qquad (104)$$

and du/dy and all terms of unsymmetrical form will be zero.

If the mass of gas is of the same temperature throughout there will be no conduction of heat, and the equation (94) will become

$$\tfrac{1}{2}\rho\beta\frac{\partial \bar{V}^2}{\partial t} - \tfrac{1}{3}\bar{V}^2\frac{\partial \rho}{\partial t} = 0 \qquad (105)$$

or

$$2\frac{\partial \rho}{\rho} = 3\beta\frac{\partial \bar{V}^2}{\bar{V}^2} = 3\beta\frac{\partial \theta}{\theta} \qquad (106)$$

or

$$\frac{\partial \theta}{\theta} = \frac{2}{3\beta}\frac{\partial \rho}{\rho} \qquad (107)$$

which gives the relation between the density and the temperature in a gas expanding without exchange of heat with other bodies. We also find

$$\frac{\partial p}{p} = \frac{\partial \rho}{\rho} + \frac{\partial \theta}{\theta}$$

$$= \frac{2+3\beta}{3\beta}\frac{\partial \rho}{\rho} \qquad (108)$$

which gives the relation between the pressure and the density.

Specific Heat of Unit of Mass at Constant Volume

The total energy of agitation of unit of mass is $\tfrac{1}{2}\beta \bar{V}^3 = \tfrac{1}{2}E$, or

$$E = \frac{3\beta}{2}\frac{p}{\rho} \qquad (109)$$

If, now, additional energy in the form of heat be communicated to it without changing its density,

$$\partial E = \frac{3\beta}{2}\frac{\partial p}{\rho} = \frac{3\beta}{2}\frac{p}{\rho}\frac{\partial \theta}{\theta} \tag{110}$$

Hence the specific heat of unit of mass at constant volume is in dynamical measure

$$\frac{\partial E}{\partial \theta} = \frac{3\beta}{2}\frac{p}{\rho\theta} \tag{111}$$

Specific Heat of Unit of Mass at Constant Pressure

By the addition of the heat ∂E the temperature was raised $\partial \theta$ and the pressure ∂p. Now, let the gas expand without communication of heat till the pressure sinks to its former value, and let the final temperature be $\theta + \partial'\theta$. The temperature will thus sink by a quantity $\partial\theta - \partial'\theta$, such that

$$\frac{\partial\theta - \partial'\theta}{\theta} = \frac{2}{2+3\beta}\frac{\partial p}{p} = \frac{2}{2+3\beta}\frac{\partial\theta}{\theta}$$

whence

$$\frac{\partial'\theta}{\theta} = \frac{3\beta}{2+3\beta}\frac{\partial\theta}{\theta} \tag{112}$$

and the specific heat of unit of mass at constant pressure is

$$\frac{\partial E}{\partial'\theta} = \frac{2+3\beta}{2}\frac{p}{\rho\theta} \tag{113}$$

The ratio of the specific heat at constant pressure to that of constant volume is known in several cases from experiment. We shall denote this ratio by

$$\gamma = \frac{2+3\beta}{3\beta} \tag{114}$$

whence

$$\beta = \tfrac{2}{3}\frac{1}{\gamma-1} \tag{115}$$

The specific heat of unit of volume in ordinary measure is at constant volume

$$\frac{1}{\gamma-1}\frac{p}{J\theta} \tag{116}$$

and at constant pressure

$$\frac{\gamma}{\gamma-1}\frac{p}{J\theta} \tag{117}$$

where J is the mechanical equivalent of unit of heat.

From these expressions Dr Rankine[†] has calculated the specific heat of air, and has found the result to agree with the value afterwards determined experimentally by M. Regnault[‡].

Thermal Effects of Diffusion

If two gases are diffusing into one another, then, omitting the terms relating to heat generated by friction and to conduction of heat, the equation (94) gives

$$\begin{aligned} &\tfrac{1}{2}\rho_1 \frac{\partial}{\partial t}\beta_1(\xi_1^2+\eta_1^2+\zeta_1^2)+\tfrac{1}{2}\rho_2 \frac{\partial}{\partial t}\beta_2(\xi_2^2+\eta_2^2+\zeta_2^2) \\ &+p_1\left(\frac{du_1}{dx}+\frac{dv_1}{dy}+\frac{dw_1}{dz}\right)+p_2\left(\frac{du_2}{dx}+\frac{dv_2}{dy}+\frac{dw_2}{dz}\right) \\ &= k\rho_1\rho_2 A_1\{(u_1-u_2)^2+(v_1-v_2)^2+(w_1-w_2)^2\} \end{aligned} \tag{118}$$

By comparison with equations (78) and (79), the right-hand side of this equation becomes

$$X(\rho_1 u_1+\rho_2 u_2)+Y(\rho_1 v_1+\rho_2 v_2)+Z(\rho_1 w_1+\rho_2 w_2)$$

$$-\left(\frac{dp_1}{dx}u_1+\frac{dp_1}{dy}v_1+\frac{dp_1}{dz}w_1\right)-\left(\frac{dp_2}{dx}u_2+\frac{dp_2}{dy}v_2+\frac{dp_2}{dz}w_2\right)$$

$$-\tfrac{1}{2}\rho_1 \frac{\partial}{\partial t}(u_1^2+v_1^2+w_1^2)-\tfrac{1}{2}\rho_2 \frac{\partial}{\partial t}(u_2^2+v_2^2+w_2^2)$$

[†] *Transactions of the Royal Society of Edinburgh.* Vol. XX (1850) [p. 147].
[‡] *Comptes Rendus* [Academie des Sciences, Paris], 1853 [36, 676].

The equation (118) may now be written

$$\begin{aligned}
&\tfrac{1}{2}\rho_1 \frac{\partial}{\partial t}\{u_1^2+v_1^2+w_1^2+\beta_1(\xi_1^2+\eta_1^2+\zeta_1^2)\} \\
&+\tfrac{1}{2}\rho_2 \frac{\partial}{\partial t}\{u_2^2+v_2^2+w_2^2+\beta_2(\xi_2^2+\eta_2^2+\zeta_2^2)\} \\
&= X(\rho_1 u_1+\rho_2 u_2)+Y(\rho_1 v_1+\rho_2 v_2) \\
&+Z(\rho_1 w_1+\rho_2 w_2)-\left(\frac{d.pu}{dx}+\frac{d.pv}{dy}+\frac{d.pw}{dz}\right)
\end{aligned} \quad (119)$$

The whole increase of energy is therefore that due to the action of the external forces *minus* the cooling due to the expansion of the mixed gases. If the diffusion takes place without alteration of the volume of the mixture, the heat due to the mutual action of the gases in diffusion will be exactly neutralized by the cooling of each gas as it expands in passing from places where it is dense to places where it is rare.

Determination of the Inequality of Pressure in different Directions due to the Motion of the Medium

Let us put
$$\rho_1 \xi_1^2 = p_1+q_1 \quad \text{and} \quad \rho_2 \xi_2^2 = p_2+q_2 \quad (120)$$

Then by equation (52),

$$\begin{aligned}
\frac{\delta q_1}{\delta t} &= -3k_1 A_2 \rho_1 q_1 - \frac{k}{M_1+M_2}(2M_1 A_1+3M_2 A_2)\rho_2 q_1 \\
&- k(3A_2-2A_1)\frac{M_1}{M_1+M_2}\rho_1 q_2 - k\rho_1\rho_2 \\
&\times \frac{M_2}{M_1+M_2}(A_2-\tfrac{2}{3}A_1)(2\overline{u_1-u_2}^2-\overline{v_1-v_2}^2-\overline{w_1-w_2}^2)
\end{aligned} \quad (121)$$

the last term depending on diffusion; and if we omit in equation (75) terms of three dimensions in ξ, η, ζ, which relate to conduction of heat, and neglect quantities of the form $\xi\eta\rho$ and $\rho\xi^2-p$, when not multiplied by the large coefficients k, k_1, and k_2, we get

$$\frac{\partial q}{\partial t}+2p\frac{du}{dx}-\tfrac{2}{3}p\left(\frac{du}{dx}+\frac{dv}{dy}+\frac{dw}{dz}\right)=\frac{\delta q}{\delta t} \tag{122}$$

If the motion is not subject to any very rapid changes, as in all cases except that of the propagation of sound, we may neglect $\partial q/\partial t$. In a single system of molecules

$$\frac{\delta q}{\delta t}=-3kA_2\rho q \tag{123}$$

whence

$$q=-\frac{2p}{3kA_2\rho}\left\{\frac{du}{dx}-\tfrac{1}{3}\left(\frac{du}{dx}+\frac{dv}{dy}+\frac{dw}{dz}\right)\right\} \tag{124}$$

If we make

$$\tfrac{1}{3}\frac{1}{kA_2}\frac{p}{\rho}=\mu \tag{125}$$

μ will be the coefficient of viscosity, and we shall have by equation (120),

$$\left.\begin{aligned}\rho\xi^2 &= p-2\mu\left\{\frac{du}{dx}-\tfrac{1}{3}\left(\frac{du}{dx}+\frac{dv}{dy}+\frac{dw}{dz}\right)\right\}\\ \rho\eta^2 &= p-2\mu\left\{\frac{dv}{dy}-\tfrac{1}{3}\left(\frac{du}{dx}+\frac{dv}{dy}+\frac{dw}{dz}\right)\right\}\\ \rho\zeta^2 &= p-2\mu\left\{\frac{dw}{dz}-\tfrac{1}{3}\left(\frac{du}{dx}+\frac{dv}{dy}+\frac{dw}{dz}\right)\right\}\end{aligned}\right\} \tag{126}$$

and by transformation of co-ordinates we obtain

$$\left.\begin{aligned}\rho\eta\zeta &= -\mu\left(\frac{dv}{dz}+\frac{dw}{dy}\right)\\ \rho\zeta\xi &= -\mu\left(\frac{dw}{dx}+\frac{du}{dz}\right)\\ \rho\xi\eta &= -\mu\left(\frac{du}{dy}+\frac{dv}{dx}\right)\end{aligned}\right\} \tag{127}$$

These are the values of the normal and tangential stresses in a simple gas when the variation of motion is not very rapid, and when μ, the coefficient of viscosity, is so small that its square may be neglected.

Equations of Motion corrected for Viscosity

Substituting these values in the equation of motion (76), we find

$$\rho \frac{\partial u}{\partial t} + \frac{dp}{dx} - \mu \left\{ \frac{d^2u}{dx^2} + \frac{d^2u}{dy^2} + \frac{d^2u}{dz^2} \right\} - \tfrac{1}{3}\mu \frac{d}{dx}\left(\frac{du}{dx} + \frac{dv}{dy} + \frac{dw}{dz}\right) = X\rho \quad (128)$$

with two other equations which may be written down with symmetry. The form of these equations is identical with that of those deduced by Poisson† from the theory of elasticity, by supposing the strain to be continually relaxed at a rate proportional to its amount. The ratio of the third and fourth terms agrees with that given by Professor Stokes.‡

If we suppose the inequality of pressure which we have denoted by q to exist in the medium at any instant, and not to be maintained by the motion of the medium, we find, from equation (123),

$$q_1 = Ce^{-3kA_2\rho t} \quad (129)$$

$$= Ce^{-t/T} \quad \text{if} \quad T = \frac{1}{3kA_2\rho} = \frac{\mu}{p} \quad (130)$$

the stress q is therefore relaxed at a rate proportional to itself, so that

$$\frac{\delta q}{q} = \frac{\delta t}{T} \quad (131)$$

We may call T the modulus of the time of relaxation.

---

† *Journal de l'École Polytechnique*, 1829, Tom. XIII, Cah. XX, p. 139.

‡ " On the Friction of Fluids in Motion and the Equilibrium and Motion of Elastic Solids ", *Cambridge Phil. Trans.* Vol. VIII (1845) p. 297, equation (2).

If we next make $k = 3$, so that the stress q does not become relaxed, the medium will be an elastic solid, and the equation

$$\frac{\partial(\rho\xi^2 - p)}{\partial t} + 2p\frac{du}{dx} - \tfrac{2}{3}p\left(\frac{du}{dx} + \frac{dv}{dy} + \frac{dw}{dz}\right) = 0 \qquad (132)$$

may be written

$$\frac{\partial}{\partial t}\left\{(p_{xx} - p) + 2p\frac{d\alpha}{dx} - \tfrac{2}{3}p\left(\frac{d\alpha}{dx} + \frac{d\beta}{dy} + \frac{d\gamma}{dz}\right)\right\} = 0 \qquad (133)$$

where α, β, γ are the displacements of an element of the medium, and p_{xx} is the normal pressure in the direction of x. If we suppose the initial value of this quantity zero, and p_{xx} originally equal to p, then, after a small displacement,

$$p_{xx} = p - p\left(\frac{d\alpha}{dx} + \frac{d\beta}{dy} + \frac{d\gamma}{dz}\right) - 2p\frac{d\alpha}{dx} \qquad (134)$$

and by transformation of co-ordinates the tangential pressure

$$p_{xy} = -p\left(\frac{d\alpha}{dy} + \frac{d\beta}{dx}\right) \qquad (135)$$

The medium has now the mechanical properties of an elastic solid, the rigidity of which is p, while the cubical elasticity is $\tfrac{5}{3}p$.†

The same result and the same ratio of the elasticities would be obtained if we supposed the molecules to be at rest, and to act on one another with forces depending on the distance, as in the statical molecular theory of elasticity. The coincidence of the properties of a medium in which the molecules are held in equilibrium by attractions and repulsions, and those of a medium in which the molecules move in straight lines without acting on each other at all, deserve notice from those who speculate on theories of physics.

---

† *Camb. Phil. Trans.* Vol. VIII (1845) p. 311, equation (29).

The fluidity of our medium is therefore due to the mutual action of the molecules, causing them to be deflected from their paths.

The coefficient of instantaneous rigidity of a gas is therefore p
The modulus of the time of relaxation is T
The coefficient of viscosity is $\mu = pT$

(136)

Now p varies as the density and temperature conjointly, while T varies inversely as the density.

Hence μ varies as the absolute temperature, and is independent of the density.

This result is confirmed by the experiments of Mr Graham on the Transpiration of Gases,† and by my own experiments on the Viscosity or Internal Friction of Air and other Gases.‡

The result that the viscosity is independent of the density, follows from the Dynamical Theory of Gases, whatever be the law of force between the molecules. It was deduced by myself§ from the hypothesis of hard elastic molecules, and M. O. E. Meyer‖ has given a more complete investigation on the same hypothesis.

The experimental result, that the viscosity is proportional to the absolute temperature, requires us to abandon this hypothesis, which would make it vary as the square root of the absolute temperature, and to adopt the hypothesis of a repulsive force inversely as the fifth power of the distance between the molecules, which is the only law of force which gives the observed result.

Using the foot, the grain, and the second as units, my experiments gave for the temperature of 62° Fahrenheit, and in dry air,

$$\mu = 0.0936$$

---

† *Philosophical Transactions*, 1846 [p. 573] and 1849 [p. 349].

‡ *Proceedings of the Royal Society*, February 8, 1866 [15, 14]; *Philosophical Transactions*, 1866, p. 249.

§ *Philosophical Magazine*, January 1860 [Vol. 1, Selection 10].

‖ Poggendorff's *Annalen*, 1865 [*Ann. Phys.* **125**, 177, 401, 564].

If the pressure is 30 inches of mercury, we find, using the same units,

$$p = 477360000$$

Since $pT = \mu$, we find that the modulus of the time of relaxation of rigidity in air of this pressure and temperature is

$$\frac{1}{5099100000} \text{ of a second}$$

This time is exceedingly small, even when compared with the period of vibration of the most acute audible sounds; so that even in the theory of sound we may consider the motion as steady during this very short time, and use the equations we have already found, as has been done by Professor Stokes.†

Viscosity of a Mixture of Gases

In a complete mixture of gases, in which there is no diffusion going on, the velocity at any point is the same for all the gases.

Putting

$$\tfrac{2}{3}\left(2\frac{du}{dx} - \frac{dv}{dy} - \frac{dw}{dz}\right) = U \qquad (137)$$

equation (122) becomes

$$\left.\begin{aligned} p_1 U = -3k_1 A_2 \rho_1 q_1 &- \frac{k}{M_1 + M_2}(2M_1 A_1 + 3M_2 A_2)\rho_2 q_1 \\ &- k(3A_2 - 2A_1)\frac{M_2}{M_1 + M_2}\rho_1 q_2 \end{aligned}\right\} \qquad (138)$$

Similarly,

$$\left.\begin{aligned} p_2 U = -3k_2 A_2 \rho_2 q_2 &- \frac{k}{M_1 + M_2}(2M_2 A_1 + 3M_1 A_2)\rho_1 q_2 \\ &- k(3A_2 - 2A_1)\frac{M_1}{M_1 + M_2}\rho_2 q_1 \end{aligned}\right\} \qquad (139)$$

---

† " On the effect of the internal Friction of Fluids on the motion of Pendulums ", *Cambridge [Philosophical] Transactions*, Vol. IX (1850), art. 79.

Since $p = p_1 + p_2$ and $q = q_1 + q_2$, where p and q refer to the mixture, we shall have

$$\mu U = -q = -(q_1 + q_2)$$

where μ is the coefficient of viscosity of the mixture.

If we put s_1 and s_2 for the specific gravities of the two gases, referred to a standard gas, in which the values of p and q at temperature θ_0 and p_0 and ρ_0,

$$\mu = \frac{p_0 \theta}{\rho_0 \theta_0} \cdot \frac{Ep_1^2 + Fp_1 p_2 + Gp_2^2}{3A_2 k_1 s_1 Ep_1^2 + Hp_1 p_2 + 3A_2 k_2 s_2 Gp_2^2} \quad (140)$$

where μ is the coefficient of viscosity of the mixture, and

$$\left. \begin{array}{l} E = \dfrac{ks_1}{s_1 + s_2}(2s_2 A_1 + 3s_1 A_2) \\[1em] F = 3A_2(k_1 s_1 + k_2 s_2) - (3A_2 - 2A_1)k\dfrac{2s_1 s_2}{s_1 + s_2} \\[1em] G = \dfrac{ks_2}{s_1 + s_2}(2s_1 A_1 + 3s_2 A_2) \\[1em] H = 3A_2 s_1 s_2(3k_1 k_2 A_2 + 2k^2 A_1) \end{array} \right\} \quad (141)$$

This expression is reduced to μ_1 when $p_2 = 0$, and to μ_2 when $p_1 = 0$. For other values of p_1 and p_2 we require to know the values of k, the coefficient of mutual interference of the molecules of the two gases. This might be deduced from the observed values of μ for mixtures, but a better method is by making experiments on the interdiffusion of the two gases. The experiments of Graham on the transpiration of gases, combined with my experiments on the viscosity of air, give as values of k_1 for air, hydrogen, and carbonic acid,

Air $\quad k_1 = \quad 4 \cdot 81 \times 10^{10}$,

Hydrogen $\quad k_1 = 142 \cdot 8 \times 10^{10}$,

Carbonic acid $k_1 = \quad 3 \cdot 9 \times 10^{10}$.

The experiments of Graham in 1863,† on the interdiffusion of air and carbonic acid, give the coefficient of mutual interference of these gases,

$$\text{Air and carbonic acid } k = 5 \cdot 2 \times 10^{10};$$

and by taking this as the absolute value of k, and assuming that the ratios of the coefficients of interdiffusion given at page 67 are correct, we find

$$\text{Air and hydrogen } k = 29 \cdot 8 \times 10^{10}.$$

These numbers are to be regarded as doubtful, as we have supposed air to be a simple gas in our calculations, and we do not know the value of k between oxygen and nitrogen. It is also doubtful whether our method of calculation applies to experiments such as the earlier observations of Mr Graham.

I have also examined the transpiration-times determined by Graham for mixtures of hydrogen and carbonic acid, and hydrogen and air, assuming a value of k roughly, to satisfy the experimental results about the middle of the scale. It will be seen that the calculated numbers for hydrogen and carbonic acid exhibit the peculiarity observed in the experiments, that a small addition of hydrogen *increases* the transpiration-time of carbonic acid, and that in both series the times of mixtures depend more on the slower than on the quicker gas.

The assumed values of k in these calculations were:

$$\text{For hydrogen and carbonic acid } k = 12 \cdot 5 \times 10^{10},$$

$$\text{For hydrogen and air } \quad k = 18 \cdot 8 \times 10^{10}$$

and the results of observation and calculation are, for the times of transpiration of mixtures, given in the table.

The numbers given are the ratios of the transpiration-times of mixtures to that of oxygen as determined by Mr Graham, compared with those given by the equation (140) deduced from our theory.

† *Philosophical Transactions*, 1863 [p. 385].

Hydrogen and Carbonic acid		Observed	Calculated	Hydrogen and Air		Observed	Calculated
100	0	·4321	·4375	100	0	·4434	·4375
97·5	2·5	·4714	·4750	95	5	·5282	·5300
95	5	·5157	·5089	90	10	·5880	·6028
90	10	·5722	·5678	75	25	·7488	·7438
75	25	·6786	·6822	50	50	·8179	·8488
50	50	·7339	·7652	25	75	·8790	·8946
25	75	·7535	·7468	10	90	·8880	·8983
10	90	·7521	·7361	5	95	·8960	·8996
0	100	·7470	·7272	0	100	·9000	·9010

Conduction of Heat in a Single Medium (γ)

The rate of conduction depends on the value of the quantity

$$\tfrac{1}{2}\beta\rho(\xi^3 + \xi\eta^2 + \xi\zeta^2)$$

where ξ^3, $\xi\eta^2$, and $\xi\zeta^2$ denote the mean values of those functions of ξ, η, ζ for all the molecules in a given element of volume.

As the expressions for the variations of this quantity are somewhat complicated in a mixture of media, and as the experimental investigation of the conduction of heat in gases is attended with great difficulty, I shall confine myself here to the discussion of a single medium.

Putting

$$Q = M(u+\xi)\{u^2 + v^2 + w^2 + 2u\xi + 2v\eta + 2w\zeta + \beta(\xi^2 + \eta^2 + \zeta^2)\} \quad (142)$$

and neglecting terms of the forms $\xi\eta$ and ξ^3 and $\xi\eta^2$ when not multiplied by the large coefficient k_1, we find by equations (75), (77), and (54),

$$\left.\begin{aligned}
&\rho\frac{\partial}{\partial t}\beta(\xi^3 + \xi\eta^2 + \xi\zeta^2) + \beta\frac{d}{dx}\cdot\rho(\xi^4 + \xi^2\eta^2 + \xi^2\zeta^2) \\
&\quad -\beta(\xi^2 + \eta^2 + \zeta^2)\frac{dp}{dx} - 2\beta\xi^2\frac{dp}{dx} \\
&\quad = -3k_1\rho^2 A_2\beta\{\xi^3 + \xi\eta^2 + \xi\zeta^2\}
\end{aligned}\right\} \quad (143)$$

The first term of this equation may be neglected, as the rate of conduction will rapidly establish itself. The second term contains quantities of four dimensions in ξ, η, ζ, whose values will depend on the distribution of velocity among the molecules. If the distribution of velocity is that which we have proved to exist when the system has no external force acting on it and has arrived at its final state, we shall have by equations (29), (31), (32),

$$\overline{\xi^4} = \overline{3\xi^2} \cdot \overline{\xi^2} = 3\frac{p^2}{\rho^2} \tag{144}$$

$$\overline{\xi^2\eta^2} = \overline{\xi^2} \cdot \overline{\eta^2} = \frac{p^2}{\rho^2} \tag{145}$$

$$\overline{\xi^2\zeta^2} = \overline{\xi^2} \cdot \overline{\zeta^2} = \frac{p^2}{\rho^2} \tag{146}$$

and the equation of conduction may be written

$$5\beta\frac{p^2}{\rho\theta}\frac{d\theta}{dx} = -3k_1\rho^2 A_2\beta\{\overline{\xi^3} + \overline{\xi\eta^2} + \overline{\xi\zeta^2}\} \tag{147}$$

[Addition made December 17, 1866.]

[Final Equilibrium of Temperature]

[The left-hand side of equation (147), as sent to the Royal Society, contained a term $2(\beta-1)p/\rho\, dp/dx$, the result of which was to indicate that a column of air, when left to itself, would assume a temperature varying with the height, and greater above than below. The mistake arose from an error† in equation (143). Equation (147), as now corrected, shews that the flow of heat depends on the variation of temperature only, and not on the direction of the variation of pressure. A vertical column would therefore, when in thermal equilibrium have the same temperature throughout.

When I first attempted this investigation I overlooked the fact that $\overline{\xi^4}$ is not the same as $\overline{\xi^2} \cdot \overline{\xi^2}$, and so obtained as a result that the temperature diminishes as the height increases at a greater rate than it does by expansion when air is carried up in mass. This

---

† The last term on the left-hand side was not multiplied by β.

leads at once to a condition of instability, which is inconsistent with the second law of thermodynamics. I wrote to Professor Sir W. Thomson about this result, and the difficulty I had met with, but presently discovered *one* of my mistakes, and arrived at the conclusion that the temperature would increase with the height. This does not lead to mechanical instability, or to any self-acting currents of air, and I was in some degree satisfied with it. But it is equally inconsistent with the second law of thermodynamics. In fact, if the temperature of any substance, when in thermic equilibrium, is a function of the height, that of any other substance must be the same function of the height. For if not, let equal columns of the two substances be enclosed in cylinders impermeable to heat, and put in thermal communication at the bottom. If, when in thermal equilibrium, the tops of the two columns are at different temperatures, an engine might be worked by taking heat from the hotter and giving it up to the cooler, and the refuse heat would circulate round the system till it was all converted into mechanical energy, which is in contradiction to the second law of thermodynamics.

The result as now given is, that temperature in gases, when in thermal equilibrium, is independent of height, and it follows from what has been said that temperature is independent of height in all other substances.

If we accept this law of temperature as the actual one, and examine our assumptions, we shall find that unless $\overline{\xi^4} = 3\overline{\xi^2} \cdot \overline{\xi^2}$, we should have obtained a different result. Now this equation is derived from the law of distribution of velocities to which we were led by independent considerations. We may therefore regard this law of temperature, if true, as in some measure a confirmation of the law of distribution of velocities.]

Coefficient of Conductivity

If C is the coefficient of conductivity of the gas for heat, then the quantity of heat which passes through unit of area in unit of time measured as mechanical energy, is

$$C\frac{d\theta}{dx} = \tfrac{5}{6}\frac{\beta}{k_1 A_2}\frac{p^2}{\rho^2\theta}\frac{d\theta}{dx} \tag{148}$$

by equation (147).

Substituting for β its value in terms of γ by equation (115), and for k_1 its value in terms of μ by equation (125), and calling p_0, ρ_0, and θ_0 the simultaneous pressure, density, and temperature of the standard gas, and s the specific gravity of the gas in question, we find†

$$C = \frac{5}{3(\gamma-1)}\frac{p_0}{\rho_0\theta_0}\frac{\mu}{s} \tag{149}$$

For air we have $\gamma = 1\cdot 409$, and at the temperature of melting ice, or $274°\cdot 6$ C. above absolute zero, $\sqrt{(p/\rho)} = 918\cdot 6$ feet per second, and at $16°\cdot 6$ C., $\mu = 0\cdot 0936$ in foot-grain-second measure. Hence for air at $16°\cdot 6$ C. the conductivity for heat is

$$C = 1172 \tag{150}$$

That is to say, a horizontal stratum of air one foot thick, of which the upper surface is kept at $17°$ C., and the lower at $16°$ C., would in one second transmit through every square foot of horizontal surface a quantity of heat the mechanical energy of which is equal to that of 2344 grains moving at the rate of one foot per second.

Principal Forbes‡ has deduced from his experiments on the conduction of heat in bars, that a plate of wrought iron one foot thick, with its opposite surfaces kept $1°$ C. different in temperature, would, when the mean temperature is $25°$ C., transmit in one minute through every square foot of surface as much heat as would raise one cubic foot of water $0\cdot 0127°$C.

Now the dynamical equivalent in foot-grain-second measure of

---

† [According to Boltzmann, Selection 2, p. 141, the numerical constant should be 5/2 rather than 5/3. See also H. Poincaré, *Compt. rend. Acad. Sci., Paris*, **116**, 1020 (1893).]

‡ "Experimental Inquiry into the Laws of the Conduction of Heat in Bars", *Edinburgh [Royal Society] Transactions*, 1861–2 [**23**, 133].

the heat required to raise a cubic foot of water 1° C. is 1.9157×10^{10}.

It appears from this that iron at 25° C. conducts heat 3525 times better than air at 16°·6 C.

M. Clausius, from a different form of the theory, and from a different value of μ, found that lead should conduct heat 1400 times better than air. Now iron is twice as good a conductor of heat as lead, so that this estimate is not far different from that of M. Clausius in actual value.

In reducing the value of the conductivity from one kind of measure to another, we must remember that its dimensions are MLT^{-3}, when expressed in absolute dynamical measure.

Since all the quantities which enter into the expression for C are constant except μ, the conductivity is subject to the same laws as the viscosity, that is, it is independent of the pressure, and varies directly as the absolute temperature. The conductivity of iron diminishes as the temperature increases.

Also, since γ is nearly the same for air, oxygen, hydrogen and carbonic oxide, the conductivity of these gases will vary as the ratio of the viscosity to the specific gravity. Oxygen, nitrogen, carbonic oxide, and air will have equal conductivity, while that of hydrogen will be about seven times as great.

The value of γ for carbonic acid is 1·27, its specific gravity is $\frac{11}{8}$ of oxygen, and its viscosity $\frac{8}{11}$ of that of oxygen. The conductivity of carbonic acid for heat is therefore about $\frac{7}{9}$ of that of oxygen or of air.

2

Further Studies on the Thermal Equilibrium of Gas Molecules *

LUDWIG BOLTZMANN

SUMMARY

According to the mechanical theory of heat, the thermal properties of gases and other substances obey perfectly definite laws in spite of the fact that these substances are composed of large numbers of molecules in states of rapid irregular motion. The explanation of these properties must be based on probability theory, and for this purpose it is necessary to know the distribution function which determines the number of molecules in each state at every time. In order to determine this distribution function, $f(x, t) =$ number of molecules having energy x at time t, a partial differential equation for f is derived by considering how it changes during a small time interval as a result of collisions among molecules. If there are no external forces, and conditions are uniform throughout the gas, this equation takes the form (equation (16)):

$$\frac{\partial f(x, t)}{\partial t} = \int_0^\infty \int_0^{x+x'} \left[\frac{f(\xi, t)}{\sqrt{\xi}} \cdot \frac{f(x+x'-\xi, t)}{\sqrt{(x+x'-\xi)}} - \frac{f(x, t)}{\sqrt{x}} \cdot \frac{f(x't)}{\sqrt{x'}} \right] \sqrt{(xx')} \, \psi(x, x', \xi) \, dx' \, d\xi$$

where the variables x and x' denote the energies of two molecules before a collision, and ξ and $(x+x'-\xi)$ denote their energies after the collision; $\psi(x, x', \xi)$ is a function which depends on the nature of the forces between the molecules.

If the velocity distribution is given by Maxwell's formula

$$f(x, t) = (\text{constant}) \sqrt{(x)} \, e^{-hx}$$

---

* [Originally published under the title " Weitere Studien über das Wärmegleichgewicht unter Gasmolekülen ", in *Sitzungsberichte Akad. Wiss.*, Vienna, part II, **66**, 275–370 (1872); reprinted in Boltzmann's *Wissenschaftliche Abhandlungen*, Vol. I, Leipzig, J. A. Barth, 1909, pp. 316–402.]

then the expression in brackets in the above equation will vanish, and the time-derivative of $f(x, t)$ will be zero. This is essentially the result already obtained in another way by Maxwell: once this velocity distribution has been reached, it will not be disturbed by collisions.

With the aid of the partial differential equation for f, we are able to go further and prove that if the distribution of states is not Maxwellian, it will tend toward the Maxwellian distribution as time goes on. This proof consists in showing that a quantity defined in terms of f,

$$E = \int_0^\infty f(x, t) \left[\log \left(\frac{f(x, t)}{x} \right) - 1 \right] dx$$

can never increase but must always decrease or remain constant, if f satisfies the above differential equation. [This statement is now known as the H-theorem.] E must approach a minimum value and remain constant thereafter, and the corresponding final value of f will be the Maxwell distribution. Since E is closely related to the thermodynamic entropy in the final equilibrium state, our result is equivalent to a proof that the entropy must always *increase* or remain constant, and thus provides a microscopic interpretation of the second law of thermodynamics.

To clarify the reasoning involved in these proofs, we replace the continuous energy variable x by a discrete variable which can take only the values ϵ, 2ϵ, 3ϵ, ... ; we then show that the same result can be derived by taking the limit $\epsilon \to 0$.

The differential equation for f is also given for the case when the velocity distribution may vary from one place to another, and all directions of velocity are not equivalent (equation (44)). For the special case of intermolecular forces varying inversely as the fifth power of the distance, this equation has a simple exact solution, and the coefficients of viscosity, heat conduction, and diffusion may be calculated. The results are essentially the same as those found by Maxwell [Selection 1].

The above results are generalized to gases composed of polyatomic molecules.

The mechanical theory of heat assumes that the molecules of a gas are not at rest, but rather are in the liveliest motion. Hence, even though the body does not change its state, its individual molecules are always changing their states of motion, and the various molecules take up many different positions with respect to each other. The fact that we nevertheless observe completely definite laws of behaviour of warm bodies is to be attributed to the circumstance that the most random events, when they occur in the same proportions, give the same average value. For the molecules of the body are indeed so numerous, and their motion is so rapid, that we can perceive nothing more than average values. One might compare the regularity of these average values with the amazing constancy of the average numbers provided by statistics, which are also derived from processes each of which is determined by a completely unpredictable interaction with many other factors. The molecules are likewise just so many individuals having the most varied states of motion, and it is only because the number of them that have, on the average, a particular state of motion is constant, that the properties of the gas remain unchanged. The determination of average values is the task of probability theory. Hence, the problems of the mechanical theory of heat are also problems of probability theory. It would, however, be erroneous to believe that the mechanical theory of heat is therefore afflicted with some uncertainty because the principles of probability theory are used. One must not confuse an incompletely known law, whose validity is therefore in doubt, with a completely known law of the calculus of probabilities; the latter, like the result of any other calculus, is a necessary consequence of definite premises, and is confirmed, insofar as these are correct, by experiment, provided sufficiently many observations have been made, which is always the case in the mechanical theory of heat because of the enormous number of molecules involved. It is only doubly imperative to handle the conclusions with the greatest strictness. If one does not merely wish to guess a few occasional values of the quantities that occur in gas theory, but rather desires to work with an exact theory, then he must first of all determine the probabilities of the various states which a given molecule will have during a very long time, or which different molecules will have at the same time. In other words, one must find the number of molecules out of the total number whose states lie between any given limits. Maxwell and I have previously treated this problem in several papers, without so far managing to obtain a complete solution of it. In fact, the problem seems to be very difficult in the case where each molecule consists of several mass-points (atoms), since one cannot integrate the equations of motion even for a complex of three atoms. Yet on closer consideration, it appears not unlikely that these probabilities can be obtained from the equations of motion alone, without having to integrate them. For the many simple laws of gases show that the expression for such probabilities must have certain general properties independent of the special nature of the gas, and such general properties can frequently be deduced from the equations of motion, without the necessity of integrating them. In fact, I have succeeded in finding the solution for gas molecules consisting of an arbitrary number of atoms. However, for the sake of a better over-all view of the subject, I will first treat the simplest case, where each molecule is a single point mass. I will then treat the general case, for which the calculation is similar.

I. Consideration of Monatomic Gas Molecules

Let a space be filled with many gas molecules, each of which is a simple point mass. Each molecule moves most of the time in a straight line with uniform velocity. Only when two molecules happen to come very close do they begin to interact with each other. I call this process, during which two molecules interact with each other, a collision of the two molecules, without implying a collision of elastic bodies; the force that acts during the collision may be completely arbitrary. Even when all the molecules initially have the same velocity, they will not retain the same velocity during the course of time. As a result of collisions, many molecules will acquire larger velocities and others will come to have smaller velocities, until finally a distribution of velocities among the molecules is established such that it is not changed by further collisions. In this final distribution, in general all possible velocities from zero up to a very large velocity will occur. The number of molecules whose velocity lies between r and $r+dr$ we shall call $F(r)dr$. The function F determines the velocity distribution completely. For the case of monatomic molecules, which we are now considering, Maxwell already found the value $Av^2e^{-Bv^2}$ for $F(r)$, where A and B are constants, so that the probability of different velocities is given by a formula similar to that for the probability of different errors of observation in the theory of the method of least squares. The first proof which Maxwell gave for this formula was recognized to be incorrect even by himself. He later gave a very elegant proof that, if the above distribution has once been established, it will not be changed by collisions. He also tries to prove that it is the only velocity distribution that has this property. But the latter proof appears to me to contain a false inference.† It has still not yet been proved that, whatever the initial state of the gas may be, it must always approach the limit found by Maxwell. It is possible that there may be other possible limits. This proof is easily obtained, however, by the method which I am about to explain, and which also has the advantage that it permits one to deal directly with polyatomic molecules and thus with the case that probably occurs in nature.

I begin by defining the problem precisely again. Suppose therefore that we have a space R in which are found many gas molecules. Each molecule is a simple point mass, which moves in the

---

† First, Maxwell should really have proved that as many pairs of molecules will change their velocities from OA, OB to OA', OB' as conversely, while he actually only discusses whether a molecule changes its velocity from OA to OA' as often as it changes from OA' to OA. He then asserts that if the velocity OA were to change to OA' more often than the converse, then OA' would have to change to OA'' more often than the converse, by the same amount, since otherwise the number of molecules with velocity OA' could not remain constant. In fact, however, one can only conclude that one or more velocities OA'', OA''', etc., exist into which OA' is transformed more often than the converse. In order to prove finally that it is not possible for the velocity of a molecule to change from OA to OA' more often than the converse, Maxwell says, that otherwise there would be a recurrent cycle of velocities OA, OA', OA'' ... OA which would be more often traversed in one direction than the opposite direction. But this cannot be, since there is no reason why a molecule should prefer to go around the cycle in one direction rather than the other. This latter assertion seems to me, however, to be one that should be proved rather than taken as already established. For if we take it as already proved that the velocity of a molecule changes from OA to OA' as often as the converse, then of course there is no reason why this cycle is more likely to be traversed in one direction than the other. If we assume, on the contrary, that the theorem to be proved is not yet known to be true, then the fact that the velocity of a molecule is more likely to change from OA to OA' than the converse, and more likely to change from OA' to OA'' than the converse, and so forth, would provide a reason why the cycle is more likely to be traversed in one direction than the other. The two processes are no less than identical. One cannot conclude therefore that they are *a priori* equally probable.

way already described. During the largest part of the time, it moves in a straight line with uniform velocity. Two molecules interact only when they come very close together. The law of the force that acts duting collisions must of course be given. However, I will not make any restrictive assumptions about this force law. Perhaps the two molecules rebound from each other like elastic spheres; perhaps any other force law may be given. As for the wall of the container that encloses the gas, I will assume that it reflects the molecules like elastic spheres. Any arbitrary force law would lead to the same formulae. However, it simplifies the matter if we make this special assumption about the container. We now set ourselves the following problem: suppose that initially ($t = 0$) the position, velocity, and velocity direction of each molecule is given. What is the position, velocity, and velocity direction of each molecule after an arbitrary time t has elapsed? Since the form of the container R as well as the force law for the collisions is given, this problem is of course completely determined. It is clear, however, that it is not completely soluble in this degree of generality. The solution would be much easier to find if, instead of this general problem, we set ourselves a rather more special one. We take account of only two conditions that pertain to the nature of the subject. First, it is clear that after a very long time each direction for the velocity of a molecule is equally likely. If it is only a question of finding the velocity distribution that will be established after a long time, then we can assume that already at the beginning each velocity direction is equally probable. In the most general case one still arrives at the same final state-distribution as in this special case. This is the first condition that we shall make. The second one is that the velocity distribution should already be uniform initially. I must next clarify what I mean by a uniform velocity distribution. For the following it will be better to use the kinetic energy of a molecule rather than its velocity. We shall now do this. Let x be the kinetic energy of our gas molecule, so that $x = mv^2/2$. R is the total space in which our gas is enclosed. We construct inside this space R a smaller space called r, whose shape is completely arbitrary but whose volume will be equal to one. We assume that in the space r there are a large number of molecules, so that its dimensions are large compared to the average distance of two neighbouring molecules; this imposes no real limitation, since we may choose the unit of volume to be as large as we like. The number of molecules in the space r whose kinetic energy at time t lies between x and $x+dx$ I will call $f(x, t)dx$. This number will in general depend on where I have chosen to construct r in R. For example, one might find the fast molecules on the right side and the slower ones on the left side of R. Then the number $f(x, t)dx$ would be different, depending on whether the space r was on the right or the left of R. When this is not the case, but rather $f(x, t)dx$ is the same at a given time no matter where r is, then I say that the distribution of kinetic energy is uniform. In other words, molecules with different kinetic energies are uniformly mixed with each other. The faster ones are not on the right, nor the slower ones on the left, nor conversely. It is again clear that after a very long time the distribution of kinetic energy will become uniform, for then each position in the gas is equivalent. The presence of the walls does not disturb this uniformity, for the molecules are reflected at them like elastic spheres; they come back from the wall as if the space on the other side of the wall were filled with another gas having the same properties. Hence, we may assume that the velocity distribution is already uniform initially. This assumption, and the equal probability of all directions of velocity initially, are the two restrictive conditions under which we shall treat the problem. It is clear that these two conditions will be satisfied for all following times, so that the state of the gas at time t will be completely determined by the function $f(x, t)$. If the state of the gas is given to us at an initial time, i.e. $f(x, 0)$, then we must find the state after an arbitrary time t has elapsed, namely $f(x, t)$. The way that we shall proceed is the same way that one always proceeds in such cases. We first calculate how much the function $f(x, t)$ changes during a very small time τ; from this we obtain a partial differential equation for the function $f(x, t)$; this must then be integrated in such a way that for $t = 0$, f takes the given value $f(x, 0)$. We have therefore a double task before us, first the establishment of the partial differential equation, and second its integration. We now turn to the first problem. $f(x, t)dx$ is the number of molecules in unit volume whose kinetic energy at time t lies between x and $x+dx$. As long as a molecule collides with no other molecules, it retains the same kinetic energy. If there were no collisions, $f(x, t)$ would not change; this function changes only because of collisions. If we wish to find the change in this function during a very short time τ, then we must consider the collisions during this time. We consider a collision, before which the kinetic energy of one of the colliding molecules is

$$\text{between } x \text{ and } x+dx$$

and that of the other is

$$\text{between } x' \text{ and } x'+dx'$$

Of course the nature of the collision is by no means completely determined yet. Depending on whether the collision is head-on or more or less glancing, the kinetic energy of one of the colliding molecules can have many different values after the collision. If we assume that this kinetic energy after the collision lies

$$\text{between } \xi \text{ and } \xi+d\xi$$

then the kinetic energy of the second molecule after the collision is determined. If we denote the latter by ξ', then according to the principle of conservation of kinetic energy,

$$x+x' = \xi+\xi' \tag{1}$$

so that the sum of the kinetic energies of the two molecules is the same before and after the collision. We can now represent the limits between which the variables characterizing our collision lie by the following scheme:

	a	b	
Before the collision	$x, x+dx$	$x', x'+dx'$	(A)
After the collision	$\xi, \xi+d\xi$		

The kinetic energy of one molecule is in the column labelled a, and that of the other in the column labelled b. We now ask, how many collisions take place in time τ in unit volume such that the kinetic energies of the colliding molecules lie between the limits (A)? We shall denote this number by dn. The determination of dn can only be accomplished properly by considering the relative velocity of the two molecules. Since this calculation, although tedious, is not at all difficult, and has no special interest, and the result is so simple that one might almost say it is obvious, I will simply state the result. It is the following: dn is, first, proportional to the time τ; the longer this time τ is, the more collisions of the specified kind will occur, as long as τ is very small, so that the state of the gas during τ does not change noticeably. Second, dn is proportional to the quantity $f(x, t)dx$; this is indeed the number of molecules in unit volume whose kinetic energy lies between x and $x+dx$; the more such molecules there are in unit volume, the more often will they collide. Third, dn is proportional to $f(x', t)dx'$; for whatever holds for one of the colliding molecules will of course hold for the other one. The product of these three quantities must be multiplied by a certain proportionality factor, which is easily seen to be an infinitesimal like $d\xi$. This factor will in general depend on the nature of the collision, and hence on x, x' and ξ. To express these properties we shall write the proportionality factor as $d\xi \cdot \psi(x, x', \xi)$ so that we have therefore

$$dn = f(x, t)dx \cdot f(x', t)dx' \, d\xi \cdot \psi(x, x', \xi) \qquad (2)$$

This is the result obtained from an exact treatment of the collision process; this treatment also gives the function ψ, of course, as soon as the force law is given, since ψ depends on the force law. Since we do not need to know this function ψ yet, it would be superfluous to determine it here. We shall now keep x constant in the expression (2) for dn, and integrate x' and ξ over all possible values of those quantities, i.e. ξ from zero to $x+x'$ and x' from zero to infinity. The result of this integration will be called $\int dn$; thus

$$\int dn = \tau f(x, t)dx \int_0^\infty \int_0^{x+x'} f(x', t)\psi(x, x', \xi)dx'd\xi$$

Since x is to be considered constant for both integrations, we can write $f(x, t)$ inside the integral signs and obtain:†

$$\int dn = \tau dx \int_0^\infty \int_0^{x+x'} f(x, t)f(x', t)\psi(x, x', \xi)dx'd\xi \qquad (3)$$

What is this quantity $\int dn$? We have kept x constant. The kinetic energy of one molecule before the collision is still between

---

† Instead of actually writing out the limits of a definite integral, one can determine them in various ways, for example through inequalities. In the definite integral of equation (3) x is to be considered constant. The two integration variables are x' and ξ; these can take only positive values, including zero, for they represent kinetic energy; and indeed we must require $x+x'-\xi \geqq 0$, since $x+x'-\xi$ is the kinetic energy of the second molecule after the collision. On the other hand, it is clear that all positive x' and ξ for which $x+x'-\xi$ also comes out to be positive represent possible collisions, and therefore lie within the limits of integration. The three inequalities

$$x \geqq 0, \ \xi \geqq 0, \ x+x'-\xi \geqq 0 \qquad (3a)$$

define therefore the limits of integration of the integral in equation (3). This method of determining the limits is recommended by the fact that it often significantly shortens the calculation. One transforms the variables of integration to rectangular coordinate axes and determines the surface over

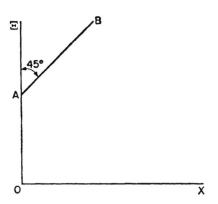

Fig. 1.

which one has to integrate. If we indicate the variable x' on the abscissa axis OX', and the variable ξ on the ordinate axis $O\Xi$, then we obtain the surface over which we have to integrate by making $OA = x$ and extending the line AB to infinity at an angle of 45° with the coordinate axes. The infinite trapezoid $XOAB$ is then the surface over which the integration is to be extended. This latter method of representing the limits has the advantage of perspicuousness.

the limits x and $x+dx$. We have integrated over all other variables. All the other variables are therefore subject to no restrictive conditions. Hence, $\int dn$ is just the number of collisions in unit volume during time τ before which the kinetic energy of one molecule lies between x and $x+dx$. By each such collision a molecule loses this kinetic energy, so that the collision diminishes the number of molecules whose kinetic energy is between x and $x+dx$ by one.† During time τ there will be in all $\int dn$ of such collisions in unit volume. Hence, this number will be decreased by $\int dn$. The number of molecules in unit volume whose kinetic energy at time t lies between x and $x+dx$ is, however, as we know, $f(x, t)dx$; during time τ, this number will decrease by $\int dn$ as a result of the collisions just considered, so that we must subtract $\int dn$ from $f(x, t)dx$.

Up to now we have only considered the collisions by which a molecule loses a kinetic energy between x and $x+dx$, so that $f(x, t)dx$ decreases. We still have to consider those collisions by which a molecule gains this kinetic energy, so that $f(x, t)dx$ thereby increases. If we denote the number of these latter collisions by $\int dv$, then $\int dv$ must be added to $f(x, t)dx$; in the sum

$$f(x, t)dx - \int dn + \int dv \qquad (4)$$

the first term is the number of molecules in unit volume whose kinetic energy is between x and $x+dx$ at time t; from this is subtracted the number of molecules that lose this kinetic energy during time τ, and to it is added the number of molecules that gain this kinetic energy during time τ. The result is clearly the number of molecules that have this kinetic energy at time $t+\tau$, i.e. $f(x, t+\tau)dx$. We thus obtain

$$f(x, t+\tau)dx = f(x, t)dx - \int dn + \int dv \qquad (5)$$

† We have excluded here those collisions for which, after the collision, the kinetic energy of one or even both molecules lies between x and $x+dx$. One sees easily that the number of such collisions—as well as the number of those before which the kinetic energies of both molecules lie between x and $x+dx$, so that two molecules simultaneously lose this kinetic energy—is a higher order infinitesimal and can be neglected. The former collisions, which we are now subtracting incorrectly, are moreover contained in $\int dv$ and hence will be added in again anyway.

We still have to determine $\int dv$. $\int dv$ is the number of collisions in unit volume during time τ, after which the kinetic energy of a molecule lies between x and $x+dx$. We must therefore choose some other notation for the kinetic energies before the collision. Therefore let dv be the number of collisions which occur in unit volume during time τ such that before the collision the kinetic energy of one molecule lies between u and $u+du$, that of the other lies between v and $v+dv$, and after the collision the kinetic energy of one molecule lies between x and $x+dx$. The kinetic energy of the other molecule after the collision is of course determined. dv is therefore the number of collisions which, corresponding to the scheme previously denoted by (A), are characterized by the following scheme:

	a	b	
Before the collision	$u, u+du$	$v, v+dv$	(B)
After the collision	$x, x+dx$		

One sees at once that the collisions characterized by (B) differ from those characterized by (A) only by having different notation for the kinetic energies before and after the collision. The number dv of the latter kind can therefore be found directly from the expression for dn by simply permuting the labels of the variables. Thus we have to write:

u instead of x, v instead of x', x instead of ξ,

and likewise (C)

du instead of dx, dv instead of dx', dx instead of $d\xi$

The number of collisions of the type considered previously was called dn and was given by equation (2). If we carry out the permutation (C), we obtain dv. Hence,

$$dv = \tau . f(u, t)duf(v, t)dvdx\psi(u, v, x)$$

Here we again keep x constant and integrate u and v over all possible values of these quantities. The result,

$$\tau dx \iint f(u, t) f(v, t) \psi(u, v, x) du dv$$

is the number of collisions in unit volume during time τ after which the kinetic energy of one molecule lies between x and $x+dx$. Thus it is the number of collisions by which one molecule gains a kinetic energy between x and $x+dx$, in other words, it is just the number which we earlier called $\int dv$.† Thus, we have

$$\int dv = \tau . dx \iint f(u, t) f(v, t) \psi(u, v, x) du dv \qquad (6)$$

The question of the limits of the double integrals now arises.‡ When $u>x$, v can run through all possible values from zero to

---

† One might think that we have forgotten the collisions after which the kinetic energy of the second of the colliding molecules lies between x and $x+dx$. For such a collision, let $u = u_1$, $v = v_1$. Since we integrate u and v over all possible values, we have also included the collision for which $u = u_1$ and $v_1 = v_1$ and the kinetic energy of the first molecule lies between x and $x+dx$ in the integral; but this is just the case that we were afraid that we had forgotten. For it makes no difference which molecule we call the first and which the second. All these collisions are therefore taken account of in our integral, if we simply write u in place of v and conversely. If one wanted to add a second integral containing collisions after which the kinetic energy of the second molecule lies between x and $x+dx$, then he would have to take the values of u and v without permutation—i.e. integrate v from zero (or $x-u$, resp.) to u, then u from zero to infinity. Only the case where the kinetic energies of both molecules lie between x and $x+dx$ after the collision has not been counted twice, as it should have been; but this is not an error, since the number of such cases is an infinitesimal of higher order.

‡ If we determine the limits according to the method indicated in the footnote on page 97, we obtain the inequalities

$$u \geqq 0, \ v \geqq 0, \ u+v-x \geqq 0$$

If we introduce now some arbitrary new variables p, q, then as is well known

$$dp \, dq = \sum \pm \frac{dp}{dx} \frac{dq}{dv} du \, dv$$

In the special case that we set $p = u+v-x$, $q = u$, the functional determinant is equal to one (it is of course to be taken positive); furthermore, in this case,

$$v = p+x-q$$

Hence, equation (6) becomes

$$\int dv = \tau \, dx \iint f(q, t) f(p+x-q, t) \psi(q, p+x-q, x) \, dp \, dq$$

The inequalities that determine the limits become

$$q \geqq 0, \ p+x-q \geqq 0, \ p \geqq 0$$

(*Footnote continued on next page*)

infinity; but if $u>x$, then v cannot be smaller than $x-u$, since otherwise $u+v-x$, which is the kinetic energy of the two molecules after the collision, would be negative. Therefore if $u<x$, v runs through all values from $x-u$ to infinity. The u integral must therefore be broken into two parts, the first from zero to x, the second from x to infinity. In the first integral one has to integrate over v from $x-u$ to infinity, and in the second from zero to infinity. The formula (6) thus becomes, when one determines the limits correctly,

$$\int dv = \tau dx \int_0^x \int_{x-u}^\infty f(u,t)f(v,t)\psi(u,v,x)dudv$$

$$+ \tau dx \int_x^\infty \int_0^\infty f(u,t)f(v,t)\psi(u,v,x)dudv \qquad (7)$$

We now introduce in place of v the new variable

$$w = u+v-x \qquad (8)$$

so that we have $v = x+w-u$. Since in the v integration, u as well as x is to be considered constant, it follows from the formula (8) that $dw = dv$. Hence, using the correct determination of limits for the w integration,

$$\int dv = \tau dx \int_0^x \int_0^\infty f(u,t)f(x+w-u,t)\psi(u,x+w-u,x)dudw$$

$$+ \tau dx \int_x^\infty \int_{u-x}^\infty f(u,t)f(x+w-u,t)\psi(u,x+w-u,x)dudv \qquad (9)$$

We can now label the variables in the integral just as we please if we also use the same labels in the inequalities. If we change the letters p, q to x', ξ, then we obtain

$$\int dv = \tau dx \iint f(\xi,t)f(x+x'-\xi,t)\psi(\xi, x+x'-\xi, x)dx' d\xi$$

The limits are determined by

$$x' \geqq 0, \ \xi \geqq 0, \ x+x'-\xi \geqq 0$$

thus we have come back to the inequalities (3a) which determine the limits in equation (3). If we attach the appropriate limits to the integral signs the last formula agrees with equation (11) in the text, which we have thus obtained here by a quicker method.

Since these integrals represent just a summation of a number of collisions, we can invert the order of integration without difficulty. The first double integral in formula (9) then becomes:

$$\int_0^\infty \int_0^x f(u, t) f(x+w-u, t) \psi(u, x+w-u, x) \, dw \, du \quad (10)$$

For the second integral, the determination of the new limits of integration is not quite so simple. We shall do it by geometric considerations (Fig. 2). We indicate the value of u on the abscissa axis OU, and the value of w on the ordinate axis OW. x is constant during this integration. We make $OA = x$ and draw through A

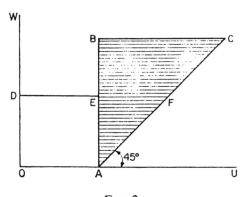

Fig. 2.

the two infinite lines AB parallel to OW, and AC inclined at a 45° angle to the coordinate axes. In the second double integral of (9) we have to integrate over u from x to infinity, i.e. from the point A to infinity; we have to integrate over w from $u-x$ to infinity, i.e. from the line AC to infinity. The total integration is thus to be extended over the unbounded triangle which is shaded in the figure. It is now easy to determine the limits if we want to integrate first over u and then over w. For a given w, say for example $w = OD$, we want to integrate u from DE to DF, hence from x to $x+w$. Thus the w integration goes from zero to infinity. The second double integral in (9) therefore becomes

$$\int_0^\infty \int_x^{x+w} f(u, t) f(x+w-u, t) \psi(u, x+w-u, x) \, dw \, du$$

This can now be combined with the first term, given in (10), into a single double integral. (The first term, incidentally, represents the integration over the infinite rectangle $WOAB$ in our figure.) The reunification of the two double integrals gives:

$$\int dv = \tau dx \int_0^\infty \int_0^{x+w} f(u, t)f(x+w-u, t)\psi(u, x+w-u, x)dwdu$$

In order to make our expression uniform with that given by formula (3) for $\int dn$, I will replace w by x' and u by ξ. As is well known, in a definite integral the variables that are integrated over can be denoted however one likes, as long as the limits remain the same. Thereby we obtain

$$\int dv = \tau dx \int_0^\infty \int_0^{x+x'} f(\xi, t)f(x+x'-\xi, t)\psi(\xi, x+x'-\xi, x)dx'd\xi \quad (11)$$

Before we substitute the two values found for $\int dn$ and $\int dv$ into equation (5), we shall transform this equation again. We expand its left-hand side according to Taylor's theorem:

$$f(x, t)dx + \frac{\partial f(x, t)}{\partial t}\tau dx + A\tau^2 dx = f(x, t)dx - \int dn + \int dv$$

where A is some finite quantity; hence

$$\frac{\partial f(x, t)}{\partial t} = \frac{\int dv}{\tau dx} - \frac{\int dn}{\tau dx} - A\tau$$

so that on substituting the values (3) and (11) for $\int dn$ and $\int dv$ we obtain

$$\frac{\partial f(x, t)}{\partial t} = \int_0^\infty \int_0^{x+x'} f(\xi, t)f(x+x'-\xi, t)\psi(\xi, x+x'-\xi, x)dx'd\xi$$

$$- \int_0^\infty \int_0^{x+x'} f(x, t)f(x', t)\psi(x, x', \xi)dx'd\xi - A\tau$$

Since everything except $A\tau$ is finite, the latter can be neglected. Furthermore, the two integrals can be combined into one, since the variables of integration and the limits are the same in both. Thus we obtain:

$$\frac{\partial f(x, t)}{\partial t} = \int_0^\infty \int_0^{x+x'} [f(\xi, t)f(x+x'-\xi, t)\psi(\xi, x+x'-\xi, x)$$
$$-f(x, t)f(x', t)\psi(x, x', \xi)]dx'd\xi \quad (12)$$

This is the desired partial differential equation, which determines the law of variation of the function f. It needs one more transformation, for which we use the properties of the function ψ expressed by the following two equations, valid for any x, x' and ξ:

$$\psi(x, x', \xi) = \psi(x', x, x+x'-\xi) \quad (13)$$

$$\sqrt{xx'}\psi(x, x', \xi) = \sqrt{\xi(x+x'-\xi)}\psi(\xi, x+x'-\xi, x) \quad (14)$$

where of course all the roots are to be taken with the positive sign; the ψ's are also essentially positive quantities. The first of these two equations can easily be proved. Let dn' be the number of collisions that take place in unit volume during the very small time interval (previously denoted by τ) in such a way that before the collision the kinetic energy of the first molecules lies between x' and $x'+dx'$, and that of the second between x and $x+dx$; and after the collision, the kinetic energy of the first molecule lies between $x+x'-\xi-d\xi$ and $x+x'-\xi$. The collision is characterized by the scheme:

	a	b	
Before the collision	x', $x'+dx'$	x, $x+dx$	(D)
After the collision	$x+x'-\xi-d\xi$, $x+x'-\xi$		

Then dn' can be found by a simple permutation of labels from the quantity earlier denoted by dn. Indeed, on comparing the schemes (D) and (A), we see that we must write:

x' instead of x, x instead of x', $x+x'-\xi-d\xi$ instead of ξ,

dx' instead of dx, dx instead of dx'.

If one makes these exchanges in equation (2), he obtains:

$$dn' = \tau f(x', t)dx'.f(x, t)dx.d\xi.\psi(x', x, x+x'-\xi-d\xi) \quad (15)$$

However, if the kinetic energy of one molecule after the collision lies between $x+x'-\xi-d\xi$ and $x+x'-\xi$, then that of the other lies between ξ and $\xi+d\xi$. Instead of the scheme (D) we can therefore characterize our collision by the following:

	a	b
Before the collision	$x', x'+dx'$	$x, x+dx$
After the collision		$\xi, \xi+d\xi$.

And now one sees that this is exactly the same collision as the one characterized by the scheme (A). For clearly it makes no difference which molecule I call the first (*a*) and which the second (*b*). Since the two kinds of collisions are the same, their numbers must be equal, hence $dn = dn'$. If we equate the two values (2) and (15) and cancel the common factors, we obtain

$$\psi(x, x', \xi) = \psi(x', x, x+x'-\xi-d\xi)$$

Here the differential $d\xi$ can be neglected compared to finite quantities, since ψ cannot be discontinuous, and thus we obtain equation (13).

The proof of equation (14) is more difficult. The proof was first given, though in a somewhat different form, by Maxwell; it was then considerably generalized by me, and was shown to be a special case of Jacobi's principle of the last multiplier; hence, I do not have to give the proof here, but can assume it to be known. I remark only that in the proof it is assumed that the force between two material points is a function of their distance, which acts in the direction of their line of centres, and that action and reaction are equal. These assumptions are also necessary for the validity of the following calculations. Taking account of (14), we can take a common factor ψ out of the brackets in equation (12) and obtain:

$$\frac{\partial f(x, t)}{\partial t} = \int_0^\infty \int_0^{x+x'} \left[\frac{f(\xi, t)f(x+x'-\xi, t)}{\sqrt{\xi}\sqrt{x+x'-\xi}} - \frac{f(x, t)f(x', t)}{\sqrt{x}\sqrt{x'}} \right]$$

$$\times \sqrt{xx'}\psi(x, x', \xi)dx'd\xi \quad (16)$$

This is the fundamental equation for the variation of the function $f(x, t)$. I remark again that the roots are to be taken positive, and that both ψ and f are essentially positive quantities. If for a moment we set
$$f(x, t) = C\sqrt{x}e^{-hx} \tag{16a}$$
where C and h are constants, so that therefore
$$f(x', t) = C\sqrt{x'}e^{-hx'}, \quad f(\xi, t) = C\sqrt{\xi}e^{-h\xi},$$
$$f(x+x'-\xi, t) = C\sqrt{x+x'-\xi}\,e^{-h(x+x'-\xi)}$$
then the expression in brackets in equation (16) vanishes; hence $\partial f(x, t)/\partial t = 0$. This is nothing more than Maxwell's proof transcribed in our present notation. If the distribution of states at any time is determined by the formula (16a), that $\partial f(x, t)/\partial t = 0$, i.e. this distribution does not change further in the course of time. This, and nothing else, is what Maxwell proved. However, we shall now consider the problem more generally. We shall assume that the distribution of kinetic energy is initially a completely arbitrary one, and we shall ask ourselves how it changes in the course of time. Its variation is determined by the partial differential equation (16). This partial differential equation can be transformed, as we shall see later on, into a system of ordinary differential equations, if one replaces the double integrals by a sum of many terms. Indeed, as is well known, such a double integral is nothing more than an abbreviated notation for a sum of infinitely many terms. The methods of calculation are then more intuitively clear for the system of ordinary differential equations. However, I will not at first undertake this exchange of summation and integration, in order that it will not appear to be necessary for the proof of our theorem. This proof can be carried out without abandoning the symbolism of integral calculus. It is only for the sake of better understanding that we use the summation formulae at the end.

We shall first give the proof of a theorem which forms the basis of our present investigation: the theorem that the quantity

$$E = \int_0^\infty f(x, t) \left\{ \log\left[\frac{f(x, t)}{\sqrt{x}}\right] - 1 \right\} dx \tag{17}$$

can never increase, when the function $f(x, t)$ that occurs in the definite integral satisfies the differential equation (16). On the right-hand side of equation (17) one has to integrate x from zero to infinity. Therefore E does not depend on x, but only on t. Since t does not appear in the limits of the integral, we can obtain the time derivative dE/dt by finding the partial derivative with respect to t of the quantity under the integral sign, keeping x constant. This differentiation, which can very easily be carried out, yields

$$\frac{dE}{dt} = \int_0^\infty \log\left[\frac{f(x, t)}{x}\right] \frac{\partial f(x, t)}{\partial t} dx$$

We assume that $f(x, t)$ satisfies equation (16). If we substitute from this equation the value for $\partial f(x, t)/\partial t$, then we obtain

$$\frac{dE}{dt} = \int_0^\infty \log\left[\frac{f(x, t)}{\sqrt{x}}\right] dx \int_0^\infty \int_0^{x+x'} \left[\frac{f(\xi, t)f(x+x'-\xi, t)}{\sqrt{\xi}\sqrt{x+x'-\xi}}\right.$$

$$\left. - \frac{f(x, t)}{\sqrt{x}}\frac{f(x', t)}{\sqrt{x'}}\right] \times \sqrt{xx'}\psi(x, x', \xi)dx'd\xi$$

Since in the integration over x' and ξ the quantity x is to be kept fixed, we can also put the logarithm under the two following integration signs and write

$$\frac{dE}{dt} = \int_0^\infty \int_0^\infty \int_0^{x+x'} \log\frac{f(x, t)}{\sqrt{x}}\left[\frac{f(\xi, t)f(x+x'-\xi, t)}{\sqrt{\xi}\sqrt{x+x'-\xi}} - \frac{f(x, t)f(x', t)}{\sqrt{x}\sqrt{x'}}\right]$$

$$\times \sqrt{xx'}\psi(x, x', \xi)dxdx'd\xi \qquad (18)$$

The real meaning of the transformation which we are now going to perform with this expression will of course become completely clear only when we replace the integrals by sums. It will then appear that all the following transformations of integrals are nothing more than changes in the order of summation; it will also be clear just why these changes in the order of summation are necessary. However, I will not discuss this any further at this point, but simply proceed as quickly as possible to the proof of the desired theorem, that E cannot increase. In equation (18) we can integrate first over x' and then over x;† we thereby obtain:

$$\frac{dE}{dt} = \int_0^\infty \int_0^\infty \int_0^{x+x'} \log\frac{f(x,t)}{\sqrt{x}} \left[\frac{f(\xi,t)}{\sqrt{\xi}} \frac{(x+x'-\xi,t)}{\sqrt{x+x'-\xi}} - \frac{f(x,t)}{\sqrt{x}} \frac{f(x',t)}{\sqrt{x'}}\right]$$
$$\times \sqrt{xx'}\psi(x,x',\xi)dx'dxd\xi \quad (19)$$

or when we substitute for $\psi(x, x', \xi)$ its value from equation (13),

$$\frac{dE}{dt} = \int_0^\infty \int_0^\infty \int_0^{x+x'} \log\frac{f(x,t)}{\sqrt{x}} \left[\frac{f(\xi,t)f(x+x'-\xi,t)}{\sqrt{\xi}\sqrt{x+x'-\xi}} - \frac{f(x,t)}{\sqrt{x}} \frac{f(x',t)}{\sqrt{x'}}\right]$$
$$\times \sqrt{xx'}\psi(x',x,x+x'-\xi)dx'dxd\xi$$

We now leave the variables x' and x unchanged, but instead of ξ we introduce the new variable $\xi' = x+x'-\xi$, so that $\xi = x+x'-\xi'$, $d\xi = -d\xi'$. Then we have

$$\frac{dE}{dt} = -\int_0^\infty \int_0^\infty \int_{x+x'}^0 \log\frac{f(x,t)}{\sqrt{x}} \left[\frac{(x+x'-\xi',t)f(\xi',t)}{\sqrt{x+x'-\xi'}\sqrt{\xi'}}\right.$$
$$\left. -\frac{f(x,t)}{\sqrt{x}} \frac{f(x',t)}{\sqrt{x'}}\right] \times \sqrt{xx'}\psi(x',x,\xi')dx'dxd\xi'$$

---

† That the change in order of integration is unconditionally allowed follows from the fact that we could have derived equations (20), (22) and (23) directly in just the same way that we have obtained (18); we have adopted the method of transformations merely in order not to have to repeat four times the arguments by which we obtained equation (18). Also, the fact that the earlier differentiation under the integral sign is not permitted, because of the discontinuity of the integrand, does not destroy the proof in the text, as one can show by excluding from the entire space over which the integration extends in equations (18), (20), (22) and (23) a very thin strip on the surface at those places where one of the quantities s, s', σ, or σ' is zero or infinite. One can prove by means of Taylor's theorem that the sum of the terms thus excluded from $4\ dE/dt$ cannot be positive, if none of these quantities is infinite or has infinitely many discontinuities.

or, when we reverse the sign and limits of the first integral,

$$\frac{dE}{dt} = \int_0^\infty \int_0^\infty \int_0^{x+x'} \log\frac{f(x,t)}{\sqrt{x}} \left[\frac{f(\xi',t)f(x+x'-\xi',t)}{\sqrt{\xi'}\sqrt{x+x'-\xi'}} - \frac{f(x,t)f(x',t)}{\sqrt{x}\sqrt{x'}} \right]$$
$$\times \sqrt{xx'}\psi(x',x,\xi')dx'dxd\xi$$

This triple integral is now constructed in the same way as that in equation (18); but the variables over which one has to integrate are labelled differently. But this is only an apparent difference. The variables of integration of a definite integral may be labelled in any way one likes, as long as the limits remain the same. Hence, we can write ξ in place of ξ' in (19), and also permute the letters x and x'. Thereby we obtain

$$\frac{dE}{dt} = \int_0^\infty \int_0^\infty \int_0^{x+x'} \log\frac{f(x',t)}{\sqrt{x'}} \left[\frac{f(\xi,t)f(x+x'-\xi,t)}{\sqrt{\xi}\sqrt{x+x'-\xi}} - \frac{f(x,t)f(x',t)}{\sqrt{x}\sqrt{x'}} \right]$$
$$\times \sqrt{xx'}\psi(x,x',\xi)dxdx'd\xi \quad (20)$$

There can be no doubt as to the identity of the two integrals (19) and (20), since they differ only by the labels by which the variables of integration are denoted.

We obtain a third expression for dE/dt in the following way. We substitute in (18) in place of $\sqrt{xx'}\,\psi(x,x',\xi)$ its value from equation (14). Thereby we obtain:

$$\frac{dE}{dt} = \int_0^\infty \int_0^\infty \int_0^{x+x'} \log\frac{f(x,t)}{\sqrt{x}} \left[\frac{f(\xi,t)f(x+x'-\xi,t)}{\sqrt{\xi}\sqrt{x+x'-\xi}} - \frac{f(x,t)f(x',t)}{\sqrt{x}\sqrt{x'}} \right]$$
$$\times \sqrt{\xi(x+x'-\xi)}\psi(\xi,x+x'-\xi,x)dxdx'd\xi$$

We will now introduce a new variable for x'. We want to do the integration over x' first, before the integration over ξ. We need to transform the double integral

$$\int_0^\infty \int_0^{x+x'} \log\frac{f(x,t)}{\sqrt{x}} \left[\frac{f(\xi,t)f(x+x'-\xi,t)}{\sqrt{\xi}\sqrt{x+x'-\xi}} - \frac{f(x,t)f(x',t)}{\sqrt{x}\sqrt{x'}} \right]$$
$$\times \sqrt{\xi(x+x'-\xi)}\psi(\xi,x+x'-\xi,x)dx'd\xi$$

In order to obtain dE/dt we then simply multiply this by dx and integrate over x from zero to infinity. We have already previously inverted the order of integration in such a double integral. By the same argument as before we can transform it to a sum of two integrals:

$$\int_0^x \int_0^\infty \log\frac{f(x,t)}{\sqrt{x}} \left[\frac{f(\xi,t)f(x+x'-\xi,t)}{\sqrt{\xi}\sqrt{x+x'-\xi}} - \frac{f(x,t)f(x',t)}{\sqrt{x}\sqrt{x'}}\right]$$

$$\times \sqrt{\xi(x+x'-\xi)}\psi(\xi, x+x'-\xi, x)d\xi dx'$$

$$+ \int_x^\infty \int_{\xi-x}^\infty \log\frac{f(x,t)}{\sqrt{x}} \left[\frac{f(\xi,t)f(x+x'-\xi,t)}{\sqrt{\xi}\sqrt{x+x'-\xi}} - \frac{f(x,t)f(x',t)}{\sqrt{x}\sqrt{x'}}\right]$$

$$\times \sqrt{\xi(x+x'-\xi)}\psi(\xi, x+x'-\xi, x)d\xi dx'$$

If we introduce in these two integrals the variable

$$\xi' = x+x'-\xi$$

in place of x', we obtain, with a correct determination of the limits:

$$\int_0^x \int_{x-\xi}^\infty \log\frac{f(x,t)}{\sqrt{x}} \left[\frac{f(\xi,t)f(\xi',t)}{\sqrt{\xi}\sqrt{\xi'}} - \frac{f(x,t)f(\xi+\xi'-x,t)}{\sqrt{x}\sqrt{\xi+\xi'-x}}\right]$$

$$\times \sqrt{\xi\xi'}\psi(\xi, \xi', x)d\xi d\xi'$$

$$+ \int_x^\infty \int_0^\infty \log\frac{f(x,t)}{\sqrt{x}} \left[\frac{f(\xi,t)f(\xi',t)}{\sqrt{\xi}\sqrt{\xi'}} - \frac{f(x,t)f(\xi+\xi'-x,t)}{\sqrt{x}\sqrt{\xi+\xi'-x}}\right]$$

$$\times \sqrt{\xi\xi'}\psi(\xi, \xi', x)d\xi d\xi'$$

These two definite integrals are to be integrated over x from zero to infinity, so that we obtain:

$$\frac{dE}{dt} = \int_0^\infty \int_0^x \int_{x-\xi}^\infty \log\frac{f(x,t)}{\sqrt{x}} \left[\frac{f(\xi,t)f(\xi',t)}{\sqrt{\xi}\sqrt{\xi'}} - \frac{f(x,t)}{\sqrt{x}}\frac{f(\xi+\xi'-x,t)}{\sqrt{\xi+\xi'-x}}\right]$$

$$\times \sqrt{\xi\xi'}\psi(\xi,\xi',x)dx\,d\xi\,d\xi'$$

$$+ \int_0^\infty \int_x^\infty \int_0^\infty \log\frac{f(x,t)}{\sqrt{x}} \left[\frac{f(\xi,t)f(\xi',t)}{\sqrt{\xi}\sqrt{\xi'}} - \frac{f(x,t)}{\sqrt{x}}\frac{f(\xi+\xi'-x,t)}{\sqrt{\xi+\xi'-x}}\right]$$

$$\times \sqrt{\xi\xi'}\psi(\xi,\xi',x)dx\,d\xi\,d\xi' \tag{21}$$

We must now change the order of integration so that we integrate first over x, then over ξ', and last over ξ.† For the purpose of determining limits of integration it is best to visualize the integration space geometrically (Fig. 3). Since the integral is

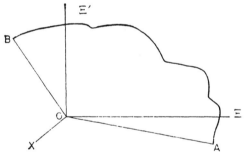

Fig. 3.

a triple one, we must use three-dimensional space. We construct three rectangular coordinate axes OX, $O\Xi$, $O\Xi'$ in space, and indicate on these the values of x, ξ, and ξ'. Further, we draw in the

(Text continued on page 114)

---

† All the rather tedious determinations of limits in the text can be very much simplified by using the method already indicated in the footnote on page 97. Then equation (18) reads as follows:

$$\frac{dE}{dt} = \iiint \log \frac{f(x,t)}{\sqrt{x}} \left[\frac{f(\xi,t)}{\sqrt{\xi}}\frac{f(x+x'-\xi,t)}{\sqrt{x+x'-\xi}} - \frac{f(x,t)}{\sqrt{x}}\frac{f(x',t)}{\sqrt{x'}}\right]$$

$$\times \sqrt{xx'}\psi(x,x',\xi)\,dx\,dx'\,d\xi \tag{18a}$$

(Footnote continued on next page)

where the integration is over all values satisfying the inequalities:
$$x \geq 0,\ x' \geq 0,\ \xi \geq 0,\ x+x'-\xi \geq 0 \tag{18b}$$

The two equations (18a) and (18b) are equivalent to the single definite integral (18), and I remark that now the order of integration is completely arbitrary; there is no requirement that we must integrate first over one variable and then another, but instead we just have to integrate over all values that satisfy (18b). If we now introduce some arbitrary new variables u, v, w, then as is well known

$$du\, dv\, dw = \sum \pm \frac{du}{dx}\frac{dv}{dy}\frac{dw}{dz}\, dx\, dy\, dz$$

If we now wish to obtain equation (20) of the text, then we need only set

$$u = x',\ v = x,\ w = x+x'-\xi$$

Then the functional determinant is equal to one, and it is clear that it is to be taken with the positive sign, if we always integrate from smaller to larger values of the variables, so that the differentials are to be considered positive. Then

$$du\, dv\, dw = dx\, dx'\, d\xi$$

and equation (18a) becomes

$$\frac{dE}{dt} = \iiint \log \frac{f(v,t)}{\sqrt{v}} \left[\frac{f(u+v-w,t)}{\sqrt{u+v-w}} \frac{f(w,t)}{\sqrt{w}} - \frac{f(v,t)}{\sqrt{v}} \frac{f(u,t)}{\sqrt{u}} \right]$$
$$\times \sqrt{uv}\, \psi(v, u, u+v-w)\, du\, dv\, dw$$

The inequalities (18b), which determine the limits, become

$$v \geq 0,\ u \geq 0,\ u+v-w \geq 0,\ w \geq 0$$

We can now again replace the letters u, v, w by x, x' and ξ (since the labels of integration variables are arbitrary) and obtain for the integral

$$\frac{dE}{dt} = \iiint \log \frac{f(x',t)}{\sqrt{x'}} \left[\frac{f(x+x'-\xi,t)}{\sqrt{x+x'-\xi}} \frac{f(\xi,t)}{\sqrt{\xi}} - \frac{f(x,t)}{\sqrt{x}} \frac{f(x',t)}{\sqrt{x'}} \right]$$
$$\times \sqrt{xx'}\, \psi(x', x, x+x'-\xi)\, dx\, dx'\, d\xi$$

and for the inequalities, which determine the limits,

$$x \geq 0,\ x' \geq 0,\ \xi \geq 0,\ x+x'-\xi \geq 0 \tag{20b}$$

If we now finally replace $\psi(x', x, x+x'-\xi)$ by $\psi(x, x', \xi)$ using equation (13), we obtain

$$\frac{dE}{dt} = \iiint \log \frac{f(x',t)}{\sqrt{x'}} \left[\frac{f(\xi,t)}{\sqrt{\xi}} \frac{f(x+x'-\xi,t)}{\sqrt{x+x'-\xi}} - \frac{f(x,t)}{\sqrt{x}} \frac{f(x',t)}{\sqrt{x'}} \right]$$
$$\times \sqrt{xx'}\, \psi(x, x', \xi)\, dx\, dx'\, d\xi \tag{20a}$$

The inequalities (20b) are identical with the inequalities (18a). If we combine the two equations (20a) and (20b) into a single one, by supposing that we integrate first over ξ, then over x', and last over x, and that the limits of integration are attached to the integral signs, then we obtain the desired equation (20) in the text.

If we want to obtain equations (22) of the text, then we set

$$u = \xi, v = x+x'-\xi, w = x$$

The determinant is again one, hence

$$du\, dv\, dw = dx\, dx'\, d\xi$$

Furthermore,

$$x' = u+v-w$$

Equation (18a) becomes, after introducing these variables:

$$\frac{dE}{dt} = \iiint \log \frac{f(w,t)}{\sqrt{w}} \left[\frac{f(u,t)}{\sqrt{u}} \frac{f(v,t)}{\sqrt{v}} - \frac{f(w,t)}{\sqrt{w}} \frac{f(u+v-w,t)}{\sqrt{u+v-w}}\right]$$
$$\times \sqrt{w(u+v-w)}\, \psi(w, u+v-w, u)\, du\, dv\, dw$$

and the inequalities (18b) become

$$w \geq 0,\ u+v-w \geq 0,\ u \geq 0,\ v \geq 0$$

If, as before, we replace the letters u, v, w by x, x', ξ, then we obtain

$$\frac{dE}{dt} = \iiint \log \frac{f(\xi,t]}{\sqrt{\xi}} \left[\frac{f(x,t]}{\sqrt{x}} \frac{f(x',t]}{\sqrt{x'}} - \frac{f(\xi,t]}{\sqrt{\xi}} \frac{f(x+x'-\xi,t]}{\sqrt{x+x'-\xi}}\right]$$
$$\times \sqrt{\xi(x+x'-\xi)}\, \psi(\xi, x+x'-\xi, x)\, dx\, dx'\, d\xi \qquad (22a)$$
$$x \geq 0,\ x' \geq 0,\ x+x'-\xi \geq 0,\ \xi \geq 0$$

We see that we must simply use (14) and write in the limits of the integrals in order to obtain equation (22) of the text from (22a). One sees therefore that if the method of determination of limits by inequalities is used, the transformations can be made with hardly any calculation at all although detailed calculations were needed in the text. I used the longer method in the text simply because the method of inequalities is somewhat unusual.

I note here that one could also use the following expression for E:

$$E_1 = \int_0^\infty f(x,t) \log \left[\frac{f(x,t)}{\sqrt{x}}\right] dx \qquad (17a)$$

This expression is different from the one given in the text for E simply by a term $\int_0^\infty f(x,t)\, dx$, which is just the total number of molecules in unit volume. Since this total number is constant, we see that E_1 is also a quantity that cannot increase.

Another transformation method could be developed by defining the function $\psi(x, x', \xi)$ to be zero whenever $x+x'-\xi < 0$; all the integrations could then be extended from zero to infinity.

plane $XO\Xi$ the line OA, which makes 45° angles with OX and $O\Xi$, and likewise in the plane $\Xi O\Xi'$ the line OB. We now consider the first triple integral in (21). Here we have to integrate ξ' from $x-\xi$ to infinity, i.e. from a point of the plane AOB to infinity. We have to integrate ξ from zero to x, i.e. from zero to a point of the line OA. The integration space of the first integral is therefore just that part of space which stands vertically on AOB (if one imagines that the $O\Xi'$ axis is vertical). Likewise one finds that the integration space of the second integral in (21) is the part of space that stands vertically on the triangle $AO\Xi$. (The triangle formed by extending the lines OA and $O\Xi$ to infinity.) The two integrals can therefore be represented as a single integration over a solid angle bounded by the four figures AOB, $AO\Xi$, $BO\Xi'$ and $\Xi O\Xi'$. Now it is easy to determine the limits if we first integrate over x. For constant ξ and ξ' we remain in the solid angle as x goes from zero to $\xi+\xi'$. Zero and $\xi+\xi'$ are therefore the limits of integration for x. The integration over ξ and ξ' simply goes from zero to infinity. With this new order of integration we can combine the two integrals into one, and we have

$$\frac{dE}{dt} = \int_0^\infty \int_0^\infty \int_0^{\xi+\xi'} \log\frac{f(x,t)}{\sqrt{x}} \left[\frac{f(\xi,t)f(\xi',t)}{\sqrt{\xi}\sqrt{\xi'}} - \frac{f(x,t)f(\xi+\xi'-x,t)}{\sqrt{x}\sqrt{\xi+\xi'-x}} \right] \times \sqrt{\xi\xi'}\psi(\xi,\xi',x)d\xi d\xi' dx$$

In this definite integral it again makes no difference how we label the variables over which we integrate. Hence, we can denote the two variables ξ and ξ' by the Latin letters x and x', and the variable x with the letter ξ. If we do this, and put a minus sign in front of the whole integral while reversing the signs inside the bracket, we obtain

$$\frac{dE}{dt} = -\int_0^\infty \int_0^\infty \int_0^{x+x'} \log\frac{f(\xi,t)}{\sqrt{\xi}} \left[\frac{f(\xi,t)f(x+x'-\xi,t)}{\sqrt{\xi}\sqrt{x+x'-\xi}} - \frac{f(x,t)f(x',t)}{\sqrt{x}\sqrt{x'}} \right]$$
$$\times \sqrt{xx'}\psi(x,x',\xi)dxdx'd\xi \quad (22)$$

If we apply to this equation the same transformation by which we obtained equation (20) from equation (18), then we obtain still a fourth expression for dE/dt. I do not need to go through this transformation in detail; one sees easily that its result is the following:

$$\frac{dE}{dt} = -\int_0^\infty \int_0^\infty \int_0^{x+x'} \log\frac{f(x+x'-\xi,t)}{\sqrt{x+x'-\xi}} \left[\frac{f(\xi,t)f(x+x'-\xi,t)}{\sqrt{\xi}\sqrt{x+x'-\xi}}\right. \\ \left. -\frac{f(x,t)f(x',t)}{\sqrt{x}\sqrt{x'}}\right] \times \sqrt{xx'}\psi(x,x',\xi)dxdx'd\xi \qquad (23)$$

I will now collect together the four expressions that we have obtained for dE/dt, for which purpose I will use the following abbreviations. I set:

$$\frac{f(x,t)}{\sqrt{x}} = s, \quad \frac{f(x',t)}{\sqrt{x'}} = s', \quad \frac{f(\xi,t)}{\sqrt{\xi}} = \sigma, \quad \frac{f(x+x'-\xi,t)}{\sqrt{x+x'-\xi}} = \sigma'$$

$$\sqrt{xx'}\psi(x,x',\xi) = r$$

The four equations (18), (20), (22), and (23) thus reduce to

$$\frac{dE}{dt} = \int_0^\infty \int_0^\infty \int_0^{x+x'} \log s(\sigma\sigma'-ss')rdxdx'd\xi$$

$$\frac{dE}{dt} = \int_0^\infty \int_0^\infty \int_0^{x+x'} \log s'(\sigma\sigma'-ss')rdxdx'd\xi$$

$$\frac{dE}{dt} = -\int_0^\infty \int_0^\infty \int_0^{x+x'} \log \sigma(\sigma\sigma'-ss')rdxdx'd\xi$$

$$\frac{dE}{dt} = -\int_0^\infty \int_0^\infty \int_0^{x+x'} \log \sigma'(\sigma\sigma'-ss')rdxdx'd\xi$$

We also obtain dE/dt when we add together all four expressions and divide by 4. Since on the right there are simply definite integrals with the same integration variables and the same limits, we can write the integral sign in front of the sum and then add the various quantities under the integral sign. Taking out the common factor, we obtain

$$\frac{dE}{dt} = \tfrac{1}{4}\int_0^\infty \int_0^\infty \int_0^{x+x'} (\log s + \log s' - \log \sigma - \log \sigma')(\sigma\sigma' - ss')r\,dx\,dx'\,d\xi$$

or, after replacing the sum of logarithms by a logarithm of a product,

$$\frac{dE}{dt} = \tfrac{1}{4}\int_0^\infty \int_0^\infty \int_0^{x+x'} \log\left(\frac{ss'}{\sigma\sigma'}\right)(\sigma\sigma' - ss')r\,dx\,dx'\,d\xi \quad (24)$$

If it is not true for all combinations of values of the variables contained in the s and σ that

$$ss' = \sigma\sigma' \quad (25)$$

then for some combinations we must have either $ss' > \sigma\sigma'$ or $ss' < \sigma\sigma'$. In the first case, $\log(ss'/\sigma\sigma')$ is positive but $\sigma\sigma' - ss'$ is negative, and in the second case the converse is true; in both cases the product $\log(ss'/\sigma\sigma')(\sigma\sigma' - ss')$ is negative. Now the quantity r is essentially positive, since ψ is always positive, and the square roots are to be taken with positive signs. Hence, the quantity under the integral sign, and therefore the entire integral, is necessarily negative. Therefore E must necessarily decrease. Only when equation (25) holds can E remain constant. Since, as we shall see later on, E cannot go to negative infinity, it must approach a minimum with increasing time, and at the minimum we have $dE/dt = 0$, hence equation (25) holds. This equation implies, when we substitute for s, s', σ, and σ' their definitions,

$$\frac{f(x,t)}{\sqrt{x}}\frac{f(x',t)}{\sqrt{x'}} = \frac{f(\xi,t)}{\sqrt{\xi}}\frac{f(x+x'-\xi,t)}{\sqrt{x+x'-\xi}}$$

In order that this equation may be satisfied for all values of the variables x, x', and ξ, it is necessary, as can easily be shown, that

$$f(x,t) = C\sqrt{x}\,e^{-hx}$$

It has thus been rigorously proved that, whatever may be the initial distribution of kinetic energy, in the course of a very long time it must always necessarily approach the one found by Maxwell. The procedure used so far is of course nothing more than a mathematical artifice employed in order to give a rigorous proof of a theorem whose exact proof has not previously been found. It gains meaning by its applicability to the theory of polyatomic gas molecules. There one can again prove that a certain quantity E can only decrease as a consequence of molecular motion, or in a limiting case can remain constant. One can also prove that for the atomic motion of a system of arbitrarily many material points there always exists a certain quantity which, in consequence of any atomic motion, cannot increase, and this quantity agrees up to a constant factor with the value found for the well-known integral $\int dQ/T$ in my paper on the "Analytical proof of the 2nd law, etc."† We have therefore prepared the way for an analytical proof of the second law in a completely different way from those previously investigated. Up to now the object has been to show that $\int dQ/T = 0$ for reversible cyclic processes, but it has not been proved analytically that this quantity is always negative for irreversible processes, which are the only ones that occur in nature. The reversible cyclic process is only an ideal, which one can more or less closely approach but never completely attain. Here, however, we have succeeded in showing that $\int dQ/T$ is in general negative, and is equal to zero only for the limiting case, which is of course the reversible cyclic process (since if one can go through the process in either direction, $\int dQ/T$ cannot be negative).

II. Replacement of Integrals by Sums

I will not consider any further the relation between E and $\int dQ/T$, but rather now show how all the previous calculations can be made much clearer when the partial differential equation (16) is transformed to a system of ordinary differential equations. This is done

---

† L. Boltzmann, *Wien. Ber.* **63**, 712 (1871).

by replacing the double integral which appears in the partial differential equation by a sum, according to the well-known formula

$$\int_0^\infty f(x,t)dx = \lim \varepsilon[f(\varepsilon,t)+f(2\varepsilon,t)+f(3\varepsilon,t)+\ldots f(p\varepsilon,t)]$$

for $\lim \varepsilon = 0, \quad \lim p\varepsilon = \infty$

We wish to replace both integrals in (16) by such sums, and first take ε and p both finite. The (16) becomes a differential equation with the following unknowns: $f(\varepsilon, t), f(2\varepsilon, t), \ldots, f(p\varepsilon, t)$. Each unknown is a function only of time. The number of unknowns is p. But equation (16) must hold for each x. If we substitute therein the values

$$x = \varepsilon, \ x = 2\varepsilon, \ldots, \ x = p\varepsilon$$

then we obtain in all p differential equations relating our p unknowns; and since the unknowns are functions only of time, the differential equations are not partial ones. This system of p ordinary differential equations in p unknowns will first be solved, and then we shall investigate the limit approached by the solution when ε becomes infinitesimal and $p\varepsilon$ becomes infinite. The limit is then the solution of the partial differential equation. The substitution of summation formulae into the partial differential equation offers no difficulty. It is transformed into the system of equations (34) below. We shall carry out explicitly the calculations which are only sketched here. I will also show how one must modify our problem in order to arrive directly at the system of p ordinary differential equations instead of first deriving a partial differential equation. The method that we shall use is by no means new. As is well known, the integral is nothing more than a symbolic notation for the sum of infinitely many infinitesimal terms. The symbolic notation of the integral calculus has the advantage of such great brevity that in most cases it would only lead to useless complications to write out the integral first as a sum of p terms and then let p become large. In spite of this, there are cases in which the latter method—on account of its generality, and especially on account of its greater perspicuousness, in that it allows the various solutions of a problem to appear—should not be completely rejected. I recall the elegant solution of the problem of string-vibrations by Lagrange in the *Miscellanea taurinensia*, where he first treated the vibrations of a system of n spheres bound together, and then obtained the vibrations of a string by letting n become large while the mass of each sphere became small.† The problem of diffusion and heat conduction was solved in a similar way by Stefan.‡ Another beautiful application of this method was made by Reimann for the differential equation§

$$\frac{d^2w}{dr\,ds} = a\left(\frac{dw}{dr}+\frac{dw}{ds}\right)$$

It appears to me that in our case also, once one has become used to some abstractions, this method very much improves clarity. We wish to replace the continuous variable x by a series of discrete values $\varepsilon, 2\varepsilon, 3\varepsilon, \ldots p\varepsilon$. Hence we must assume that our molecules are not able to take up a continuous series of kinetic energy values, but rather only values that are multiples of a certain quantity ε. Otherwise we shall treat exactly the same problem as before. We have many gas molecules in a space R. They are able to have only the following kinetic energies:

$$\varepsilon, 2\varepsilon, 3\varepsilon, 4\varepsilon, \ldots p\varepsilon \qquad (26)$$

No molecule may have an intermediate or a greater kinetic energy. When two molecules collide, they can change their kinetic energies in many different ways. However, after the collision the kinetic energy of each molecule must always be a multiple of ε. I certainly do not need to remark that for the moment we are not concerned with a real physical problem. It would be difficult to imagine an apparatus that could regulate the collisions of two bodies in such a way that their kinetic energies after a collision are always

---

† [J. L. LaGrange, *Misc. Taur.* **1**, (1759), reprinted in his *Oeuvres*, Gauthier-Villars, Paris, 1867, **1**, 37.]

‡ J. Stefan, *Wien. Ber.* **47**, 327 (1863), and R. Beez [*Z. Math. Phys.* **7**, 327 (1862), **10**, 358 (1865)].

§ B. Riemann, *Abh. K. Ges. Wiss. Göttingen* **8**, 43 (1859).

multiples of ε. That is not the question here. In any case we are free to study the mathematical consequences of this assumption, which is nothing more than an artifice to help us to calculate physical processes. For at the end we shall make ε infinitely small and $p\varepsilon$ infinitely large, so that the series of kinetic energies given in (26) will become a continuous one, and our mathematical fiction will reduce to the physical problem treated earlier.

We now assume that at time t there are w_1 molecules with kinetic energy ε, w_2 with kinetic energy $2\varepsilon, \ldots$ and w_p with kinetic energy $p\varepsilon$, in unit volume. We again assume that the distribution of kinetic energy at time t is already uniform (the w's are independent of position in space) and all directions of velocity are equally probable. In the course of time, molecules with a certain kinetic energy, for example, $k\varepsilon$, will leave unit volume; but since the distribution of kinetic energy is uniform, on the average the same amount will come in from the surroundings. Since it is only a question of average values, the w's will change only by collisions. If we want to establish a differential equation for the changes of the w's, we must subject the collisions to a closer examination. We denote by $N_{\kappa\lambda}^{kl}$ the number of collisions in unit volume that occur during a very small time τ, such that the kinetic energies of the two molecules are $k\varepsilon$ and $l\varepsilon$ before the collision, and $\kappa\varepsilon$ and $\lambda\varepsilon$ after the collision. The four quantities k, l, κ, λ are positive integers, $\leq p$; for collisions in which k, l, κ, λ have other values do not occur. Moreover we have the equation

$$k+l = \kappa+\lambda \tag{27}$$

since the sum of the kinetic energies of the two molecules must be the same before and after the collision. Since we are not at present dealing with a real physical problem, we cannot actually determine these numbers $N_{\kappa\lambda}^{kl}$; we can only make some rather arbitrary assumptions about them and study the consequences thereof. However, if we want our problem to reduce to the one treated earlier in the limit of infinitely small ε, then we must assume that the $N_{\kappa\lambda}^{kl}$ are determined in just the same way as the collision numbers previously were. We therefore assume that $N_{\kappa\lambda}^{kl}$ is proportional to the time τ, to the number of molecules with kinetic energy $k\varepsilon$ in unit volume, w_k, and to w_l. The product of these three quantities is to be multiplied by a certain proportionality factor, which still depends on the four quantities k, l, κ, λ that determine the nature of the collision, but not on the time; we denote it by $A_{\kappa\lambda}^{kl}$. Thus we write

$$N_{\kappa\lambda}^{kl} = \tau w_k w_l A_{\kappa\lambda}^{kl} \tag{28}$$

The number of collisions is analogous to what we determined earlier in equation (2). The quantity A appears here in place of what was earlier denoted by ψ. If we wish to make the analogy complete, we must attribute to A the properties of ψ. ψ satisfies the equation

$$\sqrt{xx'}\psi(x, x', \xi) = \sqrt{\xi(x+x'-\xi)}\psi(\xi, x+x'-\xi, x) \tag{29}$$

In our case the kinetic energies before the collision are $k\varepsilon$, $l\varepsilon$, and after the collision $\kappa\varepsilon$, $\lambda\varepsilon$; in our case, therefore,

$$x = k\varepsilon; \quad x' = l\varepsilon; \quad \xi = \kappa\varepsilon; \quad x+x'-\xi = \lambda\varepsilon$$

The quantity $\psi(x, x', \xi)$ corresponds to $A_{\kappa\lambda}^{kl}$, and one sees easily that the quantity $\psi(\xi, x+x'-\xi, x)$ corresponds to $A_{kl}^{\kappa\lambda}$. Equation (29) therefore reduces in our case to

$$\sqrt{kl} A_{\kappa\lambda}^{kl} = \sqrt{\kappa\lambda} A_{kl}^{\kappa\lambda} \tag{30}$$

Now the analogy is complete, and we need only to make ε infinitely small and $p\varepsilon$ infinitely large in order to obtain from the solution of this problem the solution of the physical one treated earlier. The formulae will be somewhat simpler if we denote $\sqrt{kl} A_{\kappa\lambda}^{kl}$ by $B_{\kappa\lambda}^{kl}$. Then equation (30) reduces to

$$B_{\kappa\lambda}^{kl} = B_{kl}^{\kappa\lambda} \tag{31}$$

and equation (28) becomes

$$N_{\kappa\lambda}^{kl} = \tau \frac{w_k w_l}{\sqrt{kl}} B_{\kappa\lambda}^{kl} \tag{32}$$

The square roots are of course to be taken positive, since $N_{\kappa\lambda}^{kl}$ and the w's are essentially positive numbers, and we wish to choose the B's always positive.

After these preparations, we ask what change the quantity w_1 experiences during the time τ. w_1 is the number of molecules with kinetic energy ε in unit volume. We know that this number changes only through collisions. Every time two molecules collide such that before the collision one of them has kinetic energy ε, while afterwards it no longer has that energy, this number decreases by one. Conversely, every time two molecules collide in such a way that before the collision neither, but after it one of them has kinetic energy ε, this number increases by one. If we subtract the former number from w_1 and add the latter to it, we obtain the number of molecules in unit volume which have kinetic energy ε at time $t+\tau$. We now have to find out how many collisions there are in which one molecule has kinetic energy ε before the collision. When the other one also has kinetic energy ε, then after the collision each must still have kinetic energy ε, since the total kinetic energy of the two must still be 2ε, and no other kinetic energies than those listed in (26) can occur. If one molecule has kinetic energy ε before the collision and the other has kinetic energy 2ε, then for the same reason, after the collision one must have ε and the other 2ε. None of these collisions change the number of molecules with energy ε. However, if one has ε and the other has 3ε before the collision, then after the collision both may have 2ε. Each such collision will decrease w_1 by one. There will be N_{22}^{13} of these collisions in unit volume during time τ, so that w_1 decreases by N_{22}^{13}. We have to subtract N_{22}^{13} from w_1. Likewise we have to subtract $N_{23}^{14}, N_{32}^{14}, N_{24}^{15}, \ldots, N_{p-1,2}^{1p}$ from w_1. On the other hand we have to add the numbers $N_{13}^{22}, N_{14}^{23}, \ldots, N_{1p}^{p-1,2}$, since through each of these collision the number of molecules with kinetic energy ε increases by one. Thus we obtain

$$w_1' = w_1 - N_{22}^{13} - N_{23}^{14} - N_{32}^{14} - N_{24}^{15} - \ldots,$$
$$+ N_{13}^{22} + N_{14}^{23} + N_{14}^{32} + N_{15}^{24} + \ldots \quad (33)$$

The law of formation of the series is easily grasped. We have to subtract all the N's which have an upper index 1, and add all those that have a lower index 1. Those which have this index both above and below are to be added and subtracted both, so that they may be left out entirely. (Earlier in the integral we have not eliminated these self-cancelling terms, for the sake of convenience.) We must also observe that the four indices of N must satisfy equation (27), and that two N's which are the same if one permutes the upper and lower indices simultaneously (e.g. N_{23}^{14} and N_{32}^{41}) correspond to completely identical collisions, and hence should be added (or subtracted) only once.

If we expand w_1' according to Taylor's theorem, we obtain

$$w_1' = w_1 + \tau \frac{dw_1}{dt}$$

If we substitute this as well as the value of N given by equation (32) into equation (33), then we obtain, after dividing through by τ:

$$\frac{dw_1}{dt} = -B_{22}^{13}\frac{w_1 w_3}{\sqrt{1\sqrt{3}}} - B_{23}^{14}\frac{w_1 w_4}{\sqrt{1\sqrt{4}}} - B_{32}^{14}\frac{w_1 w_4}{\sqrt{1\sqrt{4}}} - B_{24}^{15}\frac{w_1 w_5}{\sqrt{1\sqrt{5}}} - \cdots$$

$$+ B_{13}^{22}\frac{(w_2)^2}{2} + B_{14}^{23}\frac{w_2 w_3}{\sqrt{2\sqrt{3}}} + B_{14}^{32}\frac{w_2 w_3}{\sqrt{2\sqrt{3}}} + B_{15}^{24}\frac{w_2 w_4}{\sqrt{2\sqrt{4}}} + \cdots$$

which can also be written, taking account of equation (32), as:

$$\frac{dw_1}{dt} = B_{22}^{13}\left(\frac{(w_2)^2}{2} - \frac{w_1 w_3}{\sqrt{1\sqrt{3}}}\right) + (B_{23}^{14} + B_{32}^{14})\left(\frac{w_2 w_3}{\sqrt{2\sqrt{3}}} - \frac{w_1 w_4}{\sqrt{1\sqrt{4}}}\right) + \cdots$$

Likewise one finds

$$\left.\begin{aligned}
\frac{dw_2}{dt} &= 2B_{22}^{13}\left(\frac{w_1 w_3}{\sqrt{1\sqrt{3}}} - \frac{(w_2)^2}{2}\right) \\
&\quad + (B_{23}^{14} + B_{32}^{14})\left(\frac{w_1 w_4}{\sqrt{1\sqrt{4}}} - \frac{w_2 w_3}{\sqrt{2\sqrt{3}}}\right) + \cdots \\
\frac{dw_p}{dt} &= (B_{2,\,p-1}^{1,\,p} + B_{p-1,\,2}^{1,\,p})\left(\frac{w_{p-1} w_2}{\sqrt{p-1}\sqrt{2}} - \frac{w_1 w_p}{\sqrt{1\sqrt{p}}}\right) \\
&\quad + (B_{3,\,p-2}^{1,\,p} + B_{p-2,\,3}^{1,\,p})\left(\frac{w_3 w_{p-2}}{\sqrt{3}\sqrt{p-2}} - \frac{w_1 w_p}{\sqrt{1\sqrt{p}}}\right) + \cdots
\end{aligned}\right\} \quad (34)$$

All that needs to be explained is why the term

$$B_{22}^{13}\frac{w_1 w_3}{\sqrt{1}\sqrt{3}}$$

has the factor 2 in the expression for dw_2/dt. This term comes from collisions for which the kinetic energies are ε, 3ε before the collision and 2ε, 2ε after the collision; each such collision changes the number of molecules with kinetic energy 2ε by two rather than by one, since two molecules simultaneously acquire kinetic energy 2ε. Hence, all these collisions must be counted twice. Likewise in the expression for dw_3/dt, the terms

$$B_{33}^{15}\frac{w_1 w_5}{\sqrt{1}\sqrt{5}} \quad \text{and} \quad B_{33}^{24}\frac{w_2 w_4}{\sqrt{2}\sqrt{4}}$$

are counted twice, and so forth.

It would be easy to represent the system of equations (34) by summation formulae; however, I believe that this would not make it any easier to understand; the law of formation of the terms is already clear enough. One also sees that this is exactly the system of equations to which the partial differential equation (18) is transformed when one uses the Lagrange method, as previously explained, to replace it by a system of p ordinary differential equations, and denotes $f(k\varepsilon, t)$ by w_k. In order to simplify equations (34) somewhat, we set

$$w_k = \sqrt{k}u_k$$

These equations then become:

$$\left.\begin{aligned}\frac{du_1}{dt} &= B_{22}^{13}(u_2^2 - u_1 u_3) + (B_{23}^{14} + B_{32}^{14})(u_2 u_3 - u_1 u_4) + \cdots \\ \sqrt{2}\frac{du_2}{dt} &= 2B_{22}^{13}(u_1 u_3 - u_2^2) + (B_{23}^{14} + B_{32}^{14})(u_1 u_4 - u_2 u_3) + \cdots \\ &\quad\vdots \\ \sqrt{p}\frac{du_p}{dt} &= (B_{2,\,p-1}^{1,\,p} + B_{p-1,\,2}^{1,\,p})(u_2 u_{p-1} - u_1 u_p) + \cdots\end{aligned}\right\} \quad (35)$$

From these equations it can again be proved that

$$E = u_1 \log u_1 + \sqrt{2} u_2 \log u_2 + \ldots + \sqrt{p} u_p \log u_p$$

must always decrease unless $u_2^2 - u_1 u_3$, $u_2 u_3 - u_1 u_4$, ... (in other words all the expressions multiplied by the coefficients B in equation (35)) vanish. Equations (35) have the inconvenient feature that while they can be written with summation formulae they cannot be written out completely explicitly. It would undoubtedly be an aid to clarity, therefore, if we begin with the simplest case and then proceed gradually to the general case. First let $p = 3$; the molecules can only have three different kinetic energies, ε, 2ε, and 3ε. Then the system of equations (35) reduces to the following three equations:

$$\left. \begin{aligned} \frac{du_1}{dt} &= B_{22}^{13}(u_2^2 - u_1 u_3) \\[1em] \sqrt{2}\frac{du_2}{dt} &= 2B_{22}^{13}(u_1 u_3 - u_2^2) \\[1em] \sqrt{3}\frac{du_3}{dt} &= B_{22}^{12}(u_2^2 - u_1 u_3) \end{aligned} \right\} \quad (36)$$

and the expression for E reduces to

$$E = u_1 \log u_1 + \sqrt{2} u_2 \log u_2 + \sqrt{3} u_3 \log u_3$$

The differentiation gives

$$\frac{dE}{dt} = (\log u_1 + 1)\frac{du_1}{dt} + \sqrt{2}(\log u_2 + 1)\frac{du_2}{dt} + \sqrt{3}(\log u_3 + 1)\frac{du_3}{dt}$$

or, with a different arrangement of the terms,

$$\frac{dE}{dt} = \log u_1 \frac{du_1}{dt} + \sqrt{2} \log u_2 \frac{du_2}{dt} + \sqrt{3} \log u_3 \frac{du_3}{dt}$$

$$+ \frac{du_1}{dt} + \sqrt{2} \frac{du_2}{dt} + \sqrt{3} \frac{du_3}{dt}$$

The sum of the last three terms vanishes according to equations (36) so that one obtains dE/dt by multiplying the first of these equations by $\log u_1$, the second by $\log u_2$, and the third by $\log u_3$, and adding all three together. If one does this he obtains

$$\frac{dE}{dt} = B_{22}^{13}.(u_2^2 - u_1 u_3).(\log u_1 + \log u_3 - 2 \log u_2)$$

or

$$\frac{dE}{dt} = B_{22}^{13}.(u_2^2 - u_1 u_3) \log \left(\frac{u_1 u_3}{u_2^2} \right)$$

Of the two factors multiplying B_{22}^{13} on the right-hand side of this equation, the first is positive and the second is negative when $u_2^2 > u_1 u_2$, whereas the first is negative and the second is positive when $u_2^2 > u_1 u_2$. Hence their product is always negative, and since B_{22}^{13} must be positive, dE/dt is always negative or zero. The latter is true when $u_2^2 = u_1 u_3$. Now it can easily be shown that E cannot become negatively infinite. Obviously none of the three quantities u_1, u_2, u_3 may be negative or imaginary. For positive u, however, $u \log u$ cannot have a larger negative value than $-1/e$, hence E cannot have a larger negative value than

$$-\frac{1 + \sqrt{2} + \sqrt{3}}{e}$$

where e is the base of natural logarithms.

Therefore E, since its derivative cannot be positive, must continually approach a minimum for which $dE/dt = 0$, and for which $u_2^2 = u_1 u_3$. The proof cannot be carried out in just the same way when $n > 3$. I consider here only the case $n = 4$. In this case equations (35) reduce to

$$\left.\begin{aligned}
\frac{du_1}{dt} &= B_{22}^{13}(u_2^2 - u_1 u_3) + (B_{23}^{14} + B_{32}^{14})(u_2 u_3 - u_1 u_4) \\
\sqrt{2}\frac{du_2}{dt} &= 2B_{22}^{13}(u_1 u_3 - u_2^2) + (B_{23}^{14} + B_{32}^{14})(u_1 u_4 - u_2 u_3) \\
&\quad + B_{33}^{24}(u_3^2 - u_2 u_4) \\
\sqrt{3}\frac{du_3}{dt} &= B_{22}^{13}(u_2^2 - u_1 u_3) + (B_{23}^{14} + B_{32}^{14})(u_1 u_4 - u_2 u_3) \\
&\quad + 2B_{33}^{24}(u_2 u_4 - u_3^2) \\
\sqrt{4}\frac{du_4}{dt} &= (B_{23}^{14} + B_{32}^{14})(u_2 u_3 - u_1 u_4) + B_{33}^{24}(u_3^2 - u_2 u_4)
\end{aligned}\right\} \quad (37)$$

For E one finds

$$E = u_1 \log u_1 + \sqrt{2} u_2 \log u_2 + \sqrt{3} u_3 \log u_3 + \sqrt{4} u_4 \log u_4$$

$$\frac{dE}{dt} = \log u_1 \frac{du_1}{dt} + \sqrt{2} \log u_2 \frac{du_2}{dt} + \sqrt{3} \log u_3 \frac{du_3}{dt}$$

$$+ \sqrt{4} \log u_4 \frac{du_4}{dt}$$

If one substitutes here for

$$\frac{du_1}{dt}, \quad \frac{du_2}{dt}, \quad \frac{du_3}{dt}, \quad \frac{du_4}{dt}$$

their values from equations (37), he obtains, with a suitable rearrangement of terms,

$$\frac{dE}{dt} = B_{22}^{13}(u_2^2 - u_1 u_3) \log\left(\frac{u_1 u_3}{u_2^2}\right) + B_{33}^{24}(u_3^2 - u_2 u_4) \log\left(\frac{u_2 u_4}{u_3^2}\right)$$

$$+ (B_{23}^{14} + B_{32}^{14})(u_2 u_3 - u_1 u_4) \log\left(\frac{u_1 u_4}{u_2 u_3}\right)$$

I remark that the change in the order of the summands, which is necessary here, is analogous to our previous transformation of definite integrals. From the above expression one sees at once that dE/dt is again necessarily negative, unless simultaneously we have

$$u_2^2 = u_1 u_3, \quad u_3^2 = u_2 u_4, \quad u_2 u_3 = u_1 u_4$$

which can also be written

$$u_3 = \frac{u_2^2}{u_1}, \quad u_4 = \frac{u_2^3}{u_1^2}$$

Likewise one finds in the general case that dE/dt is necessarily negative so that E must decrease unless

$$u_3 = \frac{u_2^2}{u_1}, \quad u_4 = \frac{u_2^3}{u_1^2}, \ldots \tag{38}$$

Since E cannot have a larger negative value than

$$-\frac{1+\sqrt{2}+\sqrt{3}+\ldots+\sqrt{p}}{e} \tag{39}$$

it must necessarily approach a minimum value for which equations (38) hold. Thus it continually approaches the distribution of states determined by equations (38). We now have to prove that equations (38) uniquely determine the distribution of states. If we add together all the equations (35), we obtain

$$\frac{du_1}{dt}+\sqrt{2}\frac{du_2}{dt}+\sqrt{3}\frac{du_3}{dt}+\ldots+\sqrt{p}\frac{du_p}{dt}=0$$

hence

$$u_1+\sqrt{2}u_2+\sqrt{3}u_3+\ldots+\sqrt{p}u_p = a \tag{40}$$

In a similar way we find that

$$u_1+2\sqrt{2}u_2+3\sqrt{3}u_3+\ldots+p\sqrt{p}u_p = \frac{b}{\varepsilon} \tag{41}$$

where a and b are constants. The meaning of these equations is obvious. In particular,

$$w_1 + w_2 + w_3 + \ldots = u_1 + \sqrt{2}u_3 + \sqrt{3}u_3 + \ldots = a$$

is the total number of molecules in unit volume, while b is their total kinetic energy. Equations (40) and (41) therefore tell us that these two quantities are constant. Suppose that the two quantities a and b are given. Then we set the quotient u_2/u_1 equal to γ. Equations (38) then reduce to

$$u_3 = \gamma^2 u_1, \; u_4 = \gamma^3 u_1, \; \ldots, \; u_p = \gamma^{p-1} u_1$$

If one substitutes these values into equations (40) and (41), then he finds easily

$$\left(pa - \frac{b}{\varepsilon}\right)\sqrt{p}\gamma^{p-1} + \left[(p-1)a - \frac{b}{\varepsilon}\right]\sqrt{p-1}\gamma^{p-2} + \ldots \\ + \left(3a - \frac{b}{\varepsilon}\right)\sqrt{3}\gamma^2 + \left(2a - \frac{b}{\varepsilon}\right)\sqrt{2}\gamma + a - \frac{b}{\varepsilon} = 0 \quad (42)$$

Since all the u's are necessarily positive, we see immediately that $(b/\varepsilon) - a$ must be positive while $(b/\varepsilon) - pa$ must be negative. Hence b must lie between εa and $\varepsilon p a$. Hence, in equation (42) the coefficient of γ^{p-1} is positive, while the term independent of γ must be negative. The polynomial is therefore positive for $\gamma = \infty$, and negative for $\gamma = 0$; therefore there is one and only one positive root for γ, since the series of coefficients changes sign only once. Negative or imaginary values for γ are of course meaningless. But from γ we can determine uniquely all the u's and also all the w's. Hence, whatever may be the initial distribution of states, there is one and only one distribution which it approaches with increasing time. This distribution depends only on the constants a and b, the total number and total kinetic energy of the molecules (density and temperature of the gas). This theorem was proved first only for the case that the distribution of states is initially uniform. It must also hold, however, when this is not true, provided only that the molecules are distributed in such a way that they tend to become mixed as time progresses, so that the distribution becomes uniform after a very long time. This will always happen with the exception of certain special cases, for example, when the molecules move initially in a straight line and are reflected back in this straight line at the walls. Since we have established this for arbitrary p and ε, we can immediately go to the case where $1/p$ and ε become infinitesimal.† We have first:

$$w_k = \sqrt{k u_k} = u_1 \sqrt{k \gamma^{k-1}}$$

---

† For very large p, the expression (39) will be very large, of order $p^{\frac{3}{2}}$. In this case it is necessary to look for a smaller negative value that E can never exceed. The quantity denoted here by E differs by a constant from the one earlier so denoted. If we wish to obtain the quantity denoted by E_1 in equation (17a), page 113, which again differs only by a constant from the other quantities denoted by this letter, then we must add to our present E,

$$-\frac{3 \log \varepsilon}{2} (u_1 + \sqrt{2} u_2 + \ldots)$$

Therefore

$$E_1 = E - \frac{3 \log \varepsilon}{2} (u_1 + \sqrt{2}\, u_2 + \ldots) = u_1 \log \frac{u_1}{\varepsilon^{\frac{3}{2}}} + \sqrt{2}\, u_2 \log \frac{u_2}{\varepsilon^{\frac{3}{2}}} + \ldots$$

It is clear now that E_1 is a real and continuous function of the u's for all real positive values of u. Furthermore, if we say that a negative quantity is smaller, the greater its numerical value is, then E is not smaller than the expression (39), hence E_1 is not smaller than

$$-\frac{1}{2}(1 + \sqrt{2} + \ldots \sqrt{p}) - \frac{3}{2} a \log \varepsilon$$

Hence, E_1 must have a minimum if the u's run through all real positive values compatible with equations (40) and (41). One can then easily show that for this minimum none of the u's can be equal to zero, so that the minimum cannot lie on the boundary of the space formed from the u's, and consequently it can be found by applying the usual rules of differential calculus. If we add to the total differential of E_1 that of the two equations (40) and (41), multiplying the former with the undetermined multiplier λ, and the latter by μ, then we obtain

$$(\log u_1 + \lambda + \mu)\, du_1 + (\log u_2 + \lambda + 2\mu)\, \sqrt{2}\, du_2 + \ldots = 0$$

At the minimum, the factor of each differential must vanish, whence on elimination of λ and μ one obtains

$$\log u_2 - \log u_1 = \log u_3 - \log u_2 = \ldots$$

or

$$u_3 = \frac{u_2^2}{u_1}, \quad u_4 = \frac{u_3^2}{u_2} \ldots$$

(Footnote continued on next page

For infinitesimal ε we can again set

$$\varepsilon = dx, \ k\varepsilon = x, \ \gamma = e^{-h\varepsilon}, \ \frac{u_1}{\gamma \varepsilon^{\frac{3}{2}}} = C \tag{43}$$

and obtain

$$w_k = C\sqrt{x}e^{-hx}dx$$

which is again the Maxwell distribution. Likewise one can convince himself that the sum which we have here denoted by E

---

which we recognize to be the same as equations (38). These equations therefore determine the smallest value that E_1 can have when the u's take all possible values consistent with equations (40) and (41). However, since the u's are actually subject to equations (40 and (41) during the entire process, this is the smallest value of E_1 during the entire process. In order to calculate it, we set again

$$u_2 = u_1\gamma, u_3 = u_1\gamma^2 \ldots$$

We know that we then find from equations (38), (40) and (41) a unique positive value for γ, which must correspond to the actual minimum of E_1. This minimum value of E_1 is therefore

$$E = \frac{1}{\varepsilon} b \log \gamma + a \log \left(\frac{u_1}{\gamma \varepsilon^{\frac{3}{2}}}\right)$$

E_1 cannot have a smaller value than this. This value remains finite for infinitesimal ε and infinite p. Taking account of equations (43), we see that it reduces to

$$a \log C - bh$$

or, since

$$a = \frac{1}{2}\sqrt{\frac{\pi}{h^3}} \ C, \ b = \frac{3a}{2h}$$

one can write for it,

$$\frac{1}{2}\sqrt{\frac{\pi}{h^3}} \ C\left(\log C - \frac{3}{2}\right)$$

which, since the constants C and h are not infinite, is a finite quantity. Hence E_1 cannot be minus infinity. On the other hand, it may be plus infinity. We still have to show that in that case there cannot be thermal equilibrium. This proof, as well as an explicit discussion of the exceptional case where

$$\lim (\varepsilon/\tau)[f(\varepsilon, t+\tau) \log f(\varepsilon, t+\tau) + \sqrt{2} f(2\varepsilon, t+\tau) \log f(2\varepsilon, t+\tau) + \ldots$$
$$- f(\varepsilon, t) \log f(\varepsilon, t) - \sqrt{2} f(2\varepsilon, t) \log f(\varepsilon, t) - \ldots]$$

comes out to be different according as ϵ/τ or τ/ϵ vanishes, will not be discussed further here.

reduces, aside from a constant additive term, to the integral in equation (17a); we therefore obtain by this method all the results that we earlier found by transformations of definite integrals, but it is the advantage of being much simpler and clearer. One only has to accept the abstraction that a molecule may have only a finite number of kinetic energies as a transition stage.

If one sets the time derivatives in equations (35) equal to zero, he obtains the conditions that the distribution of states does not change with time but is stationary. One sees also that equations (35) have many other solutions in addition to the one we have found, but these do not represent acceptable stationary distributions since the probabilities of certain kinetic energies comes out to be negative or imaginary. The same is true when, as actually happens in nature, each molecule can have any kinetic energy from zero to infinity. The condition that the distribution be stationary is obtained by setting

$$\frac{\partial f(x,t)}{\partial t} = 0$$

in equation (16). This gives

$$0 = \int_0^\infty \int_0^{x+x'} \left[\frac{f(\xi)f(x+x'-\xi)}{\sqrt{\xi}\sqrt{x+x'-\xi}} - \frac{f(x)f(x')}{\sqrt{xx'}} \right] \sqrt{xx'}\psi(x,x',\xi)dx'd\xi$$

A solution of this equation is

$$f(x) = C\sqrt{x}e^{-hx}$$

which is the Maxwell distribution. From what has been said previously it follows that there are infinitely many other solutions, which are not useful however since $f(x)$ comes out negative or imaginary for some values of x. Hence, it follows very clearly that Maxwell's attempt to prove *a priori* that his solution is the only one must fail, since it is not the only one but rather it is the only one that gives purely positive probabilities, and therefore it is the only useful one.

III. Diffusion, viscosity, and heat conduction of a Gas

Here we shall make room for a few remarks pertaining to the case that the distribution is not completely irregular, but is still not what we have called uniform, so that not all directions of velocity are equivalent; this corresponds to the case of viscosity and heat conduction. Let

$$f(\xi, \eta, \zeta, x, y, z, t)d\xi d\eta d\zeta$$

be the number of molecules in unit volume at the position (x, y, z) in the gas for which the velocity components in the x-direction lie between ξ and $\xi + d\xi$, those in the y-direction lie between η and $\eta + d\eta$, and those in the z-direction lie between ζ and $\zeta + d\zeta$. A collision is determined by the velocity components ξ, η, ζ and ξ_1, η_1, ζ_1 of the two colliding molecules before the collision and by the quantities b and ϕ. (The latter two quantities, as well as V, k, A_2, X, etc., which will appear later, have the same meaning as in Maxwell's paper.†) The velocity components after the collision: ξ, η, ζ and ξ_1, η_1, ζ_1 are functions of these eight variables. If we write for brevity $d\omega_1$ for $d\xi_1 \, d\eta_1 \, d\zeta_1$ and denote by f the value of the function $f(\xi, \eta, \zeta, x, y, z, t)$, and by f_1, f' and f'_1 the values of this function when one substitutes for (ξ, η, ζ) the variables (ξ_1, η_1, ζ_1), (ξ', η', ζ') or $(\xi'_1, \eta'_1, \zeta'_1)$ respectively, then the function f must satisfy the differential equation

$$\left. \begin{array}{l} \dfrac{\partial f}{\partial t} + \xi \dfrac{\partial f}{\partial x} + \eta \dfrac{\partial f}{\partial y} + \zeta \dfrac{\partial f}{\partial z} + X \dfrac{\partial f}{\partial \xi} + Y \dfrac{\partial f}{\partial \eta} + Z \dfrac{\partial f}{\partial \zeta} \\ + \int d\omega_1 \int b \, db \int d\phi \, V(f f_1 - f' f'_1) = 0 \end{array} \right\} \quad (44)$$

This can easily be seen by imagining that the volume element moves with velocity (ξ, η, ζ) and considering how the distribution of states is changed by collisions. If the gas is enclosed by fixed walls, then it follows again from equation (44) that E can only be decreased by molecular motion, if one sets

$$E = \iiint\!\!\iiint f \log f \, dx \, dy \, dz \, d\xi \, d\eta \, d\zeta$$

---

† J. C. Maxwell, *Phil. Mag.* **35**, 129, 185 (1868) [Selection 1].

which expression is proportional to the entropy of the gas. In order to give just one example for the case of different boundary conditions, let the repulsion of two molecules be inversely proportional to the fifth power of their distance. X, Y, and Z will always be zero in the following. We shall set

$$f = A(1+2hay\xi+c\xi\eta)e^{-h(\xi^2+\eta^2+\zeta^2)} \qquad (45)$$

where the two constants a and c are very small. If we substitute this value into equation (44), neglect squares and products of a and c, and carry out the integration over b and ϕ just as Maxwell has done,† then we find that equation (44) is satisfied when

$$c = -\frac{2ha}{3A_2k\rho}$$

The formula (45) thus gives a possible distribution of states: one in which each layer parallel to the xz plane moves in the direction of the x axis with velocity ay, if y is the y-coordinate of that layer. This is the simplest case of viscous flow. The viscosity constant is then the momentum transported through unit surface in unit time divided by $-a$:

$$-\frac{\rho\overline{\xi\eta}}{a} = -\frac{\rho\iiint\xi\eta f d\xi d\eta d\zeta}{a \iiint f d\xi d\eta d\zeta} = \frac{1}{6A_2kh} = \frac{p}{3A_2k\rho}$$

just as Maxwell has already found. The notations are those of Maxwell. A more general expression is the following:

$$\left.\begin{aligned}f = A\Bigl[&1-\frac{2ht}{3}\left(\frac{\partial u}{\partial x}+\frac{\partial v}{\partial y}+\frac{\partial w}{\partial z}\right)(\xi^2+\eta^2+\zeta^2)\\ &+2h(u\xi+v\eta+w\zeta)+\alpha\xi^2+\beta\eta^2+\gamma\zeta^2+\alpha'\eta\zeta+\beta'\xi\zeta\\ &+\gamma'\xi\eta\Bigr]e^{-h(\xi^2+\eta^2+\zeta^2)}\end{aligned}\right\} \qquad (46)$$

---

† Maxwell, *op. cit.* p. 141–4.

This expression likewise satisfies equation (44), when u, v, and w are linear functions of x, y, z, and

$$\alpha = -\frac{2h}{3A_2 k\rho} \frac{\partial u}{\partial x}, \quad \alpha' = -\frac{2h}{3A_2 k\rho} \left(\frac{\partial v}{\partial z} + \frac{\partial w}{\partial z}\right)$$

β, γ, β' and γ' have similar values. The expression (46) represents an arbitrary motion of the gas in which the velocity components u, v, w at the point with coordinates x, y, z are linear functions of these coordinates. Unless

$$\frac{\partial u}{\partial x} + \frac{\partial v}{\partial y} + \frac{\partial w}{\partial z} = 0$$

the density and temperature change with time.

If one calculates

$$\overline{\xi^2}, \overline{\eta^2}, \overline{\xi\eta} \ldots$$

using the expression (46), he obtains again the values found by Maxwell. If $\partial^2 u/\partial x^2$ is different from zero, one obtains one more term in equation (44) which does not vanish, namely

$$\frac{\partial \alpha}{\partial x} \cdot \xi^3 e^{-h(\xi^2 + \eta^2 + \zeta^2)}$$

whose average value we can take to be approximately

$$\left.\begin{aligned}-\frac{\partial \alpha}{\partial x}(\overline{\xi^2})^{\frac{3}{2}} e^{-h(\xi^2+\eta^2+\zeta^2)} &= \frac{2h}{3A_2 k\rho}\sqrt{\frac{p^3}{\rho^3}}e^{-h(\xi^2+\eta^2+\zeta^2)}\frac{\partial^2 u}{\partial x^2}\\ &= \frac{\mu}{\rho}\sqrt{\frac{\rho}{p}\frac{\partial^2 u}{\partial x^2}}e^{-h(\xi^2+\eta^2+\zeta^2)}\end{aligned}\right\} \quad (47)$$

One sees easily that this term vanishes in comparison to the other terms in equation (44) if one substitutes therein the value of f from equation (46), so that equation (44) is still approximately satisfied. In calculating the distribution of states one should expand the quantities u, v, w in a Taylor series and retain the first powers of

x, y, and z. The first term of (44), after substituting the value of f from (46), would thus be

$$-\frac{2h}{3}\xi^2 \frac{\partial u}{\partial x} e^{-h(\xi^2+\eta^2+\zeta^2)}$$

Its average value is therefore

$$\frac{1}{3}\frac{\partial u}{\partial x} e^{-h(\xi^2+\eta^2+\zeta^2)}$$

If one calculates the ratio of (47) to this quantity, for air at 0°C and normal atmospheric pressure, he finds that it is about

$$0\cdot 00009 \text{ mm } \frac{\partial^2 u/\partial x^2}{\partial u/\partial x}$$

It is therefore negligible when

$$\frac{\partial u/\partial x}{\partial^2 u/\partial x^2}$$

is about 1 mm, in other words if values of $\partial u/\partial x$ at points 1 mm apart are in the ratio 1 : 2 on the average. Only when $\partial u/\partial x$ starts to change this much over distances of the order of a mean free path will the ratio become significant.

The value

$$f = A[1+ax+by+cz-(a\xi+b\eta+c\zeta)t]\, e^{-h(\xi^2+\eta^2+\zeta^2)}$$

also satisfies equation (44).

For a mixture of two kinds of gas, we shall indicate quantities pertaining to the second kind by an asterisk, so that p and p_* will be the partial pressures, m and m_* the masses of molecules of the two kinds. Then in place of equation (44) we have:

$$\left.\begin{aligned}&\frac{\partial f}{\partial t}+\frac{\partial f}{\partial x}+\frac{\partial f}{\partial y}+\frac{\partial f}{\partial z}+\int d\omega_1 \int b\,db \int d\phi\, V(ff_1-f'f_1')\\ &\qquad +\int d\omega_* \int b\,db \int d\phi\, V(ff_*-f'f_*')=0\end{aligned}\right\} \quad (44^*)$$

and a similar equation for the second kind of gas. The simplest case of diffusion corresponds to

$$\left. \begin{array}{l} f = \sqrt{m^3 h^3/\pi^3}\, N(1+2hmu\xi)\, e^{-hm(\xi^2+\eta^2+\zeta^2)} \\[4pt] f_* = \sqrt{m_*^3 h^3/\pi^3}\, N_*(1+2hm_* u\xi)\, e^{-hm_*(\xi^2+\eta^2+\zeta^2)} \end{array} \right\} \quad (46^*)$$

where N and N_* are functions of x, but Nu and $N_* u_*$ are constant. None of these quantities depends on time. Equation (44*) is satisfied when

$$\frac{dN}{dx} + NN_* 2hmm_*(u-u_*)A_1 k = 0$$

A similar equation must hold for N_*. We must also have $N + N_* = $ constant $=$ number of molecules of both kinds in unit volume; hence, it follows that $Nu = -N_* u_* = $ number of molecules of one kind of gas that go through a unit cross-section in unit time. The diffusion constant is

$$-\frac{Nu}{dN/dx} = \frac{1}{(N+N_*)2hmm_* A_1 k} = \frac{pp_*}{A_1 k \rho \rho_*(p+p_*)}$$

since

$$2h = \frac{N+N_*}{p+p_*} = \frac{N}{p} = \frac{N_*}{p_*}$$

To obtain the equations of motion, one multiplies equation (44) or (44*) by $m\xi\, d\omega$ (where $d\omega = d\xi\, d\eta\, d\zeta$) and integrates over all ξ, η, and ζ. The first four terms of these equations become

$$\frac{\partial(\rho u)}{\partial t} + \frac{\partial(\rho \overline{\xi^2})}{\partial x} + \frac{\partial(\rho \overline{\xi\eta})}{\partial y} + \frac{\partial(\rho \overline{\xi\zeta})}{\partial z}$$

or since

$$\frac{\partial \rho}{\partial t} + \frac{\partial(\rho u)}{\partial x} + \frac{\partial(\rho v)}{\partial y} + \frac{\partial(\rho w)}{\partial z} = 0$$

they become

$$\rho\left(\frac{\partial u}{\partial t} + u\frac{\partial u}{\partial x} + v\frac{\partial u}{\partial y} + w\frac{\partial u}{\partial z}\right) + \frac{\partial(\rho \overline{\xi'^2})}{\partial x} + \frac{\partial(\rho \overline{\xi'\eta'})}{\partial y} + \frac{\partial(\rho \overline{\xi'\zeta'})}{\partial z}$$

where we have set
$$\xi = \xi' + u, \quad \eta = \eta' + v, \quad \zeta = \zeta' + w$$

The other terms are just minus the momentum transported by molecular collisions, which is zero if no second gas is present. The momentum transport, aside from that resulting from the pressure forces,
$$\left(-\frac{\partial(\rho\overline{\xi'^2})}{\partial x} - \frac{\partial(\rho\overline{\xi'\eta'})}{\partial y} - \frac{\partial(\rho\overline{\xi'\zeta'})}{\partial z}\right)$$
is therefore equal to the acceleration multiplied by the density:
$$\left(\frac{\partial u}{\partial t} + u\frac{\partial u}{\partial x} + v\frac{\partial u}{\partial y} + w\frac{\partial u}{\partial z}\right)$$

The latter equations are valid for any force law. On the other hand, the expressions (45), (46) and (46*) are correct only when the repulsion between two molecules is inversely proportional to the fifth power of their distance. For any other force law—for example, when the molecules bounce off each other like elastic spheres—the expressions (45), (46) and (46*) do not satisfy equations (44) and (44*), so that for other force laws the velocity distribution for diffusion, viscosity, etc., does not have such a simple form. For the case of diffusion, we have to represent f in the following form:

$$A[1 + a\xi + b\xi^3 + c(\eta^2 + \zeta^2)\xi + d\xi^5 \ldots]e^{-h(\xi^2 + \eta^2 + \zeta^2)} \quad (47^*)$$

and I see no other way to solve equation (44*) than by successive determination of the coefficients $a, b, c \ldots$ For all other force laws, therefore, the velocity distribution of a diffusing gas is not the same as if it were moving alone in space with its diffusion velocity u. This is because molecules with different velocities also have different diffusion velocities, so that the velocity distribution will be continually disturbed. Since the terms in the expression (47*) with $\xi^3, \xi\eta^2, \ldots$, lead to terms in the diffusion constant that are of the same order of magnitude as the one with ξ, the diffusion constant cannot be obtained accurately by first leaving out the terms for momentum transport. Nevertheless, the error thereby incurred should scarcely be very large. The same holds for viscosity and heat conduction. Indeed it is not merely the values of the diffusion, viscosity, etc., constants that are in question here, but rather their constancy, in the case of force laws other than Maxwell's.

The case of heat conduction in the direction of the x-axis corresponds to the following value of f, assuming Maxwell's force law:

$$f = A[1 + ax(\xi^2 + \eta^2 + \zeta^2) + bx + c\xi + g\xi(\xi^2 + \eta^2 + \zeta^2)]e^{-h(\xi^2 + \eta^2 + \zeta^2)}$$

whence it follows that

$$\xi\frac{\partial f}{\partial x} + \eta\frac{\partial f}{\partial y} + \zeta\frac{\partial f}{\partial z} = \xi A e^{-h(\xi^2 + \eta^2 + \zeta^2)}[a(\xi^2 + \eta^2 + \zeta^2) + b]$$

When one substitutes this value of f into equation (44) and carries out all the integrations according to Maxwell's procedure, the last term of (44) reduces to

$$2gA_2kMN\xi\left(\xi^2 + \eta^2 + \zeta^2 - \frac{5}{2h}\right)Ae^{-h(\xi^2 + \eta^2 + \zeta^2)}$$

In order that equation (44) may be satisfied, we must take

$$a = -2gA_2kMN, \quad b = 5gA_2kMN \cdot \frac{1}{h} = -\frac{5a}{2h}$$

The mass passing through unit surface in unit time is

$$\rho\bar{\xi} = \rho\left(\frac{c}{2h} + \frac{5g}{4h^2}\right)$$

If the heat conduction is not associated with any mass motion, then we must have

$$c = -\frac{5g}{2h}$$

If we denote the absolute temperature by T, with B a constant, then

$$T = \frac{M}{2}(\overline{\xi^2} + \overline{\eta^2} + \overline{\zeta^2}) \cdot B = \frac{3MB}{h}\left(1 + \frac{ax}{h}\right)$$

hence, neglecting infinitesimal terms, we have

$$\frac{dT}{dx} = \frac{a}{h}T$$

The amount of kinetic energy passing through unit surface in unit time is

$$L = \frac{\rho}{2}(\overline{\xi^3} + \overline{\xi\eta^2} + \overline{\xi\zeta^2})$$

The mean value can easily be calculated by using the assumed value of f. One obtains, after computing all the integrals that arise of the form

$$\iiint \xi^2(\xi^2+\eta^2+\zeta^2)^n e^{-h(\xi^2+\eta^2+\zeta^2)}\,d\xi\,d\eta\,d\zeta$$

which can best be done by differentiating $N\overline{\xi^2}$ with respect to h

$$L = \frac{5}{8}\frac{\rho g}{h^3} = \frac{5MNg}{8h^3}$$

The heat conduction constant is

$$C = -\frac{L}{dT/dx} = -\frac{5}{8}\frac{MNg}{h^3} \cdot \frac{h}{aT} = \frac{5}{16h^2 T A_2 k_1}$$

Noting that

$$\frac{p}{\rho} = \overline{\xi^2} = \frac{1}{2h}$$

one obtains finally

$$C = \frac{5p^2}{4\rho^2 T A_2 k_1}$$

Since I consider the gas molecules to be simple mass-points, Maxwell's parameter β is equal to 1 and the ratio of specific heats is $\gamma = 1\frac{2}{3}$. If we denote the specific heat (of unit mass of the gas) at constant volume in the usual thermal units by w, and the mechanical equivalent of heat by $1/J$, then according to a well-known formula

$$(\gamma - 1)w = \tfrac{2}{3}w = \frac{pJ}{\rho T}$$

The heat conduction constant, measured in the usual thermal units, is therefore

$$C' = JC = \frac{5wp}{6kA_2\rho} = \tfrac{5}{2}w\mu$$

where μ is the viscosity constant. This value of the heat conduction constant is $\tfrac{3}{2}$ times as large as the one found by Maxwell, because of Maxwell's error in deriving his equation (43) from (39).

The gas molecules have been assumed to be simple mass-points, since with this assumption all the calculations can be carried out exactly. This assumption is clearly not fulfilled in nature, so that the above formulae require some modification if they are to be compared with experiments. If one includes the intramolecular motion following Maxwell's method, he obtains

$$C = \frac{5\beta p^2}{4\rho^2 T A_2 k}, \quad C = \tfrac{5}{2} w\mu$$

However, this seems very arbitrary to me, and if one includes intramolecular motion in some other way, he can easily obtain significantly different values for the heat conduction constant. It appears that an exact calculation of this constant from the theory is impossible until we know more about the intramolecular motion. Since the heat conduction constant, whose value was previously thought to be incapable of experimental measurement, has been determined so exactly by Stefan,[†] it appears that our

---

[†] [J. Stefan, *Wien. Ber.* **65**, 45 (1872); see also *Wien. Ber.* **47**, 81, 327 (1863); **72**, 69 (1876).]

experimental knowledge is here much better than our theoretical knowledge.

It seems unnecessary to explain what would happen when the function f is not a linear function of x, or when at the same time there takes place heat conduction or motion in other directions.

IV. Treatment of polyatomic gas molecules

We have assumed up to now that each molecule is a single mass-point. This is certainly not the case for the gases existing in nature. We would clearly come closer to the truth if we assumed that each molecule consists of several mass-points (atoms). The properties of such polyatomic gas molecules will be considered in the present section. Note that the previous definitions of symbols are not necessarily retained in this section.

Let the number of mass-points or atoms in a molecule be r. These may be held together by any type of force; we assume only that the force between two points depends only on the distance between them, and acts along the line between them, and that the force is such that the atoms of a given molecule can never be completely separated from each other. I will call this force the internal force of the molecule. During by far the largest part of the time, these forces act only among the atoms in the same molecule. Only when the molecule comes very close to another one will its atoms act on those of the other molecule, and conversely. I call this process, during which two molecules are so close that their atoms act on each other significantly, a collision; and the force which the atoms of different molecules exert on each other will be called the collision force. I assume that this force is also a function of distance that acts along the line of centres, and that the atoms of the two molecules are not exchanged in a collision, but rather that each molecule consists of the same atoms after the collision as before. In order to be able to define precisely the instant when the collision begins, I assume the interaction of two molecules begins whenever the distance between their centres of gravity is equal to a certain quantity l. This distance will then become smaller than l, then increase and when it is again equal to l the collision ends. Of course in reality the moment of beginning a collision is probably not so sharply defined. Our conclusions are not changed if collisions are defined as in my previous paper.† I will therefore retain the above assumption, which is otherwise not inferior in generality to any other assumption, since we have merely assumed that as long as the distance of centres is greater than l, no interaction takes place. If the distance is equal to l, then in many cases the interaction will still be zero, and will not start until later.

In order to define the state of a molecule at a certain time t, we consider three fixed perpendicular directions in space. Through the point at which the centre of our molecule finds itself at time t, we draw three rectangular coordinate axes parallel to these three directions, and we denote the coordinates of the molecule with respect to these axes by $\xi_1, \eta_1, \zeta_1, \xi_2, \ldots, \zeta_r$. Let c_1 be the velocity of the first atom, and u_1, v_1, w_1 its components in the directions of the coordinate axes; the same quantities for the second atom will be c_2, u_2, v_2, w_2; for the third, c_3, u_3, v_3, w_3, and so forth. Then the state of our molecule at time t will be completely determined when we know the values of $6r-3$ quantities

$$\xi_1, \eta_1, \zeta_1, \xi_2, \ldots \xi_{r-1}, \eta_{r-1}, \zeta_{r-1}, u_1, v_1, w_1, u_2, \ldots w_r \quad (A)$$

ξ_r, η_r, ζ_r are functions of the other ξ, η, ζ, since the centre of gravity is the origin of coordinates. The coordinates of the centre of gravity of our molecule with respect to the fixed coordinate axes do not determine its state but only its position. When our molecule is not interacting with any others, then only the internal

---

† L. Boltzmann, *Wien. Ber.* **63**, 397 (1871).

force between the atoms is present. We can therefore establish, between the time and the $6r-3$ quantities (A), as many differential equations, which we shall call the equations of motion of the molecule. These equations will have $6r-3$ integrals, through which the values of the variables (A) can be expressed as functions of time and of the values of these quantities at some initial time. If we eliminate the time, there remain $6r-4$ equations with the same number of arbitrary constants of integration. Let these be

$$\phi_1 = a_1, \; \phi_2 = a_2, \ldots, \; \phi_\rho = a_\rho$$

where the a's are constants of integration while the ϕ's are functions of the variables (A). ρ is equal to $6r-4$. We can therefore express all but one of the variables (A) as functions of this one variable and the $6r-4$ constants of integration. I will always call this one variable x; it can be either one of the ξ, η, ζ, or one of the u, v, w. As long as the molecule does not collide with another one, these variables (A) satisfy the equations of motion of the molecule, and hence the a's remain constant and the value of each of the variables (A) depends only on the value of x. I will say that $a_1, a_2, \ldots, a_\rho$ determine the kind of motion of the molecule, while x determines the phase of the motion. As long as the molecule does not collide with another one, only the variable x determining the phase will change. But when the molecule does collide with another one, then the a's change their values; the kind of motion of the molecule also changes.

We shall now assume that we again have a space $\mathscr{R}$ containing a large number of molecules. All these molecules are equivalent—i.e. they all consist of the same number of mass-points, and the forces acting between them are identical functions of their relative distances for all the molecules. If we now choose, somewhere in the space $\mathscr{R}$, a smaller space of volume R, still large compared to the distance between two molecules, then there will be RN molecules in this smaller space. Of these,

$$Rf(t, a_1, a_2 \ldots a_\rho) da_1 da_2 \ldots da_\rho$$

will be in such a state at time t that

$$\left.\begin{array}{l}\phi_1 \text{ lies between } a_1 \text{ and } a_1 + da_1 \\ \phi_2 \text{ lies between } a_2 \text{ and } a_2 + da_2 \ldots\end{array}\right\} \quad (B)$$

The constants a determine the kind of motion of a molecule; hence, if the function f is given, then the number of molecules in each of the various kinds of motion at time t in the space R is determined. Hence, we say that the function f determines the distribution of various kinds of motion among the molecules at time t. I assume again that there is already at the initial time, and hence at all subsequent times, a uniform distribution; i.e. the function f is independent of the position of the space R as long as that space is very large compared to the average distance between two neighbouring molecules. For the sake of brevity I will say that a molecule is in a space when its centre of gravity is in that space. We now assume that the value of the function f at time $t = 0$, namely $f(0, a_1, a_2, \ldots)$ is given; we wish to determine the value of f at any later time. The constants a change their values only through collisions; hence f can only change through collisions, and our problem is to establish the equations that determine the variation of f. We must again compute how many collisions occur during a certain time Δt such that before the collision the a's lie between the limits (B), and also how many occur such that after the collision the a's lie between the limits (B). If we add the first number to $f(t, a_1, a_2 \ldots)$ and subtract the latter from it, then we obtain the number of molecules for which the a's lie between the limits (B) after time Δt, namely the quantity $f(t+\Delta t, a_1, a_2 \ldots) da_1 da_2 \ldots$

We now consider some collision between two molecules; for the first molecule, the a's are assumed to lie between the limits (B) before the collision. For the second,

$$\left.\begin{array}{l}\phi_1 \text{ lies between } a_1' \text{ and } a_1' + da_1' \\ \phi_2 \text{ lies between } a_2' \text{ and } a_2' + da_2' \text{ and so forth}\end{array}\right\} \quad (C)$$

The collision is not yet completely determined, of course; the phases of the two colliding molecules, as well as their relative position at the beginning of the collision, must also be given. Let the phase of the first molecule be such that

$$x \text{ lies between } x \text{ and } x+dx, \qquad \text{(D)}$$

while for the second molecule,

$$x \text{ lies between } x' \text{ and } x'+dx' \qquad \text{(E)}$$

In order to determine the relative position of the two molecules at the beginning of the collision, we denote the angle between the line of centres and the x-axis by θ, the angle between the xy-plane and a plane through the x-axis parallel to the line of centres by ω, and assume that at the beginning the collision

$$\left.\begin{array}{l}\theta \text{ lies between } \theta \text{ and } \theta+d\theta \\ \omega \text{ lies between } \omega \text{ and } \omega+d\omega\end{array}\right\} \qquad \text{(F)}$$

All collisions that take place in such a way that the conditions (B), (C), (D), (E) and (F) are satisfied, I will call collisions of type (G). The next question is, how many collisions of type (G) occur in unit volume during a certain time Δt? We make the assumption that the internal motions are so rapid, and the collisions so infrequent, that a molecule passes through all its possible phases of motion between one collision and the next. We can then choose Δt so large that each molecule goes through all possible phases of motion during Δt, yet so small that only a few collisions take place during Δt, so that f changes only slightly. We consider some particular molecule whose kind of motion lies between the limits (B); we call it the B molecule. We assume that it passes through all possible phases several times during the time Δt; it can then be shown that the sum of all times during which it has the phase (D) in the course of time Δt is to the total time Δt in the same ratio as $s\,dx$ to $\int s\,dx$, so that the sum of all these times is

$$\tau = \Delta t \frac{s dx}{\int s dx} \qquad (48)$$

where s is given by the following equation:

$$\frac{1}{s} = \sum \pm \frac{\partial a_1}{\partial \xi_1} \frac{\partial a_2}{\partial \eta_1} \cdots \frac{\partial a_\rho}{\partial w_r}$$

The integration is over all possible values of x, i.e. all possible phases. The product $s\,dx$ is always to be taken with the positive sign.

In the functional determinant there will occur the derivatives with respect to all the variable $\xi_1, \eta_1, \ldots \zeta_{r-1}, u_1, v_1, \ldots w_r$ except for x; it can therefore be expressed as a function of x and of the integration constants a. The theorem just stated can be

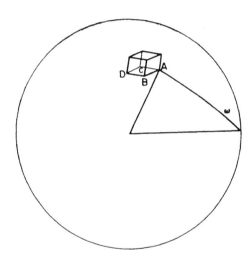

FIG. 4.

shown to be completely similar to Jacobi's principle of the last multiplier;† since I have already given the proof in a previous paper,‡ I will not repeat it here. I have also shown that the situation is not changed when one variable is no longer sufficient to determine the phase.

The total length of all paths which the centre of gravity of (B) traverses relative to the centre of a molecule with property (C),

---

† L. Boltzmann, *Wien. Ber.* **63**, 679 (1871).

‡ See especially note 1 of L. Boltzmann, *Wien. Ber.* **63**, 679 (1871).

during the time interval denoted above by τ, is equal to $\gamma\tau$, where γ is the relative velocity of the centres of the two molecules. If we now describe around the centre of the B molecule a sphere of radius l, then the set of all points of this sphere for which the angles θ and ω lie between the limits (F) form an infinitesimal rectangle $ABCD$ (see Figure 4) of surface area $l^2 \sin \theta \, d\theta \, d\omega$. If we imagine that this rectangle is rigidly connected to the centre of the B molecule, then the total length of the paths which it traverses during τ relative to the C molecule is likewise $\gamma\tau$. All these paths make the same angle with the coordinate axes (neglecting infinitesimal quantities); hence, they also make the same angle (ε) with the plane of the rectangle $ABCD$, since the velocities of all the atoms in B are enclosed within infinitesimally near limits by the conditions (B) and (D). The total volume swept out by our rectangle during τ, if it moves with its relative velocity with respect to C, is

$$V = l^2 \sin \theta \, d\theta \, d\omega \cdot \sin \varepsilon \cdot \gamma\tau \qquad (49)$$

and it is easily seen that all molecules inside this volume will collide with the B molecule in such a way that conditions (F) are fulfilled. It is now a question of finding how many molecules with the kind of motion (C) and phase (E) will lie in this volume. We know that in unit volume there will be $f(t, a'_1, a'_2, \ldots) da'_1 \, da'_2 \ldots$ molecules with the kind of motion (C); hence, since the distribution of kinds of motion is uniform, there will be $Vf(t, a'_1, a'_2 \ldots) da'_1 \, da'_2 \ldots$ in volume V. However, not all of these molecules will have phase (E). Rather, the number that have this phase will be in the same ratio to the total number as the time during which a molecule has phase (E) is to the time during which it passes through all possible phases. As in equation (48) we can write this as the ratio of $s'dx'$ to $\int s'dx'$, where

$$\frac{1}{s'} = \sum \pm \frac{\partial a'_1}{\partial \xi'_1} \frac{\partial a'_a}{\partial \eta'_1} \ldots$$

This number is therefore

$$v = Vf(t, a_1', a_2' \ldots) da_1' da_2' \ldots \frac{s'dx'}{\int s'dx'} \tag{50}$$

Here we have assumed that, among the molecules with property (C), the different phases during time τ are distributed in the same way as during time Δt, so that there is no particular correlation between the phases (D) and (E) of the two molecules. If the period of vibration of the molecule with property C is not commensurable with that of the molecule with property B, this is obvious. However, if the two periods of vibration for all or a finite number of molecule pairs are commensurable, then this must be assumed as a property of the initial distribution of states, which is then maintained for all subsequent times.

The expression (50) gives us the number of molecules in volume V for which conditions (C) and (E) are satisfied, and we know that all these, and only these will collide during Δt with the B molecules in such a way that conditions (C), (D), (E) and (F) are satisfied. There are $f(t, a_1, a_2 \ldots) da_1 da_2 \ldots$ molecules with property B in unit volume; if we multiply by this number, we obtain $dn = vf(t, a_1, a_2 \ldots) da_1 da_2 \ldots$ for the number of molecule pairs that collide in unit volume during time Δt in such a way that all five conditions (B), (C), (D), (E) and (F) are satisfied; this is therefore the number of collisions of type (G) during this time. Taking account of equations (48), (49) and (50), we obtain for this number the value:

$$\left. \begin{array}{l} dn = \dfrac{f(t, a_1, a_2 \ldots) f(t, a_1', a_2', \ldots)}{\int s dx \int s' dx'} l^2 \gamma \sin\theta \sin\varepsilon \, d\theta d\omega \Delta t \\ \quad \times s dx da_1 da_2 \ldots s' dx' da_1' da_2' \ldots \end{array} \right\} \tag{51}$$

If the values of the quantities

$$a_1, a_2, \ldots a_1', a_2', \ldots x, x', \theta, \omega \tag{J}$$

are given at the beginning of the collision, then the nature of the collision is completely characterized; the values of these quantities at the end of the collision can therefore be calculated. These values after the collision will be denoted by the corresponding capital letters. The quantities

$$A_1, A_2, \ldots A'_1, A'_2, \ldots X, X', \Theta, \Omega \qquad \text{(K)}$$

can therefore be expressed as functions of the variables (J). I will denote the set of conditions (B), (C), (D), (E) and (F) as the conditions (G). All collisions for which the values of the variables at the beginning of the collision satisfy (G) will proceed in a very similar way. Hence, at the end of the collisions the values of the variables will also lie between certain infinitesimally near limits. We will assume that for all these, and only these, collisions, the variables after the collision lie between the limits

$$\left.\begin{array}{l} A_1 \text{ and } A_1+dA_1, A_2 \text{ and } A_2+dA_2, \ldots A'_1 \text{ and } A'_1+dA'_1 \\ A'_2 \text{ and } A'_2+dA'_2, \ldots X \text{ and } X+dX, X' \text{ and } X+dX' \\ \Theta \text{ and } \Theta + d\Theta, \Omega \text{ and } \Omega + d\Omega \end{array}\right\} \text{(H)}$$

Since the variables (K) are functions of the variables (J), we can replace differentials of (J) in equation (51) by differentials of (K); for example, in place of the four differentials $dx, dx', d\theta$, and $d\omega$ we can use differentials of A_1, A_2, A'_1 and A'_2; note that we do not necessarily have to use the four quantities A_1, A_2, A'_1 and A'_2, but rather we could just as well have chosen instead four other variables from (K). We then obtain

$$\left.\begin{array}{l} dn = \dfrac{f(a_1, a_2 \ldots)f(a'_1, a'_2 \ldots)}{\int s dx . \int s' dx'} ss's_1 l^2 \gamma \sin\theta \sin\varepsilon . \Delta t \\ \times da_1 da_2 \ldots da_\rho da'_1 \ldots da'_\rho dA_1 dA_2 dA_3 dA_4 \end{array}\right\} \text{(52)}$$

where

$$\frac{1}{s_1} = \sum \pm \frac{\partial A_1}{\partial x} \cdot \frac{\partial A_2}{\partial x'} \cdot \frac{\partial A'_1}{\partial \theta} \cdot \frac{\partial A'_2}{\partial \omega}$$

This is the number of collisions that take place in unit volume during time Δt such that before the collision the quantities $\phi_1, \phi_2 \ldots \phi'_\rho$ lie between the limits (B) and (C), and after the collision A_1, A_2, A'_1 and A'_2 lie between

$$A_1 \text{ and } A_1+dA_1 \ldots A'_2 \text{ and } A'_2+dA'_2$$

If we keep $a_1, a_2 \ldots a_\rho$ fixed in this expression and integrate over $a'_1, a'_2 \ldots a'_\rho, A_1, A_2, A'_1$ and A'_2, then we obtain all the collisions that occur such that the a's for one of the colliding molecules lie between the limits (B) before the collision, with no other restriction. This number is therefore:

$$dn' = da_1 da_2 \ldots da_p \Delta t \iint \ldots \frac{f(t, a_1, a_2 \ldots) f(t, a_1', a_2' \ldots)}{\int s\, dx \cdot \int s'\, dx} \\ \times ss' s_1 l^2 \gamma \sin\theta \sin\varepsilon\, da_1' da_2' \ldots dA_1' dA_2' \quad\quad (53)$$

Each such collision decreases by one the number of molecules for which the a's lie between the limits (B), so we have to subtract the number dn' from $f(t, a_1, a_2 \ldots) da_1 da_2 \ldots$ We still have to add the number of collisions in which the a's lie between the limits (B) after instead of before the collision. Let this number be dN'. Then

$$f(t, a_1, a_2 \ldots) da_1 da_2 \ldots da_p - dn' + dN'$$

is the number of molecules for which the a's lie between the limits (B) at time $t + \Delta t$. Therefore

$$f(t, a_1, a_2 \ldots) da_1 da_2 \ldots da_p - dn' + dN' \\ = f(t + \Delta t, a_1, a_2 \ldots) da_1 da_2 \ldots da_p \quad\quad (54)$$

The number dN' can be found in the following way. We obtained the expression (51) for the number of collisions such that the conditions (G) are satisfied before the collision. For all these collisions, the variables satisfy the conditions (H) at the end of the collision. We merely have to interchange the small and large letters in the expression in order to arrive at the number of collisions for which the values of the variables satisfy at the beginning of the collision conditions that are completely identical with (H) except that the positions of the centres of the two molecules are interchanged, and their line of centres has the opposite direction. The latter number is therefore

$$dN = \frac{f(t, A_1, A_2 \ldots) f(t, A_1', A_2' \ldots)}{\int S\, dX \int S'\, dX'} SS' l^2 \Gamma \sin\Theta \sin E \\ \times d\Theta d\Omega \Delta t\, dX\, dA_1 dA_2 \ldots dA_p dX' dA_1' dA_2' \ldots dA_p' \quad\quad (55)$$

where $S, S', E \ldots$ are constructed from $s, s', \varepsilon \ldots$ by interchanging the variables (J) with the variables (K). For all these collisions, however, the values of the variables at the end of the collision will lie between the limits (G). For it is clear that a collision at the beginning of which the conditions (G) are satisfied will proceed in just the opposite way to one at the beginning of which conditions similar to (H) are satisfied. Whereas for the former collisions the conditions (H) are satisfied at the end, for the latter collisions, conversely, conditions similar to (G) are satisfied at the end. Equation (55) therefore gives us the number of collisions at the end of which the values of the variables satisfy the conditions (G).

We next replace the differentials of X, X', Θ and Ω by differentials of a_1, a_2, a'_1 and a'_2, so that we obtain

$$dN = \frac{f(t, A_1, A_2 \ldots)f(t, A'_1, A'_2 \ldots)}{\int S dX \int S' dX'} SS'S_1 l^2 \Gamma \sin \Theta \sin E$$
$$\times \Delta t dA_1 dA_2 \ldots dA_\rho dA'_1 \ldots dA'_\rho da_1 da_2 da'_1 da'_2$$

In this formula we have

$$\frac{1}{S_1} = \pm \frac{\partial a_1}{\partial X} \cdot \frac{\partial a_2}{\partial X'} \cdot \frac{\partial a'_1}{\partial \Theta} \cdot \frac{\partial a'_2}{\partial \Omega}$$

where, in forming the partial derivatives in the functional determinant, the a's must be considered to be expressed as functions of the independent quantities (G). If we now introduce, in place of the differentials of $A_3, A_4 \ldots A_\rho, A'_3 \ldots A'_\rho$ those of $a_3, a_4 \ldots a_\rho$, $a'_3 \ldots a'_\rho$, then we obtain:

$$\left. \begin{array}{l} dN = \dfrac{f(t, A_1, A_2 \ldots)f(t, A'_1, A'_2 \ldots)}{\int S dX \int S' dX'} SS'S_1 \sigma l^2 \Gamma \sin \Theta \sin E \\ \times \Delta t da_1 da_2 \ldots da'_\rho dA_1 dA_2 dA'_1 dA'_2 \end{array} \right\} \quad (56)$$

where

$$\sigma = \sum \pm \frac{\partial A_3}{\partial a_3} \cdot \frac{\partial A_4}{\partial a_4} \ldots \frac{\partial A'_\rho}{\partial a'_\rho}$$

In this functional determinant, $A_3, A_4 \ldots A'_\rho$ are to be considered as functions of the independent variables $a_1, a_2 \ldots a'_\rho, A_1, A_2, A'_1, A'_2$. In the expression (56) we shall assume that all quantities are expressed as functions of

$$a_1, a_2 \ldots a_\rho, a'_1, \ldots a'_\rho, A_1, A_2, A'_1, A'_2$$

Furthermore, we consider $a_1, a_2 \ldots a_\rho$ as fixed and integrate over $a'_1, a'_2 \ldots A'_2$. We thereby obtain the number of collisions after which one of the colliding molecules has values of the a's between the limits (B), while everything else is arbitrary; thus we obtain precisely the quantity called dN'. Therefore

$$\left. \begin{array}{l} dN' = da_1 da_2 \ldots da_\rho \Delta t \iint \ldots \dfrac{(t, A_1 \ldots) f(t, A'_1 \ldots)}{\int S dX \int S' dX'} \\ SS'S_1 \sigma l^2 \times \Gamma \sin \Theta \sin E \, da'_1 da'_2 \ldots da'_\rho dA_1 dA_2 dA'_1 dA'_2 \end{array} \right\} \quad (57)$$

We shall now expand the right-hand side of equation (54) in powers of Δt and retain only terms of first order. It then reduces to

$$\frac{\partial f(t, a_1, a_2 \ldots)}{\partial t} da_1 da_2 \ldots da_\rho \Delta t = -dn' + dN' \quad (58)$$

Before I substitute here the values of dn' and dN', I will recall a relation which I have already proved in a previous paper.† In our present notation that equation reads:

$\gamma \sin \theta \sin \varepsilon \, d\xi_1 d\eta_1 \ldots d\zeta_{r-1} du_1 dv_1 \ldots dw_r d\theta d\omega$

$\qquad = \Gamma \sin \Theta \sin E d\Xi_1 dH_1 \ldots d\Omega$

if $\Xi_1, H_1 \ldots$ denote the values of $\xi_1, \eta_1 \ldots$ at the end of the collision. If we next introduce in this equation $a_1, a_2 \ldots a_\rho, x, x'$ in place of $\xi_1, \eta_1, \ldots w_r$, it becomes

$ss'\gamma \sin \theta \sin \varepsilon \, da_1 da_2 \ldots da'_\rho dx dx' d\theta d\omega$

$\qquad = SS' \Gamma \sin \Theta \sin E dA_1 dA_2 \ldots dA'_\rho dX dX' d\Xi d\Omega$

where s and S have the same meanings as before. If we then substitute on the left-hand side dA_1, dA_2, dA'_1, dA'_2 in place of dx, dx', $d\theta$, $d\omega$, and similarly on the right-hand side, then we obtain

$ss's_1\gamma \sin \theta \sin \varepsilon \, da_1 da_2 \ldots da'_\rho dA_1 dA_2 dA'_1 dA'_2$

$\qquad = SS'S_1 \Gamma \sin \Theta \sin E dA_1 dA_2 \ldots dA'_\rho da_1 da_2 da'_1 da'_2$

If we introduce on the right-hand side of this equation, in place of the differentials of $A_3, A_4 \ldots A_\rho, A'_3 \ldots A'_\rho$ those of

---

† L. Boltzmann, *Wien. Ber.* **63**, 397 (1871).

$a_3, a_4 \ldots a_\rho, a'_3 \ldots a'_\rho$, then we can divide through by all the differentials and we have left:

$$ss's_1\gamma \sin\theta \sin\varepsilon = SS'S_1\Gamma\sigma \sin\Theta \sin E \ldots \quad (59)$$

We wish to substitute the values (53) and (57) for dn' and dN' into equation (58). If we divide through by $da_1 da_2 \ldots da_\rho$, combine the two integrals on the right into a single one, and finally take account of equation (59), then we obtain

$$\frac{\partial f(t, a_1, a_2 \ldots a_\rho)}{\partial t}$$

$$= \int\int \ldots \left[\frac{f(t, A_1, A_2 \ldots) f(t, A'_1, A'_2 \ldots)}{\int S dX \int S' dX'} - \frac{f(t, a_1, a_2 \ldots) f(t, a'_1, a'_2 \ldots)}{\int s dx \int s' dx'} \right]$$

$$\times ss's_1 l^2 \gamma \sin\theta \sin\varepsilon \, da'_1 da'_2 \ldots da'_\rho dA_1 dA_2 dA'_1 dA'_2 \quad (60)$$

and this is to be regarded as the basic equation which determines the variation of the function $f(t, a_1, a_2 \ldots)$ with increasing time. We are to assume that all quantities are expressed as functions of the $2\rho+4$ quantities $a_1 a_2 \ldots a_\rho \ldots a'_1 \ldots a'_\rho \, A_1 A_2 A'_1 A'_2$, which completely determine the collision (aside from its absolute position in space). $\int SdX$ is the quantity which one obtains when he expresses $\int sdx$ as a function of these $2\rho+4$ variables; then by permuting the small and large letters, and again expressing the $A_3 A_4 \ldots$ as functions of the $2\rho+4$ variables, one similarly obtains $\int SdX'$.

One sees at once that this variation is equal to zero whenever the expression in brackets vanishes for all values of the variables; for then the integral must also vanish, and this is nothing but the result that I have already found in a different form in the paper on thermal equilibrium already cited. It remains only to prove that the integral can only vanish when the integrand is everywhere zero. Otherwise it might happen that the integral vanishes because the integrand is sometimes positive and sometimes negative. In order to obtain this latter proof, we proceed in just the same way as before, in the theory of the properties of monatomic molecules. Since the functional determinant s is only a function of x, a_1, $a_2 \ldots a_p$, and the limits of the integral $\int s dx$ depend only on the a's, it follows that the latter integral is a function of $a_1, a_2 \ldots a_p$. Hence, $f(t, a_1, a_2 \ldots)/ \int s dx$ is a function of $t, a_1, a_2 \ldots a_p$, and we shall set:

$$\frac{f(t, a_1, a_2 \ldots a_p)}{\int s dx} = \phi(t, a_1, a_2 \ldots a_p) \tag{61}$$

Furthermore, for brevity we set

$$ss's_1 l^2 \gamma \sin \theta \sin \varepsilon = p$$

Then we can write equation (13) in the somewhat shorter form:

$$\frac{\partial \phi(t, a_1, a_2 \ldots)}{\partial t} \int s dx = \iint \ldots [\phi(t, A_1, A_2 \ldots)\phi(t, A_1', A_2' \ldots)$$
$$- \phi(t, a_1, a_2 \ldots)\phi(t, a_1', a_2' \ldots)]p da_1' da_2' \ldots da' dA_1 dA_2 dA_1' dA_2' \tag{62}$$

The quantity p has two properties that will be used later. First, it is symmetric with respect to the variables $a_1, a_2 \ldots a_p\ A_1 A_2$ and $a_1'\ a_2' \ldots a_p'\ A_1'\ A_2'$, i.e. it does not change its value when one simultaneously interchanges a_1' and a_1, a_2' and a_2, and so forth. For this means simply that one ascribes to the second of the colliding molecules the state that the first had before the permutation, and conversely; and it is clear that this does not change the relative velocities, etc., of the two molecules. That the product ss' and the quantity s_1 are likewise symmetric in this way can be verified by simply looking at the expressions for these quantities.

Second, p is necessarily positive; for the number dn represented by equation (52) must be positive, and this expression for dn includes, aside from p, only positive factors (we assume that the differentials are always positive in any case); consequently p must likewise be positive. If we now set

$$E = \iint \ldots \phi(t, a_1, a_2 \ldots) . \int s dx . \log \phi(t, a_1, a_2 \ldots) da_1 da_2 \ldots da_p \tag{63}$$

where "log" means the natural logarithm, and the integration is to be extended over all possible values of the a's, so that E is a function only of the time. Since the limits of the integrals that determine E do not depend on t, we can find dE/dt by differentiating with respect to t under the integral sign, and then only insofar as the quantities under the integral sign contain t explicitly; for the a's are just variables over which we integrate. We know that $\int s dx$ does not depend on time; only $\phi(t, a_1, a_2 \ldots)$ contains t. If we take account of the fact that

$$\iint \ldots \phi(t, a_1, a_2 \ldots) . \int s dx . da_1 da_2 \ldots da_\rho$$

is the total number of molecules in unit volume, so that its time-derivative is equal to zero, then we obtain:

$$\frac{dE}{dT} = \iint \ldots \frac{\partial \phi(t, a_1, a_2 \ldots)}{\partial t} \log \phi(t, a_1, a_2 \ldots) . \int s dx . da_1 da_2 \ldots da_\rho$$

If we substitute here for $\partial \phi(t, a_1, a_2 \ldots)/\partial t$ its value from equation (62), and write, for brevity,

$$\phi(t, a_1, a_2 \ldots) = \phi; \quad \phi(t, a_1', a_2' \ldots) = \phi'; \quad \phi(t, A_1, A_2 \ldots) = \Phi;$$
$$\phi(t, A_1', A_2' \ldots) = \Phi'$$

then we find

$$\frac{dE}{dt} = \iiint \ldots \log \phi . [\Phi \Phi' - \phi \phi'] p da_1 da_2 \ldots da_\rho da_1' \\ \ldots da_\rho' dA_1 dA_2 dA_1' dA_2' \Bigg\} \quad (64)$$

In the definite integral on the right-hand side of (64) we may again label the variables over which we integrate just as we please. We can therefore, for example, interchange $a_1 a_2 \ldots A_2$ and $a_1' a_2' \ldots A_2'$. This does not change either p or the quantity in brackets, but only changes $\log \phi$ to $\log \phi'$. We therefore find, if we also reverse the order of integration, that the variables are integrated in the same order as before:

$$\frac{dE}{dt} = \iint \ldots \log \phi' [\Phi \Phi' - \phi \phi'] p da_1 da_2 \ldots da_\rho' dA_1 dA_2 dA_1' dA_2'$$

(65)

We shall now interchange those a's which are denoted in (64) by small Latin letters with the capitals, so that we obtain:

$$\frac{dE}{dt} = \iint \ldots \log \Phi [\phi\phi' - \Phi\Phi'] P dA_1 dA_2 \ldots dA'_\rho da_1 da_2 da'_1 da_2$$

(66)

where

$$P = SS' S_1 l^2 \Gamma \sin \Theta \sin E$$

One therefore obtains P by expressing p as a function of the $2\rho + 4$ quantities $a_1 a_2 \ldots a'_\rho \, A_1 \ldots A'_2$ and then interchanging small and large letters. However, we know that $A_3 A_4 \ldots A_\rho \, A'_3 \ldots A'_\rho$ can be expressed as functions of $a_1 a_2 \ldots a'_\rho A_1 A_2 A'_1$ and A'_2. We can therefore replace the differentials of $A_3 A_4 \ldots A_\rho \, A'_3 A'_4 \ldots A'_\rho$ in equation (66) by those of $a_3 a_4 \ldots a_\rho a'_3 \ldots a'_\rho$, which again is only a purely formal change. Indeed, we may replace under the integral sign those variables over which we integrate by any arbitrary functions of the same variables, and thus also, if we wish, by those which we earlier called $a_3 a_4 \ldots A_3 A_4$. Equation (66) is thereby transformed to

$$\frac{dE}{dt} = \iint \ldots \log \Phi [\phi\phi' - \Phi\Phi'] P\sigma da_1 da_2 \ldots da'_\rho dA_1 dA_2 dA'_1 dA'_2$$

or, if we take account of equation (59), to

$$\frac{dE}{dt} = \iint \ldots \log \Phi [\phi\phi' - \Phi\Phi'] p da_1 da_2 \ldots da'_\rho dA_1 dA_2 dA'_1 dA'_2$$

(67)

In all these equations the integrations are over all possible values of the variables.

If we finally interchange $a_1 a_2 \ldots A_2$ and $a'_1 a'_2 \ldots A'_2$, we find:

$$\frac{dE}{dt} = \iint \ldots \log \Phi' [\phi\phi' - \Phi\Phi'] p da_1 da_2 \ldots da'_\rho dA_1 dA_2 dA'_1 dA'_2$$

(68)

If we now add together equations (64), (65), (67) and (68) and divide by 4, we arrive at the result

$$\frac{dE}{dt} = \frac{1}{4}\int\int \ldots \log\left(\frac{\phi\phi'}{\Phi\Phi'}\right) \cdot [\Phi\Phi' - \phi\phi']p\,da_1\,da_2 \ldots dA'_2$$

Since p is always positive, it follows from this expression immediately that dE/dt can never be positive, and therefore E itself can only decrease or remain constant. The latter can only be the case when the expression in brackets,

$$\left.\begin{array}{l}\phi(t, a_1, a_2 \ldots a_\rho) \cdot \phi(t, a'_1, a'_2 \ldots a'_\rho) \\ \quad - \phi(t, A_1, A_2 \ldots A_\rho) \cdot \phi(t, A'_1, A'_2 \ldots A'_\rho)\end{array}\right\} \quad (69)$$

vanishes for all values of the variables. Hence, the distribution of states cannot fluctuate periodically between certain limits, and for the limit that it approaches with increasing time, the expression (69) must vanish in general.

The meaning of the above transformation of equation (64) can be further clarified, as in the case of monatomic gas molecules, by breaking up the integral into a sum. We set

$$a_1 = b_1\varepsilon, \; a_2 = b_2\varepsilon, \ldots, a'_\rho = b'_\rho\varepsilon, \; A_1 = B_1\varepsilon \ldots$$

where ε is some very small quantity, and the b's are integers. Further:

$$\phi(t, a_1, a_2 \ldots) = w_{b_1 b_2 \ldots}, \quad \int s\,dx = v_{b_1 b_2 \ldots},$$

$$\varepsilon^{\rho+4}p = D^{b_1 b_2 \ldots b'_1 b'_2 \ldots}_{B_1 B_2 \ldots B'_1 B'_2 \ldots}$$

Since D depends on all the variables, it must be given $2\rho + 4$ indices: the rest of the indices must be attached to it by symmetry. The v's are constants, whose values in general depend on the indices; the w's are functions of time. The system of ordinary differential equations, which now replaces equation (64), is the following:

$$\left.v_{b_1 b_2 \ldots}\frac{dw_{b_1 b_2 \ldots}}{dt} = \Sigma D^{b_1 b_2 \ldots b'_1 b'_2 \ldots}_{B_1 B_2 \ldots B'_1 B'_2 \ldots}[w_{B_1 B_2} \ldots w_{B_1' B_2'} \ldots \\ \qquad\qquad - w_{b_1 b_2} \ldots w^{b'_1 b'_2}_{b'_1 b'_2} \ldots]\right\} \quad (70)$$

The summation is to be taken over all possible values of $b_1' b_2' \ldots B_1 B_2 \ldots$ In equation (70) itself, $b_1 b_2 \ldots b_\rho$ can have many different values; it represents a system of many differential equations. Equation (59) now reads, in our present notation:

$$D^{b_1 b_2 \ldots b_1' b_2' \ldots}_{B_1 B_2 \ldots B_1' B_2' \ldots} = D^{b_1' b_2' \ldots b_1 b_2}_{B_1' B_2' \ldots B_1 B_2} = D^{B_1 B_2 \ldots B_1' B_2' \ldots}_{b_1 b_2 \ldots b_1' b_2' \ldots}$$

and one easily finds from this equation and from the system (70) that the derivative of

$$\Sigma v_{b_1 b_2} \ldots w_{b_1 b_2} \ldots \log w_{b_1 b_2} \ldots$$

can never be positive. The summation in this last expression is over all possible values of $b_1 b_2 \ldots b_\rho'$. This quantity therefore decreases continually until

$$w_{b_1 b_2} \ldots w_{b_1' b_2'} \ldots = w_{B_1 B_2} \ldots w_{B_1' B_2'} \ldots$$

for all possible values of $b_1 b_2 \ldots b_\rho'$ and for all values of $B_1 B_2 B_1' B_2'$ consistent with these. For D cannot be zero for any group of indices that correspond to a possible state of the molecule. If it were, then some collision would have probability zero.

We can now rewrite the quantity E given by equation (63). We saw that

$$f(t, a_1, a_2 \ldots) da_1 da_2 \ldots da_\rho \left(\frac{sdx}{\int sdx}\right)$$
$$= \phi(t, a_1, a_2 \ldots a_\rho) da_1 da_2 \ldots da_\rho . sdx$$

is the number of molecules in unit volume at time t for which conditions (B) and (D) are satisfied. If we replace the differentials of $a_1 a_2 \ldots a_\rho$ by those of $\xi_1 \eta_1 \ldots \zeta_{r-1} u_1 v_1 \ldots w_r$, then this expression becomes

$$\phi(t, a_1, a_2 \ldots a_\rho) d\xi_1 d\eta_1 \ldots d\zeta_{r-1} du_1 \ldots dw_r$$

If we again suppose that $a_1 a_2 \ldots a_\rho$ are expressed in terms of $\xi_1 \eta_1 \ldots w_r$, then ϕ will be a function of the latter variables. $\phi(t, a_1, a_2 \ldots a_\rho)$ then reduces to $F(t, \xi_1, \eta_1 \ldots w_r)$, so that therefore

$$F . d\xi_1 d\eta_1 \ldots d\zeta_{r-1} du_1 \ldots dw_r$$

is the number of molecules in unit volume for which ξ_1 lies between ξ_1 and $\xi_1 + d\xi_1 \ldots w_r$ between w_r and $w_r + dw_r$. We can then write the expression for E as

$$E = \iint \ldots F \log F \, s dx da_1 da_2 \ldots da_p$$

or, if we use differentials of $\xi_1, \eta_1 \ldots w_r$,

$$E = \iint \ldots F \log F d\xi_1 d\eta_1 \ldots d\zeta_{r-1} du_1 \ldots dw_1$$

I note that the calculation can be carried out in the same way if several kinds of molecules are present in the same container. If we denote the corresponding quantities for the second kind of molecule by adding a star, and so forth, then the quantity

$$E = \iint \ldots F \log F d\xi_1 d\eta_1 \ldots dw_r$$
$$+ \iint \ldots F^* \log F^* d\xi_1^* d\eta_1^* \ldots dw_r^* \quad (71)$$

can never increase.

V. The molecules do not make a large number of vibrations between one collision and the next

In the previous section I assumed that each molecule makes a very large number of vibrations between one collision and the next. It can easily be shown also that when this is not the case the quantity E defined by equation (71) still cannot increase. It is convenient here to denote the number of molecules in unit volume for which at time t the variables $\xi_1, \eta_1 \ldots w_r$ lie between the limits

$$\xi_1 \text{ and } \xi_1 + d\xi_1 \ldots w_r \text{ and } w_r + dw_r \quad \text{(L)}$$

by $f(t, \xi_1, \eta_1 \ldots w_r) d\xi_1 d\eta_1 \ldots dw_r$; f will now be the same function which was called F in the previous section. It can next be proved that the quantity E is not changed by the internal motions of the atoms in the molecules, so that it remains constant if the molecules do not collide with each other. The variables ξ_1, $\eta_1 \ldots w_r$ will be determined by differential equations which we have called the equations of motion of a molecule. Because of these equations of motion, we know that at time $t+\delta t$ the same molecules whose variables were within the limits (L) at time t will now have their variables lying between the limits:

$$\xi'_1 \text{ and } \xi'_1 + d\xi'_1 \ldots w'_r \text{ and } w'_r + dw'_r \quad \text{(M)}$$

In my paper on thermal equilibrium among polyatomic molecules,† I showed that‡

$$d\xi'_1 d\eta'_1 \ldots dw'_r = d\xi_1 d\eta_1 \ldots dw_r$$

If no collision occurs, then the same molecules will have their variables within the limits (M) at time $t+\delta t$ as had them within (L) at time t. The numbers of molecules in these two groups will therefore be equal, since both contain exactly the same molecules. But the number in (M) at $t+\delta t$ is

$$f(t+\delta t, \xi'_1 \ldots w'_r) d\xi'_1 \ldots dw'_r$$

the number in (L) at t is

$$f(t, \xi_1 \ldots w_r) d\xi_1 \ldots dw_r$$

Therefore

$$f(t+\delta t, \xi'_1 \ldots w'_r) d\xi'_1 \ldots dw'_r = f(t, \xi_1 \ldots w_r) d\xi_1 \ldots dw_r \quad (72)$$

(We assume that the distribution of states is uniform everywhere in the gas.) Taking account of equation (72), we obtain

$$f(t+\delta t, \xi'_1 \ldots w'_r) = f(t, \xi_1 \ldots w_r)$$

therefore we have also

$$f' \log f' \, d\xi'_1 d\eta'_1 \ldots dw'_r = f \log f \, d\xi_1 \ldots dw_r \quad (73)$$

if we write f for $f(t, \xi_1 \ldots w_r)$ and f' for $f(t+\delta t, \xi'_1 \ldots w'_r)$. Since equation (73) holds for each set of differentials, we obtain a similar

---

† L. Boltzmann, *Wien. Ber.* **63**, 397 (1871).

‡ If one wishes to calculate with finite quantities, this equation reads:

$$\lim \frac{d\xi'_1 \, d\eta'_1 \ldots dw'_r}{\delta t \, d\xi_1 \, d\eta_1 \ldots dw_r} = 1$$

expression if we integrate on both sides over all possible values of the variables. Therefore

$$\iint \ldots f' \log f' \, d\xi_1' \ldots dw_r' = \iint \ldots f \log f \, d\xi_1 \ldots dw_r$$

where we have to integrate over all possible values of the variables on both sides, so that it makes no difference how we label the variables. One can therefore write also:

$$\iint \ldots f(t+\delta t, \xi_1 \ldots w_r) \log f(t+\delta t, \xi_1 \ldots w_r) d\xi_1 \ldots dw_r$$
$$= \iint \ldots f \log f \, d\xi_1 \ldots dw_r$$

The right-hand side of this equation represents the value of E at time $t+\delta t$, while the left is its value at time t. The two are equal. E therefore does not change its value if the atoms in the molecules move according to their equations of motion, as long as the molecules do not collide with each other.

It is now a question of finding how much E changes during collisions. If δt is taken to be very small, then the total variation of E will be the sum of individual variations. If we denote the increase experienced by E as a result of collisions by δE, then we have

$$\delta E = \iint \ldots \log f \, \delta f d\xi_1 \ldots dw_r \qquad (74)$$

where δf is the increase experienced by f during time δt as a result of collisions. Hence, the number of molecules (δN) which attain the state (L) through collisions in time δt is $\delta f d\xi_1 d\eta_1 \ldots dw_r$ greater than the number of molecules (δn) which leaves this state through collisions:

$$\delta N - \delta n = \delta f d\xi_1 d\eta_1 \ldots dw_r \qquad (75)$$

I assume that a collision of two molecules begins when their centres of gravity are at a distance l. The angle between the line of centres and the x-axis will again be called θ; the angle between the xy-plane and a plane parallel to the line of centres passing through the x-axis will be called ω. Then the number of collisions in unit volume during time δt for which at the beginning the variables θ and ω lie between

$$\theta \text{ and } \theta + d\theta, \quad \omega \text{ and } \omega + d\omega \qquad (N)$$

and furthermore the variables $\xi_1 \eta_1 \ldots w_r$ for one of the colliding molecules lie between the limits (L), while the variables for the other molecule may lie within any other limits such as

$$\xi_1' \text{ and } \xi_1' + d\xi_1' \ldots w_r' \text{ and } w_r' + dw_r' \qquad (P)$$

will be:

$$\int d\xi_1 \ldots dw_r \int' d\xi_1' \ldots dw_r' g l^2 \sin\theta d\theta d\omega \delta t$$

where g is the relative velocity of the centres of the molecules. For all these, and only these molecules, θ and ω will lie between the limits

$$\Theta \text{ and } \Theta + d\Theta, \ \Omega \text{ and } \Omega + d\Omega \qquad (N^*)$$

at the end of the collision. Furthermore, the variables ξ_1, $\eta_1 \ldots w_r$ for the first molecule will lie between the limits

$$\Xi_1 \text{ and } \Xi_1 + d\Xi_1 \ldots W_r \text{ and } W_r + dW_r \qquad (L^*)$$

and for the second molecule, between the limits

$$\Xi_1' \text{ and } \Xi_1' + d\Xi_1' \ldots W_r' \text{ and } W_r' + dW_r' \qquad (P^*)$$

Each of these collisions will remove a molecule from the state (L). Hence, the total depletion during time δt is

$$\delta n = \int d\xi_1 \ldots dw_r \delta t \int\!\!\int \ldots f' d\xi_1' \ldots dw_r' g l^2 \sin\theta d\theta d\omega \qquad (76)$$

The number of collisions for which initially the variables lie between the limits (L*), (N*), and (P*) is

$$\delta v = F d\Xi_1 \ldots dW_r F' d\Xi_1' \ldots dW_r' G l^2 \sin\Theta d\Theta d\Omega \delta t$$

For all these collisions, the variables lie between the limits (L), (N) and (P) afterwards, since these collisions are just the reverse of those considered earlier.† Since the variables

$$\Xi_1 \ldots W_r, \Xi_1' \ldots W_r', \Theta, \Omega \qquad (R)$$

are functions of

$$\xi_1 \ldots w_r, \xi_1' \ldots w_r', \theta, \omega \qquad (Q)$$

---

† Of course in comparing the final state of one collision with the initial state of another one, the positions of the centres of mass of the two molecules have to be interchanged, since at the beginning they move towards each other, and at the end away from each other.

one can introduce the latter variables in place of the former in the expression (76). Then we have

$$G \sin \Theta \, d\Xi_1 \ldots dW'_r d\Theta d\Omega = g \sin \theta \, d\xi_1 \ldots dw'_r d\theta d\omega$$

as follows from the general theorem proved in the second section of a previous paper.† Consequently we have

$$\delta v = FF' d\xi_1 \ldots dw'_r \sin \theta g l^2 d\theta d\omega \delta t \tag{77}$$

In the above equation we wrote F, F', f', for $f(t, \Xi_1 \ldots W_r)$, $f(t, \Xi'_1 \ldots W'_r)$ and $f(t, \xi'_1 \ldots w'_r)$. G is the quantity that one obtains when he interchanges the variables (Q) and (R) in g. In the expression (77) one is supposed to consider that the variables (R) are expressed as functions of the variables (Q). If we integrate $\xi'_1 \ldots w'_r$, θ, ω over all possible values, then we obtain the number of molecules which attain the state (L) by collisions in unit time, and thus the number earlier denoted by δN:

$$\delta N = d\xi_1 \ldots dw_r \delta t \iint \ldots FF' g l^2 \sin \theta \, d\xi'_1 \ldots dw'_r d\theta d\omega$$

If we substitute these values for δn and δN into equation (75), we obtain

$$\delta f = \delta t \iint \ldots (FF' - ff') g l^2 \sin \theta \, d\xi'_1 \ldots dw'_r d\theta d\omega$$

and if we substitute this into equation (74), we find

$$\delta E = \delta t \iint \ldots \log f \, (FF' - ff') g l^2 \sin \theta \, d\xi_1 \ldots dw_r d\xi'_1 \ldots dw'_r d\theta d\omega$$

By interchanging the labels of the two colliding molecules, we obtain just as before

$$\delta E = \delta t \iint \ldots \log f' \, (FF' - ff') g l^2 \sin \theta \, d\xi_1 \ldots dw_r d\xi'_1 \\ \ldots dw'_r d\theta d\omega$$

By replacing the variables (Q) by (R) and then changing the labels of the variables, we find

$$\delta E = -\delta t \iint \ldots \log F \, (FF' - ff') g l^2 \sin \theta \, d\xi_1 \ldots dw'_r d\theta d\omega$$

$$\delta E = -\delta t \iint \ldots \log F' \, (FF' - ff') g l^2 \sin \theta \, d\xi_1 \ldots dw'_r d\theta d\omega$$

---

† L. Boltzmann. *Wien. Ber,* **63**, 397 (1871).

(Note that the variables (R) are the same functions of (Q) as the (Q) are of the (R).) Adding together the four equations, we find

$$\delta E = \tfrac{1}{4}\delta t \iint \ldots \log\left(\frac{ff'}{FF'}\right).(FF'-ff')gl^2 \sin\theta\, d\xi_1 \ldots dw_r d\theta d\omega$$

whence it follows again the E can only decrease as a result of collisions, and since it does not change through the motions of atoms in the molecules, it follows that in general it can only decrease.

If the distribution of states at the initial time were not uniform, then f would also depend on the position (x, y, z) in the gas. We would then have instead of E a more general expression. If

$$f(t, x, y, z, \xi_1, \eta_1 \ldots w_r) dxdydzd\xi_1 \ldots dw_r$$

is the number of molecules in the volume element $dxdydz$ at (x, y, z) at time t, for which the variables lie between the limits (L), then the quantity

$$E = \iint \ldots f \log f . dxdydzd\xi_1 \ldots dw_r \tag{78}$$

cannot increase. In order to prove this, we shall make the problem even more general. Suppose that we have many systems of mass-points (molecules). Each consists of r mass-points $m_1 m_2 \ldots m_r$ (the m's are the actual masses of the points as well as their labels). Let the mass m_1 be the same for all systems; likewise the mass m_2, etc. Let x_1, y_1, z_1 be the coordinates and u_1, v_1, w_1 the velocity components of m_1. Let $x_2, y_2 \ldots$ have similar meanings; and indeed it makes no difference whether the origin of coordinate is the same or different for the different systems. The force that acts on any of the mass-points will be assumed to be a function of the coordinates $x_1, y_1, z_1, x_2 \ldots z_r$ such that a potential function exists, and we assume that this potential is the same function of $x_1 y_1 \ldots z_r$ for all the systems. If we denote by $f(t, x_1, y_1, z_1 \ldots w_r)$ the number of systems for which the variables $x_1, y_1, z_1, x_2 \ldots z_r$, $u_1 \ldots w_r$ lie between the limits

$$x_1 \text{ and } x_1 + dx_1 \ldots w_r \text{ and } w_r + dw_r \tag{S}$$

and set
$$E = \iint \ldots f \log f \, dx_1 dy_1 \ldots dw_r \qquad (79)$$

then it can be proved just as before that E does not change as a result of the motion of the mass-points of the system as long as only the internal forces of that system act between the points. Now assume that there are interactions between the points of different systems, such that the force between two points is a function of their distance and acts in the directions of the line of centres. The conditions shall be such that when the interaction of two systems begins, there is never (or almost never) a third system simultaneously interacting with these two. (For example, the intertion of two systems might occur whenever a point of one of them comes unusually close to a point of the other.) The number of pairs of systems that interact during time δt such that the variables determining the state of one system lie initially within the limits (S), and those of the other between

$$x'_1 \text{ and } x'_1 + dx'_1 \ldots w'_r \text{ and } w'_r + dw'_r$$

is then
$$f(t, x_1 \ldots w_r) f(t, x'_1 \ldots w'_r) dx_1 \ldots dw_r dx'_1 \ldots dw'_r \delta t \cdot \phi \qquad (80)$$

ϕ is a function of the relative distances and the velocities of the atoms of the two systems. If one takes account of the fact that this function must have the general property expressed by equation (19) of my paper on the thermal equilibrium of polyatomic molecules,† then one can prove as before that E can only decrease as a result of interactions of different systems. The proof can also be carried through when the systems are not all the same but rather are of two or more kinds, as long as a large number of systems of each kind are present. (One sees at once that a mixture of gas molecules with non-uniform velocity distributions is only a special case of this.) When the potential function has the value gz, one finds for monatomic gas molecules

$$f = Ae^{-h(gz + \frac{1}{2}mc^2)}$$

---

† L. Boltzmann, *Wien. Ber.* **63**, 397 (1871).

The well-known formula for barometric height measurements as well as all aerostatic formulae thus follow as special cases from the formulae for thermal equilibrium.

VI. Solution of equation (81) and calculation of the entropy

We have proved that for polyatomic gas molecules in the case of equilibrium of kinetic energy, the expression (69) must vanish:

$$\phi(a_1, a_2 \ldots)\phi(a'_1, a'_2 \ldots) - \phi(A_1, A_2 \ldots)\phi(A'_1, A'_2 \ldots) = 0 \tag{81}$$

must hold for all possible values of the variables

$$a_1 a_2 \ldots a'_1 a'_2 \ldots A_1 A_2 \ldots A'_1 A'_2 \ldots A'_\rho \tag{82}$$

We now have to find a function ϕ that satisfies this equation. It is clear that if λ is the total kinetic energy and work contained in a molecule, then the value $\phi = Ae^{-h\lambda}$ (where A and h are constants) must satisfy (81). This is the solution of the problem which I have already found in my previous paper.† It remains only to prove that this is the only possible solution of (81). While this proof was the easiest step in the case of monatomic gas molecules, here it is the most difficult, because it is not actually possible to give in general the various equations which relate the values of the variables before and after the collision. It can still be proved at least in the case of diatomic molecules under the assumption of a particular kind of interaction of the molecules during the collision. We assume that each molecule consists of two atoms. Let r be their distance, and let $m(\chi'(r))/2$ be the force of attraction between them when the molecule is not interacting with another one, so that $m(\chi'(r))/2$ is the potential function. In order not to make the formulae too complicated, I assume that all the atoms have equal mass (the more general case can be treated in a similar way). The sum of the values of the potential and kinetic energies of the two atoms of a molecule remains constant from one collision to the next. This sum divided by half the mass of an atom will be denoted

---

† L. Boltzmann, *Wien. Ber.* **63**, 397 (1871).

by a; four times the square of the velocity of the centre of mass of the molecule will be denoted by b; four times the surface swept out by the radius vector from the centre of mass to one of the atoms in unit time will be denoted by c. Thus a, b, and c are the only constants of integration determining the nature of the atomic path. The others determine only its position in space. One can place a plane through the line of centres of two molecules in such a way that the velocity component w perpendicular to this plane is the same for both atoms. This plane will be called the orbital plane. Let α be the angle it makes with any fixed plane, and let β be the angle between its line of intersection with the fixed plane and a fixed line drawn in the fixed plane. Let γ be the angle which that velocity component of the centre of mass of the molecule which is parallel to the orbital plane makes with the line of intersection of the orbital plane and the fixed plane; let δ be the angle which the line of apsides which the orbital curve of the atom makes with the same line by its motion around the centre of mass. We assume that the gas has the same properties in all directions; between one collision of the molecule and the next, there are many maxima and minima of the distance between its atoms (apside positions). Finally, the angle between two successive apside lines is not a rational fraction of π (with the exception of an infinitesimal number of special path shapes). Then a, b, c, w, α, β, γ, δ are the constants of integration denoted by $a_1 \, a_2 \ldots a_\rho$ in the previous section, r is equal to 2, $\rho = 6r - 4 = 8$, and the quantity $\phi(a_1, a_2 \ldots a_\rho) da_1 \ldots da$ must have the form

$$\phi(a, b, c, w) \sin \alpha \, da\,db\,dc\,dw\,d\alpha\,d\beta\,d\gamma\,d\delta$$

Because of the equivalence of all directions in space, ϕ cannot depend on the angles α, β, γ. Because of our assumption about the apside lines, all values of δ are equivalent. We shall now consider only those collisions for which w and the orbital planes for the two colliding molecules are identical. Equation (81) must hold for all collisions and hence also for this kind. However, α is not changed by the collision, so that equation (81) reduces to

$$\phi(a, b, c, w)\phi(a', b', c', w') = \phi(A, B, C, w)\phi(A', B', C', w') \quad (83)$$

We shall now make a special assumption about the interaction between two molecules during the collision. The collision of two molecules shall be such that an atom of one molecule rebounds from one of the other molecules like an elastic sphere. (We label the colliding atom as the first one of its molecule.) We now construct, parallel to the line of centres of the colliding atoms at the beginning of the collision, a fixed x-axis (any line of centres shall fall in the orbital plane) and perpendicular to this but parallel to the orbital plane, a fixed y-axis. At the beginning of the collision, let u, v be the velocity components of the first atom of the first of the two colliding molecules in the directions of these two axes; let ξ, η be the coordinates of the same with respect to a system whose origin is at the centre of mass of the molecule, and whose axes are parallel to our fixed ones. Let u_1, v_1 be the velocity components of the second atom of the first molecule. The notation for the other molecule, and for quantities at the end of the collision, will be chosen in the usual way. Then

$$\left.\begin{aligned}
a &= u^2 + v^2 + u_1^2 + v_1^2 + 2w^2 + \chi \\
b &= (u+u_1)^2 + (v+v_1)^2 + 4w^2 \\
a' &= u'^2 + v'^2 + u_1'^2 + v_1'^2 + 2w'^2 + \chi' \\
b' &= (u'+u_1')^2 + (v'+v_1')^2 + 4w'^2
\end{aligned}\right\} \quad (84)$$

The collision reverses the x-components of the velocities of the colliding atoms hence $U = u'$, $U' = u$; all other capital letters have the same values as the corresponding small ones. Therefore

$$\left.\begin{aligned}
A &= u'^2 + v^2 + u_1^2 + v_1^2 + 2w^2 + \chi \\
B &= (u'+u_1)^2 + (v+v_1)^2 + 4w^2 \\
A' &= u^2 + v'^2 + u_1'^2 + v_1'^2 + 2w'^2 + \chi' \\
B' &= (u+u_1')^2 + (v'+v_1')^2 + 4w'^2
\end{aligned}\right\} \quad (85)$$

We shall now show that, if arbitrary values of the quantities

$$a, b, c, a', b', c', w, w' \quad (86)$$

are given, then the quantities $u, v, \xi \ldots$, which are not hereby determined, can also be chosen such that given values of

$$A, B, C, A', B', C'_l \tag{87}$$

will occur after the collision, provided only that the quantities (86) and (87) satisfy the single condition

$$a + a' = A + A' \tag{88}$$

which is just the equation for conservation of energy; thus there are no other relations between the quantities (86 and (87).

We now set, for the sake of brevity,

$$b - a - B + A = g, \quad b' - a' - B' + A' = g'$$

so that g and g' are the given quantities.

We find:

$$g = 2u_1(u - u'), \quad g' = 2u'_1(u' - u)$$

hence

$$u' = u - \frac{g}{2u_1}, \quad u'_1 = -\frac{g'}{g} u_1 \tag{89}$$

and consequently

$$b - B = (u + u_1)^2 - \left(u - \frac{g}{2u_1} + u_1\right)^2 \tag{90}$$

$$b' - B' = \left(u - \frac{g}{2u_1} - \frac{g'}{g} u_1\right)^2 - \left(u - \frac{g'}{g} u_1\right)^2$$

From the two latter equations it follows that

$$b - B + b' - B' = g + g'$$

which is just the equation for conservation of energy. This must be satisfied; one can then chose u_1 at will. The other equations (90) determine u quadratically, while equations (89) determine u' and u'_1. Equations (89) and (90) completely replace four of equations (84) and (85), There remain still the other four. These involve the equations for the c's. They read:

$$c = \xi(v-v_1)-\eta(u-u_1), \quad c' = \xi'(v'-v_1')-\eta'(u'-u_1')$$
$$C = \xi(v-v_1)-\eta(u'-u_1), \quad C' = \xi'(v'-v_1')-\eta'(u-u_1') \quad (91)$$

whence it follows that

$$\eta = \frac{c-C}{u'-u}, \quad \eta' = -\frac{c'-C'}{u'-u}$$

which again replaces two of equations (91), and determines η and η'. There remain still four of equations (84) and (85), and two of equations (91) to be satisfied, thus for example

$$\left. \begin{array}{l} v^2 + v_1^2 = a - u_1^2 - 2w^2 - \chi \\ (v+v_1)^2 = n - (u+u_1)^2 - 4w^2 \\ v - v_1 = \dfrac{1}{\xi}[c+\eta(u-u_1)] \end{array} \right\} \quad (92)$$

$$\left. \begin{array}{l} v'^2 + v_1'^2 = a' - u'^2 - u_1'^2 - 2w'^2 - \chi' \\ (v'+v_1')^2 = b' - (u'+u_1')^2 - 4w'^2 \\ v' - v_1' = \dfrac{1}{\xi'}[c'+\eta(u-u_1)] \end{array} \right\} \quad (93)$$

In these equations, the u's and η"s are to be considered as given, since we expressed them in terms of given quantities. If one eliminates the v's by using equations (92), there remains only a single equation for ξ, which then also determines the v's. Likewise one can determine ξ', v' and v_1' from equations (93). If equation (88) is now satisfied, then we can express the variables $\xi, \eta, u, v \ldots$ separately in terms of the given quantities

$$a, b, c, w, a', b', c', w', A, B, C, A', B', C' \quad (T)$$

The only equation relating these variables is (88). Equation (85) must therefore be satisfied for all values of the variables (T) which satisfy (88). Therefore ϕ must have the form $Ae^{-h\lambda}$. That w does not also appear in ϕ can easily be proved from the other collisions. Since already for the collisions considered, u_1 is completely arbitrary, and by considering all collisions naturally still more arbitrary quantities would be introduced, it does not seem likely that for some other force law other solutions would be possible. Yet I know of no other means of proof, at present, than to treat each force law by itself.

Inasmuch as we take it to be very probable that for the case of thermal equilibrium the function ϕ always has the form $Ae^{-h\lambda}$, we may now calculate E for any body for which thermal equilibrium has been established among its atoms. We use the expression (79) as a generally valid definition of E (ignoring constant quantities that may arise on account of the special nature of the problem).

If we wish to call E the entropy, we run up against the difficulty that the total entropy of two bodies would differ by a constant from the sum of the entropies of the individual bodies. We therefore prefer to consider the following expression, which differs from (79) only by a constant:

$$E^* = \iint \ldots f \log\left(\frac{f}{N}\right) dx_1 \ldots dw_r$$

Here N is the total number of molecules in the gas, while $f dx_1 \ldots dw_r$ is the number of those for which $x_1, y_1 \ldots w_r$ lie between the limits

$$x_1 \text{ and } x_1 + dx_1 \ldots w_r \text{ and } w_r + dw_r \tag{S}$$

If we set

$$dx_1 dy_1 \ldots dz_r = d\sigma, \; du_1 dv_1 \ldots dw_r = ds, \; \frac{f}{N} = f^*$$

then f^* also has a simple meaning. $f^* ds d\sigma$ is the probability that a molecule has the state (S) (the time during which it has that state divided by the total time during which it moves).

We then have

$$E^* = N\iint f^* \log f^* \, ds d\sigma \tag{94}$$

For monatomic gases, if N is the total number of molecules in the gas, V the volume of the container, m the mass, and T the average kinetic energy of an atom, we have

$$f^* = \frac{1}{V\left(\frac{4\pi T}{3m}\right)^{\frac{3}{2}}} e^{-\frac{3m}{4T}(u^2+v^2+w^2)}$$

hence,

$$E^* = N\iint \ldots f^* \log f^* \, dx dy dz du dv dw$$
$$= -N \log\left[V\left(\frac{4\pi T}{3m}\right)^{\frac{3}{2}}\right] - \tfrac{3}{2}N$$

which, since m and N are constant, agrees up to a constant factor and constant additive term with the expression for the entropy of a monatomic gas. For gases with r-atomic molecules, we have

$$f^* = Ae^{-h(\chi + \Sigma mc^2/2)}$$

where χ is the potential; $\Sigma mc^2/2$ is the total kinetic energy of a molecule. Since $\iint f^* \, ds d\sigma = 1$,

$$T = \frac{3}{2h}, \quad A = \frac{1}{\left(\frac{2\pi}{mh}\right)^{\frac{3}{2}} \int e^{-h\chi} d\sigma}$$

One therefore finds that

$$E^* = N \log A - hN \frac{\int \chi e^{-h\chi} d\sigma}{\int e^{-h\chi} d\sigma} - \tfrac{3}{2} rN \tag{95}$$

In order to find the relation of the quantity E^* to the second law of thermodynamics in the form $\int dQ/T < 0$, we shall interpret the system of r mass points not, as previously, as a gas molecule, but rather as an entire body. (We shall call it the system A.) During a certain period of time it interacts with a second system (B), i.e. with a second body. The two bodies may have the same or different properties. Theoretically the effect of the interaction should depend not only on the nature of the force between A and B but also on the phases of both bodies at the time when the interaction begins. However, experience shows that this is not noticeable, doubtless because the effect of the phase is masked by the effect of the large number of molecules that are interacting. (A similar opinion has already been expressed recently by Clausius.†) In order to eliminate the effects of the phase, we shall replace the single system (A) by a large number (N) of equivalent systems distributed over many different phases, but which do not interact with each other. Let $f(t, x_1 \ldots w_r)dsd\sigma$ again be the number of systems with state (S), and set f/N equal to f^*. Similarly we assume that there are many systems of type (B). Their distribution is determined by a function $f'(t, x'_1 \ldots w'_r)$ similar to f. The functions f^* and f' may also be discontinuous, so that they have large values when the variables are very close to certain values determined by one or more equations, and otherwise vanishingly small. We may choose these equations to be those that characterize visible external motion of the body and the kinetic energy contained in it. In this connection it should be noted that the kinetic energy of visible motion corresponds to such a large deviation from the final equilibrium distribution of kinetic energy that it leads to an infinity in E^*, so that from the point of view of the second law of thermodynamics it acts like heat supplied from an infinite temperature. One of the (B) systems will now act on each of the (A) systems, and thus the beginning of the interaction will coincide with all different phases. All effects that do not depend on phase will then appear just as if only one (A) system acted on one (B) system in an arbitrary phase, and we know that thermal phenomena do not in fact depend on phase. The function f can therefore be chosen arbitrarily, insofar as it is not restricted by the conditions of total kinetic energy or visible motion of the

---

† [R. Clausius, *Ann. Phys.* **142**, 433 (1871); *Phil. Mag.* **42**, 161 (1871).]

body. The probability that an (A) system in state (S) interacts with a (B) system whose state is given by a similar condition is given by a formula similar to (52). From this it can be proved that E^* can only decrease. After a long-continued interaction (to establish temperature equilibrium) E^* attains its minimum, which occurs in general when $ff' = FF'$. If the bodies are at rest, then the solution of this equation is

$$f^* = Ae^{-h(\chi + \Sigma mc^2/2)}$$

where $f^* ds d\sigma$ is the probability that an (A) system has state (S). The quantity E, which is proportional to the entropy of all N (A) systems, is again given by equation (95). The entropy of a single (A) system is therefore $1/N$ of this, and is therefore proportional to

$$E^* = \iint f^* \log f^* \, ds d\sigma = \log A - h \frac{\int \chi e^{-h\chi} d\sigma}{\int e^{-h\chi} d\sigma} - \frac{3r}{2} \qquad (96)$$

which agrees (up to a constant factor and addend) with the expression which I found in my previous paper.†

---

† L. Boltzmann, *Wien. Ber.* **63**, 712 (1871), Eqn. 18.

3

The Kinetic Theory of the Dissipation of Energy *

WILLIAM THOMSON

SUMMARY

The equations of motion in abstract dynamics are perfectly reversible; any solution of these equations remains valid when the time variable t is replaced by $-t$. Physical processes, on the other hand, are irreversible: for example, the friction of solids, conduction of heat, and diffusion. Nevertheless, the principle of dissipation of energy is compatible with a molecular theory in which each particle is subject to the laws of abstract dynamics.

Dissipation of energy, such as that due to heat conduction in a gas, might be entirely prevented by a suitable arrangement of Maxwell demons, operating in conformity with the conservation of energy and momentum. If no demons are present, the average result of the free motions of molecules will be to equalize temperature-differences. If we allowed this equalization to proceed for a certain time, and then reversed the motions of all the molecules, we would observe a disequalization. However, if the number of molecules is very large, as it is in a gas, any slight deviation from absolute precision in the reversal will greatly shorten the time during which disequalization occurs. In other words, the probability of occurrence of a distribution of velocities which will lead to disequalization of temperature for any perceptible length of time is very small. Furthermore, if we take account of the fact that no physical system can be completely isolated from its surroundings but is in principle interacting with all other molecules in the universe, and if we believe that the number of these latter molecules is infinite, then we may conclude that it is impossible for temperature-differences to arise spontaneously. A numerical calculation is given to illustrate this conclusion.

* Originally published in the *Proceedings of the Royal Society of Edinburgh*, Vol. 8, pp. 325–34 (1874); reprinted in Thomson's *Mathematical and Physical Papers*, Cambridge University Press, 1911, Vol. V, pp. 11–20.]

In abstract dynamics the instantaneous reversal of the motion of every moving particle of a system causes the system to move backwards, each particle of it along its old path, and at the same speed as before, when again in the same position. That is to say, in mathematical language, any solution remains a solution when t is changed into $-t$. In physical dynamics this simple and perfect reversibility fails, on account of forces depending on friction of solids; imperfect fluidity of fluids; imperfect elasticity of solids; inequalities of temperature, and consequent conduction of heat produced by stresses in solids and fluids; imperfect magnetic retentiveness; residual electric polarization of dielectrics; generation of heat by electric currents induced by motion; diffusion of fluids, solution of solids in fluids, and other chemical changes; and absorption of radiant heat and light. Consideration of these agencies in connection with the all-pervading law of the conservation of energy proved for them by Joule, led me twenty-three years ago to the theory of the dissipation of energy, which I communicated first to the Royal Society of Edinburgh in 1852, in a paper entitled " On a Universal Tendency in Nature to the Dissipation of Mechanical Energy."

The essence of Joule's discovery is the subjection of physical phenomena to dynamical law. If, then, the motion of every particle of matter in the universe were precisely reversed at any instant, the course of nature would be simply reversed for ever after. The bursting bubble of foam at the foot of a waterfall would reunite and descend into the water; the thermal motions would reconcentrate their energy, and throw the mass up the fall in drops re-forming into a close column of ascending water. Heat which had been generated by the friction of solids and dissipated by conduction, and radiation with absorption, would come again to the place of contact, and throw the moving body back against the force to which it had previously yielded. Boulders would recover from the mud the materials required to rebuild them into their previous jagged forms, and would become reunited to the mountain peak from which they had formerly broken away. And if also the materialistic hypothesis of life were true, living creatures would grow backwards, with conscious knowledge of the future, but no memory of the past, and would become again unborn. But the real phenomena of life infinitely transcend human science; and speculation regarding consequences of their imagined reversal is utterly unprofitable. Far otherwise, however, is it in respect to the reversal of the motions of matter uninfluenced by life, a very elementary consideration of which leads to the full explanation of the theory of dissipation of energy.

To take one of the simplest cases of the dissipation of energy, the conduction of heat through a solid—consider a bar of metal warmer at one end than the other, and left to itself. To avoid all needless complication of taking loss or gain of heat into account, imagine the bar to be varnished with a substance impermeable to heat. For the sake of definiteness, imagine the bar to be first given with one-half of it at one uniform temperature, and the other half of it at another uniform temperature. Instantly a diffusion of heat commences, and the distribution of temperature becomes continuously less and less unequal, tending to perfect uniformity, but never in any finite time attaining perfectly to this ultimate condition. This process of diffusion could be perfectly prevented by an army of Maxwell's " intelligent demons,"† stationed at the surface, or interface as we may call it with Professor James Thomson, separating the hot from the cold part of the bar. To see precisely how this is to be done, consider rather a gas than a solid, because we have much knowledge regarding the molecular motions of a gas, and little or no knowledge of the molecular motions of a solid. Take a jar with the lower half occupied by cold air or gas, and the upper half occupied with the air or gas of the same kind, but at a higher temperature; and let the mouth of the jar be closed by an air-tight lid. If the containing vessel were perfectly impermeable to heat, the diffusion

---

† The definition of a demon, according to the use of this word by Maxwell, is an intelligent being endowed with free-will and fine enough tactile and perceptive organization to give him the faculty of observing and influencing individual molecules of matter.

of heat would follow the same law in the gas as in the solid, though in the gas the diffusion of heat takes place chiefly by the diffusion of molecules, each taking its energy with it, and only to a small proportion of its whole amount by the interchange of energy between molecule and molecule; whereas in the solid there is little or no diffusion of substance, and the diffusion of heat takes place entirely, or almost entirely, through the communication of energy from one molecule to another. Fourier's exquisite mathematical analysis expresses perfectly the statistics of the process of diffusion in each case, whether it be " conduction of heat," as Fourier and his followers have called it, or the diffusion of substance in fluid masses (gaseous or liquid), which Fick showed to be subject to Fourier's formulas. Now, suppose the weapon of the ideal army to be a club, or, as it were, a molecular cricket bat; and suppose, for convenience, the mass of each demon with his weapon to be several times greater than that of a molecule. Every time he strikes a molecule he is to send it away with the same energy as it had immediately before. Each demon is to keep as nearly as possible to a certain station, making only such excursions from it as the execution of his orders requires. He is to experience no forces except such as result from collisions with molecules, and mutual forces between parts of his own mass, including his weapon. Thus his voluntary movements cannot influence the position of his centre of gravity, otherwise than by producing collision with molecules.

The whole interface between hot and cold is to be divided into small areas, each allotted to a single demon. The duty of each demon is to guard his allotment, turning molecules back, or allowing them to pass through from either side, according to certain definite orders. First, let the orders be to allow no molecules to pass from either side. The effect will be the same as if the interface were stopped by a barrier impermeable to matter and to heat. The pressure of the gas being by hypothesis equal in the hot and cold parts, the resultant momentum taken by each demon from any considerable number of molecules will be zero; and therefore he may so time his strokes that he shall never move to any considerable distance from his station. Now, instead of stopping and turning all the molecules from crossing his allotted area, let each demon permit a hundred molecules chosen arbitrary to cross it from the hot side; and the same number of molecules, chosen so as to have the same entire amount of energy and the same resultant momentum, to cross the other way from the cold side. Let this be done over and over again within certain small equal consecutive intervals of time, with care that if the specified balance of energy and momentum is not exactly fulfilled in respect to each successive hundred molecules crossing each way, the error will be carried forward, and as nearly as may be corrected, in respect to the next hundred. Thus, a certain perfectly regular diffusion of the gas both ways across the interface goes on, while the original different temperatures on the two sides of the interface are maintained without change.

Suppose, now, that in the original condition the temperature and pressure of the gas are each equal throughout the vessel, and let it be required to disequalize the temperature, but to leave the pressure the same in any two portions A and B of the whole space. Station the army on the interface as previously described. Let the orders now be that each demon is to stop all molecules from crossing his area in either direction except 100 coming from A, arbitrarily chosen to be let pass into B, and a greater number, having among them less energy but equal momentum, to cross from B to A. Let this be repeated over and over again. The temperature in A will be continually diminished and the number of molecules in it continually increased, until there are not in B enough of molecules with small enough velocities to fulfil the condition with reference to permission to pass from B to A. If after that no molecule be allowed to pass the interface in either direction, the final condition will be very great condensation and very low temperature in A; rarefaction and very high temperature in B; and equal pressures in A and B. The process of disequalization of temperature and density might be stopped at any time by changing the orders to those previously specified, and so permitting a certain degree of diffusion each way across the interface while maintaining a certain uniform difference of temperatures with equality of pressure on the two sides.

If no selective influence, such as that of the ideal "demon," guides individual molecules, the average result of their free motions and collisions must be to equalize the distribution of energy among them in the gross; and after a sufficiently long time, from the supposed initial arrangement, the difference of energy in any two equal volumes, each containing a very great number of molecules, must bear a very small proportion to the whole amount in either; or, more strictly speaking, the probability of the difference of energy exceeding any stated finite proportion of the whole energy in either is very small. Suppose now the temperature to have become thus very approximately equalized at a certain time from the beginning, and let the motion of every particle become instantaneously reversed. Each molecule will retrace its former path, and at the end of a second interval of time, equal to the former, every molecule will be in the same position, and moving with the same velocity, as at the beginning; so that the given initial unequal distribution of temperature will again be found, with only the difference that each particle is moving in the direction reverse to that of its initial motion. This difference will not prevent an instantaneous subsequent commencement of equalization, which, with entirely different paths for the individual molecules, will go on in the average according to the same law as that which took place immediately after the system was first left to itself.

By merely looking on crowds of molecules, and reckoning their energy in the gross, we could not discover that in the very special case we have just considered the progress was towards a succession of states, in which the distribution of energy deviates more and more from uniformity up to a certain time. The number of molecules being finite, it is clear that small finite deviations from absolute precision in the reversal we have supposed would not obviate the resulting disequalization of the distribution of energy. But the greater the number of molecules, the shorter will be the time during which the disequalizing will continue; and it is only when we regard the number of molecules as practically infinite that we can regard spontaneous disequalization as practically impossible. And, in point of fact, if any finite number of perfectly elastic molecules, however great, be given in motion in the interior of a perfectly rigid vessel, and be left for a sufficiently long time undisturbed except by mutual impact and collisions against the sides of the containing vessel, it must happen over and over again that (for example) something more than $\frac{9}{10}$ths of the whole energy shall be in one-half of the vessel, and less than $\frac{1}{10}$th of the whole energy in the other half. But if the number of molecules be very great, this will happen enormously less frequently than that something more than $\frac{6}{10}$ths shall be in one-half, and something less than $\frac{4}{10}$ths in the other. Taking as unit of time the average interval of free motion between consecutive collisions, it is easily seen that the probability of these being something more than any stated percentage of excess above the half of the energy in one-half of the vessel during the unit of time from a stated instant, is smaller the greater the dimensions of the vessel and the greater the stated percentage. It is a strange but nevertheless a true conception of the old well-known law of the conduction of heat, to say that it is very improbable that in the course of 1000 years one-half of the bar of iron shall of itself become warmer by a degree than the other half; and that the probability of this happening before 1,000,000 years pass is 1000 times as great as that it will happen in the course of 1000 years, and that it certainly will happen in the course of some very long time. But let it be remembered that we have supposed the bar to be covered with an impermeable varnish. Do away with this impossible ideal, and believe the number of molecules in the universe to be infinite; then we may say one-half of the bar will never become warmer than the other, except by the agency of external sources of heat or cold. This one instance suffices to explain the philosophy of the foundation on which the theory of the dissipation of energy rests.

Take, however, another case, in which the probability may be readily calculated. Let an hermetically sealed glass jar of air contain 2,000,000,000,000 molecules of oxygen, and 8,000,000,000,000 molecules of nitrogen. If examined any time in the infinitely distant future, what is the number of chances against one that all the molecules of oxygen and none of nitrogen shall be found in one stated part of the vessel equal in volume to $\frac{1}{5}$th of the whole? The number expressing the answer in the Arabic notation has about 2,173,220,000,000 of places of whole numbers. On the other hand, the chance against there being exactly $\frac{2}{10}$ths of the whole number of particles of nitrogen, and at the same time exactly $\frac{2}{10}$ths of the whole number of particles of oxygen in the first specified part of the vessel, is only 4021×10^9 to 1.

Appendix

Calculation of probability respecting Diffusion of Gases

For simplicity, I suppose the sphere of action of each molecule to be infinitely small in comparison with its average distance from its nearest neighbour; thus, the sum of the volumes of the spheres of action of all the molecules will be infinitely small in proportion to the whole volume of the containing vessel. For brevity, space external to the sphere of action of every molecule will be called free space: and a molecule will be said to be in free space at any time when its sphere of action is wholly in free space; that is to say, when its sphere of action does not overlap the sphere of action of any other molecule. Let A, B denote any two particular portions of the whole containing vessel, and let a, b be the volumes of those portions. The chance that at any instant one individual molecule of whichever gas shall be in A is $a/(a+b)$, however many or few other molecules there may be in A at the same time; because its chances of being in any specified portions of free space are proportional to their volumes; and according to our supposition, even if all the other molecules were in A, the volume of free space in it would not be sensibly diminished by their presence. The chance that of n molecules in the whole space there shall be i stated individuals in A, and that the other $n-i$ molecules shall be at the same time in B, is

$$\left(\frac{a}{a+b}\right)^i \left(\frac{b}{a+b}\right)^{n-i}, \text{ or } \frac{a^i b^{n-i}}{(a+b)^n}$$

Hence the probability of the number of molecules in A being exactly i, and in B exactly $n-i$, irrespectively of individuals, is a fraction having for denominator $(a+b)^n$, and for numerator the term involving $a^i b^{n-i}$ in the expansion of this binomial; that is to say, it is

$$\frac{n(n-1)\ldots(n-i+1)}{1.2\ldots i} \left(\frac{a}{a+b}\right)^i \left(\frac{b}{a+b}\right)^{n-i}$$

If we call this T_i, we have

$$T_{i+1} = \frac{n-i}{i+1} \frac{a}{b} T_i$$

Hence T_i is the greatest term, if i is the smallest integer which makes

$$\frac{n-i}{i+1} < \frac{b}{a}$$

this is to say, if i is the smallest integer which exceeds

$$n\frac{a}{a+b} - \frac{b}{a+b}$$

Hence if a and b are commensurable, the greatest term is that for which

$$i = n\frac{a}{a+b}$$

To apply these results to the cases considered in the preceding article, put in the first place

$$n = 2 \times 10^{12}$$

this being the number of particles of oxygen; and let $i = n$. Thus, for the probability that all the particles of oxygen shall be in A, we find

$$\left(\frac{a}{a+b}\right)^{2 \times 10^{12}}$$

Similarly, for the probability that all the particles of nitrogen are in the space B, we find

$$\left(\frac{b}{a+b}\right)^{8 \times 10^{12}}$$

Hence the probability that all the oxygen is in A and all the nitrogen in B is

$$\left(\frac{a}{a+b}\right)^{2 \times 10^{12}} \times \left(\frac{b}{a+b}\right)^{8 \times 10^{12}}$$

Now by hypothesis

$$\frac{a}{a+b} = \frac{2}{10}$$

and therefore

$$\frac{b}{a+b} = \frac{8}{10}$$

hence the required probability is

$$\frac{2^{26 \times 10^{12}}}{10^{10^{18}}}$$

Call this $1/N$, and let log denote common logarithm. We have

$$\log N = 10^{13} - 26 \times 10^{12} \times \log 2 = (10 - 26 \log 2)$$
$$\times 10^{12} = 2173220 \times 10^{6}$$

This is equivalent to the result stated in the text above. The logarithm of so great a number, unless given to more than thirteen significant places, cannot indicate more than the number of places of the whole numbers in the answer to the proposed question, expressed according to the Arabic notation.

The calculation of T_i, when i and $n-i$ are very large numbers, is practicable by Stirling's theorem, according to which we have approximately

$$1.2\ldots i = i^{i+\frac{1}{2}}\varepsilon^{-i}\sqrt{2\pi}$$

and therefore

$$\frac{n(n-1)\ldots(n-i+1)}{1.2\ldots\ldots i} = \frac{n^{n+\frac{1}{2}}}{\sqrt{2\pi}i^{i+\frac{1}{2}}(n-i)^{n-i+\frac{1}{2}}}$$

Hence for the case

$$i = n\frac{a}{a+b}$$

which, according to the preceding formulae, gives T_i its greatest value, we have

$$T_i = \frac{1}{\sqrt{2\pi nef}}$$

where

$$e = \frac{a}{a+b} \text{ and } f = \frac{b}{a+b}$$

Thus, for example, let $n = 2 \times 10^{12}$,

$$e = \cdot 2, \quad f = \cdot 8$$

we have

$$T_i = \frac{1}{800000\sqrt{\pi}} = \frac{1}{1418000}$$

This expresses the chance of there being 4×10^{11} molecules of oxygen in A, and 16×10^{11} in B. Just half this fraction expresses the probability that the molecules of nitrogen are distributed in exactly the same proportion between A and B, because the number of molecules of nitrogen is four times greater than of oxygen.

If n denote the number of the molecules of one gas, and n' that of the molecules of another, the probability that each shall be distributed between A and B in the exact proportion of the volume is

$$\frac{1}{2\pi ef\sqrt{nn'}}$$

The value for the supposed case of oxygen and nitrogen is

$$\frac{1}{2\pi \times \cdot 16 \times 4 \times 10^{12}} = \frac{1}{4021 \times 10^{9}}$$

which is the result stated at the conclusion of the text above.

4

On the Relation of a General Mechanical Theorem to the Second Law of Thermodynamics *

LUDWIG BOLTZMANN

SUMMARY

Loschmidt has pointed out that according to the laws of mechanics, a system of particles interacting with any force law, which has gone through a sequence of states starting from some specified initial conditions, will go through the same sequence in reverse and return to its initial state if one reverses the velocities of all the particles. This fact seems to cast doubt on the possibility of giving a purely mechanical proof of the second law of thermodynamics, which asserts that for any such sequence of states the entropy must always increase.

Since the entropy would decrease as the system goes through this sequence in reverse, we see that the fact that entropy actually increases in all physical processes in our own world cannot be deduced solely from the nature of the forces acting between the particles, but must be a consequence of the initial conditions. Nevertheless, we do not have to assume a special type of initial condition in order to give a mechanical proof of the second law, if we are willing to accept a statistical viewpoint. While any individual non-uniform state (corresponding to low entropy) has the same probability as any individual uniform state (corresponding to high entropy), there are many more uniform states than non-uniform states. Consequently, if the initial state is chosen at random, the system is almost certain to evolve into a uniform state, and entropy is almost certain to increase.

* Originally published under the title: " Über die Beziehung eines allgemeine mechanischen Satzes zum zweiten Hauptsatze der Warmetheorie ", *Sitzungsberichte Akad. Wiss.*, Vienna, part II, **75**, 67–73 (1877); reprinted in Boltzmann's *Wissenschaftliche Abhandlungen*, Vol. 2, Leipzig, J. A. Barth, 1909, pp. 116–22.]

In his memoir on the state of thermal equilibrium of a system of bodies with regard to gravity, Loschmidt has stated a theorem that casts doubt on the possibility of a purely mechanical proof of the second law. Since it seems to me to be quite ingenious and of great significance for the correct understanding of the second law, yet in the cited memoir it has appeared in a more philosophical garb, so that many physicists will find it rather difficult to understand, I will try to restate it here.

If we wish to give a purely mechanical proof that all natural processes take place in such a way that

$$\int \frac{dQ}{T} \leq 0$$

then we must assume the body to be an aggregate of material points. We take the force acting between these points to be a function of the relative positions of the points. When this force is known as a function of these relative positions, we shall say that the law of action of the force is known. In order to calculate the actual motion of the points, and therefore the state variations of the body, we must know also the initial positions and initial velocities of all the points. We say that the initial conditions must be given. If one tries to prove the second law mechanically, he always tries to deduce it from the nature of the law of action of the force without reference to the initial conditions, which are unknown. One therefore seeks to prove that—whatever may be the initial conditions—the state variations of the body will always take place in such a way that

$$\int \frac{dQ}{T} \leq 0$$

We now assume that we are given a certain body as an aggregate of certain material points. The initial conditions at time zero shall be such that the body undergoes state variations for which

$$\int \frac{dQ}{T} \leq 0$$

We shall show that then, without changing the law of force, other initial conditions can be found for which conversely

$$\int \frac{dQ}{T} \geqq 0$$

Consider the positions and velocities of all the points after an arbitrary time t_1 has elapsed. We now take, in place of the original initial conditions, the following: all the material points† shall have the same initial positions at time zero that they had after time t_1 with the original initial conditions, and the same velocities but in the opposite directions. For brevity we shall call this state the one opposite to that previously found at time t_1.

It is clear that the points will pass through the same states as before but in the reverse order. The initial state which they had previously had at time zero, will now be reached after time t_1 has elapsed. Whereas previously we found

$$\int \frac{dQ}{T} \leqq 0$$

this quantity is now $\geqq 0$. The sign of this integral therefore does not depend on the force law but rather only on the initial conditions.‡ The fact that this integral is actually $\leqq 0$ for all processes in the world in which we live (as experience shows) is not due to the nature of the forces, but rather to the initial conditions. If, at time zero, the state of all material points in the universe were just the opposite of that which actually occurs at a much later time t_1, then the course of all events between times t_1 and zero would be reversed, so that

$$\int \frac{dQ}{T} \geqq 0$$

---

† By this we mean all the points of all bodies interacting with the one considered, either directly or indirectly. Strictly speaking one has to include all the points in the universe, since a complex of bodies that does not interact at all with the other bodies in the universe cannot actually be found, even though we can imagine it.

‡ It need not be mentioned that if the forces act in such a way that this is not true, for example if they are dynamical, then the following also loses its applicability.

Thus any attempt to prove from the nature of bodies and of the the force law, without taking account of initial conditions, that

$$\int \frac{dQ}{T} \leqq 0$$

must necessarily be futile. One sees that this conclusion has great seductiveness and that one must call it an interesting sophism. In order to locate the source of the fallacy in this argument, we shall imagine a system of a finite number of material points which does not interact with the rest of the universe.

We imagine a large but not infinite number of absolutely elastic spheres, which move in a closed container whose walls are completely rigid and likewise absolutely elastic. No external forces act on our spheres. Suppose that at time zero the distribution of spheres in the container is not uniform; for example, suppose that the density of spheres is greater on the right than on the left, and that the ones in the upper part move faster than those in the lower, and so forth. The sophism now consists in saying that, without reference to the initial conditions, it cannot be proved that the spheres will become uniformly mixed in the course of time. For the initial conditions which we originally assumed, the spheres will be almost always uniform at time t_1, for example. We can then choose in place of the original initial conditions the distribution of states which is just the opposite of the one which would occur (in consequence of the original initial conditions) after time t_1 has elapsed. Then the spheres would sort themselves out as time progresses, and at time t_1 they would acquire a completely non-uniform distribution of states, even though the initial distribution of states was almost uniform.

We must make the following remark: a proof, that after a certain time t_1 the spheres must necessarily be mixed uniformly, whatever may be the initial distribution of states, cannot be given. This is in fact a consequence of probability theory, for any non-uniform distribution of states, no matter how improbable it may be, is still not absolutely impossible. Indeed it is clear that any individual uniform distribution, which might arise after a certain time from some particular initial state, is just as improbable as an individual non-uniform distribution; just as in the game of Lotto, any individual set of five numbers is as improbable as the set 1, 2, 3, 4, 5. It is only because there are many more uniform distributions than non-uniform ones that the distribution of states will become uniform in the course of time. One therefore cannot prove that, whatever may be the positions and velocities of the spheres at the beginning, the distribution must become uniform after a long time; rather one can only prove that infinitely many more initial states will lead to a uniform one after a definite length of time than to a non-uniform one. Loschmidt's theorem tells us only about initial states which actually lead to a very non-uniform distribution of states after a certain time t_1; but it does not prove that there are not infinitely many more initial conditions that will lead to a uniform distribution after the same time. On the contrary, it follows from the theorem itself that, since there are infinitely many more uniform than non-uniform distributions, the number of states which lead to uniform distributions after a certain time t_1 is much greater than the number that leads to non-uniform ones, and the latter are the ones that must be chosen, according to Loschmidt, in order to obtain a non-uniform distribution at t_1.

One could even calculate, from the relative numbers of the different state distributions, their probabilities, which might lead to an interesting method for the calculation of thermal equilibrium.† In just the same way one can treat the second law. It is only in some special cases that it can be proved that, when a system goes over from a non-uniform to a uniform distribution of states, then $\int dQ/T$ will be negative, whereas it is positive in the opposite case. Since there are infinitely many more uniform than non-uniform distributions of states, the latter case is extraordinarily improbable and can be considered impossible for practical purposes; just as it may be considered impossible that if one starts with oxygen and nitrogen mixed in a container, after a month one will find chemi-

---

† [Following up this remark, Boltzmann developed soon afterward his statistical method for calculating equilibrium properties, based on the relation between entropy and probability: *Wien. Ber.* **76**, 373 (1877).]

cally pure oxygen in the lower half and nitrogen in the upper half, although according to probability theory this is merely very improbable but not impossible.

Nevertheless Loschmidt's theorem seems to me to be of the greatest importance, since it shows how intimately connected are the second law and probability theory, whereas the first law is independent of it. In all cases where $\int dQ/T$ can be negative, there is also an individual very improbable initial condition for which it may be positive; and the proof that it is almost always positive can only be carried out by means of probability theory. It seems to me that for closed paths of the atom, $\int dQ/T$ must always be zero, which can therefore be proved independently of probability theory. For unclosed paths it can also be negative. I will mention here a peculiar consequence of Loschmidt's theorem, namely that when we follow the state of the world into the infinitely distant past, we are actually just as correct in taking it to be very probable that we would reach a state in which all temperature differences have disappeared, as we would be in following the state of the world into the distant future. This would be similar to the following case: if we know that in a gas at a certain time there is a non-uniform distribution of states, and that the gas has been in the same container without external disturbance for a very long time, then we must conclude that much earlier the distribution of states was uniform and that the rare case occurred that it gradually became non-uniform. In other words: any non-uniform distribution evolves into an almost uniform one after a long time t_1. The one opposite to this latter one evolves, after the same time t_1, into the initial non-uniform one (more precisely, into the opposite of it). The distribution opposite to the initial one would however, if chosen as an initial distribution, likewise evolve into a uniform distribution after time t_1.

If perhaps this reduction of the second law to the realm of probability makes its application to the entire universe appear dubious, yet the laws of probability theory are confirmed by all experiments carried out in the laboratory.

5

On the Three-body Problem and the Equations of Dynamics *

HENRI POINCARÉ

SUMMARY

Poisson attempted to show that a mechanical system is stable, in the sense that it will eventually return to a configuration very close to its initial one. He did not think that all solutions of the equations of dynamics would be stable: stability may depend on the initial conditions.

It is proved that there are infinitely many ways of choosing the initial conditions such that the system will return infinitely many times as close as one wishes to its initial position. There are also an infinite number of solutions that do not have this property, but it is shown that these unstable solutions can be regarded as "exceptional" and may be said to have zero probability.

§ 1. Notations and definitions

We consider a system of differential equations

$$\frac{dx_1}{dt} = X_1, \frac{dx_2}{dt} = X_2, \ldots, \frac{dx_n}{dt} = X_n \qquad (1)$$

where t represents the independent variable which we shall call the time, $x_1, x_2 \ldots, x_n$ are the unknown functions, and $X_1, X_2, \ldots, X_n$

---

* [Originally published under the title: " Sur le problème des trois corps et les équations de dynamique ", *Acta Mathematica* **13**, pp. 1–270 (1890); reprinted in the *Oeuvres de Henri Poincaré*, Gauthier-Villars, Paris, 1952, 7, pp. 262–479. The extracts given here are translated from pp. 8, 10 and 67–73 of the original; the definition of integral invariants, summarized in an editor's note, is taken from pp. 52–5.]

are given functions of $x_1, x_2, \ldots, x_n$. We assume in general that the functions $X_1, X_2, \ldots, X_n$ are analytic and uniform for all real values of the $x_1, x_2, \ldots, x_n$.

. . . .

We consider in particular the case $n = 3$; we can then regard x_1, x_2 and x_3 as the coordinates of a point P in space. . . . When we vary the time t, the point P will describe a certain curve in space which we call a *trajectory*. To each particular solution of equations (1) there corresponds a trajectory, and conversely.

If the functions X_1, X_2 and X_3 are uniform, one and only one trajectory will pass through each point in space. The only exception occurs when one of these three functions becomes infinite, or when all three are zero.

. . . .

[A relation of the form

$$F(x_1, x_2, \ldots, x_n) = \text{constant}$$

where the $x_1, x_2, \ldots, x_n$ are solutions of the system of equations (1), is called an integral of those equations. More generally we may have a relation such as

$$F_1(x_1, x_2, \ldots, x_n, x'_1, x'_2, \ldots, x'_n) = \text{constant}$$

where $x'_1, x'_2, \ldots, x'_n$ are another solution of the same system of equations; or we may have such a relation involving any number of such solutions. A relation which involves an integral over an infinite number of such solutions is known as an *integral invariant* of the system. The most important example is an integral over all the solutions contained in a certain volume of space:

$$\iiint dx\,dy\,dz = \text{constant}$$

In other words, the volume occupied by a specified set of initial coordinates, corresponding to a region of possible positions of the point P, remains invariant as each of these points moves along the trajectory determined by the system of equations. This " conservation of volume in phase space " is also known as Liouville's theorem in statistical mechanics.]

. . . .

§ 8. Use of integral invariants

The interest of integral invariants results from the following theorems of which we shall make frequent use.

We have defined stability above [§1] by saying that the mobile point P should stay at a finite distance from its starting place; one sometimes understands stability in a different sense, however. In order to have stability, it is necessary that the point P should return after a sufficiently long time, if not to its initial position, at least to a position as close as one wishes to this initial position.

It is in this latter sense that Poisson[†] means stability. When he showed that, if one takes account of the second powers of the masses, the major axes remain invariant, he was only interested in showing that the series expansions of the major axes contain only periodic terms of the form $\sin \alpha t$ or $\cos \alpha t$, or mixed terms of the form $t \sin \alpha t$ or $t \cos \alpha t$, without containing any secular terms of the form t or t^2. This does not mean that the major axes cannot exceed a certain value, for a mixed term $t \cos \alpha t$ can increase beyond any limit; it means only that the major axes return infinitely many times to their original values.

Does stability, in the sense of Poisson, apply to all solutions? Poisson did not think so, for his proof assumed explicitly that the average motions are not commensurable; it therefore does not apply for any arbitrary initial conditions of motion.

The existence of asymptotic solutions, which we will establish later on, is a sufficient proof that if the initial position of the point P is appropriately chosen, the point P will not pass infinitely many times as close as one wishes to this initial position.

But I propose to establish that in the particular case of the three-body problem, one can choose the initial position of P in an infinite number of ways such that it does return infinitely many times as close as one wishes to its initial position.

[†] [S. D. Poisson, *Nouveau Bulletin des Sciences d la Société Philomatique de Paris* **1,** 191 (1808); *Mémoires de l'Academie Royale des Sciences de l'Institut de France* **7,** 199 (1827).]

In other words, there will be an infinity of particular solutions of the problem which do not possess stability in the second sense of the word, that of Poisson; but there will be an infinity that do. I will add that the former can be regarded as " exceptional "; I will attempt later on to clarify the exact meaning that I attach to this word.

Let $n = 3$ and imagine that x_1, x_2, x_3 represent the coordinates of a point P in space.

THEOREM I. Suppose that the point P remains at a finite distance, and that the volume $\int dx_1 dx_2 dx_3$ is an integral invariant; if one considers any region r_0 whatever, no matter how small may be this region, there will be trajectories which traverse it infinitely many times.

In particular, suppose that P does not leave a bounded region R. I call V the volume of this region R.

Now let us imagine a very small region r_0, and call v the volume of this region. Through each point of r_0 there passes a trajectory which one may regard as being traversed by a mobile point following the law defined by our differential equations. Let us then consider an infinity of mobile points filling the region r_0 at time 0 and afterwards moving in accordance with this law. At time τ they fill a certain region r_1, at time 2τ a region r_2, etc., and at time $n\tau$ a region r_n. I may assume that τ is so large, and r_0 so small, that r_0 and r_1 have no point in common.

The volume being an integral invariant, the various regions $r_0, r_1, \ldots, r_n$ will have the same volume v. If these regions have no points in common, the total volume will be greater than nv; but since all these regions are inside R, the total volume is smaller than V. If one has

$$n > \frac{V}{v}$$

it is necessary that at least two of our regions must have a part in common. Let r_p and r_q be these two regions $(q>p)$. If r_p and r_q have a common point, it is clear that r_0 and r_{p-q} must have a common point.

More generally, if one cannot find k regions having a common part, then any point in space can belong to no more than $k-1$ of the regions $r_0, r_1, \ldots, r_n$. The total volume occupied by these regions will then be greater than $nv/(k-1)$. If then one has

$$n > (k-1)\frac{V}{v}$$

there must be k regions having a common part. Let

$$r_{p_1}, r_{p_2}, \ldots, r_{p_k}$$

be these regions. Then

$$r_0, r'_{p_2-p_1}, r'_{p_3-p_1}, \ldots, r'_{p_k-p_1}$$

will have a common part.

We now consider the question from another viewpoint. By analogy with the terminology of the previous section we agree to say that the region r_n is the nth consequent of r_0 and r_0 is the nth antecedent of r_n.

Suppose then that r_p is the first of the successive consequents of r_0 which has a part in common with r_0. Let r'_0 be this common part; let s'_0 be the pth antecedent of r'_0 which will also be part of r_0 since its pth consequent is part of r_p.

Finally let r'_{p_1} be the first of the consequents of r'_0 which has a part in common with r'_0; let r''_0 be this common part; its p_1th antecedent will be part of r'_0 and consequently of r_0, and its $p+p_1$ antecedent, which I will call s''_0, will be part of s'_0 and therefore of r_0.

Thus s''_0 will be part of r_0 as well as of its pth and $(p+p_1)$th consequents.

Continuing in the same way, we can form r'''_0 from r''_0 just as we formed r''_0 from r'_0, and r'_0 from r_0; so we likewise form $r^{IV}_0, \ldots, r^n_0, \ldots$

I will assume that the first of the successive consequents of r^n_0 which has a part in common with r^n_0 will be that of order p_n.

I will call s^n_0 the antecedent of order $p+p_1+p_2+p_n{}_1$ of r^n_0.

The s_0^n will be part of r_0 as well as of its n consequents of order:

$$p, p+p_1, p+p_1+p_2, \ldots, p+p_1+p_2+ \ldots +p_{n-1}$$

Moreover, s_0^n will be part of s_0^{n-1}; s_0^{n-1} will be part of s_0^{n-2}, and so forth.

There will then be points that belong at the same time to regions $r_0, s_0', s_0'', \ldots, s_0^n, s_0^{n+1}, \ldots$ ad infinitum. The collection of these points will form a region σ which may be reduced to one or several points.

Then the region σ will be part of r_0 as well as its consequents of order $p, p+p_1, \ldots, p+p_1+ \ldots +p_n, p+p_1+ \ldots +p_n+p_{n+1}, \ldots$, ad infinitum.

In other words, all trajectories starting from one of the points of σ will go through the region r_0 infinitely many times.

Q. E. D.

COROLLARY. It follows from the preceding that there exists an infinite number of trajectories which can pass through the region r_0 infinitely many times; but there may also exist others that traverse it only a finite number of times. I now propose to explain why these latter trajectories may be considered " exceptional ".

This expression not having by itself a precise sense, I am first obliged to complete its definition.

We shall agree to say that the probability that the initial position of the mobile point P belongs to a certain region r_0 is to the probability that this initial position belongs to another region r_0' in the same ratio as the volume of r_0 to the volume of r_0'.

Probabilities being defined thus, I propose to establish that the probability that a trajectory starting from a point of r_0 does not go through this region more than k times is zero, however large k may be, and however small the region r_0. This is what I mean when I say that the trajectories that traverse r_0 only a finite number of times are exceptional.

I assume that the initial position of the point P belongs to r_0 and I propose to calculate the probability that the trajectory starting from this point does not pass through r_0 as many as $k+1$ times in the interval between time 0 and time $n\tau$.

We saw that if the volume v of r_0 is such that

$$n > \frac{kV}{v}$$

one can find $k+1$ regions which I call

$$r_0, r_{\alpha_1}, r_{\alpha_2}, \ldots, r_{\alpha_k}$$

and which will have a common part. Let s_{α_k} be this common part, let s_0 be its antecedent of order α_k; and designate by s_p the pth consequent of s_0.

I say that if the initial position of P belongs to s_0, the trajectory starting from this point traverses $k+1$ times at least the region r_0 between time 0 and time $n\tau$. In particular, the mobile point which describes this trajectory will be in the region s_0 at time 0, in s_p at $p\tau$, and in s_n at $n\tau$. It therefore must go through the following regions between 0 and $n\tau$:

$$s_0, s_{\alpha_k-\alpha_{k-1}}, s_{\alpha_k-\alpha_{k-2}}, \ldots, s_{\alpha_k-\alpha_2}, s_{\alpha_k-\alpha_1}, s_{\alpha_k}$$

Now I say that all these regions are part of r_0. In particular, s_{α_k} is part of r_0 by definition; s_0 is part of r_0 because its α_kth consequent s_{α_k} is part of r_{α_k}; and in general $s_{\alpha_k-\alpha_i}$ will be part of r_0 because its α_ith consequent s_{α_k} is part of r_{α_i}.

Then the mobile point traverses the region r_0 at least $k+1$ times.

Q. E. D.

Now let σ_0 be the part of r_0 which belongs neither to s_0 nor to any similar region, so that the trajectories starting from the various points of σ_0 do not go through r_0 at least $k+1$ times between times 0 and $n\tau$. Let w be the volume of σ_0.

The probability sought, that is to say the probability that our trajectory does not traverse r_0 as many as $k+1$ times during this time-interval, will then be w/v.

Now by hypothesis any trajectory starting from σ_0 does not traverse r_0 as many as $k+1$ times, and *a fortiori* does not traverse σ_0 as many as $k+1$ times. One then has

$$w < \frac{kV}{v}$$

and our probability will be smaller than

$$\frac{kV}{nv}$$

However large k may be, and however small v may be, one can always take n so large that this expression will be as small as one wishes. Then the probability that our trajectory, which starts from a point of r_0, does not traverse this region more than k times between times 0 and ∞, is zero.

Q. E. D.

EXTENSION OF THEOREM I. We assumed that
1. $n = 3$.
2. The volume is an invariant integral.
3. The point P is required to remain at a finite distance.

The theorem is still true if the volume is not an invariant integral, provided there still exists any other positive invariant:

$$\int M dx_1 dx_2 dx_3$$

It is still true if $n > 3$, if there exists a positive invariant

$$\int M dx_1 dx_2 dx_3$$

and if $x_1, x_2, \ldots, x_n$, the coordinates of the point P in the n-dimensional space, are required to remain finite.

There is an even stronger statement which is valid.

Suppose that $x_1, x_2, \ldots, x_n$ are not required to remain finite, but the positive integral invariant

$$\int M dx_1 dx_2 \ldots dx_n$$

extended over the entire n-dimensional space has a finite value. The theorem will still be true then.

Here is a case that frequently arises.

Suppose that one knows an integral of equations (1):

$$F(x_1, x_2, \ldots, x_n) = \text{constant}$$

If $F = $ constant is the general equation of a system of closed surfaces in the n-dimensional space, if in other words F is a uniform function which becomes infinite whenever one of the variables $x_1, x_2, \ldots, x_n$ ceases to be finite, then it is clear that $x_1, x_2, \ldots x_n$ will always remain finite since F retains a finite constant value; one then has satisfied the conditions of the theorem.

But suppose that the surfaces $F = $ constant are not closed; it may happen nevertheless that the invariant positive integral

$$\int M dx_1 dx_2 \ldots dx_n$$

extended over the entire set of values of the x's such that

$$C_1 < F < C_2$$

has a finite value; the theorem will then still be true.

6

Mechanism and Experience *

HENRI POINCARÉ

SUMMARY

The advocates of the mechanistic conception of the universe have met with several obstacles in their attempts to reconcile mechanism with the facts of experience. In the mechanistic hypothesis, all phenomena must be reversible, while experience shows that many phenomena are irreversible. It has been suggested that the apparent irreversibility of natural phenomena is due merely to the fact that molecules are too small and too numerous for our gross senses to deal with them, although a " Maxwell demon " could do so and would thereby be able to prevent irreversibility.

The kinetic theory of gases is up to now the most serious attempt to reconcile mechanism and experience, but it is still faced with the difficulty that a mechanical system cannot tend toward a permanent final state but must always return eventually to a state very close to its initial state [Selection 5]. This difficulty can be overcome only if one is willing to assume that the universe does not tend irreversibly to a final state, as seems to be indicated by experience, but will eventually regenerate itself and reverse the second law of thermodynamics.

Everyone knows the mechanistic conception of the universe which has seduced so many good men, and the different forms in which it has been dressed.

Some represent the material world as being composed of atoms which move in straight lines because of their inertia; the velocity and direction of this motion cannot change except when two atoms collide.

Others allow action at a distance, and suppose that the atoms

---

* [Originally published under the title: " Le mécanisme et l'expérience ", *Revue de Metaphysique et de Morale* **1**, pp. 534–7 (1893).]

exert on each other an attraction (or a repulsion) which depends on their distance according to some law.

The first viewpoint is clearly only a particular case of the second; what I am going to say will be as true of one as of the other. The most important conclusions apply also to Cartesian mechanism, in which one assumes a continuous matter.

It would perhaps be appropriate to discuss here the metaphysical difficulties that underlie these conceptions; but I do not have the necessary authority for that. Rather than discussing with the readers of this review that which they know better than I do, I prefer to speak of subjects with which they are less familiar, but which may interest them indirectly.

I am going to concern myself with the obstacles which the mechanists have encountered when they wished to reconcile their system with experimental facts, and the efforts which they have made to overcome or circumvent them.

In the mechanistic hypothesis, all phenomena must be *reversible*; for example, the stars might traverse their orbits in the retrograde sense without violating Newton's law; this would be true for any law of attraction whatever. This is therefore not a fact peculiar to astronomy; reversibility is a necessary consequence of all mechanistic hypotheses.

Experience provides on the contrary a number of irreversible phenomena. For example, if one puts together a warm and a cold body, the former will give up its heat to the latter; the opposite phenomenon never occurs. Not only will the cold body not return to the warm one the heat which it has taken away when it is in direct contact with it; no matter what artifice one may employ, using other intervening bodies, this restitution will be impossible, at least unless the gain thereby realized is compensated by an equivalent or large loss. In other words, if a system of bodies can pass from state A to state B by a certain path, it cannot return from B to A, either by the same path or by a different one. It is this circumstance that one describes by saying that not only is there not *direct reversibility*, but also there is not even *indirect reversibility*.

There have been many attempts to escape this contradiction; first there was Helmholtz's hypothesis of "hidden movements". Recall the experiment made by Foucault and Panthéon with a very long pendulum. This apparatus seems to turn slowly, indicating the rotation of the earth. An observer who does not know about the movement of the earth would certainly conclude that mechanical phenomena are irreversible. The pendulum always turns in the same sense, and there is no way to make it turn in the opposite sense; to do that it would be necessary to change the sense of rotation of the earth. Such a change is of course impractical, but for us it is conceivable; it would not be so for a man who believed our planet to be immobile.

Can one not imagine that there exist similar motions in the molecular world, which are hidden from us, which we have not taken account of, and of which we cannot change the sense?

This explanation is seductive, but it is insufficient; it shows why there is not *direct* reversibility; but one can show that it still requires *indirect* reversibility.

The English have proposed a completely different hypothesis. To explain it, I will make use of a comparison: if one had a hectolitre of wheat and a grain of barley, it would be easy to hide this grain in the middle of the wheat; but it would be almost impossible to find it again, so that the phenomenon appears to be in a sense irreversible. This is because the grains are small and numerous; the apparent irreversibility of natural phenomena is likewise due to the fact that the molecules are too small and too numerous for our gross senses to deal with them.

To clarify this explanation, Maxwell introduced the fiction of a "demon" whose eyes are sharp enough to distinguish the molecules, and whose hands are small and fast enough to grab them. For such a demon, if one believes the mechanists, there would be no difficulty in making heat pass from a cold to a warm body.

The development of this idea has given rise to the kinetic theory of gases, which is up to now the most serious attempt to reconcile mechanism and experience.

But all the difficulties have not been overcome.

A theorem, easy to prove, tells us that a bounded world, governed only by the laws of mechanics, will always pass through a state very close to its initial state. On the other hand, according to accepted experimental laws (if one attributes absolute validity to them, and if one is willing to press their consequences to the extreme), the universe tends toward a certain final state, from which it will never depart. In this final state, which will be a kind of death, all bodies will be at rest at the same temperature.

I do not know if it has been remarked that the English kinetic theories can extricate themselves from this contradiction. The world, according to them, tends at first toward a state where it remains for a long time without apparent change; and this is consistent with experience; but it does not remain that way forever, if the theorem cited above is not violated; it merely stays there for an enormously long time, a time which is longer the more numerous are the molecules. This state will not be the final death of the universe, but a sort of slumber, from which it will awake after millions of millions of centuries.

According to this theory, to see heat pass from a cold body to a warm one, it will not be necessary to have the acute vision, the intelligence, and the dexterity of Maxwell's demon; it will suffice to have a little patience.

One would like to be able to stop at this point and hope that some day the telescope will show us a world in the process of waking up, where the laws of thermodynamics are reversed.

Unfortunately, other contradictions arise; Maxwell made ingenious efforts to conquer them. But I am not sure that he succeeded. The problem is so complicated that it is impossible to treat it with complete rigour. One is then forced to make certain simplifying hypotheses; are they legitimate, are they self-consistent? I do not believe they are. I do not wish to discuss them here; but there is no need for a long discussion in order to challenge an argument of which the premises are apparently in contradiction with the conclusion, where one finds in effect reversibility in the premises and irreversibility in the conclusion.

Thus the difficulties that concern us have not been overcome, and it is possible that they never will be. This would amount to a definite condemnation of mechanism, if the experimental laws should prove to be distinctly different from the theoretical ones.

7

On a Theorem of Dynamics and the Mechanical Theory of Heat *

ERNST ZERMELO

SUMMARY

Poincaré's recurrence theorem [Selection 5] shows that irreversible processes are impossible in a mechanical system. A simple proof of this theorem is given.

The kinetic theory cannot provide an explanation of irreversible processes unless one makes the implausible assumption that only those initial states that evolve irreversibly are actually realized in nature, while the other states, which from a mathematical viewpoint are more probable, actually do not occur. It is concluded that it is necessary to formulate either the second law of thermodynamics or the mechanical theory of heat in an essentially different way, or else give up the latter theory altogether.

In the second chapter of Poincaré's prize essay on the three-body problem,† there is proved a theroem from which it follows that the usual description of the thermal motion of molecules, on which is based for example the kinetic theory of gases, requires an important modification in order that it be consistent with the thermodynamic law of increase of entropy. Poincaré's theorem says that *in a system of mass-points under the influence of forces that depend only on position in space, in general any state of motion (charac-*

---

* [Originally published under the title: " Uber einen Satz der Dynamik und die mechanische Warmetheorie ", *Annalen der Physik* **57,** pp. 485–94 (1896).]

† Poincaré, " Sur les équations de la dynamique et le problème des trois corps ", *Acta Mathematica* **13,** pp. 1–270 (1890); the theorem referred to is on pp. 67–72. [Selection 5]

terized by configurations and velocities) must recur arbitrarily often, at least to any arbitrary degree of approximation even if not exactly, provided that the coordinates and velocities cannot increase to infinity. Hence, in such a system *irreversible processes are impossible* since (aside from singular initial states) no single-valued continuous function of the state variables, such as entropy, can continually increase; if there is a finite increase, then there must be a corresponding decrease when the initial state recurs. Poincaré, in the essay cited, used his theorem for astronomical discussions on the stability of sun systems; he does not seem to have noticed its applicability to systems of molecules or atoms and thus to the mechanical theory of heat, although he has taken especial interest in fundamental questions of thermodynamics, and by another method he has tried to show that irreversible processes cannot always be explained by Helmholtz's theory of "monocyclic systems.[†],[‡] In order not to have to assume an acquaintance with Poincaré's long and—to many physicists—difficultly accessible work, I will sketch here the simplest possible proof of the theorem mentioned above.

Let N be the number of mass-points; there will be $n = 6N$ state quantities, i.e. the $3N$ coordinates and $3N$ velocity components, which we denote by $x_1, x_2, \ldots, x_n$. The time-derivatives of the coordinates will be identical with the corresponding velocities; the derivatives of the velocities, i.e. the accelerations, will be the forces, which according to our assumption are single-valued continuous functions of the coordinates. The former will therefore be independent of the coordinates, and the latter of the velocities, and the differential equations for the motion are of the form

$$\frac{dx_\mu}{dt} = X_\mu(x_1, x_2, \ldots x_n) \qquad (1)$$

$$(\mu = 1, 2, \ldots n)$$

[†] Poincaré, *Compt. rend. Acad. Sci., Paris* **108**, pp. 550–2 (1889); *Vorles. über Thermodynamik*, pp. 294–6.

[‡] [H. v. Helmholtz, *Sitzber. K. Preuss. Akad. Wiss., Berlin*, **159**, 311, 755 (1884); *J. f. reine und angew. Math.* **97**, 111, 317 (1884).]

where none of the X_μ functions depends on the corresponding variable $x\mu$; thus we have the relations:

$$\frac{\partial X_1}{\partial x_1}+\frac{\partial X_2}{\partial x_2}+ \ldots +\frac{\partial X_n}{\partial x_n} = 0 \qquad (2)$$

In such a system (1) of differential equations of the first order, there corresponds to an arbitrary initial state P_0:

$$x_1 = \xi_1, x_2 = \xi_2 \ldots x_n = \xi_n, (t = t_0)$$

a definite transformed state P at time t, expressed by the integral of (1):

$$x_\mu = \phi_\mu(t-t_0, \xi_1, \xi_2, \ldots \xi_n) \qquad (3)$$

$$(\mu = 1, 2, \ldots n)$$

where the ϕ_μ are singled-valued continuous functions of all their arguments, which, regardless of the choice of initial time t_0, are determined only by the functions X_μ. These relations are equally valid for previous or later times, i.e. for either negative or positive values of $t-t_0$; the initial state P_0 is an arbitrarily chosen phase of the motion, which does not have to precede any other phase. Likewise there corresponds to any *region* g_0 filled with initial states, describable by means of relations of the form:

$$F(\xi_1, \ldots \xi_n) < 0$$

a definite transformed region $g = g_t$ at time t. To these regions there correspond the n-fold integrals,

$$\gamma_0 = \int d\xi_1 d\xi_2 \ldots d\xi_n$$

which we shall call the " extension " of g_0, and in general another extension of g.

$$\gamma = \int dx_1 dx_2 \ldots dx_n$$

However, in the special case where the functions X_μ satisfy the condition (2), Liouville's theorem† says that the second integral

---

† Jacobi, *Dynamik*, p. 93; Kirchhoff, *Theorie der Wärme*, pp. 142-4.

is equal to the first, and indeed is independent of time, however one may choose the regions g_0 or g (either of which determines the other). Thus one can write:

$$dy = dx_1 dx_2 \ldots dx_n = d\gamma_0 = \text{constant} \qquad (4)$$

"The future states that correspond to the initial states in any region will fill at a given time a region of equal extension."

An arbitrary region g_0 of states therefore goes continuously into a new region $g = g_t$, the "phases" of its transformation, which always has the same extension γ. All these "*later*" phases g_t ($t > 0$) taken together form another continuous region G_0, the "*future*" of g_0, i.e. the aggregate of all states that arise from g_0 sometime in the future, in a finite time. This region $G = G_0$ will be finite and will have a finite extension $\Gamma \geq \gamma$, if we assume that the quantities $x_1, x_2 \ldots, x_n$ can never exceed certain finite limits, for all initial states in g_0. While the region g changes with time, at the same time all its "later phases" will change into the following ones, and G also changes so that it represents at any instant t the "future" of the corresponding phase g_t. According to the definition of future, this change must take place in such a way that while earlier states may leave G, no new ones can enter: each phase of G includes all the later ones in itself, so that the extension Γ may *only decrease*. But since according to (4) this extension must remain constant, the states that leave G cannot fill a region of finite extension, so that their number is vanishingly small compared to the remaining ones; we therefore call them *singular*. Now if g_0 is contained in G_0, then it is also completely or almost completely contained in each following phase G_τ, the future of g_τ, for an arbitrarily large time interval τ. But this means that there are always states in g_τ that will later transform to states in g_0, and, conversely, states in g_0 that after time τ will again return to g_0. These latter states are found in any part of the region, no matter how small, to which the same conditions apply as to g_0; and they will be connected together, since every recurrent state must be surrounded by a neighbourhood of states that also are recurrent; hence, these states fill the entire region g_0 with the exception of singular states of total extension zero. If one excludes all these singular states, then for *any* finite time τ there will be a region g', which no longer need be continuous but still contains the overwhelming majority of the states of g_0. These states of g' will always recur once after an arbitrary time, and therefore will recur infinitely often and come back arbitrarily close to their initial values if one chooses g_0 sufficiently small.

We have proved Poincaré's theorem in complete generality, although for our present purpose it suffices to show that the states in g_0 will return at least *once* to g_0. From this it follows directly that *there can be no single-valued continuous function* $S = S(x_1, x_2 \ldots x_n)$ *of the states that always increases for all initial states in some region, no matter how small the region.* For then S would have to increase from a value less than R to a value greater than R during time τ, for some initial state P_0; and the same would also have to hold for all states in a certain neighbourhood g of P_0; and for the states in this region that recur, the function would have to decrease again.

The same result can also be proved directly very simply. If the function S continually increased for all initial states in g, then it would also do so for all states in the larger region G, the future of g, and then according to (4) the n-fold integral over G,

$$\int S dx_1 dx_2 \ldots dx_n$$

must *continually increase*. But this is impossible since the region of integration, G, can change only by losing singular states that have no finite extension, so that the value of the integral must remain *constant*.

The interpretation and proof of this theorem become very clear for the case $n = 3$, when one takes the variables x_1, x_2, x_3 to be the coordinates of a mass-point in space. Then equations (1), in connection with (2) or (4), determine a *stationary flow of an incompressible fluid* in a *closed container*, if the quantities x_μ cannot increase without limit. A definite " state " corresponds here to a point in space, and a state varying in time corresponds to a moving point. The paths described by these fluid-points, the "streamlines", taken together form "stream-tubes" or "stream-filaments" according to whether they proceed from closed curves or from surface sections, and in the case of stationary motion they remain unchanged. Now it is intuitively obvious that in this case all the stream-filaments must run back on themselves, since the fluid streaming through the filament can neither break through the sides of the tube nor accumulate somewhere inside it. From this it follows that every finite fluid particle must always return again as closely as one likes to any position, if one makes the fluid filaments thin enough and waits a sufficiently long time. Of course there are also non-recurring singular stream-lines, for example those that *asymptotically approach* an immersed solid body or cavity between the other streamlines passing on either side and avoiding each other; but these cannot form a stream-filament of finite thickness. On the other hand, if the flow has a *velocity-potential*, then in a completely enclosed container this must necessarily be many-valued, while the function S mentioned above has to be single-valued. In the more general case $n > 3$, the same analogy is often of heuristic value; one can use the same terminology and interpret equations (1) and (2) or (4) as a "stationary flow of an incompressible fluid in a space of n dimensions".

Our conclusion is therefore the following:

In a system of arbitrarily many mass-points, whose accelerations depend only on their position in space, there can be no " irreversible " processes for all initial states that occupy any region of finite extension, no matter how small, provided that the coordinates and velocities of the points can never exceed finite limits.

The theorem is even more general than this, for it holds in particular for an *arbitrary* mechanical system with generalized coordinates q_μ and corresponding momenta p_μ, whose equations of motion can be written in Hamiltonian form:

$$\frac{dp_\mu}{dt} = \frac{\partial H}{\partial q_\mu}, \frac{dq_\mu}{dt} = -\frac{\partial H}{\partial p_\mu}$$

which we may call a " conservative " system since all forces are derivable from a potential, and hence the mechanical energy is conserved. In such a system it is obvious that

$$\frac{\partial}{\partial p}\frac{dq_\mu}{dt}+\frac{\partial}{\partial q}\frac{dq_\mu}{dt}=0$$

and by analogy with the relation (2) we see that all the consequences following from that relation retain their validity.

According to the mechanical theory, in its usual atomistic version, all of nature can be represented as a system of this type: all natural processes are nothing but motions of atoms or molecules which are treated either as points without extension, or as aggregates of such points, and which interact only with " central forces " derivable from a potential, which are independent of the velocities. It is precisely this assumption that one tries to use in the " kinetic theory of gases ", in that one treats the molecules of an " ideal gas " as centres of repulsion, or as elastic spheres, or (with Boltzmann) as elastic bodies of some other form, but in any case as a " conservative " system in the sense mentioned above. However, one also restricts oneself to short-range repulsive forces between pairs of molecules.

With these assumptions it would be impossible, on the basis of the foregoing considerations, to have " irreversible " processes, unless the molecules could be dispersed to infinite distances, or could eventually attain infinitely large velocities. (We exclude the possibility of a uniform progressive motion of the centre of mass of the system.) The former possibility is excluded by the special assumption that the system is surrounded by a solid container; and the latter is excluded by the principle of conservation of energy. For in order to attain an infinitely great kinetic energy, an infinite amount of work must be done, which could only occur in the case of very close approach of two attracting force-centres. However, we must assume on physical grounds that there can be only repulsive forces at very small distances between atoms.

Suppose that we have a gas enclosed in a solid container with elastic sides that are impenetrable to heat. In general there will indeed be an infinite manifold of initial states of the molecules for which the gas will undergo *permanent* changes of state, such as viscosity, heat condition, or diffusion. However, there will also be a much larger number of possible initial states, which can be reached by arbitrarily small displacements from the former states, and these states, instead of undergoing irreversible changes, will come back periodically to their initial states as closely as one likes in the sense described above. The same must also be true when the actual *physical* state—e.g. the temperature and the entropy—is not defined by an instantaneous *state of motion*, but rather by a finite *sequence of motions*, which in any case is determined by the initial state of motion, and must recur along with it.

Hence, in order to establish the general validity of the second law of thermodynamics, one would have to assume that only those initial states that lead to irreversible processes are actually realized in nature, despite their smaller number, while the other states, which from a mathematical viewpoint are more probable, actually *do not occur*.

Though such an assumption would be irrefutable, it would hardly correspond to our requirement for causality; and in any case the spirit of the mechanical view of nature itself requires that we should always assume that all *imaginable* mechanical initial states are physically *possible*, at least within certain limits, and certainly we must allow those states that constitute an overwhelming majority and deviate by an arbitrarily small amount from the ones that actually occur. Remember that, strictly speaking, all our laws of nature refer not to definite precise quantities or processes, which can never be observed with complete accuracy, but rather only to certain ranges of values, approximations, and probabilities, whereas singularities exist only as abstract limiting cases. The assumption being discussed here would therefore be quite unique in physics, and I do not believe that anyone would be satisfied with it for very long.

That *not all conceivable* initial states can correspond to the second law follows already from the fact that by reversing the velocity directions of all the molecules at an arbitrary time, the entire temporal course of a process must be reversed. In fact, this criticism of the mechanical explanation of irreversible processes has already been made before, and in the winter of 1894–95 there was an extended discussion of this question in *Nature*, stimulated by a remark of Culverwell, though it seems to me that no satisfactory solution was found.† It was not even proved that the physical state of a gas (which is the only significant quantity) must be the same for equal and opposite velocities of all the molecules; unless this is true we cannot speak here of an actual reversal of the process, and the possibility would still remain that, at least for an extended *region* of initial states, a continual increase of entropy could occur. Both of these are paradoxes against the argument given above, and they can be disposed of by applying Poincaré's theorem.

It is now necessary to formulate either the Carnot–Clausius principle or the mechanical theory in an essentially different way, or else decide to give up the latter theory altogether. Minor changes would not serve the purpose, it seems to me. For example, if one made the interatomic or intermolecular forces velocity-dependent, so that our theorem would no longer be applicable, then it would also be necessary to introduce (in order not to violate the conservation of energy) additional forces which do no work, and whose *direction* therefore is determined by the velocities. But then the force would no longer be one that acts between points, according to the law of action and reaction, which is essential to the entire atomic theory.

Regardless of whether it may be possible, by suitable alterations of the assumptions (as for example in Hertz's *Prinzipien der Mechanik*‡) to escape these contradictions, it is in any case *impossible* on the basis of the *present* theory to carry out a mechanical derivation of the second law without specializing the

---

† [E. P. Culverwell, *Nature* **50**, 617, **51**, 105, 246 (1894), **51**, 581 (1895); S. H. Burbury, *Nature* **51**, 78, 175, 320 (1894], **52**, 104 (1895); J. Larmor, *Nature* **51**, 152 (1894); L. Boltzmann, *Nature* **51**, 413, 581 (1895).]

‡ The Helmholtz theory of " cyclic systems " in its original form, on the other hand, would be affected by the consequences of Poincaré's theorem, since in the last analysis it also, if in another way, reduces to Hamilton's equations.

initial state. It is likewise impossible to prove that the well-known velocity distribution will be reached as a stationary final state, as its discoverers Maxwell and Boltzmann wished to do. I have not given a detailed examination of the various attempts at such a proof by Boltzmann and Lorentz,† since because of the difficulties of the subject I would rather explain as clearly as possible what can be proved rigorously and what seems to be of greatest importance, and thereby contribute to a renewed discussion and final solution of these problems.

---

† Recently collected in Boltzmann's *Vorlesungen über Gastheorie* I, 1896. [English translation, *Lectures on Gas Theory*, University of California Press, Berkeley, 1964; see also H. A. Lorentz, *Wien. Ber.* **95**, 115 (1887).]

8

Reply to Zermelo's Remarks on the Theory of Heat *

LUDWIG BOLTZMANN

SUMMARY

Poincaré's theorem, on which Zermelo's remarks are based [Selection 7], is clearly correct, but Zermelo's application of it to the theory of heat is not. The nature of the H-curve (entropy vs. time) which can be deduced from the kinetic theory is such that if an initial state deviates considerably from the Maxwell distribution, it will tend toward that distribution with enormously large probability, and during an enormously long time will deviate from it by only vanishingly small amounts. Of course if one waits long enough, the initial state will eventually recur, but the recurrence time is so long that there is no possibility of ever observing it.

In contradiction to Zermelo's statement, the singular initial states which do not approach the Maxwell distribution are very small in number compared to those that do. Consequently there is no difficulty in explaining irreversible processes by means of the kinetic theory.

According to the molecular-kinetic view, the second law of thermodynamics is merely a theorem of probability theory. The fact that we never observe exceptions does not prove that the statistical viewpoint is wrong, because the theory predicts that the probability of an exception is practically zero when the number of molecules is large.

Clausius, Maxwell and others have already repeatedly mentioned that the theorems of gas theory have the character of

---

* [Originally published under the title: " Entgegnung auf die wärmetheoretischen Betrachtungen des Hrn. E. Zermelo ", *Annalen der Physik* 57, pp. 773–84 (1896).]

statistical truths. I have often emphasized as clearly as possible[†] that Maxwell's law of the distribution of velocities among gas molecules is by no means a theorem of ordinary mechanics which can be proved from the equations of motion alone; on the contrary, it can only be proved that it has very high probability, and that for a large number of molecules all other states have by comparison such a small probability that for practical purposes they can be ignored. At the same time I have also emphasized that the second law of thermodynamics is from the molecular viewpoint merely a statistical law. Zermelo's paper[‡] shows that my writings have been misunderstood; nevertheless it pleases me for it seems to be the first indication that these writings have been paid any attention in Germany.

Poincaré's theorem, which Zermelo explains at the beginning of his paper, is clearly correct, but his application of it to the theory of heat is not.

I have based the proof of Maxwell's velocity distribution law on the theorem that according to the laws of probability a certain quantity H (which is some kind of measure of the deviation of the prevailing state from Maxwell's) can only decrease for a stationary gas in a stationary container. The nature of this decrease will become most clear when one draws a graph (as I have done[§]) with time as abscissa and the corresponding values of H as ordinates, thus giving the so-called H-curve. (One may subtract off the minimum value H_{min} from all values of H.)

If one first sets the number of molecules equal to infinity and allows the time of the motion to become very large, then in the overwhelming majority of cases one obtains a curve which asymptotically approaches the abscissa axis.[||] The Poincaré theorem is not applicable in this case, as can easily be seen.

[†] L. Boltzmann, *Wien. Ber.* **75,** 67 (1877), **76,** 373 (1877), **78,** 740 (1878). " Der zweite Hauptsatz der Wärmetheorie ", lecture delivered 29 May 1866; *Almanach d. Wien. Akad.* [**36,** 225 (1886)], *Nature* **51,** 413 (1895). *Vorlesungen über Gastheorie* **1,** 42 (1896) [p. 58 in the English translation].

[‡] Zermelo, *Ann. Physik* **57,** 485 (1896).

[§] L. Boltzmann, *Nature* **51,** 413 (1895). [See also Figure on p. 244.]

[||] *Vorlesungen über Gastheorie* **1,** §5.

However, if one takes the time of the motion to be infinite, while the number of molecules is very large but not actually infinite, then the H-curve has a different appearance. As I have already shown (footnote §, page 219), it almost always runs very close to the abscissa axis. Only very rarely does it rise up above this axis; we call this a peak, and indeed the probability of a peak decreases very rapidly as the height of the peak increases. At those times when the ordinate of the H-curve is very small, Maxwell's distribution holds almost exactly; but significant deviations occur at high peaks of the H-curve. Zermelo thinks that he can conclude from Poincaré's theorem that it is only for certain singular initial states, whose number is infinitesimal compared to all possible initial states, that the Maxwell distribution will be approached, while for most initial states this law is not obeyed. This seems to me to be incorrect. It is just for certain singular initial states that the Maxwell distribution is never reached, for example when all the molecules are initially moving in a line perpendicular to two sides of the container. For the overwhelming majority of initial conditions, on the other hand, the H-curve has the character mentioned above.

If the initial state lies on an enormously high peak, i.e. if it is completely different from the Maxwellian state, then the state will approach this velocity distribution with enormously large probability, and during an enormously long time it will deviate from it by only vanishingly small amounts. Of course if one waits an even longer time, he may observe an even higher peak, and indeed the initial state will eventually recur; in a mathematical sense one must have an infinite time duration infinitely often.

Zermelo is therefore completely correct when he asserts that the motion is periodic in a mathematical sense; but, far from contradicting my theorem, this periodicity is in complete harmony with it. One should not forget that the Maxwell distribution is not a state in which each molecule has a definite position and velocity, and which is thereby attained when the position and velocity of each molecule approach these definite values asymptotically. For a finite number of molecules the Maxwell distribution can never be true exactly, but only to a high degree of approximation. It is in no way a special singular distribution which is to be contrasted to infinitely many more non-Maxwellian distributions; rather it is characterized by the fact that by far the largest number of possible velocity distributions have the characteristic properties of the Maxwell distribution, and compared to these there are only a relatively small number of possible distributions that deviate significantly from Maxwell's. Whereas Zermelo says that the number of states that finally lead to the Maxwellian state is small compared to all possible states, I assert on the contrary that by far the largest number of possible states are " Maxwellian " and that the number that deviate from the Maxwellian state is vanishingly small.†

For the first molecule, any position in space, and any values of its velocity comqonents consistent with conservation of total energy, are equally probable.

If one combines all states of all molecules, then he obtains in almost every case the Maxwell distribution, to a high degree of approximation. Only a few combinations give a completely different distribution of states.

An analogy for this is provided by the theory of the method of least squares, where one assumes that each elementary error is equally likely to have a positive or equal negative value; it is then proved that if one combines all possible values of the elementary errors in all possible ways, the great majority of combinations will obey the Gaussian law of errors, and for relatively few combinations will there be significant deviations; the deviations are not impossible, but they are very unlikely.

An even simpler example is provided by the game of dice. In 6000 throws with the same dice one might obtain 1000 one's, 1000 two's, and so forth, not because any such random sequence of throws is more probable than a series of 6000 one's, but rather because there are many more possible combinations corresponding

---

† For the definition of equally probable states, see my papers cited on previous pages.

to an equal number of one's, two's, etc., than corresponding to all one's.

The theory of probability therefore leads to the result (as is well known) that a recurrence of an initial state is not mathematically impossible, and indeed is to be expected if the time of the motion is sufficiently long, since the probability of finding a state very close to the initial state is very small but not zero. The consequence of Poincaré's theorem—that, apart from a few singular initial states, a state very close to the initial state must eventually occur after a very long time—is therefore in complete agreement with my theory.

It is only the conclusion that the mechanical viewpoint must somehow be changed or even given up that is incorrect. This conclusion would be justified only if the mechanical viewpoint led to some consequence that was in contradiction to experience. This would only be the case, however, if Zermelo could prove that the duration of the period of time after which the previous state of the gas must recur according to Poincaré's theorem has an observable length. It should indeed be obvious that if a trillion tiny spheres, each with a high velocity, are initially collected together in one corner of a container with absolutely elastic walls, then in a very short time they will be uniformly distributed throughout the container; and that the time required for all their collisions to have compensated each other in such a way that they all come back to the same corner, must be so large that no one will be present to observe it. Though it seems unnecessary, I have estimated the magnitude of this time in the appendix, and the value obtained is comfortingly large. Though this calculation makes no pretense to accuracy, it still shows that it cannot be proved from Poincaré's theorem that the theoretical existence of a recurrence time involves any contradiction with experience, since the length of this time makes any attempt to observe it ridiculous. The states that we observe all fall in the intermediate time between the beginning and end of the cycle, so that Poincaré's theorem does not exclude states that approximate with arbitrary accuracy the Maxwellian state.

Zermelo's case is therefore only one of many cases (and indeed one that does exceptionally little harm to gas theory) where a state that is theoretically only very improbable must be considered as never occuring in practice. Thus for example in oxyhydrogen gas at ordinary temperatures there must be occasional collisions of two or three molecules with very high velocities; if these were not excluded, oxyhydrogen gas would turn into water at ordinary temperatures.

To give another example, the case that during one second no molecule of a gas collides with a piston is only very improbable but not impossible.

The time that one must wait for a measurable amount of water to be produced from oxyhydrogen gas at ordinary temperature, or for the pressure on a piston to decrease by a measurable amount from its average value, is not as long as a recurrence time, but it is still sufficiently long to preclude observation. An argument against the kinetic theory can be derived from such considerations only when such phenomena fail to appear in a period of time for which calculation indicates that they should appear. This does not seem to be the case; on the contrary, for temperatures lower than the conversion temperature, actual traces of chemical conversion can be found; likewise, it is observed that very small particles in a gas execute motions which result from the fact that the pressure on the surface of the particles may fluctuate.

Thus when Zermelo concludes, from the theoretical fact that the initial states in a gas must recur—without having calculated how long a time this will take—that the hypotheses of gas theory must be rejected or else fundamentally changed, he is just like a dice player who has calculated that the probability of a sequence of 1000 one's is not zero, and then concludes that his dice must be loaded since he has not yet observed such a sequence!

The foregoing remarks are intimately connected with my interpretation of the second law of thermodynamics in the papers cited above. According to the molecular-kinetic view, this law is merely a theorem of probability theory. According to this view, it cannot be proved from the equations of motion that all phenomena must evolve in a certain direction in time. For all phenomena where only visible motion occurs, so that the body always moves as a whole, both directions must be equivalent. On the other hand, when the motion involves a very large number of very small molecules, then there must be (aside from a small number of exceptional cases) a progression from less probable to more probable states, and therefore a continual change in a definite direction, such as, in a gas, the evolution toward a Maxwellian distribution. On the other hand, when it is a question of the motions of individual molecules, this would no longer be expected.

The first and second cases are confirmed by experience; the third case has not yet been realized. Its possibility is hence neither proved nor disproved. Famous scientists, such as Helmholtz,† have believed this, and as I have tried to indicate in my book on gas theory,‡ the opinion that the second law is merely a statistical law is not only not contradicted by the facts but agrees rather well with them. Gibbs§ also arrived, by considering purely empirical facts, at the following conclusion: " The impossibility of an incompensated decrease of entropy seems to be reduced to an improbability ".

We therefore arrive at the following result: if one considers heat to be molecular motion which takes place according to the general equations of mechanics, and assumes that the complexes of bodies that we observe are at present in very improbable states, then he can obtain a theorem which agrees with the second law for phenomena observed up to now.

Of course as soon as one observes bodies of such small size that they contain only a few molecules, the theorem will no longer be valid. However, since no experiments have yet been done on such small bodies, the assumption does not contradict our present

---

† Helmholtz, *Berlin Ber.* **17**, 172 (1884).

‡ L. Boltzmann, *Vorlesungen über Gastheorie* 1, p. 61 [p. 75 in the English translation].

§ Gibbs, *Trans. Conn. Acad.* **3**, 229 (1875); p. 198 in Ostwald's German edition.

experience; indeed, the experiments that have been done on small particles in gases are favourable to the assumption, although we can hardly say that we have an experimental proof of it yet.

When the bodies in question contain many molecules, there must occur very small deviations from this theorem, since the number of molecules is not infinite. But these deviations could only add up to an observable value in a very long period of time, so that this consequence of atomistics cannot be tested by experiment. This is all the more true since gas theory claims to give only an approximate description of reality. Perturbations experienced by the molecules as a result of the aether or the electrical properties of the molecules, etc., must be left out of the theory because of our complete ignorance concerning such effects. There is no such thing as an absolutely smooth wall; on the contrary, every gas is really interacting with the entire universe, and hence the validity of the kinetic theory is not destroyed by small deviations from experience.

An answer to the question—how does it happen that at present the bodies surrounding us are in a very improbable state—cannot be given, any more than one can expect science to tell us why phenomena occur at all and take place according to certain laws.

Gas theory is not to be confused with the theory of central forces—i.e. with the hypothesis that all natural phenomena can be explained by means of central forces between mass points—since gas theory does not assume that either the properties of the aether or the internal constitution of molecules can be explained by centres of force, but only that for the interaction of two molecules during a collision the Lagrange equations of motion are valid with sufficient accuracy for the explanation of thermal phenomena.

A consequence of the Poincaré theorem may still be used against the theory of central forces with respect to the properties of the entire universe. One may say that according to Poincaré's theorem the entire universe must return to its initial state after a sufficiently long time, and hence there must be times when all processes take place in the opposite direction. How shall we decide, when we leave the domain of the observable, whether the age of the universe, or the number of centres of force which it contains is infinite? Moreover, in this case the assumption that the space available for the motion, and the total energy, are finite, is questionable. The assumption of the unlimited validity of the irreversibility principle, when applied to the universe for an infinitely long period of time, leads (as is well known) to the scarcely more attractive consequence that, when all irreversible processes have been played out, the universe will continue to exist without any events, or all events will gradually disappear. Just as it would be wrong to deduce from this the incorrectness of the irreversibility principle, so it would also be wrong to suppose that it proves anything against atomistics.

All the paradoxes raised against the mechanical viewpoint are therefore meaningless and based on errors. However, if the difficulties offered by the clear comprehension of gas-theoretic theorems cannot be overcome, then we should in fact follow the suggestion of Zermelo and decide to give up the theory entirely.

Appendix

We assume a container of volume 1 cc. In this container there will be about a trillion ($= n$) molecules of air at ordinary density. The velocity of each molecule will initially be 500 metres per second. The average distance between the centres of two neighbouring molecules is about 10^{-6} cm.

We now construct around the midpoint of each molecule a cube of edge-length 10^{-7} cm, which we call the initial space of the molecule in question. We also construct a velocity diagram by representing the velocity of each molecule by a line from the origin with the appropriate magnitude and direction. The end-point of this line is called the velocity point of the molecule. Here we divide the entire infinite space into cubes of 1 metre edge length, which we call the elementary cubes. The elementary cube in which the velocity point of a molecule is found initially will be called the initial space of its velocity point.

We now ask after how long a time, according to Poincaré's theorem, will the centres and velocity points of all the molecules return simultaneously to their initial spaces? Note that we do not require exact recurrence, since we accept the velocity state of a molecule as being the same as its initial state if its velocity components return to values that differ by no more than 1 metre from their original values.

We assume that each molecule experiences 4.10^9 collisions per second. It then follows that there will be in all about $b = 2.10^{27}$ collisions per second in the gas. In such a collision, the velocity points of two molecules will generally be displaced to different elementary cubes. According to Poincaré's theorem the original state does not have to recur until the velocity points have gone through all possible combinations of the elementary cubes.

The first molecule can have all possible velocities from zero up to $(500.10^9 = a)$ m/sec. If it has velocity v_1 m/sec, then the second can have all possible velocities from zero up to $\sqrt{a^2 - v_1^2}$ m/sec, and so forth.

The number of possible combinations of all the velocity points in the different elementary cubes is therefore:

$$N = (4\pi)^{n-1} \int_0^a v_1^2 dv_1 \int_0^{\sqrt{a^2-v_1^2}} v_2^2 dv_2 \ldots \int_0^{\sqrt{a^2-v_1^2\ldots v_{n-2}^2}} v_{n-1}^2 dv_{n-1}$$

$$= \frac{\pi^{(3n-3)/2} a^{3(n-1)}}{2.3.4.\ldots[3(n-1)/2]}$$

or

$$\frac{2.(2\pi)^{(3n-4)/2} a^{3(n-1)}}{3.5.7\ldots 3(n-1)}$$

according as n is odd or even.

Since each of these combinations lasts on the average $1/b$ seconds, all of them will be gone through in N/b seconds. After this time all molecules except one must have come back to their original velocity state. The velocity direction of this last molecule is not restricted, nor is the position of the centre of any of the molecules. In order to make the state the same as the original one, the midpoint of each molecule must also return to its initial space, so that the above number must again be multipled by another number of similar magnitude.

Though the number N/b is enormous, one can obtain some idea of its magnitude by noting that it has many trillions of digits. For comparison, suppose that every star visible with the best telescope has as many planets as does the sun, and on each planet live as many men as are on the earth, and each of these men lives a trillion years; then the total number of seconds that they all live will still have less than 50 digits.

If the gas molecules were initially distributed uniformly throughout the container, and all of them had the same velocity, then after only a hundred-millionth of a second they would already have nearly a Maxwellian velocity distribution. Comparison of these numbers shows, on the one hand, how small a fraction of the total number of possible state distributions is made up of those that deviate noticeably from the Maxwell distribution; and on the other hand, how certain are such theorems that theoretically are merely probability laws but in practice have the same significance as laws of nature.

Vienna, March 20, 1896.

9

On the Mechanical Explanation of Irreversible Processes *

ERNST ZERMELO

SUMMARY

Boltzmann has conceded [Selection 8] that the commonly accepted version of the second law of thermodynamics is incompatible with the mechanical viewpoint. Whereas the author holds that the former, a principle that summarizes an abundance of established experimental facts, is more reliable than a mathematical theorem based on unverifiable hypotheses, Boltzmann wishes to preserve the mechanical viewpoint by changing the second law into a " mere probability theorem ", which need not always be valid.

Boltzmann's assertion, that the statistical formulation of the second law is really equivalent to the usual one, is based on postulated properties of the H-curve which he has not proved, and which seem to be impossible. His argument that any arbitrarily chosen initial state will probably be a maximum on the H-curve, if it were valid, would prove that the H-curve consists entirely of maxima, which is nonsense.

The only way that the mechanical theory can lead to irreversibility is by the introduction of a new physical assumption, to the effect that the initial state always corresponds to a point at or just past the maximum on the H-curve; but this would be assuming what was supposed to be proved.

My paper in the March issue of this Journal, " On a theorem of dynamics and the mechanical theory of heat,"† has drawn from Herr Boltzmann an immediate reply,‡ in which I find a confirma-

---

* [Originally published under the title: " Ueber mechanische Erklärungen irreversibler Vorgänge ", *Annalen der Physik* **59**, 793–801 (1896).]

† E. Zermelo, *Ann. Physics* **57**, 485 (1896). [Selection 7]

‡ L. Boltzmann, *Ann. Physik* **57**, 773 (1896). [Selection 8]

tion of my own views rather than a contradiction. Not only does Herr Boltzmann recognize that the basic theorem of Poincaré is "obviously correct", but he also concedes that it is applicable to a closed system of gas molecules in the sense of the kinetic theory. Indeed, in such a system all processes are *periodic* from a mathematical viewpoint, hence *not irreversible* in the strict sense, so that one may not assert that there is an actual progressive increase of entropy as the second law, in its usual meaning, would require. To prove this, and thereby to obtain a firm basis for the discussion of the principal questions, was the purpose of my paper; at the time I was not familiar with Herr Boltzmann's investigations of gas theory, but I still think that this general clarification was not at all superfluous.

The " necessity of making a fundamental modification either in the Carnot–Clausius principle or the mechanical viewpoint " which I asserted is therefore conceded, and it remains a matter of personal opinion which of these possibilities is to be chosen. As for me (and I am not alone in this opinion), I believe that a single principle summarizing an abundance of established experimental facts is more reliable than a mathematical theorem, which by its nature represents only a theory which can never be directly verified; I prefer to give up the theorem rather than the principle, if the two are inconsistent.

Herr Boltzmann, however, will not modify the ordinary mechanical viewpoint, and instead wishes to change the second law into a " mere probability theorem " which is not valid at all times. Yet he asserts that this change, whose *principal* meaning he does not misunderstand, is really unimportant, and that " in practice " his two formulations are " completely equivalent ". Let us see how far he has succeeded in proving this.

It is undoubtedly correct, as Boltzmann emphasizes, that for a very large number of molecules in a finite volume the average duration of the Poincaré period, the time after which a state will recur, is much too large for us to expect to make a direct *observation* of the theoretical periodicity. However, his numerical estimate, which is based on a single exceptional initial state with a completely determined molecular configuration, is not conclusive. In practice one is interested in a " physical state " which can be realized by many possible combinations, and can therefore recur very much earlier. Moreover, for my purposes it is sufficient to prove the recurrence of any other state with the same or a smaller value of the entropy; the periods of recurrence of such individual values of the entropy S will of course vary, but on the whole they no longer come out to be so " comfortingly " large. Nevertheless there are functions whose periodicity is beyond observation, and the entropy function might be one of them.

For such a function it can of course happen that it *appears* to be continually increasing, since the decreasing branch of the curve, which is theoretically always present, begins so much later that it does not need to be considered. Yet it by no means follows from this that there are functions for which one *always* observes the increasing and not the decreasing part, which is the property that the mechanical analog of the entropy function must have. It is not satisfactory simply to accept this property as a fact for a particular type of initial state that we can observe at present, for it is not a question of a variable which is just observed once (as for example the eccentricity of the earth's orbit) but of the entropy of *any arbitrary* system free of external influences. How does it happen, then, that in such a system there always occurs only an *increase* of entropy and *equalization* of temperature and concentration differences, but never the reverse? And what right do we have to expect this behaviour to continue, at least for the immediate future? A satisfactory answer to this question must be given, if we are to accept a mechanical analog of the second law.

It seems to me that probability theory cannot help here, since every increase corresponds to a later decrease, and both must be equally probable or at least have probabilities of the same order of magnitude. My opinion, in agreement with Poincaré's definition,† is that the probability of occurrence of a certain property of the molecular states, for example for a definite value of the function S,

---

† H. Poincaré, *Acta Math.* **13**, 71 (1890). [Selection 5, p. 199]

can be measured only by the "extension"† γ of the "region" g of all possible states which have this property, divided of course by the total extension Γ of the region G containing all possible states. But since according to Liouville's theorem each extension γ is independent of time, any such value of a function must have the same probability at a later time as at the initial time, and no overall increase or decrease is to be expected on the grounds of probability theory.‡

Herr Boltzmann proceeds in a different way. He assumes a function H whose curve, drawn with the time t as abscissa, runs in general very close to the t-axis but occasionally has elevations or " peaks ". The larger the peaks are, the more improbable they are, and the less often they occur.§ I cannot find that he has actually *proved* this property from his other definition of the H-function. According to my definition, probability and duration of a state are not identical. Nevertheless, functions of the indicated nature may exist. He further assumes that the H-function has initially an unusually large value H_0, corresponding to a peak, but soon passes this peak and decreases almost to zero. Finally, it runs very close to the abscissa axis for a very long time. This limiting value zero of the H-function corresponds to a velocity distribution expressed by Maxwell's law, so that the properties of this H-curve provide an explanation of the probability-theoretic meaning of the distribution law, which however I do not dispute. The law does not represent a " stationary *final* state " in the strict sense, since the curve eventually rises to new peaks after a long time. Herr Boltzmann himself considers the Maxwellian state to be the " *final* state " only in an empirical or approximate sense, and it seems to me that *this* assertion does not follow sufficiently clearly from his earlier writings.

---

† E. Zermelo, *Ann. Physik* **57**, 485 (1896). [Selection 7, p. 210]

‡ [This argument is developed in more detail by Gibbs in his discussion of the generalized H-theorem: see *Elementary Principles in Statistical Mechanics* (Scribner, New York, 1902) Chapter XII.]

§ L. Boltzmann, *Ann. Physik* **57**, 773 (1896). [Selection 8, p. 220]

But it is not here a question of Maxwell's law, but of whether an analogy exists between the properties of the H-curve and the second law of thermodynamics; it is this analogy that I dispute. It is not sufficient to show that all perturbations *finally* relax to a long-lasting equilibrium state; rather it is necessary to show that changes always take place in the same sense, in the direction of equalization; that the H-function *always only* decreases during observable times, or at least that there can only be very small, practically unnoticeable increases, which will always be immediately washed out by stronger decreases. In my opinion this proof is as little possible for the H-function as for any other function. Clearly the initial state, whose probability can depend only on the initial value H_0, can just as well lie on a rising as a falling branch of the curve, and in the former case there must first be an *increase*, which can last just as long as the subsequent decrease. For this period we have $H > H_0$. Each observed decrease $H_1 \ldots H_2$ in the falling branch corresponds to an equally great increase $H_2 \ldots H_1$ in the rising branch, and the process is no more likely to begin in one way than the other. If the increase takes place in a shorter time and is hence less probable than the decrease—an assumption for which there is no basis in the theory—then it would still have to be *steeper* and therefore should be given just as much weight.

Herr Boltzmann's assertion, if I have understood it correctly,† is that the initial state has a fairly large H-value, say $H_0 > H'$, on a peak which is not too large (so that it does not have too small a probability) and as a rule must represent a *maximum*, so that of course one always observes only the decreasing branch. I cannot conceive of such a curve. Suppose for the sake of argument that the intersections of the H-curve with a line parallel to the time-axis at a height $H = H_0$ are mostly maxima, and that $H_0 > H'$. But where are the other points on the peak ($H > H'$) which are *not* maxima? Are they in fact in the minority compared to the

---

† L. Boltzmann, *Vorlesungen über Gastheorie* **1**, 44 (1896) [p. 59 in the English translation].

maxima? It is clear that this argument can make sense only if the maxima are considered not as mathematical points but as having a certain breadth, i.e. a certain time-duration. But then for any initial state the value of the function will remain constant for a longer or shorter time, thereby representing a sort of labile equilibrium; whereas according to experience, for example in the case of heat conduction, the process of equalization begins more rapidly, the greater the initial temperature differences are, that is, the further the initial state is from the stable equilibrium state.

Aside from this, I do not understand what the *initial* state has to do with the argument, except for its property of having a small probability, which it shares with the neighbouring states. Herr Boltzmann assumes that the entire H-curve, and therefore the collection of all states through which the system passes, is *given* and now asks for the probability of a certain initial state, i.e. the place on the curve where the system actually begins to move, without any external forces being present. But, as experience teaches, there is no procedure available for producing *any arbitrary* initial state by an appropriate action and then isolating the system and letting it run by itself; one cannot make any arbitrary state P_0 the initial state. If this were true, then the system would actually pass through all the states P that follow P_0 in the series, while the previous states could only be added mathematically. Now if the above argument were correct, and the initial state represents a maximum of the H-function in most cases, then the same must also be true of *all other* states for which H exceeds H', since any other state could be chosen as initial state. Moreover, the whole probability argument is just as applicable to any arbitrary state as to the initial state. All these states must therefore represent maxima, and the curve must consist purely of maxima above a certain height. This is nonsense, since the function cannot be constant. Therefore in order to obtain an approximate empirical analog of the entropy theorem, it is not sufficient to assume that the initial state is extraordinarily improbable; rather one must add the *new assumption*, that at the beginning the H-curve has a maximum or has just passed a maximum. But as long as one cannot make comprehensible the *physical origin* of the initial state, one must merely assume what one wants to prove; instead of an explanation one has a renunciation of any explanation.

I have therefore not been able to convince myself that Herr Boltzmann's probability arguments, on which " the clear comprehension of the gas-theoretic theorem "† is supposed to rest, are in fact able to dispel the doubts of a mechanical explanation of irreversible processes based on Poincaré's theorem, even if one renounces the strict irreversibility in favour of a merely empirical one. Indeed it is clear *a priori* that the probability concept has nothing to do with time and therefore cannot be used to deduce any conclusions about the *direction* of irreversible processes. On the contrary, any such deduction would be equally valid if one interchanged the initial and final states and considered the *reversed* process running in the opposite direction. Hence, the following dice game is more relevant than the example introduced by Herr Boltzmann. Two dice-players, let us suppose, have made the observation that dice they obtain from a certain source always behave in a certain way when they first start to play with them. One particular face, say the one, always comes up first. In the first 600 throws, the one comes up 200 times rather than 100 times. However, in the next 6000 throws the ones are less frequent, and after the game has continued a long time they find that one comes up on the average only 100 times out of 6000, like all the other numbers. The first player sees nothing strange in this behaviour, since the laws of probability theory are supposed to apply to very long games. But the second player says: No! This dice must be false, and it is only through long use that it gradually regains its proper condition—the latter interpretation corresponds to my own opinion.

Not only is it impossible to explain the general *principle* of irreversibility, it is also impossible to explain individual irreversible

---

† L. Boltzmann, *Ann. Physik* **57**, 773 (1896). [Selection 9, p. 226]

processes themselves without introducing new physical assumptions, at least as far as the time-direction is concerned. In particular, the differential equation for heat conduction and diffusion is

$$\frac{\partial u}{\partial t} = a^2 \frac{\partial^2 u}{\partial x^2}$$

and this equation can only represent irreversible processes. The attempt to derive this equation purely from the basic equations of mechanics, together with probability assumptions, which has been been made for example by Clausius, Maxwell, and Boltzmann, cannot reach its goal, since it is an impossible undertaking, and an apparent success can only rest on an error of deduction. The major fallacy in the methods heretofore applied seems to me to be the unprovable (because untrue) assumption that the molecular state of a gas is always, in Boltzmann's expression, " disordered " and that all possible directions and combinations are equivalent, if one can say nothing definite about the true state, which must nevertheless depend on the " ordered " initial state.† Probability theory justifies such assumptions to a certain extent for the *initial state*, at most; the probability of a later state, however, and therefore the process itself, must always first be expressed in terms of the corresponding initial state, and only then can one decide on the permissibility of such averaging assumptions. The difficulty of carrying out investigations *rigorously* from the viewpoint of probability theory may be very great, but they do not seem to me to be insurmountable. In any case such investigations cannot by themselves correct the errors of the " statistical method " used up to now; questions of principle, such as those under discussion here, require arguments whose mathematical validity is beyond question. For the present I must restrict myself to these remarks; I hope later to return to a more explicit treatment of these methodological questions.

The great successes of the kinetic theory of gases in the explanation of *equilibrium* properties do not entail its applicability to

---

† Boltzmann, *Gastheorie* **1**, 21 (1896) [p. 40 in the English translation].

time-dependent processes also, for the two are separate subjects; while in the former case the theory frequently gives us a correct and valuable picture, in the latter case, especially where it is a question of the explanation of irreversible processes, it must necessarily fail unless completely new assumptions are added to it.

Berlin, 15 September 1896.

10

On Zermelo's Paper "On the Mechanical Explanation of Irreversible Processes" *,†

LUDWIG BOLTZMANN

SUMMARY

The second law of thermodynamics can be proved from the mechanical theory if one assumes that the present state of the universe, or at least that part which surrounds us, started to evolve from an improbable state and is still in a relatively improbable state. This is a reasonable assumption to make, since it enables us to explain the facts of experience, and one should not expect to be able to deduce it from anything more fundamental.

The applicability of probability theory to physical situations, which is disputed by Zermelo, cannot by rigorously proved, but the fact that one never observes those events that theoretically should be quite rare is certainly not a valid argument against the theory.

One may speculate that the universe as a whole is in thermal equilibrium and therefore dead, but there will be local deviations from equilibrium which may last for the relatively short time of a few eons. For the universe as a whole, there is no distinction between the "backwards" and "forwards" directions of time, but for the worlds on which living beings exist, and which are therefore in relatively improbable states, the direction of time will be determined by the direction of increasing entropy, proceeding from less to more probable states.

I will be as brief as possible without loss of clarity.

§1. The second law will be explained mechanically by means of of assumption A (which is of course unprovable) that the universe,

---

* [Originally published under the title: " Zu Hrn. Zermelo's Abhandlung Über die mechanische Erklärung irreversibler Vorgange ", *Annalen der Physik* **60**, pp. 392–8 (1897).]

† E. Zermelo, *Ann. Physik* **59**, 793 (1896). [Selection 9]

considered as a mechanical system—or at least a very large part of it which surrounds us—started from a very improbable state, and is still in an improbable state. Hence, if one takes a smaller system of bodies in the state in which he actually finds them, and suddenly isolates this system from the rest of the world, then the system will initially be in an improbable state, and as long as the system remains isolated it will always proceed toward more probable states. On the other hand, there is a very small probability that the enclosed system is initially in thermal equilibrium, and that while it remains enclosed it moves far enough away from equilibrium that its entropy decrease is noticeable.

The question is not what will be the behaviour of a completely arbitrary system, but rather what will happen to a system existing in the present state of the world. The initial state precedes the later states, so that Zermelo's conclusion that all points of the H-curve must be maxima is invalid. Hence, it turns out that entropy always increases, temperature and concentration differences are always equalized, that the initial value of H is such that during the time of observation it almost always decreases, and that initial and final states are not interchangeable, in contradiction to Zermelo's assertions. Assumption A is a comprehensible physical explanation of the peculiarity of the initial state, consistent with the laws of mechanics; or better, it is a unified viewpoint corresponding to these laws, which allows one to predict the type of peculiarity of the initial state in any special case; for one can never expect that the explanatory principle must itself be explained.

On the other hand, if we do not make any assumption about the present state of the universe, then of course we cannot expect to find that a system isolated from the universe, whose initial state is completely arbitrary, will be in an improbable state initially rather than later. On the contrary it is to be expected that at the moment of separation the system will be in thermal equilibrium. In the few cases where this does not happen, it will almost always be found that if the state of the isolated system is followed either backwards or forwards in time, it will almost immediately pass to a more probable state. Much rarer will be the cases in which the state becomes still more improbable as time goes on; but such cases will be just as frequent as those where the state becomes more improbable as one follows it backwards in time.

§2. The applicability of probability theory to a particular case cannot of course be proved rigorously. If, out of 100,000 objects of a certain kind, about 100 are annually destroyed by fire, then we cannot be sure that this will happen next year. On the contrary, if the same conditions could be maintained for $10^{10^{10}}$ years, then during this time it would often happen that all 100,000 objects would burn up on the same day; and likewise there will be entire years during which not a single object is damaged. Despite this, every insurance company relies on probability theory.

It is even more valid, on account of the huge number of molecules in a cubic millimetre, to adopt the assumption (which cannot be proved mathematically for any particular case) that when two gases of different kinds or at different temperatures are brought in contact, each molecule will have all the possible different states corresponding to the laws of probability and determined by the average values at the place in question, during a long period of time. These probability arguments cannot replace a direct analysis of the motion of each molecule; yet if one starts with a variety of initial conditions, all corresponding to the same average values (and therefore equivalent from the viewpoint of observation), one is entitled to expect that the results of both methods will agree, aside from some individual exceptions which will be even rarer than in the above example of 100,000 objects all burning on the same day. The assumption that these rare cases are not observed in nature is not strictly provable (nor is the entire mechanical picture itself) but in view of what has been said it is so natural and obvious, and so much in agreement with all experience with probabilities, from the method of least squares to the dice game, that any doubt on this point certainly cannot put in question the validity of the theory when it is otherwise so useful.

It is completely incomprehensible to me how anyone can see a refutation of the applicability of probability theory in the fact that some other argument shows that exceptions must occur now and then over a period of eons of time; for probability theory itself teaches just the same thing.

§3. Let us imagine that a partition which separates two spaces filled with different kinds of gas is suddenly removed. One could hardly find another situation (at least one in which the method of least squares is applicable) where there are so many independent causes acting in such different ways, and in which the application of probability theory is so amply justified. The opinion that the laws of probability are not valid here, and that in most cases the molecules do not diffuse, but instead a large part of the container has significantly more nitrogen, and another part has significantly more oxygen, cannot be disproved, even if I were to calculate exactly the motions of trillions of molecules in millions of different special cases. Nevertheless this opinion certainly does not have enough justification to cast doubt on the usefulness of a theory that starts from the assumption of the applicability of probability theory and draws the logical consequence from this assumption.

Poincaré's theorem does not contradict the applicability of probability theory but rather supports it, since it shows that in eons of time there will occur a relatively short period during which the state probability and the entropy of the gas will significantly decrease, and that a more ordered state similar to the initial state will occur. During the enormously long period of time before this happens, any noticeable deviation of the entropy from its maximum value is of course very improbable; however, a momentary increase or decrease of entropy is equally probable.

It is also clear from this example that the process goes on irreversibly during observable times, since one intentionally starts from a very improbable state. In the case of natural processes this is explained by the assumption that one isolates the system of bodies from the universe which is at that time in a very improbable state as a whole.

This example of two initially unmixed gases gives us incidentally a possible way of imagining the initial state of the world. For if in the example we isolate the gas found in a smaller space soon after the beginning of the diffusion from the rest of the gas, we will have the asymmetry with respect to forward and backward steps in time as in the isolated system of bodies mentioned in §1.

§4. I myself have repeatedly warned against placing too much confidence in the extension of our thought pictures beyond the domain of experience, and I am aware that one must consider the form of mechanics, and especially the representation of the smallest particles of bodies as mass-points, to be only provisionally established. With all these reservations, it is still possible for those who wish to give in to their natural impulses to make up a special picture of the universe.

One has the choice of two kinds of pictures. One can assume that the entire universe finds itself at present in a very improbable state. However, one may suppose that the eons during which this improbable state lasts, and the distance from here to Sirius, are minute compared to the age and size of the universe. There must then be in the universe, which is in thermal equilibrium as a whole and therefore dead, here and there relatively small regions of the size of our galaxy (which we call worlds), which during the relatively short time of eons deviate significantly from thermal equilibrium. Among these worlds the state probability increases as often as it decreases. For the universe as a whole the two directions of time are indistinguishable, just as in space there is no up or down. However, just as at a certain place on the earth's surface we can call " down " the direction toward the centre of the earth, so a living being that finds itself in such a world at a certain period of time can define the time direction as going from less probable to more probable states (the former will be the " past " and the latter the " future ") and by virtue of this definition he will find that this small region, isolated from the rest of the universe, is " initially " always in an improbable state. This viewpoint seems to me to be the only way in which one can understand the validity of the second law and the heat death of each individual world without invoking an unidirectional change of the entire universe from a definite initial state to a final state. The objection that it is uneconomical and hence senseless to imagine such a large part of the universe as being dead in order to explain why a small part is living—this objection I consider invalid. I remember only too well a person who absolutely refused to believe that the sun could be 20 million miles from the earth, on the grounds that it is inconceivable that there could be so much space filled only with aether and so little with life.

§5. Whether one wishes to indulge in such speculations is of course a matter of taste. It is not a question of choosing as a matter of taste between the Carnot–Clausius principle and the mechanical theory. The importance of the former, as the simplest expression of the facts so far observed, is not in dispute. I assert only that the mechanical picture agrees with it in all actual observations. That it suggests the possibility of certain new observations—for example, of the motion of small particles in liquids and gases, and of viscosity and heat conduction in very rarefied gases, etc.—and that it does not agree with the Carnot–Clausius principle on some unobservable questions (for example the behaviour of the universe or a completely enclosed system during an infinite period of time), may be called a difference in principle, if you like. In any case it provides no basis for giving up the mechanical theory, as Herr Zermelo would like to do, if it cannot be changed in principle (which one should not expect). It is precisely this difference that seems to me to indicate that the universality of our thought-pictures will be improved by studying not only the consequences of the principle in the Carnot–Clausius version but also in the mechanical version.

Appendix

§6. I have always measured the probability of a state, independently of its temporal duration, by the " extension γ " (as Zermelo calls it) of its corresponding region, and I used the Liouville theorem in this connection 30 years ago.† The Maxwellian state

---

† See especially *Wien. Ber.* **58,** 517 (1868); **63,** 679, 712 (1871); **66,** (1872); **76,** 373 (1877). I have given there the proof of the above-mentioned theorem, which there is not space to repeat here.

is simply the most probable because it can be realized in the largest number of ways. The total extension γ of the region of all those states for which the velocity distribution is approximately given by the Maxwell distribution is therefore much greater than the total extension of the regions of all other states. It was only to illustrate the relation between the temporal course of the states and their probabilities that I represented the reciprocal value of this

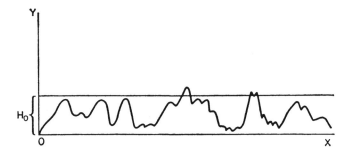

Fig. 1.

probability for the different successive states by the H-curve, in the case of a large finite number of hard gas molecules. Aside from a vanishingly small number of special initial states, the most probable states will also occur the most frequently (at least for a very large number of molecules). The ordinates of this curve are almost always very small, and these small ordinates are of course not usually maxima. It is only the ordinates with unusually large values that are mostly maxima, and indeed they are more likely to be maxima the greater they are. The fact that a very large ordinate H_0 is more often a maximum than the intersection of the line $y = H_0$ with a still higher peak is a consequence of the enormous increase in rarity of peaks with increasing height. See the above figure, which is of course to be taken with a pinch of salt. A correct figure could not be printed because the H-curve actually has a large number of maxima and minima within each finite segment, and cannot be represented by a line with continuously changing direction. It would be better to call it an aggregate of many points very close together, or small horizontal segments.†

The Poincaré theorem is of course inapplicable to a terrestrial body which we can observe, since such a body is not completely isolated; likewise, it is inapplicable to the completely isolated gas treated by the kinetic theory, if one first lets the number of molecules become infinite, and then the quotient of the time between successive collisions and the time of observation.

Vienna, 16 December 1896.

---

† *Nature* **51**, 413, 581 (1895).

Part III

*Historical Discussions
by Stephen G. Brush*

1.

GADFLIES AND GENIUSES IN THE HISTORY OF GAS THEORY

by Stephen G. Brush

1. INTRODUCTION

The history of science has often been presented as a story of the achievements of geniuses: Galileo, Newton, Maxwell, Darwin, Einstein. Recently it has become popular to enrich this story by discussing the social contexts and motivations that may have influenced the work of the genius and its acceptance; or to replace it by accounts of the doings of scientists who have no claim to genius or to discoveries of universal importance but may be typical members of the scientific community at a particular time and place. In this article I consider a different kind of story, which further research might reveal to be more common than we now suspect: progress stimulated by *gadflies* -- outspoken critics who challenge the ideas of geniuses, forcing them to revise and improve those ideas, resulting in new knowledge for which the genius gets the credit while the gadfly is forgotten.

In the history of philosophy the word 'gadfly' is associated with someone like Socrates who, rather than simply proclaiming his own position about truth and virtue, forces others to accept that position by critical dialogue and logical reasoning. I use the word in a somewhat different way: a gadfly may have ideas that turn out to be mistaken, although they are compatible with the prevailing worldview when first presented. In modern science we are often forced to accept theories that violate common sense, but it is useful for someone to articulate the conventional view in order that someone else can show exactly why a different view should be accepted.

The gadfly can even be a genius himself, like Albert Einstein whose persistent criticism of quantum mechanics (as a complete account of reality) forced other physicists to defend and test the most implausible consequences of that theory. Thus the "Einstein-Podolsky-Rosen paradox" is a remarkable phenomenon that actually occurs, contrary to the belief of Einstein and his colleagues that its *non*-existence proves the incompleteness of quantum mechanics.

Historians are beginning to take more seriously the role of gadflies in the development of science. Allen G. Debus (1998) has recently discussed Robert Fludd (1574-1637), an English mystic who criticized the new mechanical philosophy in debates with Kepler, Mersenne, and Gassendi. Debus says we should judge gadflies not by whether they were right or wrong (from a modern viewpoint) but by whether they provoked responses and debates.

The following are major landmarks in the development of gas theory: the law that the product pressure X volume is constant at a given temperature; the formula giving the average distance travelled by a gas molecule in terms of the square of its diameter; the statistical distribution of velocities of gas molecules as a function of temperature and potential energy; and the relation between entropy and probability. These achievements are generally attributed to Robert Boyle, Rudolph Clausius, James Clerk Maxwell and Ludwig Boltzmann. Rarely do we learn about the gadflies -- Franciscus Linus, C. H. D. Buys-Ballot, Francis Guthrie and Josef Loschmidt -- whose criticisms forced the geniuses to produce these results. In addition Boltzmann is known for his radical views on the direction of time and molecular randomness; these views were published as a result of debates started by Ernst Zermelo and E. P. Culverwell.[1]

2. BOYLE AND LINUS

The notion of 'air pressure' arose in the 17th century through the work of Evangelista Torricelli (1608-1647) in Italy, Blaise Pascal (1623-1662) in France, Otto von Guericke (1602-1686) in Germany, and Robert Boyle (1627-1691) in England (Webster 1965). By a combination of experiments and theoretical reasoning they persuaded other scientists that the earth is surrounded by a 'sea' of air that exerts pressure in much the same way that water does, and that air pressure is responsible for the phenomena previously attributed to 'nature's abhorrence of a vacuum'. We now view this conclusion as part of the change in attitudes that led to the mechanico-corpuscular view of nature: instead of postulating 'occult forces' or teleological principles to explain natural phenomena, scientists started to look for explanations based simply on matter and motion. But we often forget that its acceptance required a suspension of disbelief: if air pressure really accounts for suction, for example, then it must exert the enormous pressure of 15 pounds on every square inch of our bodies! Does it really make this statement any more plausible to say the reason we don't notice the pressure is that it is balanced by an equal and opposite force by the air inside? Or have we simply become so familiar with the idea of air pressure that we don't think to question it?

It was well known in the time of Galileo that a suction pump cannot lift water more than 34 feet in a pump, although Galileo himself seems to have been the first to put this fact on record in 1638. A few years later his student Torricelli devised an experiment to illustrate the same effect in the laboratory. Since mercury is about 14 times as dense as water, one might expect that it can be lifted only about 1/14 as far. This is indeed what is observed, and the fact tends to make plausible the mechanical explanation based on air pressure. Taking a glass tube about a yard long with one end closed Torricelli filled it with mercury to the top; then, placing a finger over the open end, he inverted the tube so that the open end was immersed in an open dish of mercury. When he removed his finger from the open end, the mercury in the tube fell until the top of the mercury column was about 30 inches above the level of the mercury in the open dish. (Warning: don't try this experiment at home!) Between the top of the mercury column and the upper end of the tube was an enclosed space, which, which became known as the 'Torricellian vacuum'. According to

Torricelli, it is just the mechanical pressure of the air that holds the mercury up in the tube.

Boyle, with the help of Robert Hooke (1635-1703), designed and constructed an improved air pump for Boyle, and it was with this 'pneumatic engine' or 'sucker' that Boyle performed the experiments recorded in his book *New Experiments* (Boyle 1660). Among other things, Boyle wanted to prove that air has elasticity ('spring') and can exert a mechanical pressure great enough to support a column of water or mercury.

Boyle considered that the crucial experiment in pneumatics was his No. 17, in which he enclosed the lower part of a Toricellian barometer in a container from which air could be removed by means of his pump. As the air was exhausted, the mercury in the tube fell nearly to the level of that in the dish. Thus the mercury had in fact been supported by air pressure, or rather by the difference between atmospheric pressure and the negligible pressure of the Torricellian vacuum at the top of the tube.

After publishing his account of several experiments in 1660, Boyle might well have turned to one of the many other subjects in which he was interested, without ever taking up the pressure-volume law for which he is now chiefly remembered. He would not be known as the person who 'discovered air pressure' or the person who proved that air pressure is responsible for suction phenomena; he had only added one more proof to those of his predecessors.

In 1661 Boyle's advocacy of the air-pressure theory of suction was challenged by Francis Line (1595-1675), an English Jesuit scientist who had recently returned to England after living in Belgium for many years. Line, who used the Latinized form of his name "Franciscus Linus" for his publications, was one of the most tenacious and ingenious defenders of Aristotelian philosophy against the new science of the 17th century (Reilly 1962; 1969).

In his *Tractatus de Corporum Inseparabilitate* (Treatise on the Inseparability of Bodies), Linus (1661) argued that while air may have some degree of weight and elasticity, it does not have enough to account of the behavior of the mercury in a Torricelli barometer. (Boyle's quantitative investigation of the pressure-volume relation was intended to answer this criticism, according to Shapin and Schaffer (1985, p. 168).) He disputed the conclusions Boyle had drawn from his results, and suggested an alternative theory. To prevent the formation of a vacuum, a vapor is emitted from the surface surrounding an enclosed space; this vapor forms an invisible entity, the *funiculus* (diminutive of Latin *funis*, rope), which pulls on the mercury, trying to close up the space, and preventing it from falling all the way down. Thus the funiculus enforces the law "nature abhors a vacuum."

The idea that suction is the effect of an attractive force exerted by an unseen object is not as silly as it may at first seem to the modern reader. After all, when you suck a liquid up a straw you probably think of yourself as an *active* power pulling in, rather than a passive receptacle letting air pressure do the work. But Linus could point to a simple experiment (shown by Shapin and Schaffer, p. 158) that seemed to support his theory more directly: if you fill with mercury a vertical 40-inch tube, open at both ends, and close it with your finger at the top, you will feel your finger being sucked down into the tube. (With the assistance of modern household technology you can demonstrate this effect for yourself without using mercury: put your finger on the open tube of a vacuum cleaner and turn the switch on!) . "How could this be so, Linus asked, if the pressure of the external air was, as Boyle claimed, actually pushing the column of mercury up into the tube? Should not the finger be pushed off the top of the tube, if this were really the case?" (Reilly 1969, p. 65). Even Christiaan Huygens thought Boyle's explanation of this phenomenon was unsatisfactory, though he did not like Linus's any better.

Boyle did not ignore Linus; instead, he published another book in 1662, in which he attempted to justify his theory of air pressure by further experiments and reasoning, and to reply to the objections of Linus and of Thomas Hobbes.[2] He argued that the funiculus theory was refuted by the experiments mentioned above, in particular his own Experiment No. 17, which proved a direct causal relation between the pressure of the air outside the barometer and the height of the mercury inside. Moreover, the funiculus theory is implausible because one cannot hold up a column of liquid by grabbing its top layer; only solids can be lifted this way.

To provide a quantitative assessment of the pressure or springiness of air, Boyle then described an experiment in which a certain amount of air was trapped in a J-shaped tube by a column of mercury whose height could be varied up to nearly 120 inches. He reported that when the pressure of the atmosphere was effectively doubled by adding 29 inches of mercury, the volume of the enclosed air was reduced to half its original value. He generalized this result to the statement that the pressure of air is proportional to its density (or inversely to its volume). Thus, "Almost as an afterthought, Boyle's Law would emerge from this refutation of Line's funicular hypothesis" (Reilly 1969, p. 80). Boyle was not the first to arrive at the formula now known as 'Boyle's Law' but he was the most successful in establishing and popularizing it (Webster 1965).

Back in Belgium near the end of his life, Line took on an even more formidable genius. When Isaac Newton published his theory that white light is composed of colored rays, Linus published a letter in the *Philosophical Transactions of the Royal Society of London*, stating that Newton's observations of the elongated image formed by sunlight passing through a prism were incorrect. According to his own observations such an image is formed only on cloudy days. But Line's criticism, followed by those of a swarm of other gadflies, only irritated Newton; he merely reiterated his original result.

3. BERNOULLI: GENIUS WITHOUT A GADFLY

What is the cause of the pressure or elasticity of air? Boyle discussed at length the possible atomistic interpretations, mentioning Descartes' suggestion that whirling particles might drive away neighbors who get too close, but giving his preference to a more static conception of gas structure. Boyle compared air particles to coiled-up balls of wool or springs, which would resist compression and expand into any available space. The air near the earth's surface is already compressed to a great degree by the weight of the atmosphere sitting on top of it; hence, like a squeezed spring, it exerts a strong force outwards.

Boyle's theory was cast into a simple mathematical form by Isaac Newton. In the *Principia* Newton stated the theorem: *if* a gas is composed of particles that exert repulsive forces on their neighbors, the magnitude of the force being inversely as the distance between them, *then* the pressure will be inversely as the volume. He did not claim that the converse of the theorem is true, but stated that it was a physical problem to determine whether gases are in fact made up of particles exerting such repulsive forces.

Newton's repulsive force model was the first one that gave a direct quantitative explanation of Boyle's law, and it remained in favor well into the 19th century. Chemists merged it with the caloric theory of heat; caloric was supposed to be a self-repelling fluid, which not only explained Boyle's law but also accounted for the increase of gaseous pressure with increasing temperature.

Early in the 18th century, several scientists proposed theories that attributed gas pressure to the *motion* of particles rather than to their repulsive forces. Most of these theories were based on the assumption that the particles are suspended in the ether, or consist of ether vortices, so that the type of motion considered is primarily vibrational or rotational rather that linear.

The first quantitative version of the modern kinetic theory, in which gas pressure is attributed simply to the *linear* motion of particles, was proposed by the Swiss mathematical physicist Daniel Bernoulli (1700-1782) in his book *Hydrodynamica* (1738). Bernoulli derived Boyle's law for gas pressure by computing the force exerted on a movable piston by the impacts of n particles, each having mass m and moving with speed v, in a closed space of total volume V. If V is made smaller the pressure will be greater because the particles strike the piston more frequently. He started from a general model of particles with finite diameter d, and then argued that because air can be compressed into a very small volume, d must be much smaller than the average distance D between the centers of the particles. Taking the limit $d/D \to 0$ he concluded that the pressure P should be inversely proportional to V; so, as stated by Boyle's law, the product PV is constant.

Bernoulli also showed that the pressure will be proportional to the total kinetic energy of the particles, $\frac{1}{2}nmv^2$, since the frequency of impacts is proportional to the speed v and the force [impulse] of each impact is proportional to the momentum mv. This, he remarked, explained the observed fact that increases of pressure arising from equal increases of temperature are proportional to the density, and suggested that temperature itself could be defined in terms of the pressure of air at a standard density. Although other scientists had not yet accepted the concept of an absolute temperature scale, Bernoulli's theory introduced the idea that heat or temperature could be identified with the kinetic energy of particles in an ideal gas.

Bernoulli's short chapter on kinetic theory was full of good ideas and sound reasoning. But he never followed up this promising initial effort. The reason was not that other scientists rejected his ideas; instead, his kinetic theory was simply ignored. There are several possible reasons for this lack of response (Brush 1976, pp. 20-21). Perhaps what he really needed was an obnoxious gadfly (there were plenty of those in the 18th century) who could have criticized the theory in a way that would force Bernoulli to go further.

4. HERAPATH: GADFLY WITHOUT A GENIUS

John Herapath (1790-1868) was an Englishman who might be called an amateur scientist except that the distinction between amateur and professional was not very significant at the time. But he was definitely an outsider, a clever eccentric who never enjoyed any recognition from the scientific community but lived to see some of his ideas vindicated. By temperament he was a gadfly, but unfortunately he failed to find an appropriate genius to attack. (He did publish a critique of Laplace's caloric theory of gases, but Laplace ignored him.) Instead, he developed his own version of the kinetic theory of gases; not being a genius himself, he couldn't recognize its defects.

Herapath was initially interested in developing an explanation of gravity in terms of the impacts of particles of an ethereal fluid, somewhat along the lines of the "kinetic theory of gravity" proposed by G. S. LeSage and many others. Herapath's version was somewhat different: he proposed to take account of the effect on the gravific particles of the high temperatures in the space near the Sun. In this way he came to consider the relation between temperature and particle velocity.

Herapath was puzzled by the old paradox of collisions between atoms (Scott 1970): if they meet head-on they must stop at least for an instant before rebounding. Is the collision elastic or inelastic? It can't be elastic, since an atom by definition has no smaller parts and cannot change its size, so how can it store its kinetic energy during that instant? But if it were inelastic, both atoms would have to stop; then what happens to their energy?

To avoid the paradox, Herapath decided to adopt momentum instead of energy as the fundamental measure of motion, since momentum is always conserved whether collisions are elastic or inelastic. Herapath simply assumed that the scalar momentum mv of a particle is a measure of its absolute temperature and that the total momentum of a system is conserved in collisions while individual momenta tend to be equalized. (One consequence of this unorthodox definition was that the temperature of a mixture of equal portions of hot and cold water should be somewhat lower than the average of their initial temperatures, and Herapath proposed this as a crucial test of his theory.)

Herapath submitted his first paper on kinetic theory to the Royal Society of London in 1820, hoping to get it published in the *Philosophical Transactions*. That would have given his views wide circulation in the international scientific community; moreover, as Herapath himself frankly admitted, it would have enhanced his personal reputation so that he could embark on a career of teaching and scientific research.

Humphry Davy, a well-known chemist, became president of the Royal Society shortly after the submission of Herapath's paper and was mainly responsible for its fate. Davy had earlier supported the general idea that heat is molecular motion rather than a substance, and thus he might have been expected to be receptive to a theory that gave this idea a precise mathematical formulation. But Davy failed to play the role of either genius or gadfly. He didn't have a coherent kinetic theory of his own, and couldn't accept the idea of an absolute temperature scale or a 'lowest temperature'. Moreover, it appears that his mathematical skills were probably so weak he couldn't understand the theory, though it involved no more than simple algebra.

Herapath was told that his paper would not be published in the *Philosophical Transactions*, so he sent instead to an independent scientific journal, the *Annals of Philosophy*, where it was published in 1821. Later he attacked Davy and the Royal Society in a series of letters in *The Times* of London, challenging them to solve several problems pertaining to gas properties (problems he had already solved using his own theory). There was no response. When Davy resigned the Presidency of the Royal Society, Herapath concluded that he had been forced to retire by the controversy (a conclusion for which there is no independent evidence).

Herapath later became editor of the *Railway Magazine*, a position that give him an opportunity to publish his scientific work though to a limited and perhaps unappreciative audience. Although his theory is different from the modern kinetic theory (for example, it is not consistent with Avogadro's hypothesis), it does lead to similar results in some cases. In 1836 Herapath presented a calculation of the speed of sound, which he had completed four years earlier. This was in fact the first calculation of the average speed of a molecule from the kinetic theory of gases. (J. P. Joule, was is usually credited with this accomplishment, was simply following Herapath's method.) Herapath found that the speed of sound in air at 32°F should be about 1090 feet per second, in good agreement with the experimental results available at that time.

5. CLAUSIUS AND BUYS-BALLOT

The "kinetic theory of heat" -- the old idea that heat is directly related to the kinetic energy of atomic motion -- had to be given serious consideration as soon as energy conservation and thermodynamics had been introduced in the middle of the 19th century. But there were several different kinetic hypotheses to choose from. The molecular motion might be linear, rotational, or vibrational, or a combination of all three; the molecules might be small relative to the space in which they move, or large and thus crowded together; the motion might be similar for each molecule in the system or differ according to a definite pattern. Physicists still believed that an ether is needed to transmit energy between bodies in the form of light or radiant heat; if the ether also fills the space between molecules inside a body, it should have some effect on their motion. The old idea that molecules "swim" in the ether, or are suspended by it at definite equilibrium points around which they may vibrate, was not yet dead.

Among these possibilities the kinetic theory of gases was perhaps the simplest but by no means the most plausible. In fact, it seemed too simple to be true. Like the Torricelli-Boyle hypothesis of atmospheric pressure, it required a suspension of disbelief: one had to ignore the ether and assert that molecules move through space at constant velocity, encountering no resistance except when they collide with each other or a boundary surface. The first scientist who was able to overcome the general reluctance to give serious consideration to this idea was the physicist Rudolf Clausius (1822-1888) (see biographical article by Daub 1970).

Like Bernoulli, Clausius assumed that "the space actually filled by the molecules of the gas must be infinitesimal in comparison to the whole space occupied by the gas itself." Moreover, "the influence of the molecular forces must be infinitesimal" (Clausius 1857; translation, p. 116). This means not only that the forces between molecules at their *average* distances are negligible but also that the short-range repulsive forces that cause molecules to rebound at collisions must act over a very small portion of the path of the molecule. If these conditions were not satisfied the gas would not obey the ideal gas laws. Near the end of his 1857 paper Clausius calculated the average speeds of molecules of oxygen, nitrogen, and hydrogen at the temperature of melting ice and found them to be 461 m/sec, 492 m/sec, and 1,844 m/sec respectively.

That was the cue for our next gadfly. Christoph Hendrik Diederik Buys-Ballot (1817-1890), a Dutch meteorologist, is known for 'Buys-Ballot's law': "when you place yourself in the direction of the wind ... you will have at your left the least atmospheric pressure" (Burstyn 1970). In addition to his efforts to establish an international network for reporting weather observations, Buys-Ballot had developed his own atomic theory which conflicted with that of Clausius (Snelders 1968).

Soon after the paper by Clausius appeared, Buys-Ballot (1858) pointed out that if it were really true that the molecules of a gas move at speeds of several hundred meters per second, the diffusion and mixing of gases should be much more rapid than we observe it to be. For example, if you release an odorous gas like ammonia or hydrogen sulfide at one end of a room it may take a minute or so before it is noticed at the other end; yet according to the kinetic theory all the molecules should have traversed the length of the room several times by then. Moreover, carbon dioxide may remain for a long time in an open vessel.

Buys-Ballot apparently thought he had refuted the new theory by pointing out an obvious contradiction between its predictions and the real world. To meet this objection, Clausius had to make an important change in the theory. Abandoning his earlier postulate that the gas molecules have infinitesimal size, he now assumed that they have a large enough diameter or 'sphere of action' so that a molecule cannot move very far without hitting another one.

Clausius defined a new parameter: the *mean free path* (L) of a gas molecule, to be computed as the average distance a molecule may travel before interacting with another molecule. He argued that L may be large enough compared with molecular diameters so that the basic concepts of kinetic theory used in deriving the ideal gas law are unimpaired, yet small enough so that a molecule must change its direction many times every second, and may take a fairly long time to escape from a given macroscopic region of space. In this way the *slowness* of ordinary gas diffusion, compared with molecular speeds, could be explained.

The mean free path is inversely proportional to the probability that a molecule will collide with another molecule as it moves through a small portion of the gas. For spheres of diameter d this probability is proportional to the collision cross section (πd^2) and to the number of molecules per unit volume (n/V). Thus the mean free path is determined by the formula

$$L = k\, V/nd^2$$

where k is a numerical constant of order of magnitude 1 (its precise value was a matter of dispute for some time).

When Clausius introduced the mean free path in 1858 it may have looked like only an ad hoc hypothesis invented to save the theory, since he did not have any independent method for estimating the parameters n/V and d in the above formula. But before anyone had a chance to criticize it on those grounds, Maxwell incorporated the mean free path into his own kinetic theory and showed that it could be related to gas properties such as viscosity. As a result it soon became a valuable concept, not only for interpreting experimental data, but also for determining the size of molecules and thus justifying its own existence. Of course Clausius gets the credit for introducing the mean free path, while Buys-Ballot is forgotten.

6. MAXWELL, LOSCHMIDT, BOLTZMANN, AND GUTHRIE

The kinetic theory of Clausius was quickly taken up and developed into a powerful mathematical research instrument by the Scottish physicist James Clerk Maxwell (1831-1879) (see the biography by Everitt 1975). Maxwell greatly extended the scope of the kinetic theory by showing how it could be used to calculate not only the thermal and mechanical properties of gases in equilibrium, but also their "transport properties": diffusion, viscosity, and heat conduction. In his first paper, published in 1860, he used the Clausius mean-free-path idea to obtain unexpected results for the viscosity of a gas.

Maxwell calculated the viscosity of a gas by estimating the mutual friction of neighbouring layers of gas moving at different speeds. One might expect, on the basis of experience with liquids, that a fluid will have higher viscosity (will flow less freely) at lower temperatures, and that a denser fluid will be more viscous than a rarer fluid, since in both cases the motion will be more strongly obstructed by intermolecular forces. Maxwell showed that if the kinetic theory of gases is correct, both expectations will be wrong, because the mechanism that produces viscosity is different. In a gas, viscous force originates not in the forces between neighboring molecules but in the transfer of momentum that occurs when a molecule from a faster-moving stream wanders over to a slower-moving stream and collides with a molecule there. His analysis showed that the viscosity coefficient μ increases with temperature and is independent of density.

When Maxwell published these two conclusions about gas viscosity, little empirical evidence was available, and what there was seemed to indicate that the viscosity increases with density. Maxwell seems initially to have expected that his theoretical predictions would be refuted, but instead they were confirmed, by his own experiments and those of others.

In the 1860s, when the mere existence of an atomic structure of matter was no more than a plausible hypothesis, Maxwell's theory was used to accomplish a major advance: the first reliable estimate of the *size* of an atom. This was done by Josef Loschmidt (1821-1895) (see biographical article by Böhm). He pointed out in 1865 that Maxwell's theory allowed one to estimate the product nd^2 from the measured viscosity coefficient. Thus one could determine the molecular diameter d if one other relation between n and d were known. In particular, he suggested that the volume occupied by the gas molecules themselves, if they were closely packed, should be approximately the volume of the substance condensed to the liquid state,

$$V_{liq} \approx nd^3.$$

If the density of a substance is known in both the liquid and gaseous states, the ratio or 'condensation coefficient' $V/V_{liq} = V/nd^3$ could be combined with the value of nd^2 from viscosity measurements to obtain a value for d.

In this way Loschmidt concluded that the diameter of an "air molecule" is about $d \approx 10^{-7}$ cm. This value is about four times too large according to modern data, but considerably better than any other well-founded estimate available at the time.[3]

During the next few years, other scientists (the most influential being William Thomson, Lord Kelvin) made similar estimates of atomic sizes and other parameters with the help of the kinetic theory of gases. As a result, the atom came to be regarded as no longer a merely hypothetical concept but a real physical entity, subject to quantitative measurement, even though it could not be "seen." This was one of the most important contributions of the kinetic theory to 19th-century science, a result of the independent efforts of two geniuses (Clausius and Maxwell), one gadfly (Buys-Ballot), and a respectable scientist who was soon to return to our story in the role of gadfly (Loschmidt).

Clausius had used probability concepts in his derivation of the mean-free-path formula, but it was Maxwell who converted the kinetic theory of gases into a fully statistical doctrine. Clausius and earlier kinetic theorists had assumed that all molecules in a a homogeneous gas at a given temperature have the same speed, but Maxwell asserted that the collisions among molecules will instead produce a statistical distribution of speeds. To describe this distribution he borrowed a mathematical formula from the social sciences: the 'normal distribution law' or 'law of errors' (now notorious as the 'bell curve'). According to Maxwell, at any specified absolute temperature T the average value of v^2 is proportional to T; a few molecules have much small velocities and a few have much larger velocities. This became known as the *Maxwell velocity distribution law* for molecules.

In his early papers Maxwell simply postulated that the law applies to molecules, and did not succeed in giving a convincing proof of this postulate. In a long paper on kinetic theory in 1866, Maxwell argued, by considering collisions between molecules, that this is the only possible distribution law that can hold in thermal equilibrium.

But in preparing this paper, Maxwell encountered considerable difficulty in applying his theory to answer the following question: if a column of air is in equilibrium under the influence of gravitational forces, will its temperature change with height? He tells us that the original manuscript of his 1866 paper, as sent to the Royal Society, contained an equation that implied that a column of air would assume a temperature *varying* with height. At various times he had decided that the temperature might either increase or decrease with height, but finally discovered that these conclusions were the result of calculational errors. He finally concluded that any variation of temperature with height is inconsistent with the Second Law of Thermodynamics (Garber, Brush & Everitt 1995, pp. 105-120). .

In 1868 there began a fruitful dialogue between Maxwell and Ludwig Boltzmann (1844-1906) (see Klein 1973). Boltzmann complained that Maxwell's derivation of the velocity distribution law was hard to understand because of its brevity. He devoted the first part of a 44-page memoir to filling in and illustrating with concrete examples the steps that Maxwell had glossed over. He then considered a situation in which the molecules are subject to an external force such as gravity, and showed in a rather special case that Maxwell's velocity function should be multiplied by another function involving potential energy corresponding to the force. In modern terms, the generalized distribution law is written as a function proportional to

$$e^{-E/kT}$$

where e is the base of natural logarithms (= 2.718 ...), E is the total energy, kinetic plus potential; and k is now called 'Boltzmann's constant'. This function is called the *Maxwell-Boltzmann distribution law* or simply the *Boltzmann factor*.

At this point Maxwell and Boltzmann have agreed (although they say it in very different ways) that a column of gas can be in equilibrium with the same temperature everywhere; the effect of the gravitational field is not to make the *temperature* higher at the bottom, but to make the *density* greater at the bottom.

Here we have yet another implausible result from gas theory. Everyone knows that the air gets colder as you go higher above the earth's surface. Moreover, if you believe in a kinetic theory of heat, in which temperature is directly proportional to the square of molecular velocity, you can easily see that when a molecule moves upward it has to do work against gravity and must therefore slow down; hence the molecules at greater heights will have lower velocities and the temperature will be lower. So Maxwell's and Boltzmann's conclusion seems contrary not only to common experience but also to the basic principles of the kinetic theory itself.

A younger Herapath could have found his true gadfly calling by criticizing this conclusion (which contradicted his theory), but unfortunately he died in 1868. The role was instead played by Francis Guthrie (1831-1898), professor of mathematics at the Graaff-Reinet College, Cape Colony, South Africa. When Maxwell repeated the result that temperature is independent of height in his textbook on the theory of heat in 1872, Guthrie published a letter in *Nature*, in which he wrote (1873):

> it seems obvious that thermal equilibrium can only subsist according to the kinetic theory, where the molecules encounter each other with equal average

amounts of *work* or *vis viva*, and in order that this may be the case, the velocity of the molecules (and consequent temperature) of any upper layer must be less than that of the molecules in the layer next below; since, in order to encounter each other, the former must descend, and acquire velocity, while the latter must ascend and lose it. This would establish a diminution of temperature from the bottom to the top of a column of air at the rate (in the absence of any counteracting cause) of 1°F. for 113 ft. of height ...

Maxwell (1873a) replied that Guthrie's argument, which had already occurred to him in 1866 and "nearly upset my belief in calculation," it is not conclusive because it "assumes that, of the molecules which have encounters in a given stratum, those projected upward have the same mean energy as those projected downwards. This, however, is not the case, for since the density is greater below than above, a greater *number* of molecules come from below than from above to strike those in the stratum" and hence, since the stratum is assumed to remain at rest, "a smaller number of the molecules projected upwards must have a greater initial velocity than the larger number projected downwards". But this line of reasoning is not conclusive either, and Maxwell has developed another approach, which bypasses Guthrie's objection rather than answering it directly. "It is well, however, that he has directed attention to it, and challenged the defenders of the kinetic theory to clear up their ideas..."

Maxwell (1873b), forced by Guthrie to re-examine a problem he thought he had solved, turned instead to Boltzmann's 1868 paper mentioned above. The result was a new concise derivation of the generalized Maxwell-Boltzmann distribution law, from which the thermal equilibrium of a column of gas followed as a special case (Maxwell 1873c). This paper offers little enlightenment about the Earth's atmosphere (the real-world problem that motivated Bernoulli, Herapath and Guthrie to analyze the effect of gravity on molecular collisions); instead it provided a secure foundation for the new science of *statistical mechanics*. The most important applications of the Maxwell-Boltzmann distribution law are not to situations in which the entire gas is subjected to an external force like gravity, but rather to the analysis of the effect of intermolecular forces in dense gases, liquids and solids.

Guthrie's gadfly contribution to the composition of Maxwell's paper has been hidden by the way that paper got into the literature of physics. The original version, published in a British Association report, gave full credit at the beginning to Guthrie for raising the issue and giving a *prima facia* proof that temperature does vary in a column of gas. But when the paper was reprinted in *Nature* the paragraphs mentioning Guthrie were omitted. When W. D. Niven edited Maxwell's *Scientific Papers* he used the *Nature* version, and completely omitted Maxwell's two other notes in *Nature* that mentioned Guthrie (Garber, Brush & Everitt 1995, pp. 121-123, 144-145), thus concealing from the reader the history of this fundamental paper and the role of the gadfly in provoking the genius.

7. BOLTZMANN AND LOSCHMIDT

Having provided a satisfactory molecular interpretation of the First Law of Thermodynamics, kinetic theorists soon turned their attention to the Second. As formulated by Clausius, the Second Law of Thermodynamics implied that useful mechanical work cannot be obtained from heat except when heat flows from a high temperature to a low temperature in the special way described by Sadi Carnot; that heat spontaneously flows from hot to cold but not the reverse; and that a mysterious quantity called "entropy" tends to increase to a maximum. For the universe as a whole this state of maximum entropy was called by Clausius the "Heat Death": all energy is uniformly diffused throughout space at a low temperature, so that no mechanical work can be done and life cannot exist.

In 1872 Boltzmann took a major step toward a molecular interpretation of the Second Law when he developed his version of kinetic theory, equivalent to Maxwell's theory insofar as it leads to the same formulas for the coefficients of diffusion, viscosity and heat conduction, but more convenient for some other applications. Boltzmann proceeded by analyzing the changes in the velocity distribution function in a non-uniform gas composed of colliding molecules. We will write this function $f(x,v,t)$ to indicate that it depends not only on the molecular velocity (v) but also on spatial position (x) and time (t). Boltzmann computed the change in $f(x,v,t)$ resulting from all relevant physical parameters, including especially the collisions that changed the numbers of particles having specified velocities. The result, published in 1872, was an integrodifferential equation for $f(x,v,t)$, now called *Boltzmann's transport equation* or simply 'the Boltzmann equation'. It plays a major role in 20th century kinetic theory, including theories of ionized gases (plasma physics) and in calculations of neutron flow in nuclear reactors.

Boltzmann proved that (if his equation is correct) collisions always push $f(x,v,t)$ toward the equilibrium Maxwell distribution (the exponential form given above). In particular, the quantity $H = \int f(x,v,t) \log f(x,v,t)\, dx\, dv$ always decreases with time unless f is the Maxwell distribution, in which case H maintains a fixed minimum value. This statement is now known as Boltzmann's H-theorem.[4]

For a gas in thermal equilibrium, Boltzmann's H is proportional to minus the entropy as defined by Clausius in 1865. While the entropy in thermodynamics is defined only for equilibrium states, Boltzmann suggested that his H-function could be considered a generalized entropy having a value for any state of the gas. Then the H-theorem is equivalent to the statement that the entropy always increases or remains constant, which is one version of the Second Law of Thermodynamics. The justification for Maxwell's distribution law is then based on the assertion of a general tendency for systems to pass irreversibly toward thermal equilibrium.

There is a well-known objection to this line of argument, the "reversibility paradox" usually attributed to Loschmidt. The analysis of collisions used by Maxwell and Boltzmann in developing the kinetic theory of gases assumes that each individual collision is governed by Newtonian mechanics; yet Newton's laws are clearly time-reversible: if you reverse the time direction, you will get a sequence of motions and collisions that is also compatible with Newton's laws. So how can the evolution of a system of many molecules be irreversible?

The paradox had been discussed extensively by Maxwell with his friends Peter Guthrie Tait and William Thomson; this discussion is documented by correspondence going back to 1867 (Garber, Brush & Everitt 1995, pp. 178, 187-188, 192-193, 205). The culmination of this discussion was Thomson's publication of a paper (1874) in which he distinguished between "abstract dynamics" which is perfectly reversible, and "physical dynamics" which is not. He considered a thought experiment in which, starting from an initial unequal distribution of temperature in a gas, we allow diffusion to occur until after a finite time interval the temperature is very nearly equal throughout the gas; then we instantaneously reverse the motion of each molecule. Each molecule will then retrace its former path, until at the end of a second time interval equal to the first, "every molecule will be in the same position, and moving with the same velocity, as at the beginning; so that the given unequal distribution of temperature will again be found, with only the difference that each particle is moving in the direction reverse to that of its initial motion" (Thomson 1874; Brush 1966, p. 181). While it might appear that this process is contrary to the Second Law of Thermodynamics, Thomson pointed out, first, that if the reversed motion continues, there will be an "instantaneous subsequent commencement of equalization," so that the unequal distribution of temperature will be short-lived. Second, if we looked at a gas in thermal equilibrium, there would be no way to pick out the particular arrangement that could evolve into a nonequilibrium state if the velocities were reversed. It is true that if any gas be left for a sufficiently long time in a perfectly rigid vessel with no external influences, it will inevitably happen that, for example, more than 90% of the energy will be in one half of the vessel. But the probability of this happening at any particular time is enormously smaller than the probability of a roughly equal distribution.

Thomson was not a gadfly attacking Maxwell's kinetic theory; on the contrary, Maxwell himself had already argued, with his famous 'Demon' thought experiment, that irreversibility is not an absolute law of nature but only a statistical law, valid with very high probability. Thomson was simply illustrating this conclusion in a different way.

Loschmidt's statement of the reversibility paradox is only one sentence in a series of four papers published in 1876-77, although there may also have been a verbal discussion between Loschmidt and Boltzmann at a meeting of the Vienna Academy.[5] The published papers are concerned primarily with the cosmic aspects of the Second Law of Thermodynamics (as summarized in Loschmidt 1878) and the problem of the equilibrium of a column of gases subject to gravitational forces. Loschmidt did not accept the Maxwell-Boltzmann conclusion that the column would have constant temperature throughout, but claimed instead that thermal equilibrium is possible without equality of temperature. In this way he hoped to demonstrate that the heat death of the universe is not inevitable. He claimed that the Second Law could be correctly formulated as a mechanical principle without reference to the sequence of events in time; he thought he could thus "destroy the terroristic nimbus of the second law, which has made it appear to be an annihilating principle for all living beings of the universe; and at the same time open up the comforting prospect that mankind is not dependent on mineral coal or the sun for transforming heart into work, but may have available forever an inexhaustible supply of transformable heat" (Loschmidt 1876, p. 133; see also 1877, p. 293).

After proposing a model that supposedly violated Maxwell's constant-temperature theorem, Loschmidt noted that in any system "the entire course of events will be retraced if at some isnatnt the velocities of all its parts are reversed" (1876, p. 139). His application of this reversibility principle to the validity of the Second Law was somewhat obscurely stated but Boltzmann quickly got the point (see also Boltzmann 1896-1898, §6) and published a reply (1877a). Boltzmann also discussed the thermal equilibrium of a vertical column at some length (Boltzmann 1876, 1878; Boltzmann and Nabl 1905, p. 516).

Boltzmann conceded it is impossible to prove that the entropy of a system always increases without taking account of the initial conditions. Moreover, such a statement cannot be true for *all* initial conditions, since it is certainly possible to find a special initial state (obtaining by reversing all the molecular velocities of a system which has evolved from a nonuniform one) for which succeeding states will have lower entropy. The crucial point, however, is that "since there are infinitely many more uniform that non-uniform distributions, the number of states which lead to uniform distributions after a certain time t_1 is much greater than the number that lead to non-uniform ones, and the latter are the ones that must be chosen, according to Loschmidt, in order to obtain a non-uniform distribution at t_1" (Boltzmann 1877a; translation, p. 192).

There follows the very important remark (ibid.):

> One could even calculate, from the relative numbers of the different state distributions, their probabilities, which might lead to an interesting method for the calculation of thermal equilibrium.

Following up his own suggestion, Boltzmann developed soon afterward (1877b) his famous statistical relation between entropy and probability, articulated in this discussion of the reversibility paradox. In modern terminology, a *macrostate* (determined by specified values of pressure, temperature, and other relevant variables) can correspond to many different *microstates* (determined by specified values of the positions and velocities of all the atoms). The entropy of a macrostate, according to Boltzmann, is proportional to the logarithm of the number (W) of corresponding microstates; this is the relation

$$S = k \log W$$

engraved on Boltzmann's tombstone (Brush 1976, p. 609). Thermal equilibrium is the macrostate that has the greatest probability, i.e. the largest value of W. Boltzmann's statistical interpretation of the Second Law is that a system starting in a macrostate which has low probability will almost certainly evolve to one that has high probability, because the overwhelming majority of microstates correspond to the macrostate for thermal equilibrium. "Irreversibility" is simply a tendency to go from less probable to more probable states.[6]

The distinction between macro- and microstates is crucial in Boltzmann's theory.

Like Maxwell's Demon, an observer who could deal directly with microstates would not perceive irreversibility as an invariable property of natural phenomena. It is only when we decide to group together certain microstates and call them, collectively, "disordered" or "equilibrium" macrostates, that we can talk about going from "less probable" to "more probable" states. This is an irreversible process in the same sense that shuffling the deck after dealing a grand-slam hand in bridge is an irreversible process; the rules of the game single out certain distributions of cards as "ordered" (all the same suit or all aces, kings and queens in the same hand), and we call these "rare" distributions although in fact each of the possible distributions of 52 cards among four hands of 13 each has exactly the same probability.

8. BOLTZMANN, CULVERWELL, AND BURBURY

The problem of irreversibility was revived in Britain in the 1890s. Edward Parnall Culverwell (1855-1931) raised, in 1890, what might be called the "reversibility objection to the H theorem," not to be confused with the "reversibility paradox" discussed by William Thomson, Loschmidt, and Boltzmann in the 1870s. Culverwell, at that time a Fellow of Trinity College Dublin, was later Professor of Education at Dublin and published a book on the Montessori method.

Using the example of a system of particles interacting with forces proportional to their distance [the standard 'harmonic oscillator' model of modern theoretical physics], he claimed that -- since in the case the motion is strictly periodic -- it is impossible to prove *in general* that a set of particles will tend to the "Boltzmann configuration, in which the energy is equally distributed among all the degrees of freedom". Appealing to the reversibility principle, he asserted that "for every configuration which tends to an equal distribution of energy, there is another which tends to an unequal distribution." In order to explain the fact that temperature equilibrium does nevertheless occur, he suggested that there must be some kind of interaction of molecules with the aether. After all, "one of the most important purposes for which the existence of the aether is required," he wrote, is heat transfer leading to thermal equilibrium. Conversely, if a system of particles in a vessel *not* containing aether did in fact attain equilibrium in all cases, "it would be to my mind a proof that the ultimate particles of matter did not individually obey those laws which they are known to obey when collected in the enormous numbers which compose the bodies for which the laws of motion have been experimentally proved" (Culverwell 1890a).

Explaining his views further at a British Association meeting the same year, he argued that molecular motions might be inherently irreversible, yet obey the Newtonian laws of motion when taken *en masse*. It is also conceivable, he said, that there might be periodic deviations from Newton's laws, "the period being so short that no observations could detect it" (Culverwell 1890b).

Culverwell's objection was discussed at meetings of the British Association and in the columns of Nature during the next few years. In 1894, somewhat annoyed by the endless technical arguments about Boltzmann's H-theorem, he tried to get the discussion back to the main issue as he saw it: "I do not know Boltzmann's proof, but while I suppose it is all right, I find it very hard to understand how any proof can exist." Finally he demanded: "Will some one say exactly what the H-theorem proves?" (Culverwell 1894).

It was S. H. Burbury (1831-1911), a mathematician-lawyer in London, who pointed out a plausible solution to the puzzle. He argued that the proof of the H-theorem depends on the Maxwell-Boltzmann assumption that colliding molecules are statistically uncorrelated. While this would seem a reasonable assumption to make before the collision, one might suppose that the collision itself introduces a correlation between the molecules that have just collided, so that the assumption would not be valid for later collisions. Burbury suggested that the assumption might be justified by invoking some kind of "disturbance from without [the system], coming at haphazard" (Burbury 1894, p. 78).

Culverwell then summed up the answers given by Burbury and other British scientists to his question; he concluded that if the proof of the H-theorem "does not somewhere or other introduce some assumption about averages, probability, or irreversibility, it cannot be valid" (Culverwell 1895a).

Boltzmann replied with a long letter published in *Nature*, addressing the questions "(1) Is the Theory of Gases a true physical theory as valuable as any other physical theory? (2) What can be demand from any physical theory?" (Boltzmann 1895, p. 413). In response to Culverwell, who "wishes to refute" the H-theorem, Boltzmann wrote that Culverwell was mistaken about his assumptions, which were probabilistic in nature. In fact it seems to me that Culverwell did understand that, and he complained later "it is a little hard that Dr. Boltzmann should represent me as endeavoring to *disprove* his theorem" when he had himself insisted that probabilistic assumptions were needed to prove it (Culverwell 1895b, p. 581). But the most interesting part of Boltzmann's letter is the last two paragraphs, where he presents "an idea of my old assistant, Dr. Schuetz":

> We assume that the whole universe is, and rests for ever, in thermal equilibrium. The probability that one (only one) part of the universe is in a certain state, is the smaller the further this state is from thermal equilibrium; but this probability is greater, the greater is the universe itself. If we assume the universe great enough, we can make the probability of one relatively small part being in any given state (however far from the state of thermal equilibrium), as great as we please. We can also make the probability great that, though the whole universe is in thermal equilibrium, our world is in its present state ...
>
> If this assumption were correct, our world would return more and more to thermal equilibrium; but because the whole universe is so great, it might be probable that at some future time some other world might deviate as far from thermal equilibrium as our world does at present. Then the afore-mentioned H-curve would form a representation of what takes place in the universe. The summits of the curve would represent the world where visible motion and life exist.
>
> (Boltzmann 1895, p. 415).

The outcome of Boltzmann's participation in the British discussions of the H - theorem was his acceptance of Burbury's conclusion that an additional assumption was needed, and called it the hypothesis of "molecular disorder." He argued that it could be justified by assuming that the mean free path in a gas is large compared with the mean distance of two neighboring molecules, so that a given molecule would rarely encounter again a specific molecule with which it had collided, and thus become correlated (see Boltzmann 1896-1898, §3).

"Molecular disorder" is not merely the hypothesis that states of individual molecules occur completely at random; rather it amounts to an exclusion of special ordered states of the gas that would lead to violations of the Second Law. In fact such ordered states would be generated by a random process.

In modern terminology, one makes a distinction between "random numbers" and "numbers generated by a random process" -- in preparing a table of random numbers for use in statistical studies, one rejects certain subsets, for example pages on which the frequencies of digits depart too greatly from 10%, because they are inconveniently-nonrandom products of a random process. Boltzmann recognized that the hypothesis of molecular disorder was needed to derive irreversibility, yet at the same time he admitted that the hypothesis itself may not always be valid in real gases, especially at high densities, and that recurrence may actually occur (Boltzmann 1896-1898, §4). .

9. BOLTZMANN AND ZERMELO

In section 7 an 'irreversible' process was compared to shuffling a deck of cards after a grand slam hand in bridge. If you play bridge long enough you will eventually get another grand-slam hand, not once but several times. The same is true with mechanical systems governed by Newton's laws, as the French mathematician Henri Poincaré (1854-1912) showed with his recurrence theorem in 1890: if the system has fixed total energy and is restricted to a finite volume, it will eventually return as closely as you like to any given initial set of molecular positions and velocities. If the entropy is determined by these variables, then it must also return to its original value, so if it increases during one period of time it must decrease during another.

This apparent contradiction between the behavior of a deterministic mechanical system of particles and the Second Law of Thermodynamics became known as the "Recurrence Paradox." Poincaré, in a short paper addressed to philosophers (1893), saw it as a conflict between mechanism and experience. Mechanism implies that all phenomena must be reversible, yet experience shows that many irreversible phenomena exist in nature. To escape the contradiction, physicists have postulated 'hidden movements': for example, if we did not know that the earth rotates we would regard the motion of the Foucault pendulum as 'irreversible' but having discovered that the earth does rotate, we can imagine that it might just as well be rotating in the opposite direction. Hence we do not consider this a contradiction of the principle of reversibility in Newtonian mechanics. Similarly one might suppose that there are motions in the molecular world that account for macroscopic irreversibility, and which are 'in principle' reversible.

Poincaré then pointed out that his recurrence theorem, which would seem to aply to the entire world if the kinetic theory is valid, contradicts the 'heat death' theory. According to that theory, if one ascribed absolute validity to the Second Law, then the universe, instead of returning to its initial state, would tend toward a final state of uniform temperature. One could reconcile the two theories, he suggested, by assuming that the heat death is not permanent but only lasts a very long time, so that the universe, after slumbering for millions of millions of centuries, will eventually awaken. Then "to see heat pass from a cold body to a warm one, it will not be necessary to have the acute vision, the intelligence, and the dexterity of Maxwell's demon; it will suffice to have a little patience" (Poincaré 1893, translation in Brush 1966, p. 206).

The gadfly in this case was not Poincaré but the German mathematician Ernst Zermelo (1871-1953).[7] After receiving his Ph.D. in 1894 at Berlin, where he studied with Max Planck among others, Zermelo moved to Göttingen. In 1905, "shortly after his sensational proof of the well-ordering theorem," he was appointed to a professorship there. In addition to showing that every set can be well-ordered, he is known for his axiom system for Cantor's set theory (Rootselaar 1976; Peckhaus 1990). In 1910 he accepted a chair at Zurich, where he argued with Einstein about Brownian motion (Klein 1970, p. 295).

In 1896 Zermelo, apparently unaware of Poincaré's 1893 paper, claimed that the recurrence theorem makes in impossible for the mechanistic worldview to explain irreversible processes. He argued that the Second Law is an absolute truth, so any theory that leads to predictions inconsistent with it must be false. This refutation would apply not only to the kinetic theory of gases but to any theory based on the assumption that matter is composed of particles moving in accordance with the laws of mechanics.

Boltzmann had previously denied the possibility of such recurrences and might have continued to deny their certainty by rejecting the determinism postulated in the Poincaré-Zermelo argument. Instead, having been forced to reconsider the subject by his debate with Culverwell and Burbury, he admitted quite frankly that recurrences are completely consistent with the statistical viewpoint, as the card-game analogy suggests; they are fluctuations, which are almost certain to occur if you wait long enough. So determinism leads to the same qualitative consequence that would be expected from a random sequence of states! In either case the recurrence time is so inconceivably long that our failure to observe it cannot constitute an objection to the theory.

Boltzmann could decisively refute the contention that the mechanical viewpoint contradicts 'experience' because the term 'experience' had been improperly extended by Zermelo (and Poincaré) to include *theoretical predictions* about what will happen to the universe in the remote future. The heat death is not a fact of experience but only an extrapolation from the observation that heat 'always' flows from hot to cold. From the kinetic theory Boltzmann could estimate the time needed for an approximate recurrence of the positions and velocities of all the molecules in 1 cubic centimeter of gas at ordinary density; it is a number so large that it would take trillions of digits even to write it down.

Thus the recurrence paradox has nothing to do with the behavior of gases in the laboratory

> when Zermelo concludes, from the theoretical fact that the initial states ina gas must recur, -- without having calculated how long a time this will take -- that the hypotheses of gas theory must be rejected or else fundamentally changed, he is just like a dice player who has calculated that the probability of a sequence of 1000 one's is not zero, and then concludes that his dice must be loaded since he has not yet observed such a sequence!
> (Boltzmann 1896; translation in Brush 1966, p. 223).

But Boltzmann's theory of the *H-curve* (the overall behavior of H as a function of time) was still only qualitative, and somewhat unsatisfactory from a mathematical viewpoint. Boltzmann asserted that the curve runs along very close to its minimum value most of the time, with occasional peaks corresponding to significant deviations from the equilibrium state. The probability of a peak decreases rapidly as the height of the peak decreases, and if the initial state lies on a very high peak, the state of the system will drop down toward the equilibrium state (minimum value of H) "with enormously large probability, and during an enormously long time it will deviate from it by only vanishingly small amounts". On the other hand if one waits an even longer time the initial state will eventually recur. Yet Boltzmann insisted that for any state with a value of H above the minimum, H is more likely to decrease than increase. No evidence was presented for any of these statements other than the original (qualitative) H-theorem.

In a second paper, Zermelo (1896b) protested that the properties attributed to the H-curve by Boltzmann are not only unproved, but incompatible with the laws of mechanics; and that probability theory cannot resolve this contradiction. First, the overall periodicity of the system implies that every decrease in H must be balanced by an increase at some other time. Second, the probability of occurrence of a certain value of H should be measured by the volume in phase space of all states having this value; but from the equations of motion it can be shown that this volume is independent of time (this is called "Liouville's theorem" by physicists). Hence there cannot be any tendency for H to increase or decrease.[8] While these objections might apply to an ensemble of systems over a long period of time, Zermelo realized that Boltzmann had based his case for the H-theorem on more specific assumptions about the short-time behavior of the H curve for individual systems, and so he must also attack these assumptions.

If we assume that H has occasional peaks, and we choose the initial state to have a value of H ($= H_0$) greater than its minimum value, then it would seem that H_0 can just as well lie on a rising as a falling part of the curve and therefore can either increase or decrease. If we assume that the increasing branch occupies a smaller time interval, so that the probability of landing there is smaller, it would still appear that the increase observed when one *does* land there is steeper and thus must be given a correspondingly greater weight. Zermelo interpreted Boltzmann's argument as an attempt to avoid this objection by postulating that H_0 is always at a *maximum* of the H-curve, so that one only observes it to decrease. But Zermelo says he "cannot conceive of such a curve" which consists only of maxima, nor can anyone else. As he says, it would make sense only if the maxima are not mathematical points but flat portions of the curve; but this again contradicts our experience of rapid dissipations of temperature inequalities or other ordered states.

Zermelo also revived the reversibility argument, which he contended makes it impossible ever to derive irreversibility. Any alleged deduction must depend on errors or fallacious assumptions, in particular the "unprovable (because untrue) assumption that the molecular state of a gas is always, in Boltzmann's expression, 'disordered'".[9] According to Zermelo, only the initial state may be assumed to be disordered; the probability of a later state must depend on the initial state.

It is not surprising that when Zermelo refers to the "mechanical view of nature" he still has in mind a deterministic mechanical system in which randomness plays no role in the molecular motions themselves, but only in the observer's description of these motions. What is more remarkable is that Boltzmann, after having introduced the molecular disorder" postulate, does not challenge this view; as we noted earlier, molecular disorder is for Boltzmann an assumption that may or may not be true -- not a postulate.

Boltzmann's reply (1897) to Zermelo's second paper is in one sense a reiteration of his earlier arguments, but it is at the same time a retreat from his contention that irreversibility follows *in general* from the kinetic theory. We are concerned, he says, with what will happen in the present state of the world, which happens to be a state of low entropy; therefore we can say that H is a maximum in the initial state without having to claim (as Zermelo suggested) that *all* points of the H-curve are maxima. If, however, we selected a completely arbitrary state of the universe, there are four possibilities. First, and most likely, the state is one of thermal equilibrium, so there will be no significant change of H at all from its minimum value. Second, H is above its minimum, and will "almost immediately" decrease if we follow it either forwards or backwards in time. Third, H is above its minimum, on an increasing branch, so the system passes to more improbable states as one goes forward in time. Fourth, H is above its minimum, on a decreasing branch, so the system passes to more improbable states as one goes backwards in time. The third and fourth cases have equal probability but both are "much rarer" than the second, which is in turn much rarer than the first.

From the description of the second case we can perhaps see why Boltzmann persists in saying that most parts of the H-curve above its minimum are maxima, even though he admits that this cannot be literally true. The key word is "almost" -- if the peak is very narrow in time, then even if the initial state is slightly to the left of the actual maximum, one will quickly get over the top and further down the other side in a short time interval (Boltzmann 1898; P. & T. Ehrenfest, 1959 translation, p. 34). Thus it is a maximum with respect to finite differences but not with respect to infinitesimal differences. Some of the confusion might have been avoided if Boltzmann had stated this more explicitly.

But the second case cannot be taken as the "initial state" in laboratory experiments, for if we look at the value of H in 1896 and follow it backwards in time we expect it to increase not decrease. Thus Boltzmann is really forced into accepting the fourth case as the typical one, which means that the third case is equally likely to be found somewhere, sometime, elsewhere in the universe. He therefore proposes that one should really *define the direction of time* as the direction in which one goes from less to more probable states. This would make the direction of time dependent on the individual observer, and would be different for different parts of the universe at different epochs:

> This viewpoint seems to me to be the only way in which one can understand the validity of the Second Law and the Heat Death of each individual world without invoking a unidirectional change of the entire universe from a definite initial state to a final state
> (quoted from translation in Brush 1966, p. 242; see also Boltzmann 1896-1898, part II, §90).

Boltzmann's conception[10] of alternating time directions in the universe, and the idea that the direction of time is determined by human experience, has been revived in recent years by philosophers and cosmologists (Reichenbach 1956; other references in Brush 1976, pp. 639-640). But it is more significant to note that the proposal was motivated by Boltzmann's desire to push the deterministic (though statistical) mechanical world view to its furthest extreme, perhaps not entirely seriously. In this debate he chose *not* to make the alternative assumption that "molecular disorder" is continually maintained by random or external causes acting at the molecular level, as had been suggested by Burbury, though he accepted it elsewhere. While it is true that the statistical interpretation, like Poincaré's deterministic calculation, predicts recurrences, it is not hard to conceive of a postulate of continual or repeated randomization that would enforce irreversibility much more strongly; in fact this is just what Wolfgang Pauli's proof of the quantum-mechanical H-theorem involves (Pauli 1928; Tolman 1938, p. 455). With such a postulate, recurrence is not impossible but neither is it certain. One must distinguish between a statistical and a stochastic explanation of the Second Law (Kohler 1931).

At the same time Boltzmann was willing to speculate in another direction, more in line with the "hidden variables" interpretation which would attribute randomness in molecular motion to determinism at a still lower level:

> Since today it is popular to look forward to the time when our view of nature will have been completely changed, I will mention the possibility that the fundamental equations for the motion of individual molecules will turn out to be only approximate formulas which give average values, resulting according to the probability calculus from the interactions of many independent moving entities forming the surrounding medium-as for example in meteorology the laws are valid only for average values obtained by long series of observations using the probability calculus. These entities must of course be so numerous and must act so rapidly that the correct average values are attained in millionths of a second.

(Boltzmann 1896-1898; from the 1964 translation, p. 449)

In view of Boltzmann's partial abandonment of determinism on the molecular level, and similar actions by Maxwell and Max Planck (Brush 1983, pp. 89-94) we must reconsider the view that 19th-century physicists always assumed determinism and used statistical methods only for convenience ('epistemological randomness'). By the early decades of the 20th century, thanks in part to the debates provoked by gadflies like Loschmidt, Zermelo, and Culverwell, physicists were familiar with the notion that a molecular interpretation of the Second Law of Thermodynamics might involve a more fundamental ('ontological') randomness. In this meandering journey away from determinism Boltzmann, the genius most often stung by gadflies, led the way.[11]

NOTES

[1] Loschmidt and Zermelo did become part of physicists' history of their discipline as propounders of the reversibility and recurrence paradoxes, respectively, in part because of the influential article by Paul and Tatiana Ehrenfest (1911) which introduced these terms. But it is somewhat misleading to think of Loschmidt as an opponent of kinetic theory since he made a spectacularly-successful application of it, to obtain the first good quantitative estimate of molecule diameters (see below).

[2] Thomas Hobbes (1588-1679), the well-known English political philosopher, thought a Cartesian ether rather than a vacuum is in the space above the mercury in Torricelli's barometer. In this episode he may also be considered a gadfly. The debate between Hobbes and Boyle has been discussed at considerable length by Steven Shapin and Simon Schaffer (1985). See also the earlier account of James Bryant Conant (1950) of Boyle's experiments and his response to critics.

[3] The corresponding value of the number of molecules in a cubic centimeter of an ideal gas at standard conditions (0°C, 1 atm pressure) would be

$$N_L \approx 2 \times 10^{18}.$$

Although Loschmidt himself did not give this result explicitly in his 1865 paper, it can easily be deduced from his formula, and so this number is now sometimes called "Loschmidt's number." Its modern value is 2.687×10^{19}. It should not be confused with the related constant, "Avogadro's number," defined as the number of molecules per gram-mole, equal to

$$N_A = N_L/V_0 = 6.02 \times 10^{23},$$

where ($V_0 = 22420.7$ cm^3 atm mole^{-1}. Avogadro himself did not give any estimate of this number, but only postulated that it should have the same value for all gases. See also Boltzmann (1979), Floderer (1971); Porterfield & Krause (1995), and other works cited by Brush (1976, pp. 75-76). [4]In his 1872 paper Boltzmann used E for this function; the letter H was introduced by S. H. Burbury in 1890 (Brush 1976, pp. 619, 626).

[5]There is also an embellishment of the story circulating among modern physicists, to the effect that when Loschmidt told Boltzmann that his system would simply run backwards if all the molecular velocities were reversed, Boltzmann replied, "Well, *you* just try to reverse them!" (Mayer 1964, p. 10; Kac 1959, p. 61). For Loschmidt's earlier views on this subject see Daub (1970). On the (mostly-friendly) personal relations between Boltzmann and Loschmidt see Blackmore (1995, Book One, pp. 16-17)

[6]In classical mechanics W is infinite because positions and velocities are continuously variable, so Boltzmann actually postulated only that the difference between the entropies of two macrostates is proportional to the logarithm of the ratio of their probabilities; W includes a factor depending on the size of an elementary cell $\Delta x\, \Delta v$, which cancels out in the ratio. In quantum theory the cell size depends on Planck's constant, which therefore appears in the (Sackur-Tetrode) formula for the absolute entropy.

[7] In his report on the journal of a German professor to 'El Dorado' [California], Boltzmann refers to Zermelo as a *Pestalutz* [an evil figure in Friedrich Schiller's *Wallensteins Tod*] (Broda 1955, p. 93, translation p. 91; Blackmore 1995, Book 1, p. 174). Jagdish Mehra (1975) reported: "Professor G. E. Uhlenbeck once told me that he had heard from Ehrenfest that Boltzmann always referred to Zermelo as 'Dieser Halunke' (rogue or villain)".

[8]The same argument was developed in more detail by Gibbs (1902, chapter XII); see also Tolman (1938, pp. 165-179).

[9]As mentioned in the previous section, Boltzmann adopted Burbury's suggestion that because of random external perturbations, the velocities of a pair of colliding molecules will be statistically uncorrelated after as before they collide, and defined this as a state of 'molecular disorder'. But he also acknowledged, in a brief remark (Boltzmann 1895) that a gas can also pass to a state of molecular order, i.e. a recurrence of a previous ordered state.

[10]It is ironic that Boltzmann has now adopted the viewpoint of another of his critics, Ernst Mach, who in 1894 wrote: "If we could really determine the entropy of the world it would represent a true, absolute measure of time. In this way is best seen the utter tautology of a statement that the entropy of the world increases with time. Time, and the fact that certain changes take place only in a definite sense, are one and the same thing" (see the 1943 reprint, p. 178). The idea was also anticipated by G. J. Stoney (1887) who rediscovered the reversibility paradox and concluded that the Second Law is not a "true dynamical law" but suggested that time itself does not exist apart from events in the universe.

[11]For a comprehensive survey of Boltzmann's scientific and philosophical situation and his relations with contemporaries *circa* 1900 see Blackmore (1995). Several retrospective evaluations of Boltzmann's ideas in the light of 20th century physics have been published in conference proceedings and elsewhere; see for example Sexl & Blackmore (1982), papers by Martin Curd and Alfred Wehrl; Battimelli, Ianniello & Kresten (1993), especially the papers by Carlo Cercignani; Stiller (1988).

In 1911 Boltzmann's theory sustained yet another criticism, too late for him to reply (he committed suicide in 1905). Paul and Tatiana Ehrenfest questioned the validity of what they called the *ergodic hypothesis*, the assumption by Boltzmann (and Maxwell) that a gas eventually passes through all microstates with the same energy. This would be one way to justify the postulate ('equipartition theorem') that all mechanical degrees of freedom have the same energy. Soon afterward two mathematicians, Artur Rosenthal and Michel Plancherel, confirmed the Ehrenfests' suspicion that it is mathematically impossible for the system to go through *every* microstate (i.e., every point in a 6N-1 dimensional phase space); see Brush (1971; 1976, pp. 363-385).

REFERENCES

[The abbreviation *Wien. Ber.* is used for the frequently-cited journal *Sitzungsberichte der kaiserlichen Akademie der Wissenschaften, Wien, Mathematisch-Naturwissenschaftliche Klasse* (Teil II or IIa).]

Battimelli, G.M. G. Ianniello and O. Kresten (eds.): 1993, *Proceedings of the International Symposium on Ludwig Boltzmann (Rome, February 9-11, 1989)*, Verlag der Österreichischen Akademie der Wissenschaften, Wien.

Bernoulli, D.: 1738, *Hydrodynamica, sive de viribus et motibus fluidorum commentarii*, Dulseckeri, Argentorati. Translation/summary of section on kinetic theory in Brush (1965, pp. 57-65).

Blackmore, J. (ed).: 1995, *Ludwig Boltzmann: His Later Life and Philosophy, 1900-1906*, Book One: *A Documentary History*; Book Two: *The Philosopher (Boston Studies in the Philosophy of Science*, **168** and **174**), Kluwer Academic Publishers, Dordrecht/Boston/London.

Böhm, W.: 1973, 'Loschmidt, Johann Joseph (b. Putschirn, near Carlsbad, Bohemia [now Karlovy Vary, Czechoslovakia], 15 March 1821; d. Vienna, Austria, 8 July 1895), *physics, chemistry*', in C. C. Gillispie (ed.), *Dictionary of Scientific Biography*, Charles Scribner's Sons, New York, vol. 8, pp. 507-511.

Boltzmann, L.: 1868, 'Studien über das Gleichgewicht der lebendiges Kraft zwischen bewegten materiellen Punkten', *Wien, Ber.* **58**, 517-560.

Boltzmann, L.: 1872, 'Weitere Studien über das Wärmegleichgewicht unter Gasmolekülen', *Wien Ber.* **66**, 275-370. English translation in Brush (1965, pp. 88-175).

Boltzmann, L.: 1876, 'Über die Aufstellung und Integration der Gleichungen, welche die Molecularbewegung in Gasen bestimmen', *Wien. Ber.* **74**, 503-552.

Boltzmann, L.: 1877a, 'Bemerkungen über einige Probleme der mechanischen Wärmetheorie', *Wien. Ber.* **75**, 62-100. English translation of the second section (his reply to Loschmidt) in Brush (1966, pp. 189-193).

Boltzmann, L.: 1877b, 'Über die Beziehung zwischen des zweiten Hauptsatze der mechanischen Wärmetheorie und der Wahrscheinlichkeitsrechnung, respectiv den

Satzen über das Wärmegleichgewicht', *Wien. Ber.* **76**, 373-435.

Boltzmann, L.: 1878, 'Weitere Bemerkungen über einige Probleme der mechanischen Wärmetheorie', *Wien. Ber.* **78**, 7-46.

Boltzmann, L.: 1886, 'Der zweite Hauptsatze der mechanische Wärmetheorie', *Almanach der kaiserlichen Akademie der Wissenschaften in Wien* **36**, 225-259. Translation in McGuinness (1974, pp. 13-32).

Boltzmann, L.: 1895, 'On certain Questions of the Theory of Gases', *Nature* **51**, 413-415.

Boltzmann, L.: 1896, 'Entgegnung auf die wärmetheoretischen Betrachtungen des Hrn. E. Zermelo', *Annalen der Physik* [3] **57**, 773-784. English translation in Brush (1966, pp. 218-228).

Boltzmann, L.: 1896-1898, *Vorlesungen über Gastheorie*, 2 vols., Barth, Leipzig. English translation with introduction and notes by S. G. Brush, *Lectures on Gas Theory*, University of California Press, Berkeley, 1964.

Boltzmann, L.: 1897, 'Zu Hrn. Zermelo's Abhandlung Über die mechanische Erklärung irreversibler Vorgänge', *Annalen der Physik* [3] **60**, 392-398. English translation in Brush (1966, pp. 238-245).

Boltzmann, L.: 1898, 'Über die sogennante H-Curve', *Mathematische Annalen* **50**, 325-332.

Boltzmann, L.: 1905, *Populäre Schriften*, J. A. Barth, Leipzig. New edition, selected and edited by E. Broda, Vieweg, Wiesbaden, 1979.

Boltzmann, L.: 1979, 'Zur Erinnerung an Josef Loschmidt', in E. Broda (ed.), Boltzmann's *Populäre Schriften*, Vieweg, Wiesbaden, Germany.

Boltzmann, L. and J. Nabl, 1905: 'Kinetische Theorie der Materie', *Encycklopädie der mathematischen Wissenschaften*, vol. 5, part 1, p. 494-558.

Boltzmann Gesamtausgabe, see Sexl (1981-)

Boyle, R.: 1660, *New Experiments Physico-Mechanical, Touching the Spring of the Air, and its Effects; made, for the most part in a new Pneumatical Engine; Written by Way of Letter to the Right Honorable Charles Lord Viscount of Dungarvan, eldest Son to the Earl of Corke*. Oxford. Reprinted in *The Works of the Honourable Robert Boyle*, London, 1772. Extensive extracts are quoted in Conant (1950).

Boyle, R.: 1662, *A Defence of the Doctrine touching the Spring and Weight of the Air, proposed by Mr. R. Boyle in his New Physico-Mechanical Experiments; Against the Objections of Franciscus Linus, wherewith the Objector's Funicular Hypothesis is also Examined*. Oxford.

Broda, E.: 1955, *Ludwig Boltzmann: Mensch, Physiker, Philosoph*. Deutscher Verlag der Wissenschaften, Berlin. Translation by Larry Gay and the author, *Ludwig Boltzmann: Man--Physicist--Philosopher*, Ox Bow Press, Woodbridge, CT (1983).

Brush, S. G.: 1965, *Kinetic Theory*, Vol. 1, *The Nature of Gases and of Heat*, Pergamon Press, Oxford.

Brush, S. G.: 1966, *Kinetic Theory*, Vol. 2, *Irreversible Processes*, Pergamon Press,

Oxford.
Brush, S. G.: 1971, 'Proof of the Impossibility of Ergodic Systems: The 1913 Papers of Rosenthal and Plancherel', *Transport Theory and Statistical Physics* **1**, 287-311.
Brush, S. G.: 1976, *The Kind of Motion We Call Heat: A History of the Kinetic Theory of Gases in the 19th Century*, North-Holland
Burbury, S. H.: 1894, 'Boltzmann's Minimum Function', *Nature* **51**, 78, 320.
Burstyn, H. L.: 1970, 'Buys Ballot, Christoph Hendrik Diederik (b. Kloetinge, Netherlands, 10 October 1817; d. Utrecht, Netherlands, 3 February 1890), meteorology, physical chemistry', in C. C. Gillispie (ed.), *Dictionary of Scientific Biography*, Vol. 2, Scribner's Sons, New York, pp. 628.
Buys Ballot, C. H. D.: 1858, 'Ueber die Art von Bewegung, welche wir Wärme und Electricitat nennen', *Annalen der Physik* [2] **103**, 240-259.
Clausius, R., 1857: 'Ueber die Art der Bewegung, welche wir Wärme nennen', *Annalen der Physik* [2] **100**, 353-380. English translation in Brush (1965, pp. 111-134).
Clausius, R.: 1858, 'Ueber die mittlere Länge der Wege, welche bei Molecularbewegung gasförmigen Körper von den einzelnen Molecülen zurücklegen werden, nebst anderen Bemerkungen über der mechanischen Wärmetheorie', *Annalen der Physik* [2] **105**, 239-258. English translation in Brush (1965, pp. 135-147).
Conant, J. B: 1950, 'Robert Boyle's Experiments in Pneumatics', in J. B. Conant (ed.), *Harvard Case Histories in Experimental Science*, Harvard University Press, Cambridge, Mass., pp. 1-63.
Culverwell, E. P.: 1890a, 'Note on Boltzmann's Kinetic Theory of Gases, and on Sir W. Thomson's Address to Section A, British Association, 1884', *Philosophical Magazine* [5] **30**, 95-99.
Culverwell, E. P.: 1890b, 'Possibility of Irreversible Molecular Motions', *Report of the 60th Meeting of the British Association for the Advancement of Science*, 744.
Culverwell, E. P.: 1894, 'Dr. Watson's Proof of Boltzmann's Theorem on Permanence of Distributions', *Nature* **50**, 617.
Culverwell, E. P.: 1895a, 'Boltzmann's Minimum Theorem', *Nature* **51**, 246.
Culverwell, E. P.: 1895b, 'Professor Boltzmann's Letter on the Kinetic Theory of Gases', *Nature* **51**, 581.
Daub, E. E.: 1970, 'Maxwell's Demon', *Studies in History and Philosophy of Science* **1**, 213-227.
Debus, A. G.: 1998, 'Chemists, Physicians, and Changing Perspectives on the Scientific Revolution', *Isis* **89**, 66-81.
Dugas, R.: 1959, *La Théorie Physique au Sens de Boltzmann*, Griffon, Neuchatel.
Ehrenfest, P., and T. Ehrenfest: 1911, 'Begriffliche Grundlagen der statistischen Auffassung in der Mechanik', *Encyklopädie der mathematische Wissenschaften*, vol. 4, part 32. English translation by M. J. Moravcsik, *The Conceptual Foundations of the Statistical Approach in Mechanics*, Cornell University Press, Ithaca, NY, 1959.
Everitt, C. W. F.: 1975, *James Clerk Maxwell: Physicist and Natural Philosopher*,

Scribner, New York.
Floderer, A.: 1971, 'Avogadro'sche oder Loschmidt'sche Zahl?', *Bohemia: Jahrbuch des Collegium Carolinum* **12**, 377-384.
Garber, E., S. G. Brush and C. W. F. Everitt: 1986, *Maxwell on Molecules and Gases*, The MIT Press, Cambridge, MA.
Garber, E., S. G. Brush and C. W. F. Everitt: 1995, *Maxwell on Heat and Statistical Mechanics*, Lehigh University Press, Bethlehem, PA.
Gibbs, J. W.: 1902, *Elementary Principles in Statistical Mechanics*, Scribner, New York.
Guthrie, F.: 1873, 'Kinetic Theory of Gases', *Nature* **8**, 67, reprinted in Garber, Brush & Everitt (1995, pp. 120-121).
Guthrie, F.: 1874, 'Molecular Motion', *Nature* **10**, 123; reprinted in Garber, Brush & Everitt (1995, pp.143-44).
Hann, J.: 1896, Obituary of Josef Loschmidt, *Almanach der kaiserlichen Akademie der Wissenschaften, Wien* **46**, 258-262.
Herapath, J.: 1821, 'A Mathematical Inquiry into the Causes, laws and Principal Phenomenae of Heat, Gases, Gravitation, etc.', *Annals of Philosophy* [2] **1**, 273-293, 340-351, 401-406. Reprinted in 1972 with Herapath (1847).
Herapath, J.: 1847, *Mathematical Physics*. Whittaker, London. Reprinted with Selected Papers by John Herapath, edited by S. G. Brush, Johnson Reprint Corp., New York, 1972.
Kac, M.: 1959, *Probability and related Topics in Physical Sciences*, Interscience, New York.
Klein, M. J.: 1970, *Paul Ehrenfest*, vol. 1, North-Holland, Amsterdam.
Klein, M. J.: 1973, 'The Maxwell-Boltzmann Relationship', in J. Kestin (ed.), *Transport Phenomena-1973*, American Institute of Physics (Conference Proceedings, No. 11), New York, pp. 297-308.
Kohler, W.: 1931, 'Zur Boltzmannschen Theorie des zweiten Hauptsatzes', *Erkenntnis* **2**, 336-353.
Linus, F.: 1661, *Tractatus de Corporam Inseparabilitate, in quo Experimenta de Vacuo tam Torricelliana quam Magdeburgica et Boyliana examinatur, veraque eorum causa detecta, ostenditur, vacuum naturaliter dari non posse: unde et Aristotelica de Rarefactione sententia tam contra Assertores Vacuitatum, quam Corpusculorum demonstratur*, Martin, London.
Loschmidt, J.: 1865, 'Zur Grösse der Luftmolecüle' *Wien. Ber.* **52**, 395-413.
Loschmidt, J.: 1876, 'Über den Zustand des Wärmegleichgewichtes eines Systemes von Körpern mit Rucksicht auf die Schwerkraft', *Wien. Ber.* **73**, 128-142, 366-372
Loschmidt, J.: 1877, 'Über den Zustand des Wärmegleichgewichtes eines Systemes von Körpern mit Rucksicht auf die Schwerkraft', *Wien. Ber.* **75**, 287-298; **76**, 209-225.
Loschmidt, J.: 1878, 'Cosmical Results of the Modern Heat Theory', *Nature* **18**, 184-185.
Mach, E., 1894: 'On the Principle of Conservation of Energy', *The Monist* **5**, 22-54. Reprinted in *Popular Scientific Lectures*, Open Court, LaSalle, Ill., 5th ed., 1943,

pp. 137-185..
Maxwell, J. C.: 1860, 'Illustrations of the Dynamical Theory of Gases', *Philosophical Magazine*, series 4, **19**, 19-32; **20**, 21-37; reprinted (with additional manuscript material and correspondence) in Garber, Brush & Everitt (1986, pp. 277-336).
Maxwell, J. C.: 1866, 'On the Dynamical Theory of Gases', *Philosophical Transactions of the Royal Society of London* **157**, 49-88. Reprinted in Brush (1966, pp. 23-87) and (with additional manuscript material and correspondence) in Garber, Brush & Everitt (1986, pp. 387-472)
Maxwell, J. C.: 1873a, 'Clerk Maxwell's Kinetic Theory of Gases', *Nature* **8**, 84. Reprinted in Garber, Everitt & Brush (1995, pp. 122-3).
Maxwell, J. C.: 1873b, 'On the Equilibrium of Temperature of a Gaseous Column subjected to Gravity', *Nature* **8**, 527-528, reprinted in Garber, Brush & Everitt (1995, 125-26).
Maxwell, J. C.: 1873c, 'On the Final State of a System of Molecules in Motion Subject to Forces of any Kind', *Report of the 42nd Meeting of the British Association for the Advancement of Science*, Transactions of Section A, 29-32; reprinted in Garber, Brush & Everitt (1995, 138-43); shorter version in *Nature* **8** (1873), 537-538.
Maxwell, J. C.: 1874, Reply to Guthrie, 'Molecular Motion', *Nature* **10**, 124, reprinted in Garber, Brush & Everitt (1995, 144-45).
Mayer, J. E.: 1964, 'The Death of an Ogre', in H. Craig, S. L. Miller and G. J. Wasserburg (eds.), *Isotopic and Cosmic Chemistry*, North-Holland, Amsterdam, pp. 8-15.
McGuinness, B. (ed.): 1974, *Ludwig Boltzmann: Theoretical Physics and Philosophical Problems*, Reidel, Boston.
Mehra, J.: 1975, 'Einstein and the Foundation of Statistical Mechanics', *Physica* **79A**, 447-477.-
Newton, I.: 1687: *Philosophia Naturalis Principia Mathematica*. London. An extract of the section on the repulsive model for gas pressure (Motte-Cajori translation of the third edition, 1726) is reprinted in Brush (1965, pp. 54-56).
Peckhaus, V.: 1990, ''Ich habe mich wohl gehutet, alle Patronen auf einmal zu verschiessen': Ernst Zermelo in Göttingen', *History and Philosophy of Logic* **11**, 19-58.
Poincaré, H.: 1890, 'Sur le Problème de trois Corps et les Équations de Dynamique', *Acta Mathematica* **13**, 1-270. An excerpt on the recurrence theorem is translated in Brush (1966, pp. 194-202).
Poincaré, H.: 1893, 'Le Mécanisme et l'Expérince', *Revue de Metaphysique et de Morale* **1**, 534-537. English translation in Brush (1966, pp. 203-207).
Porterfield, W. W. and W. Kruse: 1995, 'Loschmidt and the Discovery of the Small', *Journal of Chemical Education* **72**, 870-875.
Reichenbach, H.: 1956, *The Direction of Time*, University of California Press, Berkeley.
Reilly, C.: 1962, 'Francis Line, Peripatetic (1595-1675)', *Osiris* **15**, 222-253.
Reilly, C.: 1969, *Francis Line, S. J.: An Exiled English Scientist, 1595-1675*, Institutum Historicum S. J., Rome.

Rootselaar, B. Van: 1976, 'Zermelo, Ernst Friedrich Ferdinand (b. Berlin, Germany, 27 July 1871; d. Freiburg im Breisgau, Germany, 21 May 1953), *mathematics*', in C. C. Gillispie (ed.), *Dictionary of Scientific Biography*, vol. 14, Scribner's Sons, New York, pp. 613-616.

Scott, W. L.: 1970, *The Conflict between Atomism and Conservation Theory 1644-1860*, American Elsevier, New York.

Sexl, R. U. (ed.): 1981-, *Ludwig Boltzmann Gesamtausgabe*, Akademische Druck-u. Verlagsanstalt, Graz and Friedr. Vieweg & Sohn, Braunschweig/Wiesbaden.

Sexl, R. and J. Blackmore (eds.): 1982, *Ludwig Boltzmann, Internationale Tagung anlässlich des 75. Jahrestages seines Todes, 5.-8. September 1981 (Ludwig Boltzmann Gesammtausgabe, 8)*, Akademische Druck-u. Verlagsanstalt, Graz and Friedr. Vieweg & Sohn, Braunschweig/Wiesbaden.

Shapin, S. and S. Schaffer: 1985, *Leviathan and the Air Pump: Hobbes, Boyle, and the Experimental Life*, Princeton University Press, Princeton, NJ.

Snelders, H. A. M.: 1968, 'Het Materiebegrip bij Buys Ballot', *Scientiarum Historia* **10**, 154-172.

Stiller, W.: 1988, *Ludwig Boltzmann: Altmeister der klassischen Physik, Wegbereiter der Quantenphysik und Evolutionstheorie*, J. A. Barth, Leipzig.

Stoney, G. J.: 1887, 'Curious Consequences of a well-known Dynamical Theorem', *Philosophical Magazine* [5] **23**, 544-547.

Thomson, W.: 1874, 'The Kinetic Theory of the Dissipation of Energy', *Proceedings of the Royal Society of Edinburgh* **8**, 325-334. Reprinted in Brush (1966, pp. 176-187.

Tolman, R. C.: 1938, *The Principles of Statistical Mechanics*, Oxford University Press, London.

Webster, C.: 1965, 'The Discovery of Boyle's Law, and the Concept of the Elasticity of Air in the Seventeenth Century', *Archive for History of Exact Sciences* **2**, 441-502.

Zermelo, E.: 1896a, 'Ueber ein Satz der Dynamik und die mechanische Wärmetheorie', *Annalen der Physik* [2] **57**, 485-494. English translation in Brush (1966, pp. 208-217).

Zermelo, E.: 1896b, 'Ueber mechanische Erklärungen irreversibler Vorgänge', *Annalen der Physik* [2] **59**, 793-801. English translation in Brush (1966, pp. 229-237).

2.

Interatomic Forces and Gas Theory from Newton to Lennard-Jones

STEPHEN G. BRUSH

Abstract

A recurrent theme in the physical science of the past three centuries has been provided by the program attributed to ISAAC NEWTON: from the phenomena of nature to find the forces between particles of matter, and from these forces to explain and predict other phenomena. The success or failure of this program as a guide for scientific research can be assessed by considering some of the cases in which it has been applied: NEWTON's own theory of gas pressure, the BOSCOVICH theory of interatomic forces, the LAPLACE theory (short-range attractive forces and long-range repulsive forces), the billiard-ball model used in the elementary kinetic theory of gases, the MAXWELL r^{-5} repulsive force, and the VAN DER WAALS equation. A more detailed examination is presented of the rise and fall of the "Lennard-Jones potential" in relation to calculations and experimental data on virial coefficients and transport properties of gases, solid state properties, and the quantum theory of interatomic forces.

The history of the subject suggests that the hypothetico-deductive model of scientific method has not been followed in practice, since the reasons for adopting or rejecting new interatomic force laws are often not simply related to the success or failure of the force law in calculations of gas properties.

At present there is serious doubt as to whether it is worthwhile trying to establish a single "realistic" force law for the interaction between two atoms or molecules. It may be more fruitful to abandon this program and to choose force laws instead on the basis of their convenience in a particular mathematical theory of the properties of matter.

Can the law of force between atoms be determined by theoretical analysis of the macroscopic properties of matter? Should theories of the macroscopic properties of matter be based on atomic models using the best available force laws? These questions are still of considerable importance to many scientists today, while others may be surprised to learn that negative answers to them are being seriously contemplated. I propose to discuss them here from an historical viewpoint, with special emphasis on the interactions between certain selected atomic models and the properties of gases. However, this paper is intended primarily as a contribution to a current controversy rather than to the history of science; hence I will focus on the common features of theories published by a few major scientists during the past three centuries, ignoring both the special historical contexts of those theories and the numerous works of other scientists, the study of which might provide a better sense of historical evolution and continuity.

The basic problem is surely among the oldest in the history of science: we see around us materials having various different properties — solid, liquid, gaseous, hot, cold, slippery, sticky, hard, elastic, *etc.* We would like to explain these properties in terms of the properties of "atoms" — invisible particles of which we conceive the materials to be composed. Why? Mainly to satisfy our intellectual curiosity, although we might claim that once we can establish a valid atomic theory, we will be able to predict new properties of natural or synthetic materials, and thus extend our control over the physical world.

But these ends might also be accomplished through other kinds of theories: why an atomic one? Before the present century there was really no solid evidence for the atomic structure of matter, and as soon as such evidence did emerge, it came in such a way as to refute at the same time the old concept of the atom as an *indivisible* building block. (Since about 1910, the existence of "atoms" that can break up by natural or artificial means has been generally accepted.) Nevertheless there has obviously been a tremendously strong attraction for most scientists in the idea of a *corpuscular* theory, if not a strictly atomic one.

If we choose to limit ourselves to the quantitative theories proposed in modern times, a convenient starting point is the program advocated by ISAAC NEWTON in the *Principia*:

> We ... treat mainly those [powers] that pertain to heaviness, lightness, elastic force, the resistance of fluids, and attractive or repulsive forces of that kind: And it is because of them that we propose these our own (so to speak) mathematical principles of philosophy. Indeed, all the difficulty of Philosophy seems to lie in this, that from the phenomena of motions we should investigate the forces of nature, and thereupon from these forces we should demonstrate the rest of the phenomena. And to this point pertain the general propositions which we have treated at length in the first and second books. In the third book, moreover, we have proposed an example of this business in the unfolding of the System of the World. There, indeed, from the heavenly phenomena, through propositions demonstrated mathematically in the first two books, are derived the forces of gravity by means of which bodies strive toward the sun and the several planets. Thereupon, from these forces through propositions likewise mathematical are deduced the motions of the planets, comets, moon, and ocean. Would it were possible to derive the rest of the phenomena of nature from mechanical principles by the same kind of reasoning! For many things move me to suspect somewhat that they all may depend upon certain forces, by which the little parts of bodies through causes not yet known are either driven against one another and crowd together according to regular figures, or are driven away from each other and draw apart...[1]

---

[1] ISAAC NEWTON, *Philosophiae Naturalis Principia Mathematica* (London, 1687, same in the 3rd ed. 1726), passage translated by C. TRUESDELL. The MOTTE-CAJORI translation of this passage is somewhat misleading, especially in the last sentence where the phrase "secundum figuras regulares" is translated "in regular figures." [See *Sir Isaac Newton's Mathematical Principles of Natural Philosophy and his System of the World* (Berkeley, 1934), p. xvii–xviii, reprinted in S. G. BRUSH *Kinetic Theory*, vol. 1 (Oxford, 1965) p. 52–53.] According to TRUESDELL, NEWTON'S use of the word "figura" is deliberately vague; there is no correct English translation, since NEWTON might have meant either "geometrical figures" or "according to regular qualities" or "according to qualities specified by rules." The last of these employs the only possible sense of "regularis" in classical Latin. (Letter from C. TRUESDELL to S. BRUSH, January 30, 1970; see also footnote 7 of TRUESDELL'S paper cited in our footnote 2.)

For the historian it should be important to discern what NEWTON himself meant by these words. Thus, one should note that he does not refer specifically here to *atoms*, and indeed very little of the solid substance of NEWTON's work has any relation to atomic theory. Moreover, as TRUESDELL has pointed out, there is nothing here to imply that the forces must be central or even pairwise, or that optical and electromagnetic properties are excluded by the qualification "mechanical". In NEWTON's time, the English word "mechanical" was used much more generally to mean "working like a machine" or "practical" as opposed to something involving thought, spirit, or speculation.[2] What NEWTON does emphasize in the preface where this program is proposed is that rigorous mathematical methods should be employed in working out the consequences of any assumption about forces, whatever that assumption may be. While NEWTON did indeed write out some speculations about the relations between interatomic forces and the chemical and physical properties of matter, which may be found in the famous Queries in his *Opticks*[3] and in an unpublished "conclusion" of the *Principia*,[4] he confined himself to mathematical demonstrations in the main text of the *Principia*.

But the historian must also take account of the meaning that was read into NEWTON's words by other scientists who lacked the ability or desire to master the contents of his book in detail. They assumed he was simply restating the "mechanical philosophy" of RENÉ DESCARTES and ROBERT BOYLE.[5] According to this philosophy, the purpose of theoretical physics is to construct explanations of phenomena in terms of atoms, assuming that the atom can have only "primary" mechanical properties such as mass, size, shape, and motion. The large amount of 18th-century research and writing inspired by this supposedly Newtonian program has recently been surveyed by ROBERT SCHOFIELD,[6] and RUSSELL MCCORMMACH has argued that it provided a unifying theme for the work of HENRY CAVENDISH.[7]

Now it has been a hotly debated question whether "force" can be a legitimate mechanical property of an atom. According to the Cartesian tradition, the only admissible atomic force is a contact repulsion that prevents two atoms from

---

[2] C. TRUESDELL, Texas Quarterly, **10**, 238 (1967), reprinted in his *Essays in the History of Mechanics* (Springer-Verlag New York, 1968); see esp. p. 177–182.

[3] ISAAC NEWTON, *Opticks* (London, 4th ed. 1730, reprinted by Dover, New York 1952), p. 375–402.

[4] See A. RUPERT HALL & MARIE BOAS HALL, *Unpublished Scientific Papers of Isaac Newton* (Cambridge University Press, 1962), p. 334–344; MARIE BOAS HALL (ed.), *Nature and Nature's Laws* (New York, 1970), p. 323–333.

[5] See for example the article by MARIE BOAS, Osiris **10**, 412 (1952), or the more extended earlier treatment by E. A. BURTT, *The Metaphysical Foundations of Modern Physical Science* (New York, 1952, reprint of the second edition of 1932). Historians still disagree as to how much NEWTON himself accepted this philosophy; at least he made it clear in the LEIBNIZ-CLARKE debate that he rejected the clockwork-universe concept of BOYLE and LEIBNIZ. See H. G. ALEXANDER (ed.), *The Leibniz-Clarke correspondence with extracts from Newton's Principia and Opticks* (New York, 1956). A comprehensive analysis of the relations between NEWTON's and DESCARTES' metaphysical positions may be found in ALEXANDRE KOYRÉ, *Newtonian Studies* (Cambridge, Mass., 1965).

[6] ROBERT E. SCHOFIELD, *Mechanism and materialism: British Natural Philosophy in an age of Reason* (Princeton, 1970).

[7] RUSSELL MCCORMMACH, Isis **60**, 293 (1969).

occupying the same space; any kind of long-range attraction or repulsion is inconceivable as an inherent property of atoms, and must ultimately be explained in terms of contact actions (perhaps propagated through an "ether").[8] Whether you are willing to agree with this as a philosophical position or not, you have to recognize that it has been quite congenial to many scientists throughout history; it helps to account for the popularity of such speculations as the kinetic theory of gravity[9] and the vortex atom.[10]

At the other extreme is the theory of ROGER BOSCOVICH (1711–1787), which eliminates mass, size, and shape as inherent properties of atoms and replaces them by a rather complicated force law for the interaction of point atoms.[11] BOSCOVICH's ideas had some influence on later scientific theories, and a few of the more metaphysical thinkers of the early 19th century were quite content to reduce matter entirely to force, in some cases even eliminating the atom altogether.[12]

From a metaphysical viewpoint it might appear that the Cartesian and Boscovichean theories are mutually exclusive: the first denies the existence of force (except in a trivial sense as a label for that which prevents pieces of matter from interpenetrating), while the second denies the existence of matter (except in a trivial sense as the locus of a strongly repulsive force). But most successful scientists pay little attention to such metaphysical issues, and so we must take seriously a model that seems to be a compromise between the two extremes: an atom with finite size and mass and a definite shape (usually spherical) which can also exert both long and short-range attractive and/or repulsive forces on other atoms. Given such a model, how can we use it to explain the properties of matter, and what do we gain by doing so?

Let us sketch a procedure that corresponds to a common conception of scientific method, as it ought to be applied in this case. Certain features of this

---

[8] For general surveys see MAX JAMMER, *Concepts of Force* (Cambridge, Mass., 1957); MARY B. HESSE, *Forces and Fields* (London, 1961); A. KOYRÉ, *Newtonian Studies* (Cambridge, Mass., 1965).

[9] This is the theory that gravity results from the bombardment of matter by streams of invisible particles; adjacent pieces of matter "shield" each other so the net effect is the same as an attractive force between them. The theory was popularized by G. L. LE SAGE in Geneva at the end of the 18th century; a large number of periodical articles published during the 19th century are listed in the *Royal Society Catalogue of Scientific Literature* (see index).

[10] The theory that atoms are stable vortex motions in a continuous fluid was proposed by WILLIAM THOMSON (Lord KELVIN) and was popular among British physicists in the last quarter of the 19th century. See R. SILLIMAN, Isis 54, 173 (1963).

[11] R. J. BOSCOVICH, *Philosophiae Naturalis Theoria redacta ad unicam legem virium in natura existentium* (Vienna, 1758); English translation, *A Theory of Natural Philosophy* (Chicago, 1922, reprinted by MIT Press, Cambridge, 1966).

[12] Such ideas may be found in the writings of the German *Naturphilosophen*, SCHELLING, GOETHE, *etc.*, and rather explicitly in a paper by I. H. FICHTE, Z. Philos. & Philos. Kritik 24, 24 (1854). I have discussed the relation between this tradition and the later attacks on atomism by MACH and others in my paper in Synthese 18, 192 (1968). The case for BOSCOVICH's influence on later scientists is presented in a book of essays edited by L. L. WHYTE, *Roger Joseph Boscovich* (New York: Fordham University Press 1961). The specific influence of BOSCOVICH on FARADAY, claimed by L. PEARCE WILLIAMS in his biography *Michael Faraday* (New York: Basic Books, 1965) has recently been challenged by J. BROOKES SPENCER, Arch. Hist. Exact Sci. 4, 184 (1967).

procedure have been popularized in recent years by the philosopher of science KARL POPPER,[13] but the basic ideas are quite old.

We could expect to start by knowing property P_1 of a substance; we then guess a plausible atomic model A_1, and by carrying out mathematical operation M_1 try to calculate P_1. If the theoretical and experimental values of P_1 agree sufficiently well — there is unfortunately no objective or generally accepted criterion for "sufficiently well," as KUHN has pointed out[13a] — we declare that a *prima facie* case has been made for the model A_1. We then use another operation M_2 on the same A_1 to calculate another property P_2. If the agreement is still satisfactory we continue the process, and our confidence in the validity of the model increases, though if we are good Popperians, we keep telling ourselves all the while that we have not *proved* that the model represents reality, and indeed can never hope to do so; it is only a conjecture that has so far survived all attempts at refutation.[13] As soon as one property P_i calculated from the model fails to agree with experiment we must at once reject A_1, pick another model A_2 and start all over again.

To apply this procedure to gases, we recall that the first quantitative properties of air to be studied in the 17th century were its pressure (or weight) and its volume. Following the earlier work of TORRICELLI, PASCAL, and GUERICKE, ROBERT BOYLE (1627–1691) provided us with the simple relation: pressure X volume = constant. (I am here speaking of BOYLE's significance, retrospectively, as the modern scientist sees it; at the time, the *quantitative* "Boyle's law", suggested by HENRY POWER, RICHARD TOWNELEY, and others, was less important than the *qualitative* fact that the suction phenomena previously attributed to nature's abhorrence of a vacuum could be explained mechanically in terms of air pressure.[14])

BOYLE's law, which I will call property P_1, was published in 1662. Twenty-five years later ISAAC NEWTON proposed a simple atomic model, A_1: gases are composed of particles that repel their neighbors with a force inversely proportional to the distance. NEWTON does not mention here any motion of the atoms or any thermal property of matter; his analysis presumes tacitly that only adjacent atoms exert any force upon each other. He states and proves the following theorem (*Principia*, Book II, Sect. V, Prop. XXIII):

> If a fluid be composed of particles fleeing from each other, and the density be as the compression, the centrifugal forces of the particles will be inversely proportional to the distances of their centres. And, conversely, particles fleeing from each other, with forces that are inversely proportional to the distances of their centres, compose an elastic fluid, whose density is as the compression.

From the viewpoint of the hypothetico-deductive procedure outlined above, NEWTON has only accomplished the mathematical demonstration that $A_1 \rightarrow P_1$; the converse implication $P_1 \rightarrow A_1$ is valid if and only if one accepts the hypothesis that gases *are* composed of repelling atoms, and this had not yet been

---

[13] KARL R. POPPER, *Logik der Forschung* (Vienna, 1934), English translation *The Logic of Scientific Discovery* (London, 1959); *Conjectures and Refutations* (New York & London, 1962).

[13a] T. S. KUHN, Isis **52**, 161 (1961).

[14] See S. G. BRUSH, *Kinetic Theory*, vol. 1 (Oxford: Pergamon Press 1965), p. 3–4 and 43–51 (reprint of part of BOYLE's 1660 work); C. WEBSTER, Arch. Hist. Exact Sci. **2**, 441 (1965).

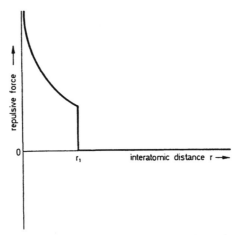

Fig. 1. Model A_1 (NEWTON, 1687). Repulsive force $F(r) \propto 1/r$ for $r \leq r_1$; $F(r) = 0$ for $r > r_1$, where r_1 is the distance of the nearest neighbor.

established. In other words, if we introduce the more general hypothesis A_1^*: "a fluid is composed of particles fleeing from (*i.e.*, repelling) each other" with the understanding that the repulsive forces act only on nearest neighbors and that no account is to be taken of the effects of atomic motion, then the first part of NEWTON's theorem reads: $(P_1 + A_1^*) \to A_1$, where A_1 is a special case of A_1^* in which the forces are inversely as the distance.

But NEWTON, unlike some of his followers, was careful to point out that A_1^* should not be taken for granted:

> But whether elastic fluids truly consist in little parts fleeing each other mutually, is a physical question. We have demonstrated mathematically a property of fluids consisting in little parts of this kind, so as to furnish philosophers a handle for taking up that question.[15]

Nor did NEWTON consistently adopt the model A_1 in his other writings; later, in his *Opticks*, he leaned toward the hypothesis that atoms are hard and impenetrable but have short-range *attractive* forces diminishing with distance.[3]

The model A_1 acquired considerable historical importance as the first quantitative interatomic force law to be inferred from the properties of gases, in spite of the fact that NEWTON himself did not take it very seriously. Judged (unfairly, of course) by modern standards, it must be regarded as a complete failure. Not only is it wrong in a quantitative sense; it is implausible since it implies that the force between two atoms depends on whether a third is in between them or they are neighbors. (If one does not make that stipulation, the theory no longer gives the correct P_1; the pressure-volume relation depends on the size and shape of the container, as NEWTON himself pointed out, if it is assumed that $1/r$ forces act between all pairs of atoms.) NEWTON could justify this feature of his model

---

[15] ISAAC NEWTON, *Principia*, Book II, Prop. XXIII, Theorem XVIII, reprinted in S. G. BRUSH, *Kinetic Theory*, vol. 1 (Oxford, 1965) p. 52–56 (from the MOTTE-CAJORI translation); the last passage has been retranslated by C. TRUESDELL.

only by suggesting an analogy with magnetic forces that can be reduced by interposing an iron plate between two magnetic bodies.

The reason for the failure of NEWTON's model is that the motion of the atoms has been ignored; we now realize that the pressure of a gas at ordinary density is almost entirely due to impacts of atoms against the walls rather than to continuously acting interatomic repulsive forces. But the amazing historical fact is that it took almost two centuries for the majority of physicists to accept that conclusion. With a few significant exceptions such as LEONHARD EULER, DANIEL BERNOULLI, and JOHN HERAPATH, most of them were so subservient to NEWTON's authority that they adhered firmly to what NEWTON had suggested only as a tentative hypothesis. One might say they rejected his hypothetico-deductive *method* while adopting what appeared to be its *conclusion*.

Of course there is more to it than that: during the 18th century, the idea that *heat* is associated with the interatomic repulsion force was added to NEWTON's theory, and accounted for much of its popularity. Yet it is rather disturbing to find not only most of the textbooks but also some first-rate scientists such as JOHN DALTON and THOMAS YOUNG stating at the beginning of the 19th century that the repulsive force law for gas particles suggested by NEWTON has been definitely proved beyond any possible doubt.[16]

The first attempt to relate interatomic forces to gas properties was a disaster. What about the second? This was the 1758 theory of BOSCOVICH, already mentioned. Atoms are point centers of force, but since they have no other properties except motion, the force law must do a lot more explanatory work. It must change from repulsive to attractive and back again several times, in order to account for the following properties:

(properties)	(features of model)
P_2, impenetrability of particles	repulsive force increases without limit as $r \to 0$ ($r =$ distance between atoms)
P_3, solid has an equilibrium volume at low pressures	force $=0$ at distance r_1; force is repulsive for slightly smaller distances and attractive for slightly larger, so this is a point of stable equilibrium.
P_4, solid changes to liquid which has larger equilibrium volume	force $=0$ at r_3 (note that r_2 is point of *unstable* equilibrium)
P_5, at large distance any two pieces of matter have inverse square gravitational attraction	force $\propto 1/r^2$ as $r \to \infty$

There are many further ramifications of BOSCOVICH's theory which I will not go into. Instead there are three things to be said about it: (1) the basic idea was attractive to some scientists, including major ones like Lord KELVIN;[17] (2) from

---

[16] J. DALTON, Mem. Manchester Lit. Phil. Soc. **5**, 540 (1802); THOMAS YOUNG, article "Cohesion" written in 1816 for the *Britannica*, reprinted in *Miscellaneous Works of the Late Thomas Young* (London, 1855), vol. I, p. 454.

[17] KELVIN, Brit. Assoc. Report for 1889, p. 494, and other references given on p. 225 of WHYTE's book (*op. cit.*); R. OLSON, Isis **60**, 91 (1969); see also recent issues of the *Isis* Critical Bibliography, entries indexed under BOSCOVICH.

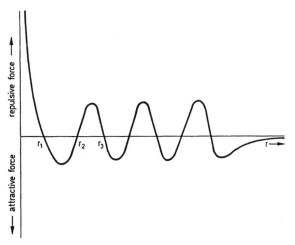

Fig. 2. Model A_2 (Boscovich, 1758). Repulsive force for $r < r_1$; attractive force between r_1 and r_2; repulsive again between r_2 and r_3, etc.; attractive force $\propto 1/r^2$ for large r.

a modern viewpoint it is definitely wrong as a *fundamental* postulate — we have now rejected the idea of explaining all these properties of matter simply by attributing a complicated system of forces to *point* atoms. Instead we prefer to start with a much simpler force law, say an electrostatic force between subatomic particles, and with the aid of more complicated assumptions about atomic structure and motions we may eventually *derive* an "effective" interatomic force law that has some resemblance to Boscovich's curve. (3) No quantitative calculations or predictions (with the possible exception of those of Kelvin) could be based on it during the period when it was entertained as a possible model of matter. The process of testing it against experiment never really got off the ground; one either accepted or rejected it for metaphysical reasons. So by the strict criteria of modern scientific methodology this was disaster number 2. (Whether those criteria are really applicable to the behavior of scientists is another question.)

On the other hand we must at least credit Boscovich with extending the range of material properties that an atomic theory should try to explain: he has proposed the rather interesting idea that the same model might account for the existence of a substance in more than one physical state. If you look at some of the other speculations published in the 18th and early 19th centuries you will realize that this is not as trivial as it seems today. Many scientists seemed to think that the nature of the atom itself would have to be different in the gaseous, liquid, and solid states, if only because of the change in the amount of heat-fluid condensed around the atomic nucleus. The modern notion that one should try to explain all three states of matter (except metals and plasmas) with the same atomic model did not really take hold until after J. D. van der Waals published his famous theory in 1873. His theory predicted not only the liquid-gas transition but also the disappearance of that transition at the critical point.

I will come back to van der Waals' theory later in its proper chronological sequence, but I mentioned the critical point because I wanted to introduce the

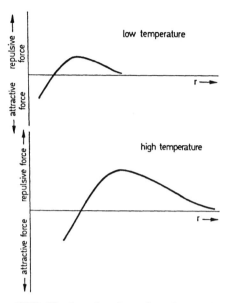

Fig. 3. Model A_3 (LAPLACE, 1822). The force law depends on temperature. As temperature rises, the atom acquires more caloric, thereby increasing its size and repulsive force. At larger distances the repulsive force is proportional to $1/r$.

third in the sequence of atomic force models. This was the one proposed by LAPLACE in 1822[18] to explain the phenomenon discovered in the same year by CAGNIARD DE LA TOUR: if you seal a liquid with its own vapor in a strong container and heat it up, you find that eventually the meniscus dividing liquid and vapor disappears. This was interpreted at the time to mean that the liquid changes to gas, even though it has remained liquid far above its normal boiling temperature because of the high pressure. (The modern explanation that gaseous and liquid states are continuous above the critical point was first hinted at by JOHN HERSCHEL in 1830[19] but definitely established only with the work of THOMAS ANDREWS in the 1860's.)

LAPLACE had already developed a modification of NEWTON's gas theory, in which he explained the repulsion of gas atoms (still generally accepted) in terms of continual radiation and emission of caloric (heat) particles. But he also postulated short-range attractive forces between atoms to account for phenomena such as surface tension, capillarity, and the cohesion of solids and liquids. The repulsive forces were supposed to operate at greater interatomic distances, in spite of NEWTON's attempt to restrict them to nearest neighbors only. With this model, A_3, LAPLACE could give a qualitative explanation of CAGNIARD DE LA TOUR's experiment: the attractive forces dominate when atoms are close together (as in a

---

[18] P. S. DE LAPLACE, Ann. Chim. Phys. **21**, 22 (1822). The historical background of similar assumptions about short-range forces has been reviewed by W. B. HARDY, Nature **109**, 375 (1922).

[19] J. HERSCHEL, *Preliminary discourse on the study of natural philosophy* (London 1830), p. 234.

solid or liquid), while the repulsive forces dominate when they are far apart (as in a gas); so by squeezing the atoms of a gas together one can shift the balance toward attraction and keep the substance in the liquid state at higher temperatures. So the boiling temperature increases with pressure, a property which we will call P_6. On the other hand, the amount of caloric in the substance increases with temperature, and eventually provides enough repulsive force to change the liquid to a gas even at high pressures: this is (apparent) property P_7.

The figures presented here are *not* taken from LAPLACE's paper, but represent my own guess as to what he had in mind.

Now LAPLACE's model is clearly wrong; in fact it is just the opposite of the one now generally accepted (short-range repulsive forces, long-range attractive forces). Even worse, it was not formulated quantitatively in such a way that it could be properly tested against experiments. Although LAPLACE did present a derivation of the ideal gas law based on his radiation-exchange theory of the repulsive force, he did not apparently realize that (as NEWTON had pointed out) if there is a long-range $1/r$ repulsive force between all pairs of atoms, the model no longer explains BOYLE's law.

The caloric theory of heat was eventually rejected, of course, and LAPLACE's model along with it, but *not* because A_3 led to an incorrect value of any particular property P_i. On the contrary, one might argue that the LAPLACE-POISSON explanation of the speed of sound, based on the concept of adiabatic compression and using empirical values for the ratio of specific heats, is somewhat more satisfactory than the 19th-century kinetic-theory explanation of this property (P_8); yet LAPLACE's theory was replaced by the kinetic theory for other reasons. (One reason, which might seem to be completely irrelevant, was that the wave theory of light won out over the particle theory around 1820–25.[20]) All in all, it looks like defeat number 3 for the program of relating interatomic forces to gas properties.

Of course I have exaggerated the failure of LAPLACE's theory; it did provide the best explanation of capillarity for many years, and thereby stimulated J. D. VAN DER WAALS in his own attempts to construct a more satisfactory atomic explanation of such phenomena; thus LAPLACE's theory did make some positive contributions to the progress of science.

Toward the "Correct Modern Theory"

Now at last we come to the kinetic theory of gases. As every schoolboy knows, if we regard a gas as being made up of billiard balls which have only contact action, and move freely at constant velocity most of the time, colliding occasionally with each other and with the walls of the container, then we can derive BOYLE's law by a simple calculation. (This was done by DANIEL BERNOULLI in 1738, and by several other scientists who independently rediscovered the theory.) This assumes completely elastic collisions and specular reflection at the walls, in the usual textbook presentation.

Next one introduces the property of thermal expansion, P_9, proposed by AMONTONS and CHARLES in the 18th century and firmly established by GAY-LUSSAC

---

[20] See S. G. BRUSH, Brit. J. Hist. Sci. (in press).

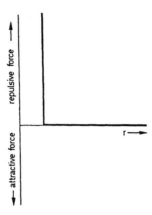

Fig. 4. Model A_4 (BERNOULLI, 1738). The atom is a billiard ball: no force except for an infinite repulsion when two atoms come in contact.

at the beginning of the 19th century: at constant pressure, equal increments of temperature produce equal increments of volume ($\Delta V \propto \Delta T$ for any T). Or, what seems to be regarded as the same law, equal increments of temperature produce equal increments of pressure if the volume is kept fixed.

It was not at all obvious at the beginning of the 19th century that the law of thermal expansion and BOYLE's law could be written together in the now-familiar form

$$PV = RT.$$

That would seem to imply that there is an "absolute" temperature scale, and that one can reduce the pressure or volume of a gas to nothing by cooling it down to "absolute zero." Even GAY-LUSSAC himself, as late as 1818, denied this.[21] Nevertheless an absolute temperature scale was finally accepted so that the ideal gas law could be written as above by about 1850.

In the kinetic theory of gases, as it was generally understood after 1850, one postulates that the absolute temperature is proportional to the average kinetic energy of the particles. Presumably if you heat up the outside of the container, the kinetic energies of the particles inside will tend to increase in the same proportion, although it is not clear from the simple model of billiard balls bouncing elastically off rigid plane walls how this energy-transfer would actually occur. I don't think anyone really worried about that problem until the 1870's when it was found to be advisable to assume explicitly that the temperature of the wall does affect the rebound velocity of the particles, in order to explain the radiometer effect. Of course the question of whether (and how fast) the equilibration of energy between wall and gas takes place is crucial to the distinction between an adiabatic and an isothermal process.

Let us backtrack a bit and ask whether it is obvious that the concept of thermal equilibrium entails that different systems tend to acquire the same average kinetic

---

[21] J. L. GAY-LUSSAC, Ann. Chim. Phys. 9, 305 (1818).

energy per particle, $\frac{1}{2}mv^2$. All one really gets from the textbook derivation is the BERNOULLI formula
$$PV = \tfrac{1}{3} N m \overline{v^2}.$$

When JOHN HERAPATH (a forgotten English scientist whom I am trying to revive [22]) rediscovered this formula in 1816, he argued that temperature should be defined as proportional to atomic *momentum* (mv) rather than kinetic energy. This might appear to be a purely verbal difference but it's not, since HERAPATH claimed that if you mix two equal weights of liquid at different temperatures, the mixture will have the average temperature calculated his way, not the usual way. In other words, defining HERAPATH's temperature as $T_H \propto \sqrt{T_K}$, where T_K is the now-standard Kelvin temperature, one has an ideal gas law of the form

$$PV \propto T_H^2,$$

and if you mix equal weights of water at temperatures T_H and T_H', the mixture should have temperature $\frac{1}{2}(T_H + T_H')$ on the HERAPATH scale.

Instead of disputing HERAPATH's collision dynamics, we could allow him to include his temperature postulate as part of the definition of his model, A_4'. (Remember that the generalized law of conservation of energy had not yet been adopted when HERAPATH published his theory in 1821.) While he can explain properties P_1 (BOYLE's law) and P_9 (thermal expansion), we expect he will run aground on P_{10}, the temperature of mixtures. But in fact the data available at the time actually agreed better with HERAPATH's predictions than with those derived from averaging kinetic energies! HERAPATH considered this the crucial experiment in support of his theory as against conventional ideas, and carried on a long feud with HUMPHRY DAVY and others in the Royal Society of London who refused to listen to him.[22]

HERAPATH was also able to calculate the speed of sound (P_8) from his theory, assuming it is about the same as the speed of an air molecule. He got very close to the right value, using slightly incorrect assumptions. When the calculation was redone later on for mechanical models of diatomic molecules, it turned out that you really can't get the correct experimental value if you insist on using the equipartition theorem of statistical mechanics (see below).

On the other hand, HERAPATH scored on the side of technological application with another result: by treating a railway locomotive as a giant air molecule he was able to derive a relation between wind resistance and the speed of the locomotive. He found that the resistance is proportional to the square of the speed (as NEWTON had claimed in Book II of the *Principia*) and was also able to calculate numerical values which he said were in good agreement with existing experimental data at low speeds. He then discussed the need for streamlining the locomotive to cut down the effects of the much greater resistance expected at high speeds. His work on gas theory led him into a number of other engineering problems which he discussed in his own rather successful magazine, *Herapath's Railway Journal*.

---

[22] J. HERAPATH, Ann. Phil. **8**, 56 (1816), [2] **1**, 273, 340, 401, **2**, 50, 89, 201, 256, 363, 434 (1821), **3**, 16 (1822). S. G. BRUSH, Ann. Sci. **13**, 188 (1957), Notes & Rec. Roy. Soc. London, **18**, 161 (1963); JOHN HERAPATH's *Mathematical Physics and other selected Papers*, reprinted with a new introduction by S. G. BRUSH (New York: Johnson Reprint Corp., in press).

To sum up HERAPATH's theory: it has definite defects, but they were corrected by CLAUSIUS and others later on; on the whole it should be counted as a victory for the program, which is rather encouraging after three straight defeats!

But does the kinetic theory depend on any particular assumption about the interatomic force law? This is a point that has often been misunderstood. You have to make *some* assumption in order to calculate anything at all, and the easiest model to use is the billiard ball. So one often hears it said that 19th-century physicists naively accepted the billiard-ball model of the atom, whereas of course we sophisticated moderns know that it's much more complicated than that. But I claim that all the first-rate 19th-century physicists also realized that it's much more complicated than that — but, being unwilling to give up theoretical atomic physics merely because they didn't know what the atom looks like, they plunged ahead with models anyway. I would further claim that it was precisely in the area of gas theory that this approach paid off, because *many* properties of gases can be calculated accurately from any one of a whole class of atomic models. Not only the *PVT* relations but also the general features of transport processes such as viscosity, diffusion, effusion, heat conduction, and sound propagation can be explained fairly well. The main exception was the specific heats of polyatomic gases, which do depend on details of atomic and molecular structure, so that one cannot predict the exact value of the speed of sound with a purely classical theory.

The struggles of 19th-century scientists with the specific heat problem throw a strange light on the procedure of hypothesis-testing. The discrepancy was already implicit in WATERSTON's theory of 1845, but he failed to realize it because of a numerical error.[23] RANKINE found it a reason for rejecting the kinetic theory in favor of his own vortex model.[24] CLAUSIUS was willing to avoid the problem by postulating an arbitrary empirical ratio between the kinetic and internal energies of molecules.[25] MAXWELL took it much more seriously, and in 1860 he told the British Association that "This result of the dynamical theory, being at variance with experiment, overturns the whole hypothesis, however satisfactory the other results may be."[26] BOLTZMANN in 1876 proposed to force the theory to agree with experiment by forbidding one mechanical degree of freedom of a molecule to gain or lose any energy in collisions,[27] but MAXWELL refused to accept this dodge.[28]

Even though the specific heat problem was never satisfactorily resolved within the framework of classical mechanics, it is not correct to say that it produced a crisis that forced physicists to abandon classical mechanics. In the debates of the 1880's and 1890's, those who rejected atomistic or mechanical models in physics did so for other reasons — at least they did not consider the specific heat discrepancy to be one of the more important issues at stake. On the other hand, those physicists who were most concerned about the problem did not conclude

---

[23] J. J. WATERSTON, Phil. Trans. A **183**, 1 (1892) (published posthumously with notes by Lord RAYLEIGH).

[24] W. J. M. RANKINE, Trans. Roy. Soc. Edinburgh **20**, 561 (1853); **25**, 559 (1869); Proc. Glasgow Phil. Soc. **5**, 128 (1864).

[25] R. CLAUSIUS, Ann. Physik [2] **100**, 353 (1857); see S. G. BRUSH, *Kinetic Theory*, vol. 1, p. 131–134.

[26] J. C. MAXWELL, Brit. Assn. Rept. **30**, 15 (1860).

[27] L. BOLTZMANN, Sitzungsber. Akad. Wiss. Wien **74**, 553 (1877).

[28] J. C. MAXWELL, Nature **16**, 242 (1877).

that it was necessary to reject the kinetic-molecular hypothesis because of this one failure. If they believed strongly enough in the basic validity of the model they did not accept the experimental data of the day as the supreme authority.

The history of Maxwell's work on kinetic theory may suggest why such an attitude is defensible. In 1860 he had before him two clear experimental refutations of the kinetic theory. His initial reaction to one of them has been quoted above. The other one was the data pertaining to his theoretical prediction that the viscosity of a gas should be independent of density. He had written to G. G. Stokes in May 1859, stating this prediction which he called "certainly very unexpected" and asking Stokes, the expert on viscosity: "Have you the means of refuting the results of the hypothesis?"[29] On the basis of Stokes' reply (which has apparently been lost) Maxwell wrote in the published paper of 1860: "Such a consequence of a mathematical theory is very startling, and the only experiment I have met with on the subject does not seem to confirm it."[30]

Despite these two disproofs of his hypothesis, Maxwell devoted several more years of his life to working out and testing the kinetic theory. It turned out eventually that in computing the viscosity of air from pendulum experiments Stokes had implicitly assumed that the viscosity decreases in proportion to the density.[31] This was a natural assumption for anyone to make before 1859; it was only with the help of Maxwell's *theory* that one could correctly interpret the *experiments* designed to test the theory! (The situation was similar but a little more complicated in the case of heat conduction experiments, where it was necessary to disentangle the effects of radiation and convection from conduction; the Dulong-Petit radiation law, based on the assumption that conduction as well as convection is negligible at low pressures so the remaining heat transfer must be due to radiation, was accepted for several decades until Maxwell's kinetic theory suggested the fallacy of the assumption.)

Special Force Laws and the Kinetic Theory

Maxwell's confidence in the kinetic theory, presumably based on his own experimental confirmation of his viscosity prediction, outweighed the skepticism which he continued to maintain in regard to specific heats. He proceeded to develop an elaborate version of the theory, taking into account the collision dynamics of particles interacting with a general r^{-n} force law, and introducing a general non-equilibrium distribution function. While he did not find a general solution of his equations, he did discover that in the special case of an r^{-5} repulsive force he could compute the transport coefficients without knowing the distribution function. According to this model, which we call A_5, the viscosity coefficient comes out to be directly proportional to the temperature, in agreement with Maxwell's own experiments.[32] (The billiard-ball model, A_4, gives a viscosity

---

[29] See S. G. Brush, *Kinetic Theory*, vol. 1, p. 26–27.

[30] J. C. Maxwell, Phil. Mag. [4] **19**, 19, 20, 21 (1860); see S. G. Brush, *Kinetic Theory*, vol. 1, p. 166.

[31] O. E. Meyer, Ann. Physik [2] **125**, 177, 401, 564 (1865); G. G. Stokes, *Mathematical and Physical Papers* (New York: Johnson Reprint Corp., 1966, reprint of the 1880–1905 edition) vol. 3, p. 76, 137.

[32] J. C. Maxwell, Phil. Trans. **157**, 49 (1867); see S. G. Brush, *Kinetic Theory*, vol. 2, p. 79.

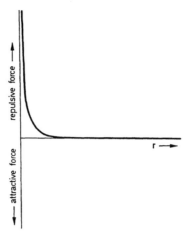

Fig. 5. Model A_5 (MAXWELL, 1867). Repulsive force proportional to r^{-5} at all distances.

proportional to the square root of the temperature.) Thus MAXWELL proposed to adopt this model, and some of his British colleagues were carried along by his enthusiasm even after MAXWELL himself realized that it could not be the true interatomic force law. German scientists such as O. E. MEYER did not think much of this new model for several reasons. First, the theory of attractive and repulsive interatomic forces was associated with a different philosophical viewpoint from the theory of billiard-ball atoms, as noted above. Second, the JOULE-THOMSON experiment had shown that long-range interatomic forces must be attractive rather than repulsive as MAXWELL's model requires. Third, the experimental data of MEYER and others suggested that the exponent of the temperature in the viscosity law is definitely less than 1, although it was also definitely greater than $\frac{1}{2}$.[33] Of course if one rejected both A_4 and A_5 on the basis of the third argument there would be nothing left; it was considered preferable to go back to the billiard ball model but to assume that the diameter of the billiard ball varies with temperature (i.e., with average collision speed).[34]

Next we come to the model of VAN DER WAALS (1873). The model, as employed by VAN DER WAALS in obtaining his famous equation of state

$$\left(p+\frac{a}{V^2}\right)(V-b)=RT$$

should probably be considered a success since the equation derived from it was successful in correlating and predicting gas properties and critical phenomena. The only question is: what *is* the model? This question is not so easily answered. There is no doubt that it involves a billiard ball of definite diameter d, since the atoms are excluded from a space $2\pi d^3/3$. It also includes an attractive force

---

[33] MEYER, Ann. Physik [2] **148**, 1, 203 (1873); J. PULUJ, Sitzungsber. Akad. Wiss. Wien **69** (2), 287 (1874), **70** (2), 243 (1875), **73** (2), 589 (1876); A. v. OBERMAYER, Sitzungsber. Akad. Wiss. Wien **71** (2), 281 (1875), **73** (2), 433 (1876).

[34] MEYER, *op. cit.*; a similar hypothesis was proposed by STEFAN in order to explain diffusion data: see J. STEFAN, Sitzungsber. Akad. Wiss. Wien **65** (2), 323 (1872).

whose effect on the pressure is represented by the term a/V^2, *i.e.* the pressure is decreased by the force. But there has been considerable confusion about the nature of the attractive force implied by the van der Waals equation, beginning with VAN DER WAALS himself. He simply asserted that the results of the JOULE-THOMSON experiment showed that there are attractive forces, but that these forces must vanish beyond some distance, whereas a repulsive force must dominate at much shorter distances. He did not propose any definite law for the attractive force, so that his result a/V^2 must be regarded as a lucky guess based on physical "intuition" rather than the consequence of a mathematical derivation.[35]

MAXWELL, in his review of VAN DER WAALS' theory, refused to accept the factor $(V-b)$ representing the effect of atomic size (or short-range repulsive forces), but he did agree that the a/V^2 term could be inferred from experiments with the help of the CLAUSIUS virial theorem:

> It appears by Dr. Andrews' experiments that when the volume of carbonic acid is diminished, the temperature remaining constant, the product of the volume and pressure at first diminishes, the rate of diminution becoming more and more rapid as the density increases. Now, the virial depends on the number of pairs of molecules which are at a given instant acting on one another, and this number in unit volume is proportional to the square of the density. Hence the part of the pressure depending on the virial increases as the square of the density, and since, in the case of carbonic acid, it diminishes the pressure, it must be of the positive sign, that is, it must arise from *attraction* between the molecules.[36]

Since at higher densities compression produces a rise of pressure, the same argument indicates that the intermolecular repulsive forces are operating.

> We have thus evidence that the molecules of gases attract each other at a certain small distance, but when they are brought still nearer they repel each other. This is quite in accordance with Boscovich's theory of atoms as massive centres of force, the force being a function of the distance, and changing from attractive to repulsive, and back again several times, as the distance diminishes. If we suppose that when the force begins to be repulsive it increases very rapidly as the distance diminishes, so as to become enormous if the distance is less by a very small quantity than that at which the force first begins to be repulsive, the phenomena will be precisely the same as those of smooth elastic spheres.[36]

This view of MAXWELL is recognizably modern, including his disdain for the philosophical distinction between the Boscovichean and Cartesian conceptions of the atom. As a *qualitative* description of the relation between the force law

---

[35] JOHANNES DIDERIK VAN DER WAALS, *Over de Continuiteit van den Gas- en Vloeistoftoestand*, Academisch Proefschrift (Leiden, 1873); see also the English translation by R. THRELFALL & J. F. ADAIR, based on the revised and expanded German edition of 1881, in *Physical Memoirs*, vol. 1, p. 333 (Physical Society of London, 1890). Some historical discussion of "van der Waals forces" may be found in the article by H. MARGENAU, Rev. Mod. Phys. **11**, 1 (1939).

[36] J. C. MAXWELL, Nature **10**, 477 (1874), reprinted in *The Scientific Papers of James Clerk Maxwell* (New York: Dover Publications 1965), vol. 2, p. 407.

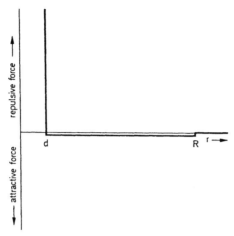

Fig. 6. Model A_6 (VAN DER WAALS, 1873, as later interpreted by BOLTZMANN, and by KAC, UHLENBECK, & HEMMER). Infinite repulsive force at $r=d$; attractive force given by the limit as $R \to \infty, c \to 0$ (keeping cR fixed) of the function
$$F(r) = c \quad \text{for } r \leq R$$
$$= 0 \quad \text{for } r > R.$$

and the properties of gases and liquids it fails on only one major point: MAXWELL seems to think that the condensation of a gas depends on attractive forces, so that a phase transition would be impossible if the interatomic forces were purely repulsive:

> In hydrogen the repulsive force seems to prevail even at ordinary pressures. This gas has never been liquefied, and it is probable that it never will be liquefied, as the attractive force is so weak.[36]

Recent computer experiments show, however, that systems of particles with purely repulsive forces can change from a gaseous to a solid phase, though they need not go through an intermediate liquid phase.[37]

Whereas VAN DER WAALS stated that the attractive force would decrease rapidly to zero at large distances, BOLTZMANN complained in 1898 that this had never been proven satisfactorily, and claimed instead that the VAN DER WAALS equation could be derived only on the assumption that the force has a small but *constant* value at large distances. When he told this to VAN DER WAALS, the latter replied, first, that he had never made any such assumption himself, and second, that he thought it was physically improbable.[38]

Some writers have stated that the a/V^2 term implies an inverse 4th power attractive force. Others, misled perhaps by the nomenclature "van der Waals forces" for the quantum-mechanical dispersion forces calculated by WANG, SLATER, EISENSCHITZ, and LONDON in the late 1920's, have talked about inverse 7th-power forces. While I hesitate to contradict some of the eminent scientists who

---

[37] B. J. ALDER & T. E. WAINWRIGHT, J. Chem. Phys. **27**, 1208 (1957), **31**, 459 (1959), **33**, 1439 (1960); S. G. BRUSH, H. L. SAHLIN & E. TELLER, J. Chem. Phys. **45**, 2102 (1966).

[38] See L. BOLTZMANN, *Lectures on Gas Theory* (Berkeley: University of California Press, 1964, translated from the German edition of 1896–98 by S. G. BRUSH), p. 219, 220, 375.

have written on this subject, I am reliably informed by my colleagues in statistical mechanics that the latest work confirms BOLTZMANN'S conclusion (at least for the cases that can be solved exactly): the VAN DER WAALS equation of state implies a very long-range attractive force that does not vary with distance but has a very small magnitude.[39] More precisely, the force law for model A_6 is the limit of a function $F(r) = -c$ for $r \leq R$ as $R \to \infty$ and $c \to 0$, keeping the product cR fixed; together with a billiard-ball repulsive core mentioned above. So we are left with a very nice theory but no physically plausible interatomic force law.

Let us pause briefly to take stock of our assets. We have recorded two successful atomic models: the kinetic theory (A_4) and the VAN DER WAALS theory (A_6). But all we have learned about interatomic forces from these models is the diameter of the billiard ball, together with the conclusion that the long-range force is attractive but very weak. (Comparison with experiment allows us to determine the constants a and b, which gives numerical values for the diameter d and for an integral of the attractive force $F(r)$ from $r=d$ to ∞, but not $F(r)$ itself.) These are significant results but a lot less than might have been expected.

The Lennard-Jones Model

I do not want to review here all the various models proposed since VAN DER WAALS; instead I will concentrate on just one, selected because many people even today seem to think it is the best "realistic" interatomic force model, at least for calculations of properties of monatomic gases and liquids by statistical mechanics. This is the LENNARD-JONES "6, 12" potential or "7, 13" force law:

$$F(r) = -\frac{a}{r^7} + \frac{b}{r^{13}}$$

(inverse 7th power attractive force dominating at large distances, inverse 13th power repulsive force dominating at short distances).

LENNARD-JONES did not simply start out with this force law; he arrived at it as the best special case of the general form $F(r) = -a/r^n + b/r^m$. Although MAX BORN and other physicists had used this type of force law in theories of solids around 1920, it was LENNARD-JONES who pioneered its application to gas theory. After CHAPMAN and ENSKOG developed their methods for calculating transport coefficients for a large class of force laws (1916–17), it seemed reasonable to try to determine the force law by fitting experimental data on viscosity and other properties. One could also use the relation between the second virial coefficient (correction to the ideal-gas equation of state) and the force law, as well as data on the compressibility of solids which could be related to the force law. This was undertaken as a definite program by J. E. JONES, a former student of CHAPMAN at Manchester, around 1924. However, JONES took the wise precaution of changing his surname at the beginning of his scientific career: after marrying Kathleen LENNARD in 1925, he called himself LENNARD-JONES. (The practice of hyphenating common names like JONES and SMITH is not unusual in Britain.) Unlike WILLIAM THOMSON, who caused considerable confusion by becoming Lord KELVIN at the age of 68, LENNARD-JONES probably enhanced the popularity of his model, although I

---

[39] M. KAC, G. E. UHLENBECK & P. C. HEMMER, J. Math. Phys. 4, 216 (1963).

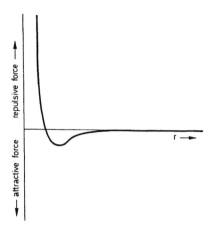

Fig. 7. Model A_7 (LENNARD-JONES, 1931). $F(r) = -\dfrac{a}{r^7} + \dfrac{b}{r^{13}}$ for all r (positive value of $F(r)$ means that the force is repulsive, negative means it is attractive).

must leave it to experts in psychology or sociology to determine whether A_7 would have been as readily accepted under the name "Jones potential."

In his first paper (1924), LENNARD-JONES proposed the general two-parameter form but immediately adopted $n=3$ for the attractive force because it simplified the calculation of collision integrals in the CHAPMAN-ENSKOG theory. The resulting formula for viscosity could be used to represent the data fairly accurately over a wide temperature range, but it did not seem to require a unique value of the exponent m (repulsive part of the force law). LENNARD-JONES presented his results for $m = 11$, $14\frac{1}{3}$, 21, and ∞, and found that $14\frac{1}{3}$ was slightly preferable for argon. For hydrogen and helium, on the other hand, the data could be explained well enough by ignoring the attractive force entirely and using only a repulsive force, $r^{-11.2}$ and $r^{-14.6}$ respectively.[40]

The inverse third power attractive force was quickly abandoned; in the second paper of the series, published at about the same time, LENNARD-JONES showed that when virial-coefficient data of HOLBORN & OTTO were combined with viscosity data for argon, a consistent fit could be obtained with $n=5$ and $m=14\frac{1}{3}$. Three months later, he was able to strengthen the case for this combination by using X-ray diffraction data on the atomic spacing of argon, recently obtained in a crystalline state by Simon and Simson. The index $14\frac{1}{3}$, originally chosen for computational convenience $\left(\dfrac{m-5}{m-1} = 0.7\right)$, was now changed to 15.[41] (Note that, just as in the Boscovich theory, the equilibrium volume of the solid is determined by the value of r that makes $F=0$.)

The following year, LENNARD-JONES extended his calculations to helium, neon, krypton, and ions having the same electronic structure as the rare-gas atoms. Transport data suggested rather high values of $m - 14$ or 15 for helium and argon,

---

[40] J. E. [LENNARD-] JONES, Proc. Roy. Soc. London A **106**, 441 (1924).
[41] J. E. LENNARD-JONES, Proc. Roy. Soc. London A **106**, 463, 709 (1924).

and 21 for neon — but information on the lattice spacing and compressibility of crystals indicated values of 9, 10, or 11 for this index. Significantly, LENNARD-JONES seemed to prefer the lower values, and came to depend more on crystal data and less on kinetic theory.[42] By 1926 he was also able to cite quantum-theoretical calculations of BORN and HEISENBERG to justify his use of the fifth power attractive force.[43] But he had not been able to compute the theoretical viscosity coefficient for a general force law with m^{th} power repulsion and n^{th} power attraction for values of n other than 3.

A further computational advance was made by HASSÉ and COOK at Bristol in 1927: they succeeded in determining the viscosity coefficient, using the CHAPMAN-ENSKOG formula for hard-spheres with inverse 5th power attractive forces.[44] They then attacked the LENNARD-JONES problem for the same attractive force, and found that the calculations could be somewhat simplified if a 9th-power repulsive force was combined with the 5th-power attractive force. These results, published in 1929, indicated that the (5, 9) model could fit the viscosity data fairly well but was not completely consistent with virial coefficient measurements. They concluded this paper with the remark:

> The trigonometrical treatment used in the calculation for $n=5$, $m=9$ could be used for $n=7$, $m=13$, which seems likely to provide interesting results, but the choice of n and m would need careful consideration.[45]

Had they followed through on this suggestion themselves, we would have had the HASSÉ-COOK model rather than the LENNARD-JONES model.

HASSÉ and COOK may have realized by this time that calculations based on the new quantum mechanics of HEISENBERG and SCHRÖDINGER (1925–26) were already pointing to an inverse 7th power (rather than 5th power) attractive force, though they did not mention these developments in their 1929 paper. What are now called "dispersion" forces were first discussed by WANG in 1927; he obtained an inverse 7th power attractive force though his numerical value for the constant was later found to be incorrect.[46] Soon afterwards, J. C. SLATER obtained an approximate formula for the force between two helium atoms, including an exponential short-range repulsion as well as an inverse 7th-power repulsion:

$$F(r) = A e^{-kr} - \frac{B}{r^7}.$$

This is usually referred to as the "exp, 6" potential since the potential energy of the attractive force is proportional to r^{-6}; it is considered the prototype for quantum-mechanical force laws though more accurate expressions have sub-

---

[42] J. E. LENNARD-JONES, Proc. Roy. Soc. London A **107**, 157 (1925), A **109**, 584 (1925); LENNARD-JONES & B. A. TAYLOR, Proc. Roy. Soc. London A **109**, 476 (1925).

[43] J. E. LENNARD-JONES & B. M. DENT, Proc. Roy. Soc. London A **112**, 230 (1926); M. BORN & W. HEISENBERG, Z. Phys. **112**, 230 (1924). A general survey of the situation at this time may be found in the chapter on intermolecular forces which LENNARD-JONES wrote for R. H. FOWLER's book *Statistical Mechanics* (Cambridge University Press, 1929).

[44] H. R. HASSÉ & W. R. COOK, Phil. Mag. [7] **3**, 978 (1927).

[45] H. R. HASSÉ & W. R. COOK, Proc. Roy. Soc. London A **125**, 196 (1929).

[46] S. C. WANG, Phys. Z. **28**, 663 (1927).

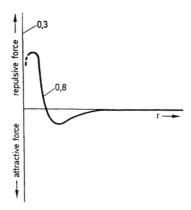

Fig. 8. Model A_8 (SLATER, 1928, usually called "exp, 6" potential). $F(r) = Ae^{-kr} - \dfrac{B}{r^7}$ for all r, except that various modifications are used to eliminate the negative infinity at $r = 0$.

sequently been obtained.[47] In 1930, EISENSCHITZ & LONDON recalculated the interaction between hydrogen atoms (originally done by WANG) and gave a comprehensive treatment of dispersion forces and their relation to molecular polarizabilities.[48]

LENNARD-JONES himself was one of the early participants in these quantum-mechanical calculations of interatomic forces, and his 1931 lecture on "Cohesion" to the Physical Society of London provides a good contemporary survey of the subject.[49] Perhaps more important in the present context, however, is the fact that this lecture includes the first presentation of the 7, 13 potential. We have seen that up to 1929, his program for determining the force law from experimental data had been executed more or less in accordance with the procedure I outlined at the beginning of this article, subject to the difficulties in doing mathematical calculations of collision integrals for most of the possible force laws. It seemed to have been fairly definitely established by LENNARD-JONES and by HASSÉ & COOK that the attractive force r^{-5} was satisfactory, while the exponent in the repulsive force was probably somewhere between 8 and 15. But now the 5th-power attraction was completely abandoned, simply because it did not agree with *another* theory, quantum mechanics. Clearly experimental data were of little importance compared to the fascination of the new theory; or rather, the connection between the data and the force law was so weak that it did not have to be taken too seriously.

The switch from r^{-5} to r^{-7} attraction, combined with more recent data on virial coefficients, crystal spacing, and heats of sublimation, now seemed to make higher values of the repulsive index m preferable. LENNARD-JONES recognized that theoretical calculations "show that the repulsive field is more complicated"

---

[47] J.C. SLATER, Phys. Rev. [2] **32**, 349 (1928); J.C. SLATER & J.G. KIRKWOOD, Phys. Rev. [2] **37**, 682 (1931). More complicated exponential forms were proposed by A. UNSÖLD, Z. Physik **43**, 563 (1927); L. PAULING, Z. Krist. **67**, 377 (1928); H. BRÜCK, Z. Physik **51**, 707 (1928).

[48] R. EISENSCHITZ & F. LONDON, Z. Physik **60**, 491 (1930); F. LONDON, Z. Physik. Chem. B **11**, 222 (1930).

[49] J.E. LENNARD-JONES, Proc. Phys. Soc. London **43**, 461 (1931).

than his original r^{-m} form "and contains terms of the form e^{-aR}, but it falls off very rapidly with distance and can (in the case of helium at any rate) be represented, over the range which is most effective in atomic collisions, by a term of the type $\lambda(\text{rep.})^{R-m}$." In order to determine m, gas properties were not sufficient, but crystal spacing data and heats of sublimation could be used. However, these were known only for neon, argon, and nitrogen.

Here is the portion of the table from LENNARD-JONES' 1931 paper on which his choice of the index $m=13$ in his force law appears to have been based:

	m	Calcd. closest distance in crystal (A.U.)	Calcd. heat of sublimation in cals/gm. mole	Observed crystal spacing	Observed heat of sublimation
Neon	10	3.14	499		
	11	3.04	542	3.20	590
	13	2.99	612		
Argon	10	3.90	1730		
	11	3.83	1847	3.84	2030
	13	3.74	2030		
Nitrogen	10	4.23	1380		
	11	4.17	1460	4.0	1860
	13	4.06	1640		

As can be seen from the above table, the 7, 13 model was a weakling at its birth; it has suffered from uncertain health ever since. The major part of the experimental evidence for the r^{-m} repulsion came from crystal data; but in 1932, MAX BORN and J. E. MAYER renounced the r^{-m} repulsive force law which BORN had used in his earlier work in crystal theory; they reported that a potential of the form e^{-ar} now fit the data better, quite apart from the question of its quantum-mechanical justification.[50] In 1933, MASSEY & MOHR showed that the viscosity data that had previously been used to infer an r^{-m} repulsive force could now be explained by the simpler hard-sphere model, provided that quantum-mechanical scattering theory was used.[51] Hence, at least for hydrogen and helium, LENNARD-JONES had to concede that "no information about interatomic forces can be derived from the viscosity by classical methods."[52] MASSEY and MOHR then showed that the viscosity of helium computed from the SLATER force law (exp, 6 potential) was in fairly good agreement with experiment except at low temperatures where deviations might be expected because of the effect of BOSE-EINSTEIN statistics.[53]

When LENNARD-JONES revised his chapter on "Interatomic Forces" for the second edition of R. H. FOWLER's *Statistical Mechanics* (1936), he included some of his theoretical values of the second virial coefficient and crystal spacing for the 7, 13 force law, but no longer claimed that 13 was the best value for the repulsive term. Instead, he reverted to his earlier preference for 9th power forces. While recognizing that the work of BORN and MAYER and of MASSEY & MOHR had under-

[50] M. BORN & J.E. MAYER, Z. Physik **75**, 1 (1932).
[51] H.S.W. MASSEY & C.B.O. MOHR, Proc. Roy. Soc. London A **141**, 434 (1933).
[52] LENNARD-JONES, in FOWLER's *Statistical Mechanics*, 2nd ed. (1936) p. 312.
[53] MASSEY & MOHR, Proc. Roy. Soc. London A **144**, 188 (1934).

mined some of his earlier arguments for using a power-law repulsion, LENNARD-JONES seemed to think that the practical difference between such a law and the theoretically preferable exponential form would not be sufficient to justify redoing all his calculations. (Moreover, the virial-coefficient integrals had not been evaluated for the exponential form.)

While LENNARD-JONES' argument that a 9th-power repulsive force is a convenient model for interpreting experimental data, and also approximates the quantum-mechanical result, is reasonable and has in fact sometimes been accepted on just this basis, it does seem surprising that in a series of papers with DEVONSHIRE on the equation of state and critical phenomena in 1937–39, the 7, 13 force law was revived, with the statement that it "has been found to represent the fields of some of the inert gases satisfactorily."[54] The only evidence for this assertion was a reference to the 1931 lecture, and the 13th power was not even justified as being computationally simpler than the 9th power for the problem at hand.

In 1938, R. A. BUCKINGHAM advocated the use of the exp, 6 potential in correlating crystal data with second virial coefficients, and joined with MASSEY in calculating the viscosity of helium with this potential.[55] But about the same time J. DE BOER in Holland, and J. O. HIRSCHFELDER's group in the U.S.A. began to make extensive use of the 7, 13 force law in their researches.

There was considerable discussion of the relative merits of the two models A_7 and A_8 (the exp, 6 potential) during the 1930's, based mainly on equilibrium properties of gases (second virial coefficients and JOULE-THOMSON coefficients) and crystal data. The transport coefficients had not yet been calculated for either model. The first calculation of these coefficients for the 7, 13 law was reported by the Japanese scientists KIHARA & KOTANI in 1943, although it did not become known to Western scientists until 1949.[56] More elaborate computations were published by HIRSCHFELDER, BIRD & SPOTZ in 1948, and by DE BOER & VAN KRANENDONK a little later in the same year.[57] Shortly after this, ROWLINSON in England reported a similar calculation.[58] The fact that four independent calculations gave essentially the same numerical results for the 7, 13 collision integrals produced some confidence in these results; but perhaps more significantly, the fact that all four groups had chosen to use the same potential created the impression that the

---

[54] J. E. LENNARD-JONES & A. F. DEVONSHIRE, Proc. Roy. Soc. London A **163**, 53 (1937); see also LENNARD-JONES, Physica **4**, 941 (1937), where it is stated that "a value between 9 and 12 fits the observations satisfactorily" but 12 is immediately adopted without explanation, for the further calculations. The 7, 13 model is also used by J. CORNER & J. E. LENNARD-JONES, Proc. Roy. Soc. London A **178**, 401 (1941). When FOWLER published a textbook version of his monograph under the title *Statistical Thermodynamics* (Cambridge, 1939) with E. A. GUGGENHEIM as co-author, he recognized the theoretical superiority of the exp-6 potential of SLATER & BUCKINGHAM (see below) but — perhaps still under the influence of LENNARD-JONES — maintained that the 7, 13 model was a sufficiently good approximation to it (see p. 280–292).

[55] R. A. BUCKINGHAM, Proc. Roy. Soc. London A **168**, 264 (1938); H. S. W. MASSEY & R. A. BUCKINGHAM, Proc. Roy. Soc. London A **168**, 378 (1938).

[56] T. KIHARA & M. KOTANI, Proc. Physico-Math. Soc. Japan [3] **25**, 602 (1943); see J. O. HIRSCHFELDER, R. B. BIRD & E. L. SPOTZ, J. Chem. Phys. **17**, 1343 (1949).

[57] J. O. HIRSCHFELDER, R. B. BIRD & E. L. SPOTZ, J. Chem. Phys. **16**, 968 (1948); J. DE BOER & J. VAN KRANENDONK, Physica **14**, 442 (1948); J. DE BOER, Physica **14**, 139 (1948); J. DE BOER & B. S. BLAISSE, Physica **14**, 149 (1948).

[58] J. S. ROWLINSON, J. Chem. Phys. **17**, 101 (1949).

7, 13 model was now the preferred one. At least it was the only "realistic" potential for which the parameters were established for a large number of gases. At the time when HIRSCHFELDER, CURTISS & BIRD published their massive treatise, *Molecular Theory of Gases and Liquids* (1954), calculations with the exp-6 potential were only just beginning,[59] so that the postwar generation of scientists who used *Molecular Theory* as their chief source of information may well have gotten the impression that the 7, 13 model had been generally accepted by the experts.

A strict comparison of results based on the exp, 6 and 7, 13 models was not possible until virial coefficients had been computed for both of them. The calculations are somewhat simpler for the 7, 13 model, so it had a head start despite the theoretical prestige of the exponential repulsion. Moreover, it was necessary to adopt some kind of short-range modification of the latter, since otherwise the inverse 7th power attraction would overcome the exponential repulsion and produce an "unphysical" negatively infinite potential energy at $r=0$.[60]

By the end of 1954, just after the publication of *Molecular Theory*, MASON & RICE had carried out direct comparisons of the 7, 13 and exp-6 models for several gases. They found that while the exp-6 was definitely better when applied to hydrogen and helium, on the whole neither model was markedly superior to the other in reproducing virial coefficients and viscosity coefficients.[61]

Thus, nearly 40 years after CHAPMAN and ENSKOG had established the basic theory of transport properties, and 30 years after LENNARD-JONES had begun the program of using this theory (and others) to determine intermolecular forces, it was finally established that gas theory could *not* be used to select the correct form of the force law.

Further humiliation was still in store for the LENNARD-JONES model. It was still generally believed that even though the LENNARD-JONES procedure could not fix the functional form of the force law, at least it could be used to calculate a consistent set of parameters when the form had been fixed by other methods. But in 1965, KINGSTON reported that the coefficient of the r^{-6} attractive energy, calculated theoretically from quantum mechanics, is only about half of the value determined from second virial coefficients and viscosity. Although the discrepancy, once it was known to exist, could be removed by reinterpretation and extension of the experimental data, KINGSTON had to conclude that "in general the LENNARD-JONES (12-6) potential is not a good representation of the true interatomic potential".[62]

The inadequacy of force laws determined from gas properties is especially apparent in dealing with properties that depend on the nature of the force law

---

[59] E. A. MASON, J. Chem. Phys. **22**, 169 (1954); W. E. RICE & J. O. HIRSCHFELDER, J. Chem. Phys. **22**, 187 (1954); E. A. MASON & W. E. RICE, J. Chem. Phys. **22**, 522 (1954). HIRSCHFELDER *et al.* state on p. 203 of *Molecular Theory* that the exp-6 potential provides "a definite improvement over the fit given by the Lennard-Jones (6-12) potential." But the statement favoring the 6-12 potential on p. 23 remained unchanged in the 1964 "corrected printing" of the book (see also the notes added to p. 181 and 1050 in this printing, at the end).

[60] R. A. BUCKINGHAM & J. CORNER, Proc. Soc. London A **189**, 118 (1947); see also RICE & HIRSCHFELDER, *op. cit.* (note 59); J. L. YNTEMA & W. G. SCHNEIDER, J. Chem. Phys. **18**, 646 (1950); E. A. MASON, *op. cit.* (note 59).

[61] E. A. MASON & W. E. RICE, J. Chem. Phys. **22**, 522, 843 (1954); E. A. MASON, J. Chem. Phys. **23**, 49 (1955).

[62] A. E. KINGSTON, J. Chem. Phys. **42**, 719 (1965).

at very short or very large distances. Thus CASIMIR & POLDER discovered in 1946 that when "retardation" effects are taken into account by means of quantum electrodynamics, the long-range attractive force between neutral atoms changes from an inverse 7th power to an inverse 8th power.[63] On the other hand, at very short distances, the dominant force is the COULOMB repulsion between the nuclei rather than an inverse 13th power or even an exponential repulsion.[64]

An even more serious challenge to the Newtonian program is the recent accumulation of evidence that the "true interatomic potential" is not a very useful concept. It is of course possible to determine the force between two atoms by scattering experiments, at least under circumstances where it can be assumed that the potential is a monotonic function of distance. A considerable amount of information of this type is available and can be represented by assuming that the short-range repulsive force has the form r^{-n}, where n ranges from 4 to 20 for various atoms and molecules.[65] But if an interatomic force law is to be of any use in explaining and predicting properties of bulk matter, it must be applicable to systems of more than two atoms, and it must give an adequate description of attractive as well as repulsive forces. Are these criteria satisfied by any of the force laws now under consideration?

Since experimental data on crystals played a significant role in the original determination of the exponents 7 and 13 in the LENNARD-JONES model, one might at least expect that this model would be useful in the theory of solids even though it does not help much in understanding gas properties. For metallic solids or ionic crystals such a model would of course be irrelevant; no one nowadays would attribute a major share of the binding energy of such solids to the forces known to act between neutral molecules. However, there is a small but interesting class of substances of low melting point, such as the solids of the rare gases, which are believed to be held together by essentially the same forces which produce the deviations from the ideal gas law at higher temperatures, the so-called "van der Waals lattices." Not surprisingly in view of the history discussed above, the lattice energy of these solids can be fitted fairly well by either A_7 or A_8. But one also expects to be able to explain theoretically the actual crystal structure of such a solid, and here we encounter another catastrophe. It turns out that any physically plausible potential of the LENNARD-JONES or (exp, 6) type leads to the prediction that a face-centered cubic lattice has a *higher* energy (*i.e.* is less stable) than a hexagonal-close-packed lattice, whereas in reality all the rare gases crystallize in the face-centered-cubic form except for helium, where the quantum zero-point energy is dominant.[66]

It now appears, as a result of research done during the last decade, that the interatomic-force concept is inadequate because the interaction energy of a

---

[63] H.B.G. CASIMIR & D. POLDER, Nature **158**, 787 (1946); Phys. Rev. [2] **73**, 360 (1948).

[64] A.A. FROST & J.H. WOODSON, J. Am. Chem. Soc. **80**, 2615 (1958); R.A. BUCKINGHAM, Trans. Faraday Soc. **54**, 453 (1958).

[65] E.A. MASON & J.T. VANDERSLICE, in *Atomic and Molecular Processes*, ed. D.R. BATES (New York: Academic Press, 1962), Chapter 17; I. AMDUR & J.E. JORDAN, Adv. Chem. Phys. **10**, 29 (1966). It has recently been asserted that none of the usual two or three parameter potential functions can fit the details of the total scattering cross-section data: F. VON BOSCH *et al.*, Z. Physik **199**, 518 (1967); R. DUEREN & C. SCHLIER, J. Chem. Phys. **46**, 4535 (1967).

[66] L. MEYER, Adv. Chem. Phys. **16**, 343 (1969).

system of three or more atoms or molecules cannot be written accurately as a sum of two-particle energies. For example, an elaborate study of the properties of the rare gas solids was published by KLEIN & MUNN, based on the potential function

$$\frac{\phi^{12}(x)}{4\varepsilon} = e^{4(1-x^2)} \sum_{i=0}^{3} C_{2i} x^{2i-6} - x^{-6} \sum_{i=0}^{3} C_{2i} \quad \text{where } x = r/\sigma.$$

Values of the parameters ε, σ, (C_{2i}) had been determined earlier by MUNN & SMITH[67] so as to fit virial coefficient and transport data for the gaseous phase. According to KLEIN & MUNN, the resulting potential function "is probably the best available approximation to the true two-body potential." They used it to calculate the lattice parameters, sublimation energies, and bulk modulus for the solids at 0 °K, and found that "if pairwise additivity is assumed for the interatomic forces in the solid, the agreement with experiment is poor." (In this case "poor" means about 20% too high.) But the agreement could be improved significantly by adding an "Axilrod-Teller triple-dipole force.[68]

Similar conclusions emerged from experiments on shock compression of argon done by Ross & ALDER. They used both the LENNARD-JONES and exp-6 potentials to calculate the shock HUGONIOT equation of state, and found that these theoretical pressures (based on the assumption of pairwise additivity) were different from the experimental pressures by up to 30% for high-density states.[69]

Other studies have shown that even in liquids and gases some kind of density-dependent 3-body interaction must be included in order to interpret experimental results.[70] The quantum mechanical basis for such a 3-body force is well understood and has been investigated in detail for simple systems as long ago as 1955.[71]

In 1967 an entire volume of the review series "Advances in Chemical Physics" was devoted to the topic *Intermolecular Forces* under the editorship of J.O. HIRSCHFELDER. According to the index there are only two references to the LENNARD-JONES potential in the entire book. On page 290, O. SINANOĞLU states rather casually that this potential function "is known not be sufficiently typical of even the rare gases." The other reference is found in the editor's introduction and may serve as an obituary for the LENNARD-JONES program:

> A few years ago, the parametrized potential energy functions of Morse, Lennard-Jones, Buckingham, Stockmayer, and Kihara were satisfactory for representing the available experimental data. With the development of a variety of ingenious experimental techniques which provide detailed information regarding individual molecular collisions, it has become necessary to provide a more realistic basis for describing the energy of interaction.

---

[67] R.J. MUNN & E.J. SMITH, J. Chem. Phys. **43**, 3998 (1965).
[68] M.L. KLEIN & R.J. MUNN, J. Chem. Phys. **47**, 1035 (1967).
[69] M. ROSS & B.J. ALDER, J. Chem. Phys. **46**, 703 (1967).
[70] T. HALICIOĞLU & O. SINANOĞLU, J. Chem. Phys. **49**, 996 (1968); D.A. COPELAND & N.R. KESTNER, J. Chem. Phys. **49**, 5214 (1968).
[71] A. SHOSTAK, J. Chem. Phys. **23**, 1808 (1955). For further discussion and review of the problem of nonadditivity see H. MARGENAU & J. STAMPER, Adv. Quant. Chem. **3**, 129 (1967).

Some Conclusions and Invidious Comparisons

One might conclude that the entire program of determining interatomic forces from gas properties was a mistake, destined to fail simply because gas properties do not depend very much on the details of the force law, and because the "atom" is not a stable entity unaffected by the presence of other nearby atoms. Of course NEWTON could not have known these facts, and even LENNARD-JONES probably did not realize them until he had been working on the problem for several years. In any case it was inevitable that the attempt would be made; what matters now is that we realize that it did in fact fail.

But the failure affects not only this particular program but also the validity of the hypothetico-deductive method which I described at the beginning of this paper. While it is true that the final collapse of the last phase of the program was due to obstinate discrepancies between theory and experiment, most of the twists and turns along the way were motivated not by the outcomes of experimental tests but by changes in atomic theory initiated in other areas of science. It is clear that MAXWELL, despite his 1860 statement at the British Association meeting, did not believe that an hypothesis must be rejected because it leads to a single prediction that is unambiguously falsified by experiments; other scientists shared his devotion to the kinetic theory without even bothering to pay lip service to the tenets of hypothetico-deductive method. On the other hand, we have seen that LENNARD-JONES was willing to abandon one form of his model (the r^{-5} attractive force) for purely theoretical reasons, even though it *did* conform to the experimental data available at the time; again, his colleagues unanimously followed him on this point.

But if we reject the program of developing and testing hypothetical models in order to find eventually the "true interatomic potential," how else should we proceed? By way of contrast let us consider some of the more positive accomplishments of gas theory in recent years. I will mention only two events, one in equilibrium statistical mechanics and the other in transport theory. The first is the breakthrough in understanding the critical-point phenomena of gases, liquids, magnets, and mixtures by refined analysis of the Ising or "lattice-gas" model, due to M.E. FISHER, G.A. BAKER, and others.[72] It appears that an accurate description of cooperative phenomena should be based *not* on a "realistic" model in which all current information about the interatomic force law is used, but rather on a model which is mathematically simple enough so that the calculations can be carried out exactly. The reason is that the cooperative phenomena seem to depend very sensitively on the long-range statistical correlations produced by short-range forces, but these correlations themselves do not depend on the details of the force law. In the ISING model (first proposed by W. LENZ in 1920[73]) the potential energy of two atoms is simply a constant if they are nearest neighbors, zero if they are farther away, and $+\infty$ if they sit on top of each other; no other possibilities

---

[72] See for example M.S. GREEN & J.V. SENGERS (eds.), *Critical Phenomena: Proceedings of a Conference held in Washington, D.C., April 1965* (National Bureau of Standards Misc. Pub. 273, 1966); M.E. FISHER, Repts. Prog. Phys. **30**, 615 (1967).

[73] W. LENZ, Physik. Z. **21**, 613 (1920); E. ISING, Z. Physik **31**, 253 (1925); the history of this model is discussed by S.G. BRUSH, Rev. Mod. Phys. **39**, 883 (1967).

are permitted. (It should also be mentioned that the impetus for some of the work on critical phenomena came from the discovery by B. ALDER & T. WAINWRIGHT[37] of a phase transition in the elastic-sphere or billiard-ball fluid; this suggested that significant progress could be made by analyzing simple though unrealistic models.)

The other event is the development of generalizations of the MAXWELL-BOLTZMANN-HILBERT-CHAPMAN-ENSKOG theory to higher densities, starting with the work of BOGOLIUBOV in 1946 and culminating in the discovery of a logarithmic divergence in the density expansions of transport coefficients by WEINSTOCK, GOLDMAN, FRIEMAN, DORFMAN, and COHEN in 1965. Before 1965, laborious computations with "realistic" molecular models had produced little of general significance; it turned out that the use of the old billiard-ball model was essential in carrying out the complicated mathematical calculations and obtaining a cleancut, reliable result.[74] The reason was similar to that for progress in understanding critical-point phenomena, except that here it is correlations over long *time* periods that must be accurately treated.

In place of the program that has now been abandoned, here is a sketch of a reasonable procedure which could be followed. (This is not a new proposal but simply a summary of what many successful scientists have done in the past.) First, if you are primarily interested in knowing the interatomic force law *per se*, decide on your general or metaphysical principles of physics and on your basic assumptions about the microstructure of matter; then develop a plausible model and check your predictions against experiment just to be sure they are not too far off. Second, if you are primarily interested in understanding the properties of matter or in predicting new properties, forget about using a "realistic" force law and use the *simplest* model that still retains the crude qualitative physical properties that might be important. In many cases this will be the billiard ball, or perhaps the Maxwellian r^{-5} repulsive force which is still employed in some modern researches on kinetic theory;[75] the decision will be made on the basis of mathematical convenience. (Let M determine A.) In other words, the major effort should be put into ensuring that the predicted properties of matter are rigorous logical consequences of the model, whatever it may be, rather than into making the model itself an accurate representation of molecular properties.

I don't think the electronic computers need fear unemployment if this procedure is generally adopted. It is true that there may no longer be as much demand

---

[74] J. WEINSTOCK, Phys. Rev. [2] **140**, A 460 (1965); E. A. FRIEMAN & R. GOLDMAN, Bull. Am. Phys. Soc. **10**, 531 (1965); J. R. DORFMAN & E. G. D. COHEN, Phys. Letters **16**, 124 (1965); J. V. SENGERS, Phys. Fluids **9**, 1333 (1966). For review and further references see M. H. ERNST, L. K. HAINES & J. R. DORFMAN, Rev. Mod. Phys. **41**, 296 (1969); S. G. BRUSH, Part 1 of *Kinetic Theory*, vol. 3 (Oxford: Pergamon Press 1971).

[75] C. S. WANG CHANG & G. E. UHLENBECK, *On the theory of the thickness of weak shock waves* (Ann Arbor: Engineering Research Institute, The University of Michigan, 1948) and other works by these authors; H. GRAD, Comm. Pure & Applied Math. **2**, 331 (1949) and later papers; E. IKENBERRY & C. TRUESDELL, J. Rational Mech. Anal. **5**, 1 (1956); C. TRUESDELL, J. Rational Mech. Anal. **5**, 55 (1956); E. P. GROSS & E. A. JACKSON, Phys. Fluids **2**, 432 (1959); R. JANCEL & T. KAHAN, Compt. Rend. Acad. Sci. Paris **254**, 2292 (1962); L. SIROVICH, Phys. Fluids **6**, 1 (1963). A survey of the mathematical results that can be obtained with special models is included in the article by H. GRAD, Handbuch der Physik **12**, 205 (1958); more recent results are summarized in his paper in *Proceedings of the International Seminar on the Transport Properties of Gases* (Brown University, Providence, R. I., 1964), p. 39.

for simple operations repeated millions of times per second; but I am confident that we can teach the younger generation of computers the skills of higher analysis needed for the mathematical physics of the future.

A preliminary version of this paper was presented at a seminar of the Mechanics Department, Johns Hopkins University, 16 January 1970. The author thanks Professor C. TRUESDELL for several suggestions which were used in preparing the final version. This research was supported by the National Science Foundation, Grant GS-2475.

3.

IRREVERSIBILITY AND INDETERMINISM: FOURIER TO HEISENBERG

by Stephen G. Brush

The First Scientific Revolution, dominated by the physical astronomy of Copernicus, Kepler, Galileo and Newton, established the concept of a "clockwork universe" or "world-machine" in which all changes are cyclic and all motions are in principle determined by causal laws. The Second Scientific Revolution, associated with the theories of Darwin, Maxwell, Planck, Einstein, Heisenberg and Schrödinger, substituted a world of process and chance whose ultimate philosophical meaning still remains obscure. This paper will sketch one aspect of the change in physical theory that occurred after 1800: the acceptance of randomness at the atomic level and its relation to the acceptance of irreversibility as a characteristic property of natural phenomena.

Kenneth Ford noted a few years ago that it is historically inaccurate to suppose that indeterminism was suddenly introduced into physical science as a result of the adoption of quantum mechanics:

The idea that the fundamental processes of nature are governed by laws of probability should have hit the world of science like a bombshell. Oddly enough. it did not. It seems to have infiltrated science gradually over the first quarter of this century. Only after the quantum theory had become fully developed in about 1926 did physicists and philosophers sit up and take note of the fact that a revolution had occurred in our interpretation or natural laws.[1]

But the roots of indeterminism can be traced back to a much earlier period without losing the thread of historical continuity. In 1932 Sir Arthur Eddington. surveying the "decline of determinism," suggested that the first inkling of the existence of "secondary laws" having statistical rather than absolute validity had come from the observation that heat flows from hot to cold; the reverse is not impossible, only extremely improbable.[2] Eddington was well aware that the theory of heat conduction, as well as the Second Law of Thermodynamics, embodied a principle of irreversibility or directionality of time, and that the attempt to explain irreversibility by the atomic-kinetic theory of matter in the nineteenth century had made physicists familiar with probabilistic modes of thought, thereby helping to create a favorable climate for the introduction of quantum indeterminism in the 1920s. But later writers on the history of modern physics have been so impressed by the remarkable discoveries in atomic and nuclear physics at the beginning of this century that they tend to discount the radical significance of developments before 1900. The image of the complacent Victorian physicist, so certain that all the fundamental laws of nature had been discovered that he thought there was nothing left to do except determine the physical constants a little bit more accurately, has infiltrated our textbooks.[3] Recent scholarship in the history of quantum mechanics has provided a wealth of information about the events in science and society immediately preceding the promulgation of Max Born's statistical interpretation of the wave function and Werner Heisenberg's indeterminacy principle,[4] but there is still a need for a longer view. Without denying the crucial importance of the newer problems posed by the emission and absorption of radiation, wave-particle dualism, and the cultural environment of Weimar Germany, I shall emphasize the continuity of the stream of thought running from Fourier to Kelvin to Maxwell to Boltzmann to Planck to Einstein to Born and Heisenberg.

In reading the casual remarks, and even what seem to be carefully considered pronouncements by nineteenth-century scientists, the twentieth-century reader often finds a puzzling ambiguity. Words like randomness, irregularity, and indeterminism may appear to imply that molecules or other entities do not move in paths that are completely determined by the positions, velocities, and forces of all particles in the system; yet the same author may express himself in a manner that is completely consistent with the view that these paths really are determined but we simply cannot obtain complete *knowledge* of them. What we might now call a crucial distinction between ontological and epistemological indeterminism is frequently blurred in these writings. This ambiguity can be used to argue that almost every one of the scientists quoted here believed only in epistemological, not ontological randomness - but such an argument would conceal a gradual but extremely significant historical shift in the meaning of concepts. As Yehuda Elkana has shown in the case of the term *Kraft*, certain concepts may be in a state of flux during the time when a new theory is being created,[5] and one misses something by trying to force a precise meaning on words like "statistical" or "probabilistic" which may acquire such meaning only after the appropriate theory has been developed. Indeed, this article might be considered a chapter in the history of the changing meaning of the word "statistical."

I do not attempt to discuss here the philosophical writings on randomness (Cournot, Peirce, etc.) or the relation between determinism and the broader concept of "causality" in science.[6] Nor do I include the influence of Darwin's "chance variations" which may be claimed to have contributed something to the breakdown of determinism in science, as his "population" analysis shifted the emphasis from essentialistic to statistical thinking in biology.[7]

According to one widely held viewpoint in the history of ideas, eighteenth-century thought was dominated by the concept of the "Newtonian worldmachine."[8] God had been assigned the role of master clockmaker who designed a universe so perfect that it could run indefinitely without any need for divine tinkering. Closer inspection of Newton's own writings shows that he was actually quite firmly opposed to this concept which had been popularized by earlier writers such as Robert Boyle.[9] Newton's objections were both physical and moral; he pointed to the existence of irreversible processes tending to dissipate motion, and gravitational perturbations that seemed to threaten the stability of the solar system, and he warned that restricting God to the creation and design of the world while denying His continual supervision was a step on the road to atheism.[10] Nevertheless, in those areas of physical science where Newton's mathematical principles of natural philosophy could be applied with decisive results - planetary and lunar astronomy, fluid dynamics, elasticity, electrostatics, etc. -

eighteenth- century theorists scored numerous successes and developed increasing confidence in the mechanistic world view.

During the same period a new type of problem began to attract considerable scientific attention: the internal heat of the earth and its connection with geological history. Attempts at quantitative treatment go back to Dortous de Mairan and Buffon, both of whom argued that the present temperature near the surface is the result of a process of cooling from an initial hot molten state.[11] Buffon, from a series of laboratory experiments with hot iron spheres, concluded that the time required for the earth to reach its present temperature was about 75,000 years. J. S. Bailly suggested that progressive cooling is a property of all objects in the solar system. but each is at a different stage. Jupiter is still too hot to allow life to exist while the moon is already too cold; but everything must eventually reach a state of inanimate cold.

Such speculations came into conflict with the "Uniformitarian" view of the earth's history developed by Hutton, Playfair, and later, Lyell. While admitting that the inside of the earth is probably very hot, these geologists did not agree that the internal heat was more intense in earlier epochs, since that would have meant allowing causes whose magnitude was substantially greater than those now observable to sneak back into geological theory. Instead, they proposed that the internal heat has remained constant, providing power for recurrent upheavals to counteract the effects of erosion; this cyclic scheme has been called the "Huttonian earth-machine."[12] It was also necessary, in order to ensure that there had been no significant long term change in physical conditions on the earth's surface, to invoke the conclusion of the mathematical astronomers that the earth's average distance from the sun would only oscillate between certain minimum and maximum values.[13]

It was the critics of uniformitarian geology - forgotten men like Richard Kirwan, John Hunter, John Murray, and George Greenough - who protested that internal heat must, like all other heat, flow from hot to cold and therefore could not be assumed to remain constant unless it was being continually replenished somehow. But these qualitative objections carried little weight until after J. B. J. Fourier developed his mathematical theory of heat conduction at the beginning of the nineteenth century. By his own account it was the problem of terrestrial temperatures that led him to develop his theory,[14] and one of its first applications was the derivation of a simple formula relating the time required for a homogeneous sphere to cool down from a specified initial temperature. the present temperature gradient at the surface, and the conductivity and heat capacity of the substance. The temperature gradient (rate of change of average temperature with depth on going down into the earth) was a frequently measured quantity at this time since it could be used to find the central temperature by extrapolation, and Fourier could give rough estimates of the other parameters in his equation. The result was such a long period of time (about 200 million years) that Fourier did not even bother to write it down explicitly. He concluded that the rate of heat flow through the surface was so slow that it could have had no significant effect on the temperature during the epochs of importance to geologists at this time.[15] Thus the initial effect of Fourier's work was to confirm the uniformitarian view; it would cause difficulty only when the geologists later expanded their own time scale.

Quite apart from its geological application Fourier's theory of heat conduction marked a major turning point in the history of physics, for two distinct reasons. First, it established a new methodology for the formulation and solution of problems, based on partial differential equations. While such equations had been developed extensively by eighteenth-century mathematicians, Fourier's exploitation of trigonometric series expansions (though at first without adequate mathematical justification) greatly enhanced the power and scope of this mode of analysis. During the past 150 years physicists have followed Fourier by expressing their theories of electricity, magnetism, optics, gravity, gases, and continua in terms of differential equations; the trigonometric expansion has grown from a mathematical convenience to a new language - spectrum or harmonic analysis - which is especially suited to the description of physical processes in terms of frequencies. (The availability of this mathematical technique also reinforced the tendency to describe processes as forms of periodic wave-like motion.) Fourier's influence can be seen in both of the forms of quantum mechanics which were developed in 1925-26: Schrödinger's wave mechanics, based on a partial differential equation for the wave function, and Heisenberg's matrix mechanics, based on the manipulation of "Fourier coefficients" for transition involving radiation of various frequencies.

The second notable feature of Fourier's heat conduction theory is that it is explicitly based on a postulate of irreversibility: heat has a tendency to flow from hotter to colder parts of a body. so as to approach an equilibrium stale of uniform temperature.[16]

Irreversibility is expressed mathematically by the fact that the time variable t appears in a first-derivative term. If we follow the system backwards in time by reversing the sign of the time variable (replacing t by $-t$) we would see heat flow from cold to hot, a process which is not allowed by the theory. Newton's second law, by contrast, contains the time variable in a second derivative: $F = ma = d^2x/dt^2$, so the "time reversal" transformation $t \to -t$ leaves the right-hand side unchanged. Thus if we see an object falling toward the earth with velocity v, we observe that it has an acceleration a toward the earth due to a gravitational force F. An observer whose time direction is reversed (e.g., by running a movie film of this process backwards) will see the object moving *away* from the earth - the velocity $v = dx/dt$ has been *reversed* by the transformation $dt \to -dt$ - a process which is allowed by the theory; but he will still observe the same acceleration *toward* the earth because $d^2x/dt^2 = d^2x/d(-t)^2$.

For this reason Newton's laws are said to be *time-reversible*. Strictly speaking this is true only if one deals solely with *forces* which are themselves time-reversible. If we had introduced an air-resistance force, acting always in a direction opposite to the velocity v, into the problem in the previous paragraph, the motion would no longer be time-reversible.) It is sometimes asked why Fourier and other scientists in the first half of the nineteenth century did not discuss the apparent contradiction between reversible Newtonian mechanics and irreversible heat conduction; but in fact there is no contradiction at the phenomenological level since Newtonian mechanics had already been successfully applied to problems involving dissipative forces (such as air-resistance or friction) which are not time-reversible. The contradiction arises only when one assumes that all forces at the atomic level must he reversible.

A few years after Fourier presented his theory of heat conduction to the Academie des Sciences, Sadi Carnot published his essay on the motive power of fire and the efficiency of steam engines.[17] Carnot's essay is the source of the Second Law of Thermodynamics, established in its modern form by Rudolf Clausius in 1850, and thus is generally regarded as the origin of the principle of irreversibility in physics. Perhaps Carnot has been treated a little too generously in this respect by later writers, overcompensating for his neglect by the scientific community in his own lifetime. Irreversibility is mentioned rather casually in Carnot's memoir, not as a new scientific law but as a common-sense basis for the prescription that contact between bodies of different temperatures should be avoided in operating steam engines; such contact produces a flow of heat which could otherwise have been utilized in expanding the steam and performing mechanical work, hence entails a loss of motive power. The well-known tendency of heat to flow from hot to cold is now identified as the cause of inefficiency in steam engines, and is the reason why real engines can never attain their theoretical maximum efficiency according to Carnot's theory.

In the geological context there was nothing regrettable about the flow of heat from hot to cold; if the earth had not cooled down it would never have become hospitable to life. The unfavorable consequences of irreversibility were still in the distant future when the continued cooling of the earth and sun would presumably freeze us to death. But when Carnot pointed out that the production of useful work in steam engines depends on the *controlled* use of heat flow, it became evident that all *uncontrolled* heat flow must be, in human terms, a *waste* or *dissipation* of potential motive power. Though Carnot in 1824 had not yet decided that heat is actually transformed into mechanical work in the steam engine - he still considered heat or "caloric" to be a conserved substance - he did succeed in attaching negative connotations to its natural behavior.

William Thomson (Lord Kelvin) combined the theory of terrestrial refrigeration and the thermodynamic analysis of steam engines into a grand statement of a new law of nature: a "universal tendency toward dissipation of energy."[18] This was in 1852, after Thomson had written several papers on Fourier's theory of heat conduction, and had used it to extrapolate the earlier thermal history of a sphere with specified present temperature distribution and uniform conductivity. Thomson's statement included the assertion that the earth had been too hot in the past for human habitation, and would be too cold in the future. He had also studied the properties of steam, proposed an absolute temperature scale based on Carnot's theory, and accepted the interconvertibility of heat and mechanical work just a few months after Rudolf Clausius used this interconvertibility as the basis for the First Law of Thermodynamics (1850).[19]

The Second Law of Thermodynamics was formulated in various ways by Clausius, Thomson, and later writers. For some reason the negative versions seem to be favored by physicists, e.g., "it is impossible to obtain mechanical work by cooling a substance below the temperature of its surroundings" or "it is impossible to cause heat to flow from a low temperature to a high temperature without some equivalent compensation." It is fairly obvious that the basis for excluding such processes is that they are incompatible with the assumed natural tendency of heat to flow from hot to cold.

Clausius, who talked about "equivalent compensations," was forced to introduce a quantitative measure of the equivalence value of a transformation" in order to be able to compare heat flows with heat conversions. His measure, proposed in 1854, was the function dQ/T where dQ is the differential heat absorbed by a substance at [absolute] temperature T. Note that if the same quantity of heat dQ is emitted by a substance at temperature T_1 and absorbed by another substance at a *lower* temperature T_2, the net change in equivalence value ($- dQ/T_1 + dQ/T_2$) is necessarily positive. Thus the statement "heat always flows from hot to cold unless another compensating process occurs" becomes "the equivalence value of an uncompensated transformation is always positive."

The function dQ/T turned out to be so useful in thermodynamic analysis that Clausius decided to give it a short, snappy name; and thus "entropy" was baptized 11 years after its birth. (The Greek root means "turning inwards, modesty, humiliation, tricks or dodges.") Since dQ/T was defined as the *change* in entropy *(dS)*, the new statement of the Second Law was: entropy always increases (or remains constant in the special case of reversible processes). But the question "What is entropy?" remained unanswered within the macroscopic science of thermodynamics..

During those eleven years. when the concept of entropy was percolating in the mind of Clausius, the concept of disordered molecular motion was being revived and systematically developed by James Clerk Maxwell, following the publication of two important papers by Clausius in 1857 and 1858. After the interconvertibility of heat and mechanical work had been established, it seemed natural to many scientists that heat itself should be considered as a form of mechanical energy belonging to invisible molecules, and so the stage was set for the rediscovery of the "kinetic theory" of gases, proposed earlier by Daniel Bernoulli (1738), John Herapath (1821, 1848). and J. J. Waterston (1845), but not accepted by other scientists before 1848 James Prescott Joule and August Krönig advocated the kinetic theory in brief communications which appeared in 1848 and 1856 respectively, and Clausius then gave a systematic exposition attempting to show how it could explain a wide range of phenomena. He also demonstrated that the theory could be based on a concrete model which attributed properties such as size and atomic structure as well as mass and speed to the individual molecules.[20]

It is generally recognized that statistical ideas and methods were first introduced into physics in connection with the kinetic theory of gases.[21] At the same time, modern writers invariably point out that in this case statistics was used only as a matter of convenience in dealing with large numbers of particles whose precise positions and velocities at any instant are unknown, or, even if known, could not be used in practice to calculate the gross behavior of the gas. Nevertheless, it is claimed, nineteenth-century physicists always assumed that the gas is really a deterministic mechanical system, so that if the super-intelligence imagined by Laplace were supplied with complete information about all the individual atoms at one time he could compute their positions and motions at any other time as well as the macroscopic properties of the gas.[22] As Laplace himself had written in 1783, "The word 'chance,' then expresses only our ignorance of the causes of the phenomena that we observe to occur and to succeed one another in no apparent order."[23] This situation is to be sharply distinguished, according to the usual accounts of the history of modern physics, from the postulate of atomic randomness or indeterminism which was adopted only in the 1920's in connection with the development of quantum mechanics. Thus, part of the "scientific revolution" that occurred in the early twentieth century is supposed to have been a discontinuous change from classical determinism to quantum indeterminism.[24]

While I think it is legitimate to say that a revolution in physical thought has occurred since about 1800, I do not believe it is accurate to localize it in the two or three decades at the beginning of the twentieth century. Some of the most dramatic events did take place during that period, but they could not have had the impact they actually did if confidence in the mechanistic world view had not already been undermined to a considerable extent by developments in the nineteenth century. This argument has been presented elsewhere;[25] here I want to focus on one particular component of the revolution, the rejection of determinism and show that there was some degree of continuity between nineteenth- and twentieth-century ideas. I do not want to overstate the case for continuity; if it were possible to quantify the causal factors responsible for the ultimate effect, I would guess that twentieth-century events (including the discovery of radioactive decay though it actually occurred just before 1900) accounted for perhaps 80% of the impetus toward atomic randomness while the nineteenth-century background accounted for the remaining 20%. That 20% is still significant in view of the fact that most historians and scientists seem to give no weight at alt to the role of the well-publicized debates on the statistical interpretation of thermodynamics of the 1890's or to the well-established use of probability methods in kinetic theory.

The claim that nineteenth-century kinetic theory was based on molecular determinism must rely heavily on the evidence of the writings of James Clerk Maxwell and Ludwig Boltzmann; though in the absence of any explicit statements one might legitimately infer that they tacitly accepted the views of their contemporaries. In fact the situation is a little more complicated: their words were ambiguous but their equations pushed physical theory very definitely in the direction of indeterminism. As in other transformations of physical science - the cases of Kepler, Fresnel. Planck. and Heisenberg might be adduced here - mathematical calculation led to results that forced the acceptance of qualitatively different concepts.

Maxwell's earliest work in kinetic theory in particular his introduction of the velocity distribution law, seems to derive from the tradition of general probability theory and social statistics (as developed by Quetelet and interpreted by John Herschel) rather than from the mechanistic analysis of molecular motions.[26] Maxwell's law asserts that each component of the velocity of each molecule is a random variable which is statistically independent of every other component of the same and every other molecule. Only in his later papers did Maxwell attempt to justify the law by relating it to molecular collisions, and even then he needed to assume that the velocities of two colliding molecules are statistically independent. On the other hand, the computation of gas properties such as viscosity and thermal conductivity, whose comparison with experimental data provided the essential confirmation of the theory, did involve the precise dynamical analysis of collisions of particles with specified velocities, positions, and force laws. Without determinism in this part of the theory Maxwell could not have achieved his most striking successes in relating macroscopic properties to molecular parameters.

The debate on the statistical interpretation of irreversibility started in 1867, just after the publication of Maxwell's major work on kinetic theory. It was around the same time that Clausius. in a public lecture in Frankfort, pointed out the cosmic consequences of his new version of the Second Law of Thermodynamics (1865), "The entropy of the universe tends to a maximum":

The more the universe approaches this limiting condition in which the entropy is a maximum, the more do the occasions of further changes diminish; and supposing this condition to be at last completely obtained, no further change could evermore take place and the universe would be in a state of unchanging death.[27]

In Scotland, Maxwell was responding to his friend P. G. Tait's request for assistance in drafting a textbook exposition of thermodynamics, by imagining a tiny gatekeeper who could produce violations of the Second Law. Maxwell's immortal Demon proved that Victorian whimsy could relieve some of the gloom of the Germanic Heat Death. He is stationed at a frictionless sliding door between two chambers, one containing a hot gas, the other a cold one. According to Maxwell's distribution law, the molecules of the hot gas will have higher speeds on the average than those in the cold gas (assuming each has the same chemical constitution), but a few molecules in the hot gas will move more slowly than the average for the cold gas, while a few in the cold gas will travel faster than the average for the hot gas. The Demon identifies these exceptional molecules as they approach the door and lets them pass through to the other side, while blocking all others. In this way he gradually *increases* the average speed of molecules in the hot gas and *decreases* that in the cold, thereby in effect causing heat to flow from cold to hot

In addition to reversing the irreversible, Maxwell's demon gives us a new model for the fundamental irreversible process: he translates heat flow into molecular mixing. The ordinary phenomenon, heat passing from a hot body to a cold one, is now seen to be equivalent (though not always identical) to the transition from a partly ordered state (most fast molecules in one place, most slow molecules in another place) to a less ordered state. The concept of *molecular order and disorder* now seems to be associated with heat flow and entropy, though Maxwell himself doesn't make the connection explicit.[28]

Maxwell's conclusion was that the validity of the Second Law is not absolute but depends on the nonexistence of a Demon who can sort out molecules; hence it is a *statistical* law appropriate only to macroscopic phenomena.[29] That conclusion would have been less significant if the Second Law had originally been deduced, like other statistical laws. from analysis of large numbers of elementary events. But, having already been accepted as a member of the category of general "laws or nature," like the First Law of Thermodynamics, Newton's laws of motion, etc., the Second Law of Thermodynamics now became a rotten apple in the barrel, though this danger was not apparent until much later. Only in the 1920's did the suspicion arise that, for example, the *First* Law of Thermodynamics, "like the Second," might also have only statistical rather than absolute validity.

To call the Second Law a "statistical law" does not of course imply *logically* that it is based on random events - to the contrary, if Maxwell's Demon could not predict the future behavior of the molecules from the observations he makes as they approach the door, he could not do his job effectively. And in some of the later discussions it appeared that a relaxation of strict molecular determinism would make complete irreversibility *more* rather than less likely. (This was indeed the effect of the Burbury-Boltzmann "molecular disorder" hypothesis mentioned below.) Nevertheless at a more superficial level of discourse the characterization "statistical" conveyed the impression that an element of randomness or disorder is somehow involved.

Maxwell himself did not consistently maintain the assumption of determinism at the molecular level, though he occasionally supported that position, for example, in his lecture on "Molecules" at the British Association meeting in 1873. Yet in the same year, in private discussions and correspondence, he began to repudiate determinism as a philosophical doctrine. A detailed exposition of his views may be found in a paper titled "Does the progress of physical science tend to give any advantage to the opinion of necessity (or Determinism) over that of the contingency of events and the Freedom of the Will?" presented to an informal group at Cambridge University.[30] The answer was *no* - based on arguments such as the existence of singular points in the trajectory of dynamical systems, whereas an infinitesimal force can produce a finite effect. The conclusion was that "the promotion of natural knowledge may tend to remove that prejudice in favor of determinism which seems to arise from assuming that the physical science of the future is a mere magnified image of that of the past."

By 1875 Maxwell was asserting that molecular motion is "perfectly irregular; that is to say, that the direction and magnitude of the velocity of a molecule at a given time cannot be expressed as depending on the present position of the molecule and the time."[31] His article "Atom" in the *Britannica,* also published in 1875, indicated that this irregularity must be present in order for the system to behave irreversibly.[32]

Ludwig Boltzmann, whose publications and students transmitted the methods and results of kinetic theory from the nineteenth century to the twentieth, proposed a quantitative measure of irreversibility in 1872. Boltzmann, in what later became known as the "*H* theorem," generalized Clausius' entropy to a function defined in terms of the velocity distribution, and showed that collisions between uncorrelated molecules would push this function toward the value corresponding to the equilibrium Maxwell distribution.[33] When his Viennese colleague Josef Loschmidt pointed out that according to Newton's laws one should be able to return to any initial state by merely reversing the molecular velocities, Boltzmann argued that entropy is really a measure of the probability of a state, defined macroscopically.[34] While each *microscopic* state (specified by giving all molecular positions and velocities) has equal probability by assumption, macroscopic states corresponding to "thermal equilibrium" are really collections of large numbers of microscopic states and thus have high probability, while macroscopic states that deviate significantly from equilibrium consist of only a few microscopic states and have very low probability. In a typical irreversible process the system passes from a non-equilibrium state (e.g., high temperature in one place, low in another) to an equilibrium state (uniform temperature), i.e., from less probable (lower entropy) to more probable (higher entropy). To reverse this process it is not sufficient to start with an equilibrium state and reverse the velocities, for that will almost certainly lead only to another equilibrium state; one must pick one of the handful of very special microscopic states (out of the immense number corresponding to macroscopic equilibrium) which has evolved from a non-equilibrium state, and reverse its velocities. Thus it is possible that entropy may decrease. but extremely improbable.

The distinction between macro- and microstates is crucial in Boltzmann's theory though he failed to explain it as clearly as one might wish. Like Maxwell's Demon, an observer who could deal directly with microstates would not perceive irreversibility as an invariable property of natural phenomena. It is only when we decide to group together certain microstates and call them, collectively, "disordered" or "equilibrium" macrostates that we can talk about going from "less probable" to more "probable" states. This is an irreversible process in the same sense that shuffling the deck after dealing a grand slam hand in bridge is an irreversible process; the rules of the game single out certain distributions of cards as "ordered" (all the same suit or all aces, kings, and queens in the same hand) and we call these "rare" distributions although in fact *each* of the possible distributions of 52 cards among 4 hands of 13 each has exactly the same probability.

If you play bridge long enough you will eventually get that grand slam hand, not once but several times - the same is true with mechanical systems governed by Newton's laws, as Henri Poincaré showed with his *recurrence theorem* in 1889: if the system has fixed total energy and is restricted to a finite volume, it will eventually return as closely as you like to any given initial set of molecular positions and velocities.[35] If the entropy is determined by these variables, then it must also return to its original value, so if it increases during one period of time it must decrease during another. This apparent contradiction between the behavior of a deterministic mechanical system of particles and The Second Law of Thermodynamics was used by Ernst Zermelo in 1896 to attack the mechanistic world view. His position was that the Second Law is an absolute truth, so any theory that leads to predictions inconsistent with it must be false. This refutation would apply not only to the kinetic theory of gases but to any theory based on the assumption that matter is composed of particles moving in accordance with the laws of mechanics.[36]

Boltzmann had previously denied the possibility of such recurrences[37] and might have continued to deny their certainty by rejecting the determinism postulated in the Poincaré-Zermelo argument. (That was in fact Planck's strategy in 1908.[38]) Instead, he admitted quite frankly that recurrences are completely consistent with the statistical viewpoint, as the card-game analogy suggests (see above); they are *fluctuations* which are almost certain to occur if you wait long enough. So determinism leads to the same (qualitative) consequence that would be expected from a random sequence of molecular states! In either case the recurrence time is so inconceivably long that our failure to observe it cannot constitute an objection to the theory.

While Boltzmann sidestepped the issue of determinism in the debate on the recurrence paradox, maintaining a somewhat ambiguous "statistical" viewpoint, he had to face the issue more squarely in another debate that came to a head at almost the same time. E. P. Culverwell in Dublin had raised, in 1890, what might be called the "reversibility objection to the *H* theorem," not to be confused with the "reversibility paradox" discussed by William Thomson, Loschmidt, and Boltzmann in the 1870s. Culverwell asked how the *H* theorem could possibly be valid as long as it was based on the assumption that molecular motions and collisions are themselves reversible, and suggested that irreversibility might enter at the molecular level, perhaps as a result of interactions with the ether.[39]

The ether was always available as a source and sink for properties of matter and energy that didn't quite fit into the framework of Newtonian physics, although some physicists were by this time quite suspicious of the tendency of their colleagues to resolve theoretical difficulties this way.

Culverwell's objection was discussed at meetings of the British Association and in the columns of *Nature* during the next few years. It was S. H. Burbury in London who pointed out, in 1894, that the proof of the *H*-theorem depends on the Maxwell-Boltzmann assumption that the velocities of colliding molecules are uncorrelated. While this would seem a plausible assumption about the velocities *before* the collision, the proof involves balancing the effects of direct and inverse collisions and therefore is valid only if there is no correlation *after* the collision. If two molecules have just collided, one supposes that certain combinations of velocities (e.g., equal and in the same direction) have a different probability than one would expect by assuming these velocities are independent random variables. The postulate that they are still independent random variables after the collision would therefore seem to imply some kind of *continual* randomization of molecular states, which Burbury thought could be attributed to "disturbance from without [the system]. coming at haphazard."[40]

Boltzmann, who participated in the British discussions of the H theorem, accepted Burbury's hypothesis, which thereafter became known as "Boltzmann's hypothesis of molecular disorder." It is not equivalent to the hypothesis that molecular states occur completely at random; rather it amounts to an exclusion of special ordered states which would lead to violations of the Second Law.[41] In fact such ordered states *would* be generated by a random process, as Boltzmann noted in his discussion of the recurrence paradox. In modern terminology, one makes a distinction between "random numbers" and "numbers generated by a random process" - in preparing a table of random numbers for use in statistical studies, one rejects certain subsets, e.g., pages on which the frequencies of digits depart too greatly from 10%. because they are inconveniently non-random products of a random process.[42] Boltzmann recognized that the hypothesis of molecular disorder was needed to derive irreversibility, yet at the same time he admitted that the hypothesis itself may not always be valid in real gases, especially at high densities, and that recurrence may actually occur.[43]

Max Planck was one of those scientists who, in the 1880s, believed that the principle of irreversibility must be regarded as an absolute law of nature, even at the cost of abandoning the atomic theory, and his student Ernst Zermelo carried this viewpoint into the 1896-97 debate on the recurrence paradox. Yet it was Planck who, in 1900 and afterwards, insisted that Boltzmann's statistical interpretation of entropy is the key to the understanding of the quantum theory of radiation, and by his popular lectures promoted the thesis that the hypothesis of molecular disorder is necessary to the explanation of irreversibility. Planck's reversal of his earlier position furnishes a singular exception to his own oft-quoted dictum that new scientific ideas win out not by rational persuasion of the opponents but only by waiting for them to die out.[44]

In the 1890s Planck developed a "theory of irreversible radiation processes" in which he attempted to show that electromagnetic theory could provide an explanation of irreversibility where mechanics had failed. This would have been a victory for the electromagnetic world view which some physicists at that time expected to replace the mechanical world view. But two circumstances turned the thrust of his efforts in an unexpected direction. The first was the task of editing Gustav Kirchhoff's lectures on heat, which happened to include an exposition of the kinetic theory of gases, and thus forced Planck (always a meticulous scholar) to become thoroughly familiar with the mathematical apparatus of kinetic theory even though he had little faith in its physical validity. The second was Boltzmann's critique (1897-98) of the technical details of Planck's derivation of irreversibility from radiation theory. The result: Planck conceded that his theory really required an assumption of elementary disorder ("natural" radiation) very similar to that which Burbury and Boltzmann had used in kinetic theory and by 1899 he was formulating an "H-theorem" for radiation which was mathematically analogous to the original one for gases.

Even without an atomistic or statistical interpretation, entropy is a crucial quantity that links equilibrium and non-equilibrium (time-dependent) properties of physical systems. Before 1900 only a handful of scientists - Willard Gibbs, Boltzmann, Planck - realized this. For Planck, the study of the entropy-energy relation for a radiating system was the key to obtaining a theoretical formula for the observable frequency distribution of black body radiation, currently the object of great interest among his experimental colleagues in Berlin.[45] Planck probably could have found the correct distribution by using only the analytical methods of macroscopic thermodynamics; his debt to Boltzmann is evident only in the step from that distribution to the quantum hypothesis. Not only did Planck adopt Boltzmann's relation between entropy and probability; he could also find in one of Boltzmann's early papers an explicit calculation of the statistical properties of an assembly of particles with quantized energies.[46] All of this reinforced Planck's belief in the validity of the statistical interpretation of entropy.

Since in his later years Planck became known as a supporter of causality and determinism,[47] it is important to remember that he was an influential spokesman for indeterminism in the period of transition from classical to quantum mechanics. In 1898, introducing his hypothesis of "natural radiation" in order to derive irreversibility, he remarked that the disorder or indeterminacy [*Unbestimmtheit*] of the radiation components "lies in the nature of the subject" - not perhaps in nature itself but at least in any rational theory we can construct about nature.[48] The following year, in a lecture to a scientific congress at Munich, he referred again to the indeterminacy associated with the fact that we cannot specify precisely the frequency of a beam of radiation but must assume that the energy is distributed irregularly over a small range of frequencies.[49]

In 1900, as is well known, Planck proposed a new frequency distribution law for black body radiation, which led directly to the introduction of the quantum hypothesis. The extreme "revolutionary" account given in some textbooks attributes this step to a crisis of classical physics: the "ultraviolet catastrophe" predicted by naive application of statistical equilibrium theory to the modes of motion of the ether. If a Victorian physicist had believed that all possible vibration frequencies of the ether must share equally in the total energy, then he would have concluded that all energy would be sucked into vibrations of indefinitely high frequency, contrary to observation as well as to common sense. As it happened, no respectable Victorian physicist did believe that, and certainly Max Planck did not. His step toward a quantum theory of radiation was not a rejection of nineteenth-century statistical physics but a more passionate embrace of that much-maligned beauty. For it was precisely Boltzmann's formula relating entropy to probability, $S = k \log W$, which Planck used to develop his quantum theory; and it was Planck who first wrote the formula in its modern form with a precise numerical value for "Boltzmann's constant" k.[50]

The new physics was most definitely a legitimate offspring of the old, though adolescence brought defiant claims of independence and denials that anything useful could be inherited from a supposedly stiff and stuffy parent. Leaders of the younger generation like Paul Ehrenfest. insisted that there was an unbridgeable gap between classical and quantum principles: whoever accepted the latter must abandon the former, though the calculational methods developed in the nineteenth century would still be of immense utility. (This was especially true of Willard Gibbs' statistical mechanics, which was found to be well suited to quantum calculations.) Planck, representing the older generation, was reluctant to admit that quantum theory entailed any real break with classical physics, especially the wave theory of light. This attitude led him to assert, in 1932, that the world view of quantum physics is just as rigidly deterministic as that of classical physics.[51]

Perhaps Planck was really a determinist at heart throughout his long career;[52] it is not my purpose to convict him of inconsistency in his innermost beliefs. I claim only that by his scientific accomplishments and by his popular lectures and essays he helped to undermine the commitment of physics to determinism and popularize the efficacy of probabilistic thinking. At Leiden in 1908, and again at Columbia University (New York) in 1909, he reminded his audience that the principle of irreversibility could be justified only by postulating microscopic disorder.[53] A more extended statement of his views was given in a lecture at Berlin in 1914, where he noted that physics since the middle of the last century has adopted a method which is statistical rather than causal. Physics, Planck asserted, "does not deal with quantities that are absolutely determined; for every number obtained by physical measurements is liable to a certain possible error."[54] Yet this same lecture can be cited as evidence that on another level Planck does not after all want to abandon determinism; having made the distinction between dynamical and statistical laws, he rejects the possibility that *all* laws are statistical, and insists that some laws must ha taken as dynamical (i.e., exact) in order to provide some foundation for deducing the statistical ones. For, after all, "the assumption of absolute determinism is a necessary basis for every scientific investigation."[55]

At the first international physics congress, held at Paris in 1900, the Director of the Puy-de-Dôme Observatory stated: "The most important question, perhaps, of contemporary scientific philosophy is that of the compatibility or incompatibility of thermodynamics and mechanics." By "thermodynamics" Bernard Brunhes meant the irreversibility of the Second Law, and by "mechanism he intended to suggest the position that all natural phenomena can be explained by the application of Newton's laws.[56] Some physicists, like H. A. Lorentz, believed that compatibility had already been established by the postulate of molecular disorder.[57] But the debate continued with new arguments and examples. Max Born (1882-1970) recalled in 1948 that when he first began to read the scientific literature there was a violent discussion raging about statistical methods in physics and the validity of the *H*-theorem.[58] According to Born, it was Paul Ehrenfest and his wife Tatiana who finally cleared up the matter beyond any doubt. The Ehrenfests published in 1912 an extensive critical review of the foundations of the statistical approach in mechanics; while it had little to say about quantum theory, this article was widely read and quoted during the following decades and seems to have been regarded as the definitive treatment of the subject.[59] In addition to discussing the need for the hypothesis of molecular chaos and similar assumptions in the kinetic theory of gases, the Ehrenfests remarked that "the last few years have seen a sudden and wide dissemination of Boltzmann's ideas" in connection with studies of electrons, colloids, and radiation.[60] They conclude at one point that the "postulate of determinism" seems to hold only for "visible" states; as soon as one includes a microscope among the instruments available to the observer, one encounters phenomena such as Brownian movement for which determinacy no longer holds.[61]

Brownian movement, the irregular motion of microscopic particles suspended in fluids, had been observed long before the botanist Robert Brown established its general character in 1828.[62] Brown's achievement was to show that the motion could not be attributed to any supposed vitality of the particles themselves, since all kinds of inorganic as well as organic substances behaved similarly. While Brownian movement was often said to be an effect of the thermal molecular motion of the surrounding fluid, no quantitative theory of this effect was successful until the work of Einstein in 1905.

Before this, as Henri Poincaré reminded his audience at the 1900 Paris congress, Léon Gouy had argued that the continual renewal of the particle's motion by contact with a medium at constant temperature could be interpreted as a violation of the Second Law of Thermodynamics.[63] Einstein, who had done extensive work in statistical physics before 1905, recognized that the motion of small particles in fluids might provide an illustration of the need for the atomistic approach to thermodynamics, a case in which *fluctuations* produce an observable and calculable effect.[64] Whereas Boltzmann had argued against Zermelo that fluctuations of sufficient magnitude to return the system to an initial ordered state would be so rare as to escape human observation,[65] Einstein's theory indicated that the combination of many random molecular impulses - which produce no net effect *on the average* - could account for the entire motion of the Brownian particles. Jean Perrin's verification of Einstein's quantitative predictions provided not only a reconfirmation of the validity of the classical atomic-statistical theory of matter, but a dramatic new proof of the existence of the atom itself.[66] From about 1910 on, Brownian movement could be cited as visible evidence of atomic randomness.[67] Later it was to be cited for a contrary purpose, as an example of an *apparently* random phenomenon that could he explained by reduction to the deterministic motion of invisible entities, thus suggesting the possibility of escaping from the indeterminism of quantum mechanics by postulating causal "hidden variables."[68] The paradox, like many others in this field, is rooted in the ambiguous use of the word "statistical" mentioned above.

Einstein, even more than Planck, is notorious for his refusal to accept indeterminism as a necessary consequence of quantum mechanics ("God does not play dice").[69] His historical role in the fall of determinism is still more remarkable. He published three important papers in 1905; each of them could later be seen as an attack on determinism, though such was certainly not Einstein's intention. I have already mentioned the paper on Brownian movement (a misnomer, strictly speaking, since that term is not used in the text); the others deal with the special theory of relativity and with the photon theory of light (often referred to as the "paper on the photoelectric effect").[70]

Modern physicists and philosophers usually assert that relativity theory, unlike quantum mechanics, retains the causal determinism of Newtonian mechanics. While this assertion may be logically correct at present, it is historically misleading. Before about 1930, the two theories were often cited together as challenges to the entire structure of classical physics, and since relativity was the first to attain a definitive formulation it was available as a model in the development of quantum theory. The best example of this influence of relativity is to be found in Heisenberg's original paper on the Indeterminacy Principle, where he compares the impossibility of talking about the simultaneity of distant events with the impossibility of talking about the precise position and momentum of a particle.[71]

Einstein's photon theory of light has been specifically credited by Max Born as the inspiration for his own statistical interpretation of the wave function. In 1926, on first introducing this interpretation, Born wrote:

I adhere to an observation of Einstein on the relationship of wave field and light quanta; he said, for example, that the waves are present only to show the corpuscular light quanta the way, and he spoke in this sense of a "ghost field." This determines the probability that a light quantum, the bearer of energy and momentum, takes a certain path...[72]

In his Nobel lecture, Born was even more generous. saying that Einstein had tried to make the duality of particles - light quanta or photons. - and waves comprehensible by interpreting the square of the optical wave amplitudes as probability density for the occurrence of photons. This concept could be carried over at once to the ψ function....[73]

It is hard to find a concrete basis for these remarks in Einstein's published writings; the 1905 paper seems to me to indicate only that the statistical methods of classical gas theory are very useful in dealing with the behavior of systems of numerous entities whose precise nature (particle or wave?) is not known. In further papers on quantum theory during the next ten years, Einstein showed the utility of the statistical approach without committing himself to any assumptions about fundamental randomness. The high point of his work in this period was a derivation of Planck's distribution law for blackbody radiation from the assumption of spontaneous as well as stimulated emission of radiation by matter. Other physicists acclaimed this derivation as the final emancipation of quantum theory from any reliance on classical physics. But for Einstein himself,

The weakness of the theory lies on the one hand in the fact that it does not get us any closer to making the connection with wave theory; on the other, that it leaves the duration and direction of the elementary processes to "chance" Nevertheless I am fully confident that the approach chosen is a reliable one.[74]

For Max Born, this derivation was the decisive step toward indeterminism, the step which "made it transparently clear that the classical concept of intensity of radiation must be replaced by the statistical concept of transition probability."[75]

Another major discovery at the beginning of the twentieth century seemed to offer direct evidence of indeterminism in nature. As Einstein noted in his 1916 paper on radiation, "the statistical law which we have assumed corresponds to that of radioactive reaction,"[76] and indeed by that time radioactive decay had become the best known example of a random process. In 1902 Rutherford and Soddy wrote:

The idea of the chemical atom in certain cases spontaneously breaking up with evolution of energy is not of itself contrary to anything that is known of the properties of atoms for the causes that bring about the disruption are not among those that are yet under our control.[77]

Physicists seemed to have no compunctions about using the word "spontaneous" to describe an event which is at the same time determined by some cause. For the progress of physics the important fact was that the atoms behave *as if* they could decide by themselves, independently of all external influences, when to explode. From that fact would follow the observed exponential decay of radioactivity, and thus Egon von Schweidler and others, starting in 1905, could construct statistical theories of radioactive processes.[78]

One could ask "was Rutherford a determinist?" and one could easily argue that Rutherford, like Maxwell, Boltzmann, Planck, and Einstein, would find it inconceivable that anything in nature could happen "by chance" without any cause at all. Hence whenever we find them using the words "probability," "chance," "statistical," or "spontaneous," we must assume that such terms only refer to our *lack of knowledge* of causes, not to the *absence of causes*. The fallacy of that interpretation is that it could apply equally well to Born and Heisenberg, or to anyone who believes in an *"uncertainty* principle" as distinct from an *"indeterminacy* principle." Heisenberg himself emphasized that his principle applies to our *knowledge* about the world. The assertion that quantum mechanics has abandoned causality means that in the proposition "If we know the present exactly we can predict the future" it is not only the conclusion but also the premise that is false, according to Heisenberg:

One might be led to the conjecture that under the perceptible statistical world there is hidden a "real" world in which the causal law holds. But it seems to us that such speculations, we emphasize explicitly, are fruitless and meaningless. Physics should only describe formally the relations of perceptions...[79]

Clearly the transition from determinism to indeterminism is linked with the positivistic-pragmatic-operationalist-instrumentalist-phenomenalist attitude that many physicists adopted in the early twentieth century, partly influenced by Ernst Mach and other critics of nineteenth-century mechanism, partly as a result of the difficulty of fitting the new phenomena of atomic physics into any consistent theoretical scheme. Positivism (as I call this complex of views) is a retreat from the aspiration to know and understand everything, an admonition to be content with the partial knowledge that can be attained at a particular stage in the development of theory and experiment. A positivist may call himself an indeterminist, meaning that his science cannot determine that which lies beyond present observation; indeterminism is then the same as "uncertainty." (This was the position of Bohr, Heisenberg, and Eddington.) Or he may call himself a determinist, meaning that his theory correlates all known or knowable facts about the observable world, and that anything beyond that is not his concern anyway. (It is in this sense that Planck and some modern physicists and philosophers say that quantum mechanics is a deterministic theory.) Conversely, an anti-positivist or realist may be a determinist in the Laplacian sense, or an indeterminist who insists that there is a fundamental randomness in nature. Since this last position is rather hard to defend, the acceptance of indeterminism was greatly facilitated by the (at least temporary) acceptance of positivism.

Yet the acceptance of indeterminism could not have occurred unless determinism had first been shown to be either untenable or inconvenient in scientific work, since positivism is not ordinarily a very attractive stance for physicists. The contribution of nineteenth-century statistical physics was to show that in many cases the use of deterministic laws is not convenient in dealing with systems of many particles even though the notion of one or two such particles can be completely determined. Developments in the first two decades of the twentieth century showed that statistical physics can also, deal with systems whose behavior does *not* seem to be determined by known causes. The positivistic attitude allowed the question of ultimate determinism to be set aside even as it encouraged the pragmatic use of statistical methods that carried the flavor of indeterminism.

A lecture by Paul Langevin in 1913 shows how statistical thermodynamics provided a link between the old and the new science. He pointed out that the physics of discontinuity must use the calculus of probabilities, which is the only possible link between the world of atoms and our observations. Thanks to Boltzmann, statistical methods permeate physics[80]

As I indicated earlier, the problem of molecular randomness was to some extent separate from the problem of the statistical nature of the Second Law (irreversibility principle). For those physicists who had not studied in detail the writings of Burbury, Boltzmann, Planck, and Ehrenfest on molecular disorder, the more important point was that the Second Law is not absolutely true, but still good enough for all practical purposes. It was no longer certain that entropy will not decrease, only highly probable.

If the Second Law of Thermodynamics - regarded by many physicists as one of the most important laws of nature to have been discovered since the time of Newton - is "only" statistical, how could one be certain that all other laws are absolutely valid? As J. M. Jauch puts it,

If statistical laws can behave nearly deterministically for systems with large degrees of freedom, could it not be true that *all* physical process laws, even the most fundamental ones, are perhaps only statistical laws and that they appear deterministic only because we cannot, with the usual observations, discern the fluctuations?[81]

Or perhaps the quest for total knowledge of the world was a symptom of the arrogance of scientists who ought to be satisfied with limited statistical knowledge. There were strong social pressures, especially in Weimar Germany, urging scientists to abandon the claim that science and only science could lead to certain knowledge. These pressures have been analyzed by Paul Forman, who presents a persuasive argument that several statements by physicists renouncing causality in the early 1920s were made in response to external criticisms of science.[82] To his account I would add the remark that the well-known statistical interpretation of irreversibility suggested a defensible fall-back position for any physicist who felt compelled to retreat from absolute causality. And in fact that was exactly the suggestion of three of the scientists discussed by Forman: Franz Exner,[83] Walther Nernst,[84] and Erwin Schrödinger.[85] Perhaps all laws are only statistical; in particular, the First Law of Thermodynamics (energy conservation).

The influence of these anti-causal statements on the later Born-Heisenberg pronouncements of indeterminism is difficult to estimate. Schrödinger's lecture was not published until 1929, and in 1931 he stated that his and Exner's ideas attracted little attention when first voiced.[86] Perhaps one can say no more than that they reflected as well as promoted a general feeling among physicists in the 1920s that classical determinism had become untenable, and that the only hope of progress was in the exploitation of statistical ideas. In any case, it would not have been enough to say "make the First Law statistical, like the Second" without a much more definite formulation of a method for calculating probabilities of atomic processes[87] and that was attained only by the Heisenberg-Schrödinger quantum mechanics in 1926.

Alfred Landé, one of the founders of quantum theory, has made probably the strongest possible statement on this point:

Starting from the Second Law as fundamental, the necessity of abandoning strict determinism in the molecular distribution in favour of irreducible laws of probability appears as a strange but necessary consequence. The peculiar indeterminacy of quantum physics, however, cannot be based on the Second Law alone, although it, too, is of a thermodynamics origin in spite of its apparently pure mechanical character.[88]

The final stages in the fall of determinism have been described in detail by Jammer, Klein, and others,[89] so only a brief summary is needed here. Niels Bohr, noting the success of Einstein's 1916 theory mentioned above, suggested in 1923 that rather than "seek a *cause* for the occurrence of radiative processes ....we simply assume that they are governed by the laws of probability."[90] The next year, with H. A. Kramers and J. C. Slater, he published a new radiation theory in which strict energy conservation was abandoned in favor of conservation on the average. Though the Bohr-Kramers-Slater theory was immediately overturned by experiments of Bothe and Geiger and of Compton and Simon, the desirability of introducing explicit randomness into quantum theory began to be widely discussed.

When Schrödinger gave de Broglie's proposal for a wave theory of matter its comprehensive mathematical formulation, the particle nature of the electron seemed to be in doubt. Born, on the basis of recent experiments on interatomic collisions by his colleague James Franck, felt it necessary to maintain the particle conception of the electron, and therefore was led to abandon his earlier belief that "the statement that a system is at a given time and place in a certain state probably has a meaning."[91] His solution was to propose that the wave function in Schrödinger's theory simply determines the probability that the electron is in a certain region of space. The wave function itself and its temporal evolution are determined by the force laws and the initial state of the system, in somewhat the same way that states are determined in Newtonian mechanics, but there seems to be no causal determinacy at the level of individual events.[92]

Werner Heisenberg has recently recalled that Bohr advised him to read Willard Gibbs' work on statistical thermodynamics in 1924.[93] (Bohr himself had been fascinated by the problems of statistical physics since the beginning of his own scientific career.[94]) Heisenberg has retrospectively identified the statistical interpretation of thermodynamics, which he attributes to Gibbs, as part of the breakdown of classical physics.[95] But Heisenberg's own formulation of the Indeterminacy Principle in 1927 showed no trace of any indebtedness to Gibbs, Boltzmann, or Maxwell. Like the original phenomenal statements of the Second Law by Clausius and Thomson, it was phrased negatively: it is impossible by any measurement to determine both the position and momentum of a particle with such an accuracy that the product of the errors of observation of these two quantities is less than an irreducible minimum value, proportional to Planck's quantum constant.

The revolution was complete (though its reverberations outside physics are only just beginning to be felt); but its history was almost obliterated![96]

Notes

[1] K. W. Ford, *Basic Physics* (Waltham, Mass., 1968), 730. A similar suggestion was made by S. Ulam, "Marian Smoluchowski and the theory of probabilities in physics," *American Journal of Physics,* **25** (1957), 475-81.

[2] A. S. Eddington, "The decline of Determinism," *Mathematical Gazette*, **16** (1932), 66-80.

[3] S. G. Brush, "Thermodynamics and History," *The Graduate Journal*, **7** (1967), 477-565; *idem*, "Romance in Six Figures," *Physics Today*, **22** (Jan. 1969), 9; L. Badash, "The Completeness of Nineteenth Century Science," *Isis*, **63** (1972) 48-58.

[4] Max Jammer, "Indeterminacy in Physics," *Dictionary of the History of Ideas*, ed. P. P. Wiener (New York, 1973), **II** , 586-94; *The Conceptual Development of Quantum Mechanics* (New York, 1967), 323-45. Paul Forman, "Weimar Culture, Causality, and Quantum Theory, 1918-1927: Adaptation by German Physicists and Mathematicians to a Hostile Intellectual Environment," *Historical Studies in the Physical Sciences*, **3** (1971), 1-115. Martin J. Klein, "The First Phase of the Bohr-Einstein Dialogue," *Historical Studies in the Physical Sciences*, **2** (1970), 1-39. E. Cassirer, *Determinism and Indeterminism in Modern Physics*, trans. O. T. Benfey (New Haven, 1956).

[5] Y. Elkana, *The Discovery of the Conservation of Energy* (London, 1974), 14-17.

[6] Loren Eisley, "The Intellectual Antecedents of *The Descent of Man*," *Sexual Selection and the Descent of Man 1871-1971*, ed. B. Campbell (Chicago, 1972), 1-16. Michael Ghiselin, *The Triumph of the Darwinian Method* (Berkeley, 1969), 54-67. A. J. Lotka, *Elements of Physical Biology* (Baltimore, 1925), 34-39.

[8] J. H. Randall, Jr., *The Making of the Modern Mind* (Boston, rev. ed., 1940).

[9] Marie Boas, "The Establishment of the Mechanical Philosophy," *Osiris*, **10** (1952), 412-541. David Kubrin, "Newton and the Cyclical Cosmos: Providence and the Mechanical Philosophy," *JHI*, **28** (1967), 325-46. H. G. Alexander, *The Leibniz-Clarke Correspondence* (Manchester, 1956).

[10] Isaac Newton, *Opticks* (London, 4th ed. 1730, rept. New York, 1952), 3970400. Alexander, *op. cit.*, **13-14**, 177-180.

[11] J. J. Dortous de Mairan, "Sur la cause generale du Froid en Hyver, & de la chaleur en Éte," *Memoires de Mathematique et de Physique tirés des Registres de l'Academie Royale des Sciences de l'Année, M.DCCXIX*, 104-35. George-Louis Leclerc, Comte deBuffon, *Introduction à l'Histoire des Mineraux* (Paris, 1774); *Oeuvres Complètes de Buffon*, ed M. Flourens, nouv. Ed. (Paris, n.d.), **IX** F. C. Haber, *The Age of the World* (Baltimore, 1959).

[12] Gordon Davies, *The Earth in Decay* (London, 1969), 170-188.

[13] John Playfair, *Illustrations of the Huttonian Theory of the Earth* (London, 1802, rept. New York, 1964), 437-38.

[14] J. B. J. Fourier, "Mémoire sur les températures du Globe terrestre et des espaces planétaires," *Mémoires de l'Académie Royale des Sciences de l'Institut de France*, **7** (1827), 570-604. *Oeuvres de Fourier* (Paris, 1890), **II**, 114.

[15] J. B. J. Fourier, "Mémoire sur le refroidissement séculaire du globe terrestre," *Annales de Chimie*, **13** (1819), 418-37.

[16] See version in the manuscript of Fourier's 1807 paper, published in I. Grattan-Guinness, *Joseph Fourier 1768-1830* (Cambridge, Mass., 1972), 33; cf. Joseph Fourier, *The Analytical Theory of Heat*, trans. A. Freeman (New York, 1955, rept. of the 1878 ed.), 14.

[17] Sadi Carnot, *Réflexions sur la Puissance Motrice de Feu et sur les Machines propres à developper cette Puissance* (Paris, 1824). E. Mendoza, *Reflections on the Motive*

Power of Fire by Sadi Carnot and other Papers on the Second Law of Thermodynamics by E. Clapeyron and R. Clausius (New York, 1960).

[18] William Thomson, "On a Universal Tendency in Nature to the Dissipation of Mechanical Energy," *Proceedings of the Royal Society of Edinburgh*, **3** (1857), 139-42; *Philosophical Magazine*, **4** (1852), 304-06.

[19] For references and further discussion: S. G. Brush, "The Development of the Kinetic Theory of Gases, VIII. Randomness and Irreversibility," *Archive for History of Exact Sciences*, **12** (1974), 1-88, #3; hereafter "Randomness and Irreversibility."

[20] S. G. Brush, "The Development of the Kinetic Theory of Gases," *Annals of Science*, **13** (1957), 188-98, 273-82; **14** (1958), 185-96; *Kinetic Theory*, Vol. 1, *The Nature of Gases and of Heat* (Oxford, 1965).

[21] K. Mendelssohn, "Probability Enters Physics," *Turning Points in Physics* (Amsterdam, 1959; New York, 1961), 45-67. Elizabeth Garber, "Aspects of the Introduction of Probability into Physics," *Centaurus*, **17** (1972), 11-39. C. C. Gillispie, "Intellectual Factors in the Background of Analysis by Probabilities," *Scientific Change*, ed. A. C. Crombie (New York, 1963), 431-53. W. Krajewski, "The Idea of Statistical Law in Nineteenth-Century Science, *Boston Studies in the Philosophy of Science*, **14** (1974), 397-405.

[22] A good example of Laplacian determinism may be found in the book by W. Stanley Jevons, *The Principles of Science* (New York, 1958, rept. of 2nd ed. 1877), 738-39. Jevons asserts that ""chance" cannot be a *cause* of events in nature; it is merely an expression of our own ignorance" (*ibid.,* 198).

[23] Quoted by C. C. Gillispie, *Proceedings of the American Philosophical Society*, 116 (Feb. 1972), 10. A similar view was stated by Benjamin Peirce on p. 120 of his article: "On the Connection of Comets with the Solar system," *Proc. of American Assoc. for the Advancement of Science*, **2** (1849), 118-22.

[24] J. von Neumann, *Mathematical Foundations of Quantum Mechanics*, trans. R. T. Beyer from German ed. of 1932 (Princeton, 1955), 207. W. Heitler, "The Departure from Classical Thought in Modern Physics," *Albert Einstein Philosopher-Scientist*, ed. P. A. Schilpp (New York, 1949), 181-98. J. B. Conant, *Modern Science and Modern Man* (New York, 1952), 84-85. George Gamow, *Biography of Physics* (New York, 1961), 258. Rudolf Carnap, *Philosophical Foundations of Physics* (New York, 1966), 378-81. A. M. Bork, "Randomness and the Twentieth Century," *Antioch Review*, 27 (1967), 40-61.

[25] S. G. Brush, *The Kind of Motion We Call Heat* (Amsterdam, 1976), #1.6.

[26] See "Randomness and Irreversibility," #4, notes 22 & 23.

[27] R. Clausius, "On the Second Fundamental Theorem of the Mechanical Theory of Heat," *Philosophical Magazine* 4, **35** (1868), 419.

[28] C. G. Knott, *Life and Scientific Work of Peter Guthrie Tait* (Cambridge, 1911), 213-15. J. C. Maxwell, *Theory of Heat* (London, 1871).

[29] Maxwell's letters are published in Knott, *op. cit.,* 115-16, 214-15. S. Amsterdamski has asserted that the impossibility of Maxwell's Demon was the first step toward discrediting the concept of an ideal observer, the final step being Heisenberg's principle of indeterminacy. *Between Experience and Metaphysics* (Dordrecht, 1975), 71.

[30] Lewis Campbell and William Garnett, *The Life of James Clerk Maxwell* (London, 1882; rept. with additions, New York, 1969), 434; *The Scientific Papers of James Clerk Maxwell*, ed. W. D. Niven (Cambridge, 1890; rept. New York, 1965), II, 760.

[31] J. C. Maxwell, "On the Dynamical Evidence of the Molecular Constitution of Bodies," *Nature*, 11 (1875), 357-59, 374-77; quotation from *The Scientific Papers of James Clerk Maxwell*, II, 436. It is not clear whether, in the context of a lecture to the Chemical Society of London, Maxwell intended to leave out the possible determination of the molecule's velocity by its past history and that of the other molecules in the system.

[32] J. C. Maxwell, "Atom," *Encyclopedia Brittanica* (Edinburgh, 8th ed., 1875), III, 36-49; *The Scientific Papers of James Clerk Maxwell*, II, 462.

[33] L. Boltzmann, "Weitere Studien über das Wärmegleichgewicht unter Gasmolekülen," *Sitzungsberichte der Mathematisch-Naturwissenschaftlichen Classe der kaiserlichen Akademie der Wissenschaften, Wien*, 66 (1872), 275-370; English translation in S. G. Brush, *Kinetic Theory* (Oxford, 1966), II, 88-175. This journal is cited hereafter as *Wien. Ber.*

[34] J. Loschmidt, "Über den Zustand des Wärmegleichgewichtes eines Systemes von Körpern mit Rucksicht auf die Schwerkraft, " *Wien. Ber.*, 73 (1876), 128-42. L. Boltzmann, "Über die Beziehung eines allgemeine mechanischen Satzes zum zweiten Haupsatze der Wärmetheorie," *ibid.*, 75 (1877), 67-73; English translation in S. G. Brush, *Kinetic Theory*, ii, 188-93.

[35] H. Poincaré, "Sur le problème des trois corps et les équations de dynamique," *Acta Mathematica*, 13 (1890), 1-270.

[36] Translations of the contributions of Poincaré, Zermelo, and Boltzmann to this debate are in S. G. Brush, *Kinetic Theory*, II, 194-245; "Randomness and Irreversibility," #7.

[37] L. Boltzmann, "Der zweite Hauptsatz der mechanische Wärmetheorie," *Almanach der kaiserliche Akademie der Wissenschaften, Wien*, 36 (1886), 225-59; quoted in "Randomness and Irreversibility," 57.

[38] Max Planck, "Die Einheit des Physikalischen Weltbildes," *Physikalische Zeitschrift*, 10 (1909), 62-75; *A survey of Physical Theory* (New York, 1960), 16.

[39] "Randomness and Irreversibility," #6, for further references and details.

[40] S. H. Burbury, "Boltzmann's Minimum Function," *Nature*, 51 (1894), 78.

[41] This distinction has been emphasized by T. S. Kuhn (private communication).

[42] M. G. Kendall and B. Babington Smith, *Tables of Random Sampling Numbers*, Tracts for Computers No. XXIV (Cambridge, 1939, rept. 1954), ix.

[43] L. Boltzmann, *Lectures on Gas Theory*, trans. S. G. Brush from the German ed. of 1896-98 (Berkeley and Los Angeles, 1964), 41-42.

[44] Max Planck, *A Scientific Autobiography and Other Papers* (London, 1950), 33-34; *Philosophy of Physics* (New York, 1936), 97.

[45] A few modern texts still repeat the myth that Planck was led to his quantum theory by the need to avoid the "ultraviolet catastrophe" predicted by classical statistical mechanics. For an accurate account of the origins of Planck's theory: Martin J. Klein, "Max Planck and the Beginnings of the Quantum Theory," *Archive for History of Exact Sciences*, 1 (1962), 461-79; Hans Kangro, *Vorgeschichte der Plankschen Strahlungsgesetzes* (Wiesbaden, 1970).

[46] L. Boltzmann, "Über die Beziehung zwischen des zweiten Hauptsatze der mechanischen Wärmetheorie und der Wahrscheinlichkeitsrechnung, respective den Satzen über das Wärmegeleichgewicht," *Wien. Ber.*, **76** (1877), 373-435.

[47] Max Planck, *Kausalgesetz und Willensfreiheit* (Berlin, 1923), 31-34, Eng. trans. *The New Science* (New York, 1959), 98-101; *Physikalische Gesetzlichkeit im Lichte neueren Forschung* (Leipzig, 1926), rept. in *Physikalische Abhandlungen und Vorträge* (Braunschweig, 1958), **III**, 159-71; *Der Kausalbegriff in der Physik* (Leipzig, 1932). quoted in Brush, "Randomness and Irreversibility," 85; Planck, "The concept of Causality," *Proceedings of the Physical Society of London*, **44** (1932), 529-39; *Determinismus oder Indeterminismus* (Leipzig, 1938).

[48] Max Planck, "Über irreversible Strahlungsvorgänge. 4. Mitteilung," *Sitzungsberichte der Preussischen Akademie der Wissenschaften, Physikal-Math. Klasse* (Berlin, 1898), 449-76. Hans Kangro, ed., *Planck's Original Papers in Quantum Physics* (London, 1972), 7, 12, 23-24, 39, 43, 54.

[49] Max Planck, "Über irreversible Strahlungsvorgänge," *Annalen der Physik*, series 4, **1** (1900), 69-122. Giulio Giorello, "Le 'Ipotesi del disordine' nell'opera di Max Planck: Caos moleculare e radiazione naturale," *Alcuni Aspetti dello Sviluppo delle teorie fisiche 1743-1911* (Pisa, 1972), 239-342.

[50] Max Planck, "Zur Theorie des Gesetzes der Energieverteilung im Normalspectrum," *Verhandlungen der Deutschen Physikalischen Gesellschaft*, **2** (1900), 237-45. Kangro, ed., *op. cit.*, 12-13, 44, 55-58. Max Planck, *The Theory of Heat Radiation*, trans. M. Masius from the 1913 German ed. (New York, 1959), 118-20.

[51] See note 47 above.

[52] Stanley Goldberg, "Max Planck's Philosophy of Nature and his response to Einstein's Special Theory of Relativity," *Historical Studies in the Physical Sciences*, **7** (Princeton, 1976), 125-60.

[53] M. Planck, *op. cit.*, (note 38).

[54] M. Planck, "Dynamische und statistische Gesetzmässigkeit" (Leipzig, 1914), rept. in *Physikalische Abhandlungen und Vorträge*, **III**, 77-90; quotation from *A Survey of Physical Theory* (New York, 1960), 57-58.

[55] *A Survey of Physical Theory*, 67-68.

[56] B. Brunhes, discussion remark, *Traveaux du Congrès International de Physique, Paris, 1900*, **4** (Paris, 1901), 29.

[57] H. A. Lorentz, "De theorie der straling en de tweede wet der thermodynamica," *Verslagen der Zittengen van de Wis- en Natuurkundige Afdeeling der Koninlijke Akademie van Wetenschappen, Amsterdam*, **9** (1900), 418-34; Lorentz's *Collected Papers*, **6** (The Hague, 1938), 270.

[58] Max Born, *Natural Philosophy of Cause and Chance* (Oxford, 1949), 58.

[59] Paul and Tatiana Ehrenfest, "Begriffliche Grundlagen der statistischen Auffassung in der Mechanik," *Encykl. der math. Wissenschaften* (1911), **IV**, 32.

[60] *Idem*, *The Conceptual Foundations of the Statistical Approach in Mechanics*, trans. M. J. Moravcsik of article in note 59 (Ithaca, 1959), 68.

[61] *Ibid.*, 36-37.

[62] S. G. Brush, "A History of Random Processes. I. Brownian Movement from Brown to Perrin," *Archive for History of Exact Sciences*, **5** (1968), 1-36.

[63] L. Gouy, "Note sur le mouvement brownien," *Journal de Physique,* series 2, **7** (1888), 561-64. H. Poincaré, "Relations entre la physique expérimentale et la physique mathématique," *Rapports présentés au Congrès International de Physique réuni a Paris en 1900* (Paris, 1900), **I**, 1-29.

[64] Einstein, "Ueber die von der molekular-kinetischen Theorie der Wärme geforderte Bewegung von in ruhended Flussigkeiten suspenierten Teilchen, " *Annalen der Physik,* series 4, **17** (1905), 549-60; Brush, *op. cit.,* (note 62).

[65] Boltzmann's 1896 article is translated in Brush, *Kinetic Theory,* **II**, 223.

[66] Brush, *op. cit.,* (note 62); Mary Jo Nye, *Molecular Reality* (London, New York, 1972).

[67] P. and T. Ehrenfest, *op. cit.* (note 60), 36-37. Planck, *A Survey of Physical Theory,* 62-63. Franz Exner, *Vorlesungen über die physikalischen Grundlagen der Naturwissenschaften* (Wien, 1919). E. Schrödinger, "Was ist ein Naturgesetz," [lecture given in 1922] *Naturwissenschaften,* **17** (1929), 9-11.

[68] David Bohm, *Causality and Chance in Modern Physics* (London, New York, 1957, rept. 1961), 81, 113; "A proposed explanation of Quantum Theory in terms of Hidden Variables at a Sub-Quantum-Mechanical Level,", *Observation and Interpretation in the Philosophy of Physics,* ed. S. Körner (London, 1957; rept. New York, 1962), 33-40.

[69] *The Born-Einstein Letters* (New York, 1971), 82, 91, 149, 155, 158, 188, 199. *Albert Einstein Philosopher-Scientist* (New York, 1949), 86-87.

[70] Einstein, "Elektrodynamik bewegter Körper," *Annalen der Physik,* **17** (1905), 891-921; "Über einen die Erzeugung und Verwandlung des Lichtes betreffenden heuristischen Gesichtspunkt," *Ibid.,* 132-48.

[71] Heisenberg, "Über den anschaulichen Inhalt der quantentheoretischen Kinematik und Mechanik," *Zeitschrift fur Physik,* **43** (1927), 172-98; 179. T. Chapman has argued that relativity theory does not imply determinism: "Special Relativity and Indeterminism," *Ratio,* **15** (1973), 107-13.

[72] Born, "Quantenmechanik der Stossvorgänge," *Zeitschrift fur Physik,* **38** (1926), 803-27; quotation from G. Ludwig, *Wave Mechanics* (Oxford, 1968), 207; Born, "Einstein's Statistical Theories," *Albert Einstein Philosopher-Scientist,* 167.

[73] Born, "Die Statistische Deutung der Quantenmechanik" (Nobelvortrag, 1954); *Nobel Lectures, Physics, 1942-1962* (Amsterdam & New York, 1964), 262.

[74] Einstein, 'Zur Quantentheorie der Strahlung," *Mitteilungen der Physikalische Gesellschaft, Zurich,* **18** (1916), 47-62; quotation from B. L. van der Waerden, *Sources of Quantum Mechanics* (Amsterdam, 1967), 76.

[75] Born, *Nobel Lectures, Physics, 1942-1962,* 258; *Albert Einstein Philosopher-Scientist,* 172. Max Jammer is skeptical about this: *Conceptual Development of Quantum Mechanics,* 113.

[76] Einstein, in van der Warerden, *Sources of Quantum Mechanics,* 66.

[77] E. Rutherford and Frederick Soddy, "The Radioactivity of Thorium Compounds. II. The Cause and Nature of Radioactivity," *Journal of the Chemical Society, Transactions.* **81** (1902), 837-60; quotation from Alfred Romer, *The Discovery of Radioactivity and Transmutation* (New York, 1964), 147.

[78] E. v. Schweidler, "Über Schwankungen der radioaktiven Umwandlung," *Premier Congrès International pour l'Étude de la Radiologie et de l'Ionisation tenu à Liege du 12 au 14 septembre 1905, Comptes Rendus* (Bruselles, 1906), "Communications

Presentées, Section de Physique, Langue allemande," 1-3 (each subsection is paginated separately). Max von Laue attributes great significance to this paper: "Here, for the first time physics encountered a process that was not accessible to causality.... The importance of the Schweidler theory lies in the fact that subsequently many other atomic processes were discovered for which the physicist can state a probability quite well without being able to determine causally the instant of their occurrence. The Schweidler viewpoint can be carried over to all these cases." *History of Physics* (New York, 1950), 111; also 140.

[79] Heisenberg, *op. cit.* (note 71), 197.

[80] P. Langevin, "La Physique du discontinu" [Conference fait à la Société francaise de Physique le 27 novembre 1913], *Les Progrès de la Physique Moléculaire* (Paris, 1914), 1-46.

[81] J. M. Jauch, "Determinism in classical and quantal physics," *Dialectica,* **27** (1973), 13-24; quotation from 19, Cf. also E. Schrödinger, "Might perhaps Energy be a merely Statistical Concept?" *Il Nuovo Cimento,* series 10, **9** (1958), 162-70.

[82] Forman, *op.cit.* (note 4). Maurice Kendall has suggested that the acceptance of inherent randomness in nature may have been part of an iconoclastic reaction to the repressive society of the 19th century, and that we may subsequently revert to an "ignorance" (epistemological) interpretation of chance. See "Chance," *Dictionary of the History of Ideas* (New York, 1972), **I**, 335-40. For a rather bombastic version of the role of entropy, probability, and radioactive decay in German thought at this time: Oswald Spengler, *Der Untergang des Abendlandes* (Munich, 1950), I. Bd., VI Kap., #14; trans. C. F. Atkinson, *The Decline of the West* (New York, abr. ed., 1932), 217-18.

[83] Exner, *op. cit.* (note 67).

[84] W. Nernst, "Zum Gültigkeitsbereich der Naturgesetze," *Naturwissenschaften,* **10** (1922), 489-95.

[85] E. Schrödinger, *op. cit.* (note 67); "Bohrs neue Strahlungshypothese und der Energiesatz," *Naturwissenschaften,* **12** (1924), 720-24.

[86] E. Schrödinger, "Indeterminism in Physics" (paper read before the Congress of the Society for Philosophical Instruction, Berlin, 16th June, 1931) in *Science Theory and Man* (New York, 1957), 52-80. Max Jammer, who discusses the ideas of Exner and C. G. Darwin in 1919, states that their proposals for indeterminism had no influence on Heisenberg ("Indeterminacy in Physics, " 588).

[87] Oskar Klein recalled that Niels Bohr gave serious consideration to abandoning the rigorous validity of the energy conservation law several years before he published the idea with H. A. Kramers and J. C. Slater in 1924, but had already come to regard it as "too cheap a way of overcoming the quantum paradox," "Glimpses of Niels Bohr as Scientist and Thinker," *Niels Bohr,* ed. S. Rosental (Copenhagen, 1964; 1st Eng. ed., New York, 1967), 74-93; quotation from 77.

[88] A. Landé, "Probability in classical and quantum theory," *Scientific Papers presented to Max Born* (New York, 1953), 59-64; quotation from 63.

[89] See note 4 above, and also J. C. Slater, *Solid State and Molecular Theory: A Scientific Biography* (New York, 1975), 8-11, 22, 44.

[90] N. Bohr, "Über die Anwerdung der Quantentheorieauf den Atombau. I. Die Grundpostulate der Quantentheorie," *Zeitschrift für Physik,* **13** (1923), 117-65.

Quotation from "On the Application of the Quantum Theory to Atomic Structure. I. The Fundamental Postulates," *Proc. Cambridge Philosophical Soc.,* Supplement (1924), 21.

[91] Born, *Problems of Atomic Dynamics* (Cambridge, Mass., 1926, rept. 1970), 129.

[92] Born, *op. cit.* (note 72); see 224 in Ludwig, *Wave Mechanics.* Born did, however, leave open a possibility "that there are other parameters not given in the theory that determine the individual event" (Ludwig, *ibid.*).

[93] Heisenberg, "Tradition in Science," *Science and Public Affairs Bulletin of the Atomic Scientists,* **29** (Dec. 1973), 4-10, quotation from 9.

[94] *Niels Bohr Collected Works,* ed. L. Rosenfeld, vol. I, *Early Work 1905-1911* (Amsterdam, 1972), e.g., in his letters to C. W. Osteen, 1 Dec. 1911, p. 430. L. Rosenfeld and E. Rüdinger, "The Decisive Years 1911-1918," *Niels Bohr,* ed. s. Rosental, 38-73. In a lecture at Copenhagen in August 1925, Bohr mentioned the success of the Clausius-Maxwell kinetic theory which describes atoms as "flying about at random," and the theory of light scattering based on "the random distribution of the atoms." Cf. "Atomic theory and Mechanics," *Atomic Theory and the Description of Nature* (Cambridge, 1934), 25-26, 27. In his Faraday Lecture in 1930, he noted the connection between entropy increase in classical statistical mechanics and the interpretation of quantum mechanics: "Chemistry and the Quantum Theory of Atomic Constitution," *Journal of the Chemical Society* (1932), 349-84.

[95] Heisenberg, *Across the Frontiers* (New York, 1974), 156 (lecture delivered in 1969). Similar views were expressed by R. von Mises, *Probability, Statistics, and Truth* (New York, 2nd ed. 1957), 204, 210.

[96] This article is based on research supported by the National Science Foundation's History and Philosophy of Science Program. I thank Paul Forman, Michael Gardner, Paul Hanle, and Thomas Kuhn for several suggestions, but recognize their disagreement with several of the interpretations here.

4.

PROOF OF THE IMPOSSIBILITY OF ERGODIC SYSTEMS:

THE 1913 PAPERS OF ROSENTHAL AND PLANCHEREL*

by Stephen G. Brush

One of the most exciting moments in a scientist's work is the sudden realization that a substantial body of knowledge or technique developed in one discipline can be directly applied to a fundamental problem in another discipline. Such an application often has a marked effect on the growth of the science which can *receive* the transplanted material; think of Max Born's discovery that Heisenberg's strange arrays of transition amplitudes were in fact *matrices* for whose manipulation an elaborate mathematical theory already existed. Less common but perhaps just as remarkable are the examples where the *donor* science gains more from the interaction than the recipient; but this is just what happened when Artur Rosenthal and Michel Plancherel independently found that existing mathematical theory could quickly dispose of a question raised by physicists. Their decisive negative answer to the question: Can a mechanical system eventually pass through every point on the energy surface in its phase space?" provoked little more than politely-stifled yawns from the physicists, but introduced a major new branch of mathematical research, ergodic theory.[1]

The work of Rosenthal and Plancherel in 1913 closed the first phase of a discussion of the foundations of statistical mechanics, in which Maxwell, Boltzmann, and Ehrenfest were the major participants, although the basic Ideas had been foreshadowed as early as 1845 in the forgotten paper of J. J. Waterston.[2] Waterston suggested that, in developing a theory of media composed of moving molecules, one should assume that the individual particles

> are for ever continually changing the velocity and direction of their individual motions; striking against and rebounding from each other, they run rapidly in their zig-zag conflict through every possible mode of concurrence, and *at each point of the medium we may thus conceive that particles are moving in every possible direction and encountering each other in every possible manner during so small an elapsed interval of time that it may be viewed as infinitesimal in respect to any sensible period.* "The simplicity of this hypothesis facilitates the application of mathematics in ascertaining the nature and properties of such media, and the study acquires much interest from the analogies that it unfolds."

This is not, of course, the ergodic hypothesis as usually understood, but the somewhat stronger assumption that all phases of the motion of single particles are realized almost immediately.

When Rudolf Clausius revived the kinetic theory of gases in 1857, he emphasized the importance of rotational and vibrational as well as translational motion of gas molecules. He argued that when thermal equilibrium has been reached, a definite ratio will have been established between the translational and internal energies of the molecules, and that this ratio will be unchanged by further collisions. Although he postulated that the ratio would be the same at different temperatures, he did not go so far as to assert that the average energy of each mechanical degree of freedom must be the same, for that would have contradicted the known experimental values for the specific heats of gases. Instead, from the fact that ratio of specific heats of most simple gases is approximately

$$\frac{c_p}{c_v} = 1.421$$

he concluded that the ratio of translational to total energy must be 0.6315.[4] From this it can be understood why Clausius had no need to make an ergodic hypothesis.

This semi-empirical approach was rejected by Maxwell, who (at least in 1860) regarded the kinetic theory as a speculative but self-consistent hypothesis which might well be refuted if its rigorous consequences could be confronted by experimental data. Maxwell asserted that in the model of rotating (but not vibrating) particles, the average kinetic energy of translation must be equal to that of rotation; he concluded that he had "proved that a system of such particles could not possibly satisfy the known relation between the two specific heats of all gases."[5] Hence, as he told the British Association meeting at Oxford in 1860,

> "This result of the dynamical theory, being at variance with experiment, overturns the whole hypothesis, however satisfactory the other results may be."

Maxwell's refutation of the kinetic theory attracted somewhat less attention than Bishop Wilberforce's refutation of the theory of evolution, announced at the same meeting. Even Maxwell himself seems to have disregarded it in his later work, for he devoted considerable effort to developing the kinetic theory of gases though he could never find a satisfactory explanation of the specific heats difficulty.[6] Indeed, one might argue that the hypothetico-deductive method which demands the rejection of a theory incompatible with a single experimental fact is not the most appropriate one for investigations in kinetic theory.[7]

In a paper written just before he died, Maxwell returned to the problem of justifying what he called "Boltzmann's theorem," i.e., the equipartition theorem which Maxwell himself had proposed in 1860, generalized by Boltzmann to mechanical systems with arbitrary conservative forces. Maxwell wrote:

> "The only assumption which is necessary for the direct proof [of the equipartition theorem] is that the system, if left to itself in its actual state of motion, will, sooner or later, pass through every phase which is consistent with the equation of energy...

> "...if we suppose that the material particles, or some of them, occasionally encounter a fixed obstacle such as the sides of the vessel containing the particles, then, except for special forms of the surface of this obstacle, each encounter will introduce a

disturbance into the motion of the system, so that it will pass from one undisturbed path into another. The two paths must both satisfy equations of energy, and they must intersect each other in the phase for which the conditions of the encounter with the fixed obstacle are satisfied, but they are not subject to the equations of momentum. It is difficult in a case of such extreme complexity to arrive at a thoroughly satisfactory conclusion, but we may with considerable confidence assert that except for particular forms of the surface of the fixed obstacle, the system will sooner or later, after a sufficient number of encounters, pass through every phase consistent with the equation of energy.[8]

Thus Maxwell proposed his hypothesis only for mechanical systems interacting with their surroundings.

A similar hypothesis had been suggested earlier by Ludwig Boltzmann, in dealing with a more specialized model: the motion of a point-mass in a plane under the influence of an attractive force described by a potential function ½ $(ax^2 + by^2)$, i.e., the compound harmonic motion which results In the so-called Lissajous figures. If the ratio of the periods of the two motions, a/b, is a rational number, then the point-mass will undergo a finite number of oscillations, then start over again and repeat the same motion. However, if a/b is irrational (the periods are incommensurable) then there is never an exact recurrence. In this case, Boltzmann asserted (1871), the point-mass goes through all possible positions within a certain rectangle.[9] He then suggested that for the general case of a gaseous system of molecules,

"The great irregularity of the thermal motion, and the multiplicity of forces that act on the body from outside, make it probable that the atoms themselves, by virtue of the motion that we call heat, pass through all possible positions and velocities consistent with the equation of kinetic energy, and that we can therefore apply the equations previously developed to the coordinates and velocities of atoms of warm bodies."

Thus the hypothesis is justified by the argument that the system will always be subjected to (random?) external forces.

Up to this point in the historical development, the word "ergodic" has not actually been used by the kinetic theorists themselves. It was introduced only in 1884 by Boltzmann, who defined the *Ergode* not as a single system that goes through every point on the energy surface but rather as a collection of systems with a certain distribution in phase space which he called ergodic.[10] While the word thus originally pertained to what we now (following Gibbs) would call an *ensemble*,[11] it subsequently acquired a completely different meaning, primarily as a result of the influential article of the Ehrenfests which we will discuss below. In order to avoid confusion, however, we will use the word in its modern sense from now on.

In another paper in 1887, Boltzmann again discussed the Lissajous figure and the more general case of gases and ensembles, asserting that all combinations of the positions and velocities consistent with the fixed total energy will be traversed in the course of time.[12] There is a slight difficulty in interpreting Boltzmann, however: it's not clear that when he says "all" he really means *all*. This doubt is raised by a paper he published in 1892, where we find in quick succession the two statements:

"all possible sets of values of x, y and θ which are consistent with the equation of *vis viva* [energy] are obtained with any required degree of approximation [*mit beliebiger Annäherung erreicht weirden*] provided the motion continues for a sufficiently long time T"

and

"In the course of an interval of time T the condition- point occupies all positions in a finite cylinder (the condition cylinder)"[13]

The first of these is of course similar to what we now call a *quasi*-ergodic hypothesis; one can, for example, get as close as one likes to any *real* number by choosing an appropriate *rational* number, but a path that intersects all the rational numbers on a line segment need not go through every point (real number) on that segment. I suspect that for Boltzmann this was not a meaningful distinction, and that his earlier statements about going through "all" combinations should not be interpreted literally in the modern mathematical sense.[14]

In order for Boltzmann to have been fully conscious of the distinction between the ergodic and the quasi-ergodic hypothesis, he would have had to read the research papers being published in the mathematical journals in the 1880's, in particular the work of Georg Cantor which had not yet found its way into the mathematics textbooks, to say nothing of the physics literature. Even though one of Cantor's major results - the proof that the class of rational numbers is denumerably infinite, in contrast to the class of real numbers which is non-denumerably infinite - was published in the *Journal für die reine und angewandte Mathmatik* in 1878, the same journal that later published some of Boltzmann's papers, we must probably assume that Boltzmann was not aware of it. We must also assume that Boltzmann did not understand another result: that an n-dimensional space could be mapped one-to-one onto a one-dimensional space,[15] - even though this mathematical fact makes *almost* plausible the strong form of the ergodic hypothesis (which requires that every point of a 6N - 1 dimensional energy surface be associated with some point on the one-dimensional time axis). In any case he never mentioned the rather sticky problem that faced mathematicians in the 1890's: how to preserve the concept of *dimension,* if dimensionality is not invariant under some mappings of point sets?

One mathematical discovery from this period did become known to a few physicists fairly soon, and was considered relevant to the problem of the ergodic hypothesis: Peano's curve, which goes through every point in a two-dimensional region.[16] This is not quite a satisfactory model for an ergodic path, since the curve goes through a few points in the region two or four times; a classical dynamical system cannot hit the same point in phase space more than once unless the motion is cyclic, and even then it can't go through some points twice as often as others.

The mathematicians soon realized that the concept of dimension could be saved if it could be proved that a point-set of a certain dimensionality *could not be mapped both one-to-one and bicontinuously* onto a point-set of different dimensionality. (In

Cantor's mappings, points that were adjacent in the n-dimensional set would not in general be adjacent in the one-dimensional set, and conversely.) The proof of this key theorem was finally found by L.E.J. Brouwer in 1910.[17] Another (equivalent) way to distinguish sets of different dimensionality was provided by the Borel-Lebesgue measure theory, developed during the same period; a point (or a denumerable set of points) has measure zero with respect to a one-dimensional line segment (i.e., all real numbers within a finite interval); a one-dimensional set has measure zero with respect to a two-dimensional set, and so forth.

It might have been thought, by some scientists in the 1890's, that refined mathematical analysis of this kind would play a role in resolving the fundamental problems of classical physics associated with the apparent failures of the equipartition theorem. But that is not what happened. Although the quantum hypothesis did dispose of the paradox of specific heats of polyatomic gases, and eliminated the possibility that ether-vibrations (having an infinite number of degrees of freedom) would drain an indefinite amount of energy out of material systems at any finite temperature, these were not the anomalies that provoked the introduction of the quantum hypothesis in the first place. Max Planck was not one of the physicists who worried about the validity of the equipartition theorem before 1900, and the myth that his distribution law for blackbody radiation was concocted merely to escape from an "ultraviolet catastrophe" predicted by the Rayleigh-Jeans law has now been thoroughly demolished.[18] It was Paul Ehrenfest who invented the ultraviolet catastrophe (eleven years *after* the publication of Rayleigh's and Planck's papers in 1900) in order to dramatize what *would have been* the consequences of the equipartition theorem If it had been valid for all classical dynamical systems (though neither Rayleigh nor Planck believed that it was).[19]

It was Paul Ehrenfest, too, who wrote (with his wife Tatyana) a massive critique of statistical mechanics, in which the ergodic hypothesis was given a prominent position.[20] It was this critique that served to transmit some of the problems of 19th-century kinetic theory to 20th-century physics, in the form that is now most frequently found in textbooks. The phrases "recurrence paradox" [kehreiwand], "reversibility paradox" [*Umkehreinwand*], "ergodic hypothesis" and "quasiergodic system" have all become part of our jargon because of this article.

Perhaps it was fortunate that the Ehrenfests exaggerated the extent to which Maxwell and Boltzmann had based their theories on the ergodic hypothesis, for despite some historical inaccuracy the article did offer a stimulating challenge to its readers to determine the validity of this assumption. Having asserted that Maxwell and Boltzmann assumed their gas models to be ergodic, and having made an explicit distinction between "ergodic" and "quasi-ergodic," the Ehrenfests wrote:

> "Now the *existence* of ergodic systems (i.e., their *selfconsistency of definition*) is *doubtful,* however: so far, not even one example is known of a mechanical system for which the single G-path *passes arbitrarily near* to every point of the corresponding 'energy surface.'

Moreover, no example is known in which the individual G-path *goes through* every point of the corresponding energy surface. Nevertheless, not only is this the *wording* of the *Maxwell-Boltzmann* definition, but precisely this feature of the definition serves as the basis of the assertion of both authors that in the gas model as an ergodic system all motions with the same total energy transverse the same ϕ-path and *therefore* give the same time average for each ϕ (p,q)."[21]

But they do not suggest any mathematical reason why ergodic systems should be impossible.

The papers of Rosenthal and Plancherel translated here were clearly written as a direct response to the Ehrenfests' statement quoted above, and they represent direct applications of the mathematical work recently completed by Brouwer and Lebesgue. Michael Plancherel was born January 16, 1885, in Bussy, Kanton Freiburg, Switzerland. He studied at Freiburg University, taking the degree Dr. ès-Sciences in Mathematics there in 1907, and was appointed Professor in 1911. He was Professor of Mathematics at the Technische Hochschule in Zürich from 1920 until his retirement in 1955; he published several papers on differential and integral equations and on expansions of functions, but wrote nothing after 1913 on the physical applications of mathematics. He died on March 4, 1967.

Artur Rosenthal was born February 24, 1887, in Fürth, Bayern, Germany; he studied in München University and received his Ph.D. degree in Mathematics there in 1909. In 1912 he was appointed to the faculty of the university, and he remained in München until 1930 when he accepted a chair of mathematics at Heidelberg. He was forced to resign from this position in 1935 and spent some time in a Nazi concentration camp before he managed to leave Germany and come to the United States in 1940. He was Professor of Mathematics at Purdue University from 1947 until his retirement in 1957. He contributed papers to various journals on Hilbert's axioms and on the singularities of curves; he also published a short paper in 1914 on the construction of gas theory with the help of the quasi-ergodic hypothesis.[22] Rosenthal died on September 9, 1959.[23]

The long-range effect of the work of Rosenthal and Plancherel was to stimulate mathematical research on dynamical systems by showing that some of the more esoteric methods of point-set and measure theory could be applied to problems that were supposedly of physical interest. Yet by 1913 most physicists had already lost interest in the ergodic hypothesis, and this topic rarely occupies more than a few paragraphs in most modern texts on statistical mechanics.[24] The reason is not that ergodic systems were proved to be nonexistent, or that classical mechanics is no longer the fundamental basis for describing systems of atoms; even the "quantum quasi-ergodic hypothesis" is unable to rouse our passions. It is simply that the 20th-century physicist is too impatient: even if the ergodic hypothesis were valid, he would not be willing to wait indefinitely long for his system to pass through all possible states in order to determine its equilibrium properties. Instead, like Waterston, he wants quick results, and prefers to sample from a statistical ensemble right away.[25] In order to get the job done, he abstains from seductive but wasteful activities such as wondering what would happen to a single system left to its own devices.

In defense of ergodic theory (against the common criticism that it is irrelevant to physics) we can only quote the remarks of a recent expositor:

> "First, the founders of statistical mechanics were not so prudish; they regarded the theory of isolated dynamical systems to be of interest. Second, if one does not understand what actually happens in such systems, one cannot be sure that the introduction of macroscopic observables or the passage to the thermodynamic limit is essential for the appearance of macroscopic determinism and irreversibility. Third, many actual systems studied in the laboratory are approximately isolated Hamiltonian systems... Such systems sometimes *do* exhibit striking features of 'irreversibility' that make them relevant for statistical mechanics..."[26]

And with that contemporary appeal to "relevance" we conclude this Introduction.

Notes

*Work supported by the U. S. Atomic Energy Commission and the National Science Foundation.

[1] Physicists have only recently begun to realize that Sinai's theorem on the ergodicity of hard sphere systems, a significant landmark in modern ergodic theory, pertains to a subject that they once considered of the greatest importance. See J. G. Sinai, in *Statistical Mechanics: Foundations and Applications*, Ed. T. Bak (Benjamin, New York, 1967), p. 559; A. S. Wightman, In *Statistical Mechanics at the Turn of the Decade*, Ed. E. G. D. Cohen (Marcel Dekker, New York, 1971), p. 1.

[2] For further details see my paper in *Arch. Hist. Exact Sci.* 4, 145 (1967), on which most of this introduction is based.

[3] Although Waterston's paper on kinetic theory was not published until 1893, an abstract announcing the equipartition theorem was printed in the *Report of the 21st Meeting of the British Association at Ipswich, 1851* (Transactions of the Sections, p. 6.)

[4] R. Clausius, *Ann. Phys.* [2] 100, 353 (1857); English translation reprinted in S. G. Brush, *Kinetic Theory*, Vol. 1 (Pergamon Press, Oxford and New York, 1965).

[5] J. C. Maxwell, *Phil. Mag.* [4] 19, 19, 20, 21 (1860). Reprinted in part in S. G. Brush, *Kinetic Theory*, Vol. 1.

[6] See *The Scientific Papers of James Clerk Maxwell* (Dover Publications, New York, 1965); S. G. Brush. *Amer. J. Phys.* 39, 631 (1971).

[7] J. Dorling, *Stud. Hist. Phil Sci. Sci.* 1, 229 (1970); S. G. Brush, *Arch. Rat. Mech. Anal.* 39, 1 (1970).

[8] J. C. Maxwell, *Trans. Cambridge Phil. Soc.* 12, 547 (1879). It may be noted that the phrase "continuity of path," which the Ehrenfests attributed to Maxwell, does not appear anywhere in this paper (or else-where in Maxwell's writings as far as I can determine). It was apparently first used by J. H. Jeans, *Phil. Mag.* [6] 4, 585 (1901).

[9] L. Boltzmann, *Sitzungsber. k. Akad. Wiss. Wien*, Kl. II, 63, 679 (1871).

[10] L. Boltzmann, *Situngsber. k. Akad. Wiss. Wien*, Kl. II, 90, 231 (1884). This paper

is discussed in a forthcoming historical study by Martin J. Klein.

[11] Boltzmann's term for ensemble was *Inbegriff*. A single system that passes through all states on the energy surface was called *isodisch*. See the footnote on page 147, Volume III, in Boltzmann's *Wissenschaftliche Abhandlungen* (Barth, Leipzig, 1909, reprinted by Chelsea Pub. Co., New York, 1968).

[12] L. Boltzmann, *J. f. reine und angew. Math.* 100, 201 (1887). This paper is sometimes cited as his first statement of the ergodic hypothesis, but as we have seen the word *Ergode* and the separate concept of an ergodic [isodic] system had been introduced earlier.

[13] L. Boltzmann, *Sitzungsber. math.-phys. Cl. k. Bay. Akad. Wiss. Munchen*, 22 (3), 329 (1892); *Phil. Mag.* [5] 35, 153 (1893).

[14] For further discussion on this point see p. 174 of my paper cited in note 2.

[15] G. Cantor, *J. f. reine und angew*, 84, 242 (1878). A convenient account of Cantor's work may be found in *Contributions to the Founding of the Theory of Transfinite Numbers*, by Georg Cantor, translated with introduction and notes by P. E. B. Jourdain (Dover Publications, New York, reprint of the 1915 edition).

[16] G. Peano, *Math. Ann.* 36, 157 (1890); P. Hertz, *Ann. Phys.* [4] 33, 225, 537 (1910).

[17] L. E. J. Brouwer, *Math. Ann.* 70, 161 (1910), 71, 97, 305 (1911), 72, 55 (1912).

[18] M. J. Klein, *Arch. Hist. Exact Sci.* 1, 459 (1962); a comprehensive account of the background of Planck's work has recently been published by Hans Kangro, *Vorgeschichte des Planckschen Strahlungsgesetzes* (Franz Steiner Verlag, Wiesbaden, 1970).

[19] P. Ehrenfest, *Ann. Phys.* 36, 91 (1911); see the discussion by M. J. Klein, *Paul Ehrenfest* (North-Holland, Amsterdam, 1970), p. 245-251.

[20] P. and T. Ehrenfest, *Encykl. d. math. Wiss.*, Vol. IV, part 2, Article IV 32 (Teubner, Leipzig, 1911); English translation by M. J.Moravcsik, *The Conceptual Foundations of the Statistical Approach in Mechanics* (Cornell University Press, Ithaca, 1959). For the circumstances of composition of this article, see Klein *Ehrenfest*, p. 81-83 and Chapter 6.

[21] *Ibid.*, p. 31-32 in the original German article, or p. 22 of the English translation (which I have revised slightly, restoring italics in the original).

[22] A. Rosenthal, *Ann. Phys.* 43, 894 (1914).

[23] For information about Plancherel and Rosenthal I am indebted to Jacqueline Walker (American Mathematical Society) and Frances Keegan (Purdue University).

[24] An exception is D. ter Haar, *Elements of Statistical Mechanics* (Rinehart, New York, 1954), p. 123-126, 333, 355-360; although he gives substantial space to a discussion of the ergodic hypothesis, ter Haar still prefers the ensemble approach, and in the second edition most of this material has been removed: see *Elements of Thermostatistics* (Holt, Rinehart and Winston, New York, 1966).

[25] This attitude is most fully articulated in the monumental volume of R. C. Tolman, *The Principles of Statistical Mechanics* (Oxford University Press, London, 1938), p. 65-70.

[26] See p. 3 of Wightman's lecture cited in note 1; he refers specifically to a statement by L. D. Landau and E. N. Lifshitz: "If one talks from the very beginning about the statistical distribution for small parts of a system (subsystems) and not for a

closed system as a whole, then one avoids the whole question of the ergodic or similar hypotheses, which are not really essential for physical statistics." See *Statistical Physics* (Addison-Wesley, Reading, Mass., 1958), p. ix; or p. xi in the second edition (1969).

PROOF OF THE IMPOSSIBILITY OF ERGODIC GAS SYSTEMS

A. Rosenthal

translated from *Annalen der Physik* 4 42, 796-806 (1913)

The Boltzmann-Maxwell formulation of gas theory is based, as P and T Ehrenfest have shown lucidly in their Encyclopedia article on statistical mechanics,[1] essentially on the hypothesis that *any gas* represents an *ergodic system*. Such a system is defined as follows:[1] "The undisturbed motion of the system takes it eventually *through every phase point* that is consistent with the given total energy."[2] In view of the fact that no example of such an ergodic system has been demonstrated with certainty P. and T. Ehrenfest doubted the *existence* of ergodic system (i.e., they doubted that their definition is not contradictory).

In the sequel it will be shown that this doubt was correct; i.e., *it will be shown* that not only *no gas* is an ergodic system, but also that in general such ergodic gas systems cannot exist.[3]

The train of thought of this proof will be as follows: first there will be marked off from the $(2\gamma N-1)$ dimensional energy surface a piece which can be reversibly uniquely and continuously referred to a $(2\gamma N-1)$ dimensional "cube" (Sec. 1). Then with the help of a simple set theoretic consideration it will be shown that for an ergodic system the mapping of the energy surface generated by the orbit of the system on the time line is in contradiction with the uniqueness and sequence of physical phenomena, i.e., with the causality principle.

For this proof we shall make no assumption about the extension of the molecules or about the kind of collisions, so that it retains all its validity when the phase space is penetrated [durchlöchert] by a finite extension of the molecule, or when the collisions change the momentum coordinates discontinuously.

The impossibility of ergodic systems is equivalent to the existence of orbits differing from each other on each energy surface; our proof provides even more, namely the demonstration of the existence of even more than a denumerably infinite number of different orbits on each energy surface.

Section I

As usual we represent a gas of N particles, each with γ degrees of freedom,[4] in a unique manner by a point G of a 2γN-dimensional (linear) space (phase space or Γ-space), where the coordinates of this Γ-space are the generalized coordinates q_s and the generalized momenta p_s of all particles. The set of all physically possible points G of the Γ-space will be denoted by Γ' (the occupied part of the phase space).[5]

Any gas and its image point G is associated with its energy in a uniquely determined way; the set of all points G of Γ'' for which the energy has a constant value E = c is called the "energy surface" E'_c.

By introducing a new (orthogonal) coordinate e into Γ-space we form a linear (2γN+1)-dimensional (linear) space, which we call the B-space; in it the Γ-space has the coordinate e = o. In B-space we associate with each point G of Γ' a point F which coincides with G in its Γ-coordinates but has as e-coordinate the value E(G) of the energy of G. It is assumed that the energy is a continuous function of q and p. Then the set F(Γ') of all points F is a reversible unique and continuous image of Γ. We now cut F(Γ ') with a 2γN-d dimensional (linear) space parallel to Γ whose equation in B-space is

$$e = c.$$

This section we denote by F_c; its parallel projection (in the direction e) on the Γ' space is just E'_c.

Now one chooses on E'_c a point G_1 with the property that G_1 does not lie on the boundary of Γ' and that at G_1, as in its neighborhood, the velocities dq_s/dt of the particles and their momentum derivatives dp_s/dt not only exist and are unique and finite but also vary continuously in the transition to neighboring systems. Such points are in each E'_c in infinite sets. If in particular the particles have finite extension, and are impenetrable and undeformable or change their momenta discontinuously in collisions, then one has to choose as G_1 a point in which no collision takes place and none of the particles touch each other. One can always find such points in the case mentioned. For if one has a system[6] G in which a collision or contact takes place, then one separates the particles in contact to a finite but very small distance such that no new contact arises.[7] The change in potential energy hereby produced can be compensated by producing a corresponding opposite change in the kinetic energy.

Around such a point one can mark off in F-space a finite 2γN-dimensional neighborhood K_1, which consists only of points having the same properties as G_1. K_1 is therefore completely contained in Γ. Let the points of F(Γ') corresponding to K_1 be denoted by $F(K_1)$; $F(K_1)$ is a reversible unique continuous image of K_1, hence it is (like K_1) 2γN- dimensional and of simple connection.[8]

Of all systems that correspond to points of K_1, we know that the velocities dq_s/dt of the particles and their momentum changes dp_s/dt exist, and are unique, finite, and continuous in their dependence on q and p. Now the canonical equations

$$\frac{dq_s}{dt} = \frac{\partial E}{\partial p_s} \quad ; \quad \frac{dp_s}{dt} = -\frac{\partial E}{\partial q_s} \quad (s = 1, 2, \ldots \gamma N)$$

are valid, and from them it follows that also all the first partial derivatives of E with respect to all the q_s and p_s, for each point of K_1 and of $F(K_1)$, exist and are unique, finite and continuous. Hence there follows the *unique existence* of a 2γNdimensional *tangent plane* T_p at each point P of the 2γN-dimensional hypersurface F(K).

At none of these points P does the tangent plane have slope zero (i.e., parallel to Γ). For were this the case, then at this point all the first partial derivatures of E would simultaneously vanish; then the system corresponding to this point would have all its velocities and momentum changes equal to zero. Such a system would generally experience no change in the course of time; it would remain completely fixed for all time. A gas cannot assume such a state. However, if our system is ergodic, then each gas system of equal energy must get into each temporal unchanging state.

We know therefore that the point G_1 on E'_c corresponds to a point P_1 in $F(K_1)$ whose tangent plane T_{P-1} is not parallel[9] to Γ-space. Hence one can certainly find one of the coordinate axes of Γ-space (let it be denoted by a) with the property that T_{P-2} is not parallel to it. T_{P-2} contains therefore no line of direction a. The line parallel to a and going through P_1. viz, a_p, has no points in common with T_p except the single point P_1, just as in general T_{P-1} is intersected by all the lines parallel to a in only one point.

Next we shall show that the part of $F(K_1)$ near to P_1 can be intersected in *no more than one* point by each of the parallels near a_{P-1}. Assume that it were not so; then there would be a sequence of parallels converging to a_{P-1}.

$$a_1, a_2, a_3, \ldots a_i, \ldots$$

such that on each of these lines a_i at least two points of $F(K_1)$ (P_i' and P_i'') lie that converge to P_1 as the index increases. Now draw through these lines a_i the usual two-dimensional planes E_i parallel to the coordinate axis e. E_i cuts out from $F(K_1)$, with which it has in common two points P_i' and P_i'' a continuous curve N_i (the lines parallel to e will be intersected only in a single point.) N_i possesses everywhere a definite tangent; for in each of its points (as, in general, in each point of $P(K_1)$) there exists the unique and finite partial derivative $\partial E/\partial a$. According to Rolle's theorem, there is at a point of N_i lying between P_i' and P_i'' a tangent a parallel to a_i. This holds for every plane E_i. Since $\partial E/\partial a$ is continuous in the neighborhood of P_1, the limit line a_p of a_i must itself be a tangent of $F(K_1)$ at the point P_1, hence it must lie in T_p, which would contradict our assumption. One can therefore mark off around P_1 a finite singly connected 2γN-dimensional neighborhood v in $F(K_1)$ such that *no* line of direction a has more than one point in common with v.

If now one draws through each point of v a line of direction a, then the set v of the points cut out of T_{P-1} by these lines is the reversible single-valued and continuous image of **V**. Hence (cf. footnote 8) V is a connected 2γN dimensional region in T_{P-1} containing P_1.

One can therefore assume in v a finite, 2γN-dimensional "cube" W, which encloses P_1 and of which a (2γN-1) dimensional "boundary cube" is parallel to the intersection of T_{P-1} with the Γ-space. The one-to-one continuous image of W in v produced by projection in the direction a is a 2γN-dimensional region γ enclosing P_1. Now one constructs through P the 2γN-dimensional plane parallel to the Γ-space, namely e = c. This cuts W in a (2γN-1) dimensional "cube" S and it cuts γ in a likewise (2γN-1) dimensional region R, which is a piece of F_c and will be mapped by the lines lying in (e = c) with direction a, reversibly uniquely and continuously on S. S and R enclose the point P_1. If one now projects R parallel to the coordinate direction e on the Γ-space then he obtains a (2γN-1) dimensional region R' which is completely contained in E_c' and encloses the point G_1.[10] Thus the *existence of such a region R' in our "energy surface", E_c' which can be mapped reversibly uniquely, and continuously on a (2γN-1) dimensional "cube" S, is proved.*

Section II

If one has a gas with fixed energy E = C, which is represented at each instant of time by a point G lying on E_c', then the representative point describes in the course of time a known curve C^{12} on E_c'. The corresponding state of the gas is completely determined by the generalized coordinates q_s and momenta p_s. Since the physical phenomena follow each other in a uniquely determined manner independent of time (causality principle) then one finds that the representative point G traverses its path C in a uniquely determined way, and that, if it is at the same place at different times, it proceeds on its path in exactly the same way each time. Thus it can never happen that two different parts of the path-curve intersect each other. The path-curve C *cannot therefore have any multiple points* (in so far as one restricts himself in the case of purely periodic paths to a single cycle.)

The *ergodic hypothesis* now requires that the representative point G of the system must pass through *all* points of the "energy surface" E_c', hence a fortiori *all* points of R'. Thereby there would be effected (by means of the ordering association of path points and time moments) a certain mapping of E_c' on the time line t and of R' (like S) on a certain part of the time line t. Certainly from what has been said above, for *finite* values of t the mapping A of R' (and of S) on any part of the time line t is reversibly unique (provided one restricts himself in the case of a purely periodic path to a single full cycle).

For finite *bounded* time intervals this mapping A is also reversibly continuous. For let:

$$t_1, t_2, \ldots, t_i \ldots$$

be a monotonic sequence of time moments, completely enclosed between the finite bounds t and t. To them corresponds the sequence

$$P_1, P_2, \ldots, P_i \ldots$$

of points lying in R'. The accumulation value t_w of the $\{t_i\}$, which also belongs to the interval t . . . t, will be associated with a definite point P_o. Assume now that P_o lay outside of R' (cf. footnote 11); then G, in approaching t_w, must pass arbitrarily many times into R' in an arbitrarily short time, which is impossible, since the velocity of G or its components q_s and j_s are continuous in R' and therefore are bounded. Hence P_o lies in R'. Then however t belongs to a time interval T_w which corresponds to a continuous connected curve segment C_w in R'. For a specified i, all t_i must lie in T_w and hence all P_i in C_w. Hence P_o is the accumulation point and indeed the only accumulation point of the $\{P_i\}$. Conversely there also corresponds to an accumulation point P_w of $\{P_i\}$ an accumulation value of $\{t_i\}$, inasmuch as the times t_i in question lie between finite limits. For if one selects from $\{P_i\}$ a subset $\{P_i'\}$ which has P_w as an accumulation point, then because of the foregoing, and because of the reversible uniqueness the relation between the accumulation value t_w' of $\{t_i\}$ and P_w is reciprocally uniquely coordinated.[13]

Our proof now amounts to showing that the mapping A is impossible.

First, it would be conceivable that the curve C does not leave R' after a certain time, hence for t = ∞ it runs somehow asymptotically within R'. In this case we would divide R' into two subregions; then certainly in at least one of these subregions the representative point does not stay permanently for very large t. We can therefore consider that we have chosen R' from the outset such that the representative point does not remain in R' after a certain time, without ever leaving R' again.

Now in the case of ergodic systems, *three cases* are conceivable:

1. The path curve C goes through all points in R', after it enters R' during a finite bounded time ($t_o \ldots t_1$) in one transit. Then the ($2\gamma N-1$)-dimensional region R' would be mapped reversibly uniquely and continuously onto the one-dimensional segment ($t_o \ldots t_1$) of the time line. This is however *impossible* on account of the *invariance of the dimension number*[14] for reversible unique continuous mappings (as long as ($2\gamma N-1$) > 1).

2. The path curve C goes through all points of R' but not in one transit; rather it returns a finite or infinite number of times to R'; between the first entrance and the last exit there intervenes a finite bounded time interval ($t_o \ldots t_1$). Then the simply connected region R' would be mapped on a nonconnected piece of the interval ($t_o \ldots t_1$) reversibly uniquely and continuously; which is *impossible* on account of the *invariance of connectivity* of closed sets under reversible unique continuous mappings.

3. The path curve C, after (or before) an *arbitrarily long time* always enters R' again and then leaves it, and thereby goes through all the points of R'. This is (by virtue of Poincaré's theorem)[16] the physically most important case which one has in mind in stating the ergodic hypothesis. For this case, however, the previous methods of proof fail to work. For here at $t = \pm\infty$, the mapping is neither unique nor continuous, rather for $t = \pm\infty$ one obtains no definite limiting position. [This indeterminacy is physically admissible since actual infinite time has no physical meaning.] Even if we restrict it to finite times, the mapping is not reversibly continuous (cf. footnote 13); on the contrary, neighboring points of R' correspond in part to widely separated moments of time.

In order to take care of case 3, we prove (with the aid of the already-known impossibility of cases 1 and 2 of the ergodic hypothesis) a completely general property of the part of the path curve of a gas system that goes through R'.

Each time our path C penetrates into R' and reaches interior points, it traverses a curve segment l_i and stays in R' during a certain finite time interval t_i; thereby l_i and t_i will be reversibly uniquely and continuously mapped on each other, in this way one obtains a series of separate intervals t_i and corresponding curve segments l_i. Now, according to a theorem of G. Cantor,[17] one can specify on a line only a most a *denumerable* set of non overlapping intervals. Hence there can exist only at most denumerably many time intervals t_i and hence only at most denumerably many such curve segments l_i. We consider now any one of these l_i and call it l_1. We can know from case 1 that l_1 cannot go through each point of R'. However, l_1 also cannot come arbitrarily close to each point of R'; for since l_1 is a closed set, it would then have to go through every point of R'. Hence there is certainly a point $Q_{(1)}$ in R' such that $Q_{(1)}$ is separated from every point of l_1 by a finite non-vanishing distance, where the lower limit $q_{(1)}$ of these distances is different from zero. One can therefore construct around $Q_{(1)}$ a (2N-1)-dimensional "sphere" with radius $q_{(1)}$, such that within this "sphere" there can be no points of l_1. These points $Q_{(1)}$ with their corresponding "spheres" are moreover *everywhere dense* in R'. For: otherwise there must be a (2 N-1)- dimensional subregion M' of R' through all of whose points l_1 passes, which according to cases 1 and 2 is impossible. Hence the points of l_1 form a set that is *nowhere dense* in R'. Hence according to the definition of R. Baire, the aggregate of all points of R' lying on the path curve represents a *set of the first category* in R'.

However, according to R. Baire[19] the continuum R' is itself a *set of the second category* in R'. Hence in our case 3 also, the path curve cannot go through each point of R'.

Thus the impossibility of ergodic gas systems is proved.

From our proof there follows at the same time even more: we have shown in general that the aggregate of all points of a path curves in R' forms a set of the first category. Since (by virtue of the definition of sets of the first category) a denumerable infinity of sets of the first category, taken together form a set which is still of the first category, and since the aggregate of points of R' is a set of the second category, then it follows further that make *more than a denumerable infinity of path curves* must *exist on the energy* surface E_c', and that in each (no matter how small) $(2\gamma N-1)$-dimensional region of E_c' *more than a denumerable infinity* of different path curves must penetrate; on the basis of completely similar considerations this must also hold for *each region* of E_c' whose dimension is greater than 1.

Notes

[1] P. and T. Ehrenfest, *Begriffliche Grundlagenagen der Statistischen Auffassung in der Mechanik,* Encyki. d. math. Wiss. 4, 211, Heft 6, No. l0a. (19); English translation by M. J. Moravcsik, *The Conceptual Foundations of the Statistical Approach in Mechanics,* Cornell University Press, 1959.

[2] The instantaneous state of a system is represented by a point of $2\gamma N$-dimensional phase space, if the gas consists of N particles, each having γ degrees of freedom. Moreover it will be assumed that the external force field does not change in the course of time; cf. P. and T. Ehrenfest, loc. cit. No. 9a and b.

[3] The present work had already been edited when I learned from a notice in *Fortschritte der Physik* that the same matter was the subject of a report of Herr Prof. M. Plancherel (Freiburg, Switzerland) to the Berne meeting of the Swiss Physical Society (March 9, 1912). The report of the meeting (Archives des sc. phys. et nat. (4) 33, p. 254-55, 1912) contains no indication of the method of proof. Herr Prof. Plancherel was so kind as to communicate to me an outline of his proof. It appears that his ideas move in part in the same direction as mine, but they are different in some essential points. [See translator's note following footnote 19].

[4] In the following it is assumed that $\gamma N > 1$.

[5] Γ' need nor coincide with Γ; for example, if the molecules have finite extension and are impenetrable then Γ' will have gaps.

[6] With non-vanishing kinetic energy.

[7] One can always do this in the case of a gas.

[8] For the conservation of the dimension number and of the region under a reversibly unique and continuous mapping, see L. E. J. Brouwer, Math. Ann. 70, 161 (1911); 71, 305, 314 (1912); 72, 55 (1912).

[9] Moreover, at no point of $F(K_1)$ can the tangent plane be perpendicular to r-space. For, in this case, at least one of the first derivatives of E must become infinite there, which, according to the above, is forbidden for all points of $F(K_1)$.

[10] The boundary of R' will be included in R'.

[11] One chooses, for the following, S and R' not as large as possible but rather such that S is contained in the interior of a larger "cube" S and accordingly R' in the interior of a larger similar region R (where the smallest distance N of the exterior surfaces of R' and R is greater than zero).

[12] This can have discontinuities at instants when collisions take place, if these collisions produce discontinuities in the momenta. The parts of C lying in R' or R are

still always continuous.

[13] It is explicitly emphasized that for finite but *unbounded* times the mapping **A** can generally have *only one-sided continuity* but *not reversible* continuity. For: neighboring time moments must to be sure correspond to neighboring points of R', but not conversely. I.e., there corresponds also in this case to each finite accumulation value t_w of $\{t_i\}$ the only accumulation point of the corresponding $\{P_i\}$; but conversely, here, an accumulation point P_w of a series $\{P_i\}$ lying in R' need *not* correspond to an accumulation point of the corresponding $\{t_i\}$.

[14] See the work of L. E. J. Brouwer first mentioned in footnote 8; for the present case it is sufficient to use the much simpler consideration given by A. Schoenflies, *Die Entwicklung der Lehre von Punktmannigfaltigkeiten II*, Jahresb. d. Dtsch. Math. -Ver., Erganz. -Bd. 2, p. 165 (1908).

[15] A. Schoenflies, loc. cit. p. 153.

[16] H. Poincaré, Acta Math. 13, 69 (1890).

[17] G. Cantor, Math. Ann. 20, 117 (1882).

[18] R. Baire, *Lecon Sur les Fonctions discontinues*, Paris, 1905, p. 78 and 106. A set of the first category in a region G is a set which can be constructed from denumerably many sets nowhere dense in G. Each set that is not of the first category in G is then a *set of the second category* in G.

[19] R. Baire, loc. cit. p. 79 and 106.

Translator's note: The proof of Plancherel mentioned in footnote 3 was published in Annalen der Physik [4] 42, 1061 (1913). The translation follows.

PROOF OF THE IMPOSSIBILITY OF ERGODIC MECHANICAL SYSTEMS[1]

M. Plancherel

translated from Annalen der Physik [4] 42, 1061-3(1913)

If we denote by $q_1, q_2, \cdots q_n$; $p_1, p_2, \cdots p_n$ the generalized coordinates and momenta of a *conservative* mechanical system of n degrees of freedom and by $E = E(q,p)$ the total energy of the same system, then Hamilton's equations of motion are:

$$\frac{dq_h}{dt} = \frac{\partial E}{\partial p_h}, \quad \frac{dp_h}{dt} = -\frac{\partial E}{\partial q_h}, \quad (h=1,2,\cdots,n), \tag{1}$$

The equation

$$E(q,p) = \text{const.} \tag{2}$$

then gives immediately a first integral of the motion.

Geometrically we can represent any phase of the system at time t by a point with coordinates (q,p) in 2n-dimensional space (*phase space*). The actual motion of the system in the course of time then corresponds in phase space to the motion of the point (q,p) along a curve $C(t)$, the path curve of the system. The path curve, or orbit, belongs to a particular surface S (*energy surface*) of the family (2).

The orbit $C(t)$ does not go through each point of the energy surface S. On the energy surface there exists an infinite, nondenumerable set of different orbits.

This would prove the incompatibility of the ergodic hypothesis with the Hamilton equations; for this hypothesis requires that the orbit should pass through every point of the energy surface. We limit ourselves here to sketching the outlines of the proof.

1. It is always possible to fix on the surface S a point A such that for a 2n-dimensional sphere Σ, which has sufficiently small radius and has A as its midpoint, the following conditions are satisfied:[2]

a) In the interior and on the boundary of Σ, all the partial derivatives of first order, $\partial E/\partial q_h$, $\partial E/\partial p_h$, are continuous functions of the point (q,p).

b) In the interior and on the boundary of Σ, at least one of these derivatives is always different from zero. For example, let $\partial E/\partial q_1 \neq 0$.

c) If we denote by G the part of the energy surface S that lies inside and on the boundary of Σ, then G forms a simple connected (2n-1)-dimensional region.

2. At every point of G, we can therefore define the content dσ of a surface element by the formula

$$d\sigma = \frac{1}{\left|\frac{\partial E}{\partial q_1}\right|} \sqrt{\Sigma\left[\left(\frac{\partial E}{\partial q_h}\right)^2 + \left(\frac{\partial E}{\partial p_h}\right)^2\right]} \, dq_2 \cdots dq_n dp_1 \cdots dp_n.$$

Thereby it will be possible to apply the Borel-Lebesque theory of measure[3] to any point set of G. Let E be any point set of G, and let f be a function defined by

$$f = \begin{cases} 1, & \text{on } E \\ 0, & \text{outside of } E \end{cases}$$

Moreover, there exists the Lebesque integral extended over G,

$$\int_{(G)} f \, d\sigma$$

so that the integral represents the measure (G) of the point set E in G. If the measure of E in G is equal to zero, then E is called, for short, a null set. The sum of a finite or denumerably infinite number of null sets is likewise a null set.

A curved arc

$$q_h = q_h(t), \; p_h = p_h(t), \; \tau < t < \tau'$$

is *rectifiable* if the integral

$$\int_\tau^{\tau'} \sqrt{\Sigma\left[\left(\frac{dq_h}{dt}\right)^2 + \left(\frac{dp_h}{dt}\right)^2\right]} \, dt$$

exists and has a finite value. If a rectifiable curved arc consists only of points on G, the it forms (if n > 1) a null set in G.

3. We now consider any orbit C(t) passing through an interior point of G. We denote by $C_G(t)$ that part of this curve which consists of interior points of G, and by T any value-set [Wertmenge](t) which corresponds to $C_G(t)$. On account of the continuity of dq_h/dt, dp_h/dt, on G one sees easily that T consists of at most a denumerable set of non-overlapping intervals from the interior points. Consequently $C_G(t)$ is composed of a finite or denumerably infinite set of rectifiable curved arcs. Since each of these curved arcs forms a null set in G, $C_G(t)$ is itself a null set. In other words: the part of any orbit curve of a system that lies in the region G forms a null set in G. Since G itself is not a null set, the theorem to be proved follows immediately.

---

[1] I first proved this impossibility in a lecture in February, 1912 at the mathematical-physical seminar at the University of Freiburg (Switzerland). Herr Dr. A. Rosenthal has given, independently, a proof (Ann. d. Phys. 42, 796 (1913)) that is similar to my original proof. The proof given here was presented at the Berne meeting (March 1912) of the Swiss Physical Society [Schweizerischen Physikalischen Gesellschaft]. It differs from my first proof and from that of Herr Rosenthal, in that it does not use the Brouwer theorem, but rather depends on the theory of measure of point sets.

[2] For the justification of these assertions, see Rosenthal, loc. cit.

[3] See H. Lebesgue, Leçons sur l'integration, Paris, 1904.

5.

Statistical Mechanics and the Philosophy of Science:
Some Historical Notes[1]

by Stephen G. Brush

Statistical mechanics, in its broad sense, means the study of properties of bulk matter, assuming it to be composed of large numbers of particles. While each particle obeys the laws of mechanics - classical, quantum, or relativistic, whichever may be appropriate - the application of statistical methods avoids the need for considering the motion of each particle in detail, and suggests that the system as a whole has certain simple regularities. Many of these regularities were first explained (or even predicted) with the help of the kinetic theory of gases, an early version of statistical mechanics in which intuitive physical ideas based on special molecular models were applied to systems of particles whose encounters are relatively infrequent. The more general theory, developed by Maxwell, Boltzmann, and Gibbs, attempts to deal with a broad class of mechanical systems, using consistent statistical postulates. It is not necessarily limited to systems in thermal equilibrium though in practice the phrase "statistical mechanics" most often refers to the theory of such systems.

The historical development of statistical mechanics has generated several problems, some solutions, and a few singular events that are or should be of interest to philosophers of science. I will begin by mentioning the issues that have most frequently been discussed by philosophers, and then try to put them in a broader historical context. There is not time to undertake a complete analysis of the various issues but I hope you will get some idea of the range of interesting problems waiting to be explored.

1. Reduction, Statistics, and Irreversibility

To non-specialists, "statistical mechanics" often seems to mean the kinetic theory of ideal gases, based on the "billiard ball model" (elastic spheres with no long-range forces or internal structure). This theory has frequently been cited as a good example of "scientific explanation."[2] That fact in itself may have some philosophical significance, for the appeal of the theory seems to reside in features that are not typical of 20th-century physical theories: a visualizable model with contact interactions governed by Newtonian mechanics, and derivations that employ only simple algebra. But even if we agree that most scientific theories cannot be expected to have those features, the kinetic theory remains as the best available example of reduction of a theory formulated in terms of observable macroscopic quantities to one formulated in terms of unobservable microscopic entities. Thus it is often used to illustrate the procedure of theory-reduction in general, or to argue whether reduction is actually possible at all.[3] Reduction as a mode of scientific progress is associated with positivistic and neo-positivistic philosophies (Suppe [159], pp. 52-56) and thus the attack on the possibility of reduction is part of the attack on such philosophies[4] although one can defend reduction without being a positivist. Reduction also threatens the autonomy of some scientific disciplines (chemistry may be reduced to physics) and the survival of some viewpoints such as holism and vitalism which postulate non-reducible explanatory concepts; thus it impinges on the boundaries of research fields or domains.[5] As Siegel notes, it arouses antagonism in the counter-culture yet is somehow essential at least in the physical sciences ([147],[148]. Roth blames the prevalence of reductionism on the male ego, and suggests: "It is likely that a more synthetic and holistic approach, which seems to be more of a feminine characteristic, will be necessary for science in the future." [134]

One often encounters the claim that "thermodynamics has been reduced to statistical mechanics." I consider this quite misleading. Thermodynamics is the science that deals with the general relations between thermal and mechanical energy of substances whose special constitutive properties are assumed to be known. By a constitutive property I mean for example the "equation of state" relating pressure, volume and temperature for the substance when it is in thermal equilibrium. Given the equation of state one may use thermodynamics to compute such quantities as the work done when a fluid is heated from one temperature to another at constant pressure. But a theoretical calculation of the equation of state from statistical mechanics is not a derivation or reduction of thermodynamics itself. Yet that is what many philosophers present as the only concrete justification for the claim that thermodynamics has been reduced to statistical mechanics: the derivation of the ideal gas law from kinetic theory, using a model of infinitesimal billiard balls.[6]

One reason for this confusion, I think, is the use of the term "law" to describe the ideal gas equation of state (as in "Boyle's law" or "Charles's law"). The "ideal gas law" is often used to illustrate the application of thermodynamics, because of its simplicity, but it is not one of the "laws of thermodynamics." Indeed by restricting oneself to ideal gases one misses some of the most important applications of thermodynamics to modern science, including almost all cases in which intermolecular forces play a significant role.[7]

Looking at the other side of this alleged reduction, I would also point out that the reducing theory in the typical textbook example is not really statistical mechanics but rather a crude kinetic theory of gases in which statistical considerations do not play any essential role. One gets the same formula for the pressure, whether one assumes that all the molecules move at constant speed at right angles to the walls of a cubical container, or that there is a statistical distribution of the directions and magnitudes of the molecular velocity vectors. There is nothing wrong with using this as a textbook example but it is too trivial to support inferences about the nature of reduction in general, and too restricted to give any insight into the relation between thermodynamics and statistical mechanics. Indeed philosophers have been hard pressed to find in it anything worth arguing about, other than the problem of whether the word "temperature" has changed its meaning during the reduction. (Feyerabend [46], pp. 77-79)

So far I have not found any philosophical discussion of the reduction of thermodynamics to statistical mechanics except for the problem of irreversibility, where the time-asymmetry implied by the Second Law of Thermodynamics is in question. Perhaps this is because, as Bunge asserts, this reduction is so far only a program, not a fact (Bumstead [15]; Bunge [16], pp. 204, 332; [17], pp. 189-191; Sklar [151]). But philosophers of science generally do not seem to be familiar with the work that has actually been done in this area, in the 19th century and more recently.[8]

Turning now to the second issue that seems to be of philosophical interest, we find that the kinetic theory of gases is usually considered to have represented the first significant application of probabilistic reasoning in physics. Thus it provides historical background for discussions of the general notion of "statistical law" in science.[9] Usually it is asserted that there is a fundamental difference between the use of statistics by 19th-century physicists, who allegedly believed as an article of faith that all molecular motions are deterministic, and the probabilistic character of quantum physics which assumes an inherent randomness or indeterminism in nature.[10]

The third philosophical problem related to statistical mechanics is irreversibility, often expressed somewhat dramatically as the "direction of time." Ernest Nagel wrote in 1939:

perhaps the greatest triumph of probability theory within the framework of nineteenth-century physics was Boltzmann's interpretation of the irreversibility of thermal processes. ([113], p. 355)

Hans Reichenbach echoed this sentiment ([131], p. 159) and devoted a major part of his last book [132] to a discussion and extension of Boltzmann's concept of alternating time directions, originally an attempt to reconcile the Second Law of Thermodynamics with reversible Newtonian mechanics through statistical mechanics. Other philosophers concerned with the nature of time have taken up the question of entropy increase and the H-theorem.[11] The related problem of whether Brownian movement constitutes a violation of the Second Law of Thermodynamics has attracted some attention.[12]

In addition to these three major issues, philosophers of science have mentioned the kinetic theory of gases and statistical mechanics in several other connections. Following Norman Campbell [19], the billiard-ball model of kinetic theory has been invoked in arguments about the use of models in science.[13] Suppes mentioned that Gibb's work in statistical mechanics [61] could easily be formulated so as to show the distinction between theories and models; it gives a "particularly good example for applying the exact notion of model used by logicians, for there is not a direct and immediate tendency to think of Gibbs's statistical mechanical theories as being the theories of the one physical universe." ([161], p. 168)

More frequently the kinetic theory has been invoked for just the opposite reason: to illustrate the triumph of an atomistic/mechanistic view of nature.[14] At the same time it is recognized that the debate between Boltzmann and his opponents (Mach, Ostwald, and others) about the role of atomism in science has not been adequately studied by philosophers.[15] There is also a certain amount of embarrassment among the followers of Mach on this point[16] and Philipp Frank has tried to minimize the extent of disagreement between Mach and Boltzmann by claiming that the latter was not really a strict mechanist either ([57]; [58], pp. 142-144). Yehuda Elkana, who has attempted to describe Boltzmann's view within the framework of Lakatos' "methodology of scientific research programmes" [84], claims that Boltzmann changed his views in the 1890s and no longer considered meaningful the question of whether atoms really exist ([39], p. 48). This revisionist interpretation of Boltzmann has now gone so far that a selection of his philosophical writings has recently been published in a series titled "Vienna Circle Collection"![17]

Does the history of statistical mechanics play any role in the current debate about paradigms and scientific revolutions? Very little, as far as I can see; Kuhn mentioned it only in passing as an example of the discussion of the nature and standards of physics that accompanied the transition from Newtonian to modern physics ([83], p. 48). Karl Popper mentioned Boltzmann as an an example of a great scientist who doesn't fit into Kuhn's normal/revolutionary dichotomy (Popper ([123], p. 54). Perhaps the impression that scientific revolutions occur as an a response to a "crisis" produced by "anomalies" in the accepted theory has been fostered by the story of the "ultraviolet catastrophe" of classical statistical mechanics - the idea that Planck invented his quantum theory in order to avoid a divergence predicted by Rayleigh's theory of black-body radiation. The story continues to be repeated in the literature.[18]

2. Beginnings of Statistical Mechanics

The early history of statistical mechanics can provide much useful material for those philosophers who investigate the process by which scientific theories are proposed and established. The most interesting period here is the half-century from 1815 to 1865. At the beginning of that period the caloric theory of heat and gases was generally accepted, and the kinetic theory received little serious consideration when it was proposed by John Herapath. At the end of the period the caloric theory had been replaced by the kinetic theory, and Maxwell and Boltzmann were able to start developing statistical mechanics in its modern form without having to worry too much about whether all its predictions agreed with the latest experiments.

What factors were mainly responsible for this success?

First, it is important to realize that the "billiard ball" kinetic theory derivation of the ideal gas law, which I mentioned earlier as the textbook example of theory reduction, was not a major factor. Other theories of gases, based on a caloric fluid or rotating molecular vortices, could account for the ideal gas law and also for many other phenomena not yet explained by the billiard-ball model. The "domain" which defined the problems to be solved by the theory (Shapere [145]) included specific heats of liquids and solids as an well as an gases, latent heat of phase changes, and the properties of radiant heat. The caloric theory of gases offered plausible explanations of all these phenomena though its proponents were still debating the merits of two different versions (see Fox [56]).

Second, experiments and theories in fields other than gas physics prepared the way for a revival of the kinetic theory after 1845. The major event here was the establishment of the principle of conservation and convertibility of energy by Mayer, Joule, Helmholtz and Colding. This made it plausible that heat, being a form of energy interconvertible with mechanical energy, is in reality just mechanical energy on the molecular level. Even before the general principle of energy conservation was adopted, the caloric (substance) theory of heat had been transmuted into a wave (motion) theory as a result of experiments on radiant heat and light ([12], Chap. 9). Both phenomena were thought to be qualitatively similar, and when the particle (substance) theory of light was replaced by a wave (motion) theory as an a result of Fresnel's work around 1820, it seemed logical to assume that radiant heat is also a form of wave motion in the ether. At that time all heat was assumed to be similar to radiant heat.

Third, physicists took as an interest in certain gas properties that had not been systematically investigated before: free expansion (Joule-Thomson effect), diffusion and effusion, viscosity, and heat conduction. In some cases experiments were stimulated by predictions of the kinetic theory itself. The result was to open up a new domain which turned out to be more congenial to the kinetic theory than to its rivals.

Fourth, the phenomena that could be explained better by caloric theory than by kinetic theory were simply excluded from the redefined domain, at least temporarily. The most striking example of this exclusion was radiant heat, which began to be considered a phenomenon separate from other kinds of heat; thus kinetic theory ignored the"ether" and assumed that molecules move through "empty" space.[19] This redefinition of the domain failed to get rid of one awkward problem, the specific heats of gases, where there was as an obstinate discrepancy between experimental values and the kinetic-theory predictions.

Fifth, the kinetic theory offered the theoretical advantage of a simple clearly-defined molecular model which could be treated mathematically; and the fact that reasonably consistent numerical values for parameters such as an the molecular diameter could be inferred from different properties made the model quite plausible. Here we have favorable conditions for the elaboration of a more sophisticated theory.

These five factors could serve as an the basis for a "rational reconstruction" of the early history of statistical mechanics, but something essential to history is missing: the personalities and social environments of the scientists who actually made the history, and the accidental events that helped or hindered them. So I would like to say something about two pioneers: J. J. Waterston (1811-1883), who never received proper recognition during his own lifetime, and James Clerk Maxwell (1831-1879), who was well known for his work in several areas of physics and held the prestigious Cavendish professorship at Cambridge University.

One might claim that the history of statistical mechanics really began in 1845 when Waterston completed his paper "On the Physics of Media that are composed of free and perfectly elastic molecules in a State of Motion" and submitted it to the Royal Society of London. (He was at that time living in Bombay but later returned home to Scotland.) Unlike most scientists (before and after him) who worked on the kinetic theory of gases, Waterston wanted only a consistent theory based on a simple model whose properties could be worked out mathematically from a few plausible postulates, rather than a comprehensive flexible theory that could explain all the data. "Whether gases do consist of such minute elastic projectiles or not, it seems worthwhile to inquire into the physical attributes of media so constituted, and to see what analogy they bear to the elegant and symmetrical laws of aeriform bodies," he wrote ([172], p. 214).

His first postulate was that heat is some kind of motion of the smallest parts of bodies. To justify this he did not cite the principle of energy conservation, still in its infancy in 1845, but the experiments on radiant heat which had suggested the wave theory of heat mentioned earlier.

Waterston's second postulate was that the molecules move so rapidly and in so many directions, with frequent "fortuitous encounters" ([172] p. 214), that one can compute the properties of the system at any time by simply averaging over all possible molecular states. A more sophisticated version of this postulate later became known as the "ergodic hypothesis."

The third postulate, which seemed to be a consequence of the first and second, was that in a mixture of elastic particles of different masses in thermal equilibrium, each kind of particle will have the same average kinetic energy. This postulate is now called the "equipartition theorem," though Waterston did not explicitly extend it to the case where some of the particles are bound together by forces to form molecules.

Waterston applied his theory to gas properties with moderate success. That is, he explained some known facts, such as the ratio of specific heats and the decrease of temperature with height in the atmosphere, by arguments that were later found to be somewhat defective. In other cases his calculations were sound. But the Royal Society refused to publish his paper, on the recommendation of two referees who called it nonsense. Even worse, they refused to give it back to him, so he was prevented from publishing it elsewhere. As an a result Waterston's work remained generally unknown until a decade after his death.

The axiomatic style that makes Waterston's paper attractive to a modern reader was probably responsible in part for its unfavorable reception by his contemporaries. He did not attempt to show that the billiard ball is a reasonable model for a molecule, and the amount of mathematics may have seemed excessive in comparison to the physical content. Add to this the circumstance that Waterston had no reputation in the scientific community and few friends to defend his work, and one may perhaps be able to understand why his theory was ignored - though it still appears to me that he was treated rather harshly by the Royal Society.[20]

Rudolf Clausius [24] gave the kinetic theory what Waterston had failed to supply: a comprehensive justification in terms of the latest knowledge of physical and chemical properties of gases, a realistic molecular model, and the endorsement of as an acknowledged authority on heat. While Clausius was thus primarily responsible for reviving the kinetic theory and persuading other scientists to take it seriously, I suppose we must criticize him for adjusting the postulates of the theory in somewhat arbitrary ways in order to escape experimental refutation. Indeed his most famous contribution to kinetic theory, the "mean free path" concept, was devised for just this purpose. He had originally assumed that the molecules have effectively zero size and simply bounce back and forth between the solid sides of the container. But a Dutch meteorologist C. H. D. Buys-Ballot, pointed out that such molecules, moving (as Clausius estimated) at speeds of several hundred meters per second, would diffuse quickly through a large room, contrary to experience. To extricate himself from this difficulty, Clausius had to attribute to his molecules a diameter large enough so they could not diffuse too rapidly, but small enough to leave unaffected the other deductions from the theory [25]. He had no independent means of estimating the diameter, so we must conclude that this was a purely ad hoc hypothesis as of 1858. Yet the way in which the hypothesis was introduced - by assuming that the molecule travels a certain distance, on the average, before hitting another one (mean free path) - turned out to be useful in the next stage of development of the theory.

James Clerk Maxwell read Clausius' paper in English translation in 1859 and figured out a clever way to refute the kinetic theory by deducing from it a falsifiable consequence: the viscosity of a gas of billiard balls must be independent of density and must increase with temperature. Everyone knows that a fluid flows more slowly as it gets thicker and colder, but since common experience does not necessarily extend to gases, Maxwell asked G. G. Stokes, the expert on hydrodynamics, if any experiments had been done on this point. He was presumably assured by Stokes that experiments on the damping of pendulum-swings in air proved that the viscosity of a gas goes to zero as an its density goes to zero; in any case he wrote that "the only experiment I have met with on the subject does not seem to confirm" the prediction that viscosity is independent of density ([10], vol. 1, p. 166).

To clinch his argument Maxwell also worked out the ratio of specific heats for a gas of diatomic molecules and found that it should be $c_p/c_v = 1.333$, whereas experimental data clearly indicated $c_p/c_v = 1.40$ or 1.41. Whereas Clausius had evaded this problem by allowing the proportion of internal energy to be determined empirically by the specific heats, Maxwell insisted that in a consistent theory each mechanical degree of freedom must have the same average energy [102].

Armed with these calculations Maxwell went off to the British Association meeting at Oxford (1860) where he announced (referring specifically to the specific heats discrepancy): "This result of the dynamical theory, being at variance with experiment, overturns the whole hypothesis, however satisfactory the other results may be" [103]. (For some reason this conclusion, published in the report of the British Association meeting, was not included in Maxwell's collected papers.)

3. Victory

Fortunately for the progress of science Maxwell did not really believe in the hypothetico-deductive method; the kinetic theory had become such as an intriguing mathematical game that he couldn't bear to abandon it just because its predictions disagreed with experiment. Moreover, in spite of Stokes, he suspected that the density-dependence of the viscosity of air was not reliably established. So he decided to measure it himself.

The year 1865 saw the triumph of the kinetic theory. Maxwell found that the viscosity of air is indeed constant over a wide range of densities. This result was fully confirmed by other physicists. It turned out that what Stokes had claimed to be as an experimental fact was not a fact at all but a theory-laden observation: in reducing the data it had been <u>assumed</u> that the viscosity of air goes to zero as an the density goes to zero. (O. E. Meyer [108]; Stokes [158], Vol. III, pp.13, 76-77, 137-141) It seems obvious that if there is no air it cannot exert any viscous resisting force. Common sense, or a principle of continuity in nature, would suggest that if there is a very small amount of air the viscosity must be very small. But the actual pressure range in which this behavior occurs happens to be below that attainable in the first half of the 19th century, so the presumed continuous decrease of viscosity to zero was not directly observed. This episode might provide a good example for analyzing the relation between theory and experiment - only with the help of the theory was it possible to arrive at the correct interpretation of as an experiment which was supposed to test the theory.[21]

The other major event of 1865 was Josef Loschmidt's determination of the size of a molecule, using the mean-free-path formula from the kinetic theory of gases together with the ratio of gas to liquid densities. Other estimates, giving somewhat similar results (diameter 10^{-8} to 10^{-7} cm), were published soon afterwards by G. J. Stoney, L. Lorenz, and William Thomson. Suddenly the atom was real - for now atoms could be measured, weighed, and counted. But not everyone was convinced, and it was another 40 years before the "real existence of the atom" could be completely established (with the help of statistical mechanics).

4. Entropy and Probability

During the 1860s and 1870s a few physicists were also interested in the problem I mentioned at the beginning, the "reduction of thermodynamics." Since the First Law of Thermodynamics was recognized to be only a special case of the general principle of conservation of energy, it seemed reasonable to look for another general principle corresponding to the Second Law. The most popular candidate was Hamilton's version of the principle of least action. But these attempts to reduce the Second Law to mechanics dealt only with the problem of identifying a quantity that could serve as an integrating factor for the heat transferred in a reversible change. In this way one could obtain a mechanical analog of entropy (defined in thermodynamics as an heat transferred, divided by absolute temperature [26]), but the irreversible increase of entropy with time implied by the generalized Second Law was still unexplained.

Ludwig Boltzmann and James Clerk Maxwell realized around 1870 that the reduction of the Second Law must somehow involve statistical considerations. Boltzmann took an important step toward the goal with his "H-theorem" for low-density gases [7]. According to this theorem, a quantity H depending on the molecular velocity distribution always decreases with time unless the gas is in thermal equilibrium; H is then proportional to minus the thermodynamic entropy. Later Boltzmann proposed a more general but very simple interpretation: the entropy of a system is proportional to its probability. In modern notation,

$$S = k \log W$$

where W is the number of molecular states, or "microstates," corresponding to a state of the macroscopic system - a "macrostate," defined by thermodynamic variables such as pressure and temperature. If one samples at random from the microstates, the probability of finding a given macrostate is proportional to its W. Conversely, if one is told that a system has a certain macrostate, one is left in the dark about its molecular nature to the extent that it might be in any one of W microstates. If W is large one therefore says the system is "disordered." According to Boltzmann, the statement "entropy tends to increase" can be interpreted as an "a system tends to become more disordered."[22]

But there is a fundamental objection to any such reduction of irreversibility to mechanics. The objection was first pointed out explicitly by William Thomson (later Lord Kelvin), the same person who gave the first general statement of the principle of irreversibility in 1852 [75]). In 1874 Thomson [76] called attention to the apparent contradiction between the irreversibility of thermodynamics (and of macroscopic processes in general) and the reversibility of abstract Newtonian dynamics.[23] This was the "reversibility paradox," later made famous by the debate between Loschmidt and Boltzmann,[24] just as the "recurrence paradox" was made famous by the debate between Zermelo and Boltzmann in the 1890s.[25]

The controversy about the origin of irreversibility and the direction of time, which I have discussed elsewhere ([12], Chap. 14), still offers interesting material for philosophical analysis. I might mention in particular Boltzmann's semi-serious suggestion that the direction of time as we experience it <u>depends</u> on the direction of entropy change in the environment, and may itself be reversed during some epochs of the history of the universe.

In the present context it seems appropriate to point out that the attempt to reduce thermodynamics to mechanics led to the introduction of statistical ideas which changed the absolute character of thermodynamics. The "reduced theory" thereby was modified but at the same time strengthened since it could now be applied with more confidence to a wider range of phenomena, in particular those involving fluctuation phenomena such as an Brownian movement. It was just this kind of application of a generalized thermodynamics that led to the final establishment of the existence of atoms at the beginning of the 20th century.[26]

But the "reducing theory" was also affected: the introduction of statistical thermodynamics and in particular the recognition that irreversibility involves some kind of randomness at the molecular level[27] helped to undermine the rigid determinism of Newtonian mechanics. The result was that physicists in the first part of the 20th century were prepared to accept a new mechanics based on indeterminism [13]. In this way one cam see a close connection between the three philosophical aspects of the history of statistical mechanics - reduction, statistical. law, and irreversibility.

5.   Phase Transitions

During the 20th century the most important developments in statistical mechanics, aside from those associated with quantum theory, have been in the theory of phase transitions. The origin of this theory, or at least the idea that such a theory is possible, may be found in the 1873 dissertation of J. D. van der Waals [171]. Van der Waals developed as an "equation of state" which took account of intermolecular forces, thereby improving on the "ideal gas law." He found that this equation of state provided a remarkably simple explanation for phenomena near the gas-liquid critical point.[28]

If a theory such as that of van der Waals can successfully account for the transition from gaseous to liquid states of matter on a molecular basis, it should have some relevance to philosophical discussions of "levels of reality" (e.g., Grene [63], Nickles [117]). One of the arguments of antireductionists is that new concepts must be invoked in going from a lower to a higher level of complexity; the laws for the higher level, being formulated in terms of those concepts, are therefore non-reducible. Now it would seem that one of the simplest examples of this is the set of concepts "gas," "liquid" and " solid" which apparently have no meaning when applied to a single atom[29] but emerge gradually as one puts together enough atoms to form a macroscopic system. (Cf. Margenau [101], p. 42) But if statistical mechanics can show exactly how this happens, and derive the qualitative differences between states of matter from quantitative properties of atomic systems, we would have to conclude that the distinction between macroscopic and microscopic levels is not absolute but only a matter of convenience. So I am suggesting that the success of the statistical mechanical theory of phase transitions might be taken as an important example of reduction, much more significant than the derivation of the ideal gas law and more clearcut than the explanation of irreversibility in terms of randomness.

While van der Waals' theory suggested the possibility of explaining the gas-liquid transition in terms of intermolecular forces, it was not really an application of statistical mechanics. The first and simplest example of a phase transition derivable from statistical mechanics was discovered by Einstein in 1924 [37]. This is the famous condensation of a Bose-Einstein gas[30] at very low temperatures, to a peculiar substance in which a finite fraction of the particles are in the lowest quantum state. The physical significance of this mathematical result was not appreciated for another decade, but it was eventually established that the Bose-Einstein condensation is a first approximation to the "lambda transition" between the normal and "superfluid" states of liquid helium at 2.2° K (London [93], [94], [95]), and that the anomalous flow properties below the transition can be understood in this way (Tisza [164]).[31]

The lambda transition is the gateway to a spectacular realm of quantum phenomena - the unusual behavior of superfluid helium is easily visible on the macroscopic level. There is also a macroscopic quantum theory of super-fluidity, developed by L. D. Landau [86], which succeeded in explaining and predicting some (but not all) of the properties of helium below the lambda temperature, without using Bose-Einstein statistical mechanics. Thus one can even discuss the reduction of a macroscopic quantum theory to quantum statistical mechanics (Putterman [126]). Yet those philosophers of science who have written extensively on the meaning of quantum mechanics seem to have little interest in what seems to me to be one of its most striking manifestations.

Shortly after Einstein announced his discovery of the condensation of a Bose-Einstein gas, Ernst Ising published a short paper [71] on as an idealized model that has become the basis for a large amount of contemporary research on the theory of phase transitions [11]. The so-called "Ising model" was actually proposed in 1920 by Ising's mentor Wilhelm Lenz [88]. It consists of a system of atoms arranged on a regular crystalline lattice. Each atom is fixed in position and has only one property: a magnetic moment of "spin" which can have one of two possible values, +1 or -1. There is a force between nearest neighbors only, tending to make the atomic spins align in the same direction; in addition there may be an externally imposed magnetic field. The energy of a microstate of the system depends on the number of nearest-neighbor pairs of atoms having the same or opposite spins, and on the total number of spins having each value.

The basic question about the Ising model is: will the system display spontaneous magnetization, i.e.. more than half of the atomic spins have the same value in the absence of an external field, when the temperature goes below some critical temperature T_c (the "Curie temperature")?

Ising himself calculated the properties of his model only in the one-dimensional case, and found that there is no phase transition.[32] From this result he jumped to the erroneous conclusion that there would be no transition in a two or three dimensional model either.[33]

In the 1930s there was considerable interest among solid-state physicists and physical chemists in models similar to Ising's, applied to alloys and liquid mixtures. By a simple change of algebraic sign in the energy formula one can make the force such that nearest-neighbor spins tend to be unlike rather than like, and one can translate "spin +1 or -1" into "lattice site occupied by atom of type A or type B.' Thus at low temperatures microstates with an alternating ABABAB... arrangement would be favored, the so-called "superlattice" ordering. On the other hand if the force is such as to favor like pairs rather than unlike pairs (as an in the ferromagnetic case), the system may split into two phases at low temperatures, one phase containing more A atoms and the other more B atoms. Thus the Ising model might be able to explain order-disorder transitions in alloys and partial insolubility of liquids in each other as well as ferromagnetism.

The concept of the "Ising model" as a mathematical object existing independently of any particular physical application seems to have been developed by the Cambridge University group led by R. H. Fowler in the 1930s. Rudolph Peierls [120] was apparently the first to point out explicitly the equivalence of the Ising theory of ferromagnetism to the theories of other physical systems mentioned above.

The major breakthrough in the theory of phase transitions was Lars Onsager's exact solution of the zero-field two-dimensional Ising model, announced in 1942 and published in 1944 [119]. I consider this one of the most important events in theoretical physics of the 20th century.[34] For the first time it was possible to determine precisely the physical properties of a many-particle system in which interparticle forces play a significant role. Previously the only such results available were based on approximations such as series expansions valid only in temperature-density regions far away from suspected phase transitions. Although such approximations did predict that there would be a transition to an ordered state at low temperatures, they failed to describe the nature of the phase transition itself. For example, the so-called "critical exponents," which characterize the rate at which certain properties go to zero or infinity at the transition point, can only be computed accurately from mathematically rigorous theories like that of Onsager (or approximations whose error bounds are precisely established).

Although it has not yet been possible to solve the three-dimensional Ising model by finding a closed analytical formula for the partition function as Onsager did for the two-dimensional case, various methods have been refined to the point where it may be said that all the properties of the model can be computed to any desired degree of accuracy.[35] The Ising model is thus the first non-trivial case in which a phase transition has been completely explained in terms of interatomic forces. It is a much more interesting example of the reduction of thermodynamics to statistical mechanics than the textbook derivation of the ideal gas law from billiard-ball kinetic theory. In both cases one is talking about the computation of thermodynamic properties as predicted for a particular system from an atomic model rather than the derivation of laws of thermodynamics from laws of atomic physics. But a phase transition is a qualitative change in macroscopic properties that suddenly occurs at a definite temperature and pressure - spontaneous magnetization, or phase separation in a mixture - and therefore the ability to explain it seems a much more spectacular achievement.

During the last 30 years it has been shown that the Ising model applies to an even wider class of phase transitions. Around 1950 it was pointed out that the model can represent ordinary gases, liquids and solids as mixtures of atoms and "holes" - the spin variable is defined as an +1 if as an atom occupies the lattice site and as an -1 if it is empty. This is known as the "lattice gas" model.[36] With attractive forces between neighboring atoms the system may condense into a two-phase arrangement with the atoms clustered together in one region and the holes in another. With repulsive forces the system may crystallize into an ordered state with atoms and holes alternating on adjacent sites. Thus the gas-liquid transition would be analogous to the onset of spontaneous magnetization at the Curie temperature, while the liquid-solid transition would be analogous to superlattice formation in alloys.

The lattice gas model might seem to be very unrealistic, especially for gases. But M. E. Fisher [52] pointed out that it gives a fairly good estimate of the critical exponents for the gas-liquid transition - better than the van der Waals equation - and the further development of this model has been stimulated by the intense interest in critical-point experiments during the past 15 years.

While it has long been thought that the gas-liquid transition is in some way related to attractive forces, the idea that the liquid-solid transition is associated with repulsive forces is relatively recent. The possibility of crystallization of the billiard-ball gas (with no attractive forces) was suggested by Kirkwood and Monroe [79] and by I. Z. Fisher [51]; it was definitely established by B. J. Alder and T. E. Wainwright [3].[37] Systems with "softer" repulsive forces such as the inverse square also crystallize at low temperatures and high densities.[38]

The fact that the Ising model gives a reasonably good description of several different kinds of phase transitions shows that the precise nature of the interatomic force law is less important than the long-range correlations produced by these forces. Theorists now speak of a "universal" theory of phase transitions.[39] Near the transition there is a delicate balance between the ordering effects of the forces and the disordering effects of thermal agitation. As the temperature drops there is a sudden "landslide" into an ordered state; the correlations that previously influenced only nearest neighbors now extend over macroscopic distances. (This point is completely missed in the only philosophical discussion I have seen on this subject, by Girill [62].)

The success of the Ising model vindicates the approach advocated by Waterston and Maxwell. An atomic model should not incorporate all the physical factors that are believed to influence the system; instead it must be kept simple enough to permit accurate mathematical calculation of its properties.[40]

Throughout this paper I have used the term reduction without giving it as an explicit definition. This omission is probably quite annoying to a philosophical audience, but I will now try to make amends in my concluding remarks. In fact none of the definitions known to me when I started to write the paper now seem satisfactory. The difficulty is that one cannot simply say "A is reduced to B" because in all the nontrivial historical examples with which I am familiar, A and B both undergo significant changes as a result of the reduction. The attempt to reduce the Second Law of Thermodynamics changed it from an absolute to a statistical law; and a significant part of the "reducing theory," statistical mechanics, came into existence because of this attempt. Similarly the phenomenological theory of phase transitions was substantially modified as the statistical mechanical theory developed, and the current theory contains surprising new features on both macroscopic and microscopic levels. This phenomenon has been noted by Schaffner ([141], p. 116) and others.

In my view "reduction" refers to what scientists do whenever they try to replace an arbitrary or complicated empirical description by a simpler, more plausible theory, and in this sense the logical positivists were right in seeing reduction as a central feature of progress in science. But their use of "reduction" to describe other kinds of theory change (e.g., the replacement of the phlogiston theory by the oxygen theory) is unsatisfactory to me; it should be restricted to vertical motion along a hierarchy of sciences or levels, so that the reducing theory is in some way more fundamental than the reduced one.[41] (For a summary of the role of reduction in logical positivism see Darden [28].)

At the same time the antagonism toward reduction voiced by many scientists and nonscientists is also justified insofar as they have been led to accept the literal connotation of the word: as an entity <u>loses</u> something if it is "reduced." But the history of science shows that the "reduced" theory usually gains something else to replace the arbitrary or incomprehensible aspects it has lost. Often this gain will include a more precise knowledge of the limits of validity of the original theory (so that it can be used with more confidence within those limits) and indications of what has to be done to extend it beyond those limits.[42] In other cases the discovery that phenomena previously thought to be qualitatively different, such as condensation and magnetization, are essentially identical in some respects will stimulate new developments in the reduced theory that are independent of its reduction. But the diversity of historical examples makes it unlikely that these outcomes can be specified by a neat formula. I suspect that philosophical discussions of reduction, based on idealized examples taken out of their historical context, have unnecessarily impoverished the concept. Reduction is such an important process in science that it deserves better treatment than being reduced to an algorithm.

Notes

[1] This paper is based on research supported by the National Science Foundation's History and Philosophy of Science program. I thank Lindley Darden and Frederick Suppe for a number of useful comments.

The major works discussed in this parer are cited in the bibliography at the end. For further details and references to the extensive literature see [12]. Reprints and translations of the basic works may be found in [10]. A useful introduction to technical aspects of statistical mechanics is the recent textbook by Thompson [163].

[2] Campbell [19], pp. 126-140; [20], pp. 81ff); Meyerson ([109], pp. 36. 205-219, 308); Hempel ([68], pp. 68-73); Toulmin ([166], pp. 39-40); Feyerabend ([50], pp. 74-78); Kattsoff ([74], pp. 149ff); Clifford ([27], pp. 146-147).

[3] Bergmann ([6], pp. 168-169); van Fraassen ([170]. p. 90); Nagel ([112], pp. 107-128; [114], Chap. 11); Schlesinger ([143], p. 53); Lambert & Brittan ([85], pp. 61-62, 102); Kemeny and Oppenheim [78], p. 7).

[4] Feyerabend ([46], pp. 76-84); Smart ([154], p. 79); Feigl ([44.], pp. 138-139); Achinstein ([2], pp. 70-71, 104).

[5] Madden ([99], pp. 113-115). Popper ([124], pp. 96-97) notes his own and Schrödinger's ambivilance about the prospect that biology may be reduced to physics. Theobald [162] has recently argued strongly against reducing chemistry to physics. Elsasser has given a mathematical-physical argument against reducing biology to physics ([40] and many earlier publications). Schaffner ([138], [139], [140], [141]) has given a thorough analysis of reduction in molecular biology.

[6] Bergmann ([6], pp. 168-169); van Fraassen ([170], p. 90). Suppes ([160], p. 271) claims that more substantive examples exist but "are too complicated to include in the

text." He gives no references to them.

[7] I think most physicists would agree with the following assessment of Boyle's law: "Although exactly true, by definition. for as an ideal gas, it is obeyed only approximately by real gases and is not a fundamental law like Newton's laws or the law of conservation of energy" ([144], p. 247). One may agree with the statement of Andrews ([4], p. 26), that "in some branches of science very minor empirical correlations are called laws, but in thermodynamics the term is kept for the most important generalizations on which the whole subject is based" - though that author also refers to the "ideal-gas law" (p. 117). Margenau is one of the few physicists who specifically include the ideal gas law among the laws of thermodynamics - but he clearly distinguishes such laws from the more general "principles" or "Laws of Thermodynamics" ([100], pp. 208-212).

[8] On the 19th century reduction attempts see Daub [29] and Klein [82]. Some examples of noteworthy recent efforts are Tisza and Quay [165], Dyson and Lenard [35], Ruelle [135]. For the current state of ergodic theory see the lectures published in [111]. I think Sklar [149] goes too far in downgrading the importance of ergodic theory.

[9] Cassirer [22]; Reichenbach ([130]; [131]. pp. 159-164); Kantor ([73], p. 89); Carnap ([21], p. 280); Hempel ([68]. pp. 66, 68-69); Dingle ([31], p. 234); Sklar ([150], p. 386); Smart ([152]. p. 58).

[10] Feigl ([42], p. 411); Carnap ([21], pp. 278, 280); Werkmeister ([173], p. 275); Nagel ([114], p. 335). Frank ([59], pp. 295-296) suggests rather tentatively that since in Newtonian physics we can derive statistical laws from causal laws by using the ergodic theorem, causal and statistical laws may not be irreconcilable but rather any statistical law might be ultimately derivable from a causal law.

[11] Grünbaum [64] and several earlier publications cited or incorporated in that book; Sklar ([150], pp. 383-411); Smart ([152], pp. 383-411). For a survey of the physical aspects see Gal-Or [60]; Davies [30].

[12] Popper [122]; Feyerabend ([46], p. 65; [47], Sect. VI; [48]).

[13] Hesse ([69], pp. 8ff; [70], pp. 213-217); Losee ([97], pp. 136-137, 144); Toulmin ([166], pp. 165-168). Benrath [5] showed how this could be carried to as an anti-atomistic extreme.

[14] Meyerson ([110], pp. 89-91, 405); Reichenbach ([131], p. 168); Johnson ([72], pp. 28-29).

[15] Feigl ([45], p. 10); Cassirer ([23], p. 97).

[16] Cf. Mehlberg ([106], p. 339).

[17] McGuinness [105]; see S. R. De Groot's remarks on p. xi of this book.

[18] Post ([125], p.222); Salmon ([136], .p. 321). As an accurate historical account of the origin of Planck's theory has been given by Klein [81].

[19] This was true of the theory but not the theorists, who believed that the ether was still needed to explain optical and perhaps electromagnetic phenomena.

[20] Rayleigh, who discovered Waterston's paper and arranged for its publication in 1893, commented at that time: "The history of this paper suggests that highly speculative investigations, especially by as an unknown author, are best brought before the world through some other channel than a scientific society, which naturally hesitates to admit into its printed records matter of uncertain value. Perhaps one may go further and say that a young author who believes himself capable of great things would usually do well to secure the favourable recognition of the scientific world by work whose scope is limited, and whose value is easily judged, before embarking on greater flights" ([172], pp. 209- 210).

[21] According to Maxwell's theory the coefficient of heat conduction should also be independent of density. This was likewise confirmed experimentally, though with somewhat more difficulty. The heat-transfer data of Dulong and Petit [34], on which they had based a formula for the temperature-dependence of radiation that was accepted for half a century, now had to be reinterpreted. They had assumed that no heat was transferred by conduction at the lowest densities involved in their experiment, and used that assumption to separate the amounts of heat transfer due to radiation and conduction at other densities. The result of the reinterpretation by Josef Stefan [157], using Maxwell's theory of heat conduction, was the modern T^4 radiation law (Stefan-Boltzmann law). This law in turn provided part of the basis for Planck's researches on black-body radiation. (See [12], Chap. XIII.)

[22] I like to think of the entropy-disorder relation in terms of bridge hands, evaluated by a point-count system, e.g. $A = 4$, $K = 3$, $Q = 2$, $J = 1$. The $52! / 13! \, 39!$ microstates - possible hands - are grouped into macrostates defined by the number of high-card points. A 10-point hand has relatively high probability and is very "disordered" compared to a 30-point hand, because we lump together a large number of hands that have the same value. More precisely, to approximate a closed system, one should consider the states of all four hands in a deal (all $52! \, (13!)^4$ microstates). It is then clear that "shuffling" is almost as an irreversible process since a low-entropy freak deal on which one pair can make notrumps is almost always followed by a deal with more nearly equal distribution of high cards. It is just in this sense that Boltzmann could interpret the Clausius version of the Second Law ("entropy tends toward a maximum") by the statement, "a system goes from less probable to more probable states."

For an introduction to some of the philosophical problems involved in this definition of entropy see Shimony [146].

[23] It is sometimes asked why this contradiction was not obvious much earlier, for example to Fourier who stressed the irreversibility of heat conduction In 1807 [55]. The answer is that Fourier and many other physicists denied that heat theory should be reducible to Newtonian dynamics, hence for them the contradiction was irrelevant. The establishment of the kinetic theory of gases created the presumption that such a reduction should be possible.

[24] Thomson resolved the reversibility paradox in much the same way Boltzmann did when he was confronted with it by Josef Loschmidt. If microstate m_1 evolves to microstate m_2 in the course of time, then on reversing all the molecular velocities we should be able to return to m_1 by a dynamically permissible path. If m_1 belongs to

macrostate M_1 which has lower entropy than macrostate M_2 to which m_2 belongs, then it would seem that the reverse process corresponds to decreasing the entropy, violating the Second Law. But we cannot get our hands on individual microstates; even if we could, the smallest error in reversing m_2 would rapidly accumulate so we would not get back anywhere near to m_1. More realistically, since we can deal only with macrostates, the chance that we would draw the particular microstate m_2 that had evolved from such an m_1 is extremely small, whereas almost all microstates in M_1 will, like m_1, evolve to a microstate belonging to a macrostate with higher entropy. It is remarkable that Kelvin captured the qualitative essence of this argument without using the concept of entropy or its relation to probability.

[25] The recurrence paradox was based on Henri Poincaré's recurrence theorem. A deterministic mechanical system, restricted to a finite space, must eventually return as an closely as you like to any initial microstate and thus, it would seem, to the corresponding macrostate with initial entropy. As Poincaré put it, in order to observe a violation of the Second Law one does not need the nimble fingers of Maxwell's Demon - one simply needs a little patience ([10], vol. 2, p. 206).

[26] Einstein [36] showed that there is an intermediate situation - small particles suspended in a fluid - where the fluctuations and reversals predicted by molecular theory are observable. The experimental confirmation of Einstein's theory by Jean Perrin was regarded by many skeptical scientists (e.g., Wilhelm Ostwald) as convincing evidence for the real existence of atoms ([12], Chap. 15; Nye [118]). The claim that matter consists of discrete atoms had previously been regarded as a metaphysical proposition by many scientists; surely the historical events that led to this proposition being accepted as a scientific truth deserve the attention of philosophers of science!

[28] "…. the fortuitous concourse of atoms is the sole foundation in Philosophy on which can be founded the doctrine that it is impossible to derive mechanical effect from heat otherwise than by taking heat from a body at a higher temperature, converting at most a definite proportion of it into mechanical effect, and giving out the whole residue to matter at a lower temperature." (Kelvin [77], p. 464, italics in original) The role of a randomness assumption in the derivation of Boltzmann's H-theorem was pointed out by Burbury [18]; see the discussion by Boltzmann [8], pp. 40-42).

[29] One assumes implicitly that the individual atom or molecule is the same entity whether it is in a gas, liquid, or solid but that is not really obvious. In the 1880s some scientists (e.g., Ramsay [127], p. 204) proposed that when a gas changes to a liquid the molecules combine chemically to form larger complexes, so that "liquid molecules" are different from "gas molecules." This possibility was later rejected on the basis of detailed study of the properties of gases and liquids near the critical point (Ramsay and Young [128]; a detailed study of this episode has been undertaken by Levelt Sengers [89]). Tresca [167] concluded from his results on the flow of solids that the behavior of substances in different states can be explained in a unified way as special cases of the action of forces on molecules.

[30] In quantum mechanics one has to take account of the fact that permutations of identical particles give rise to states that cannot be physically distinguished from each

other. This condition can be satisfied by requiring that the wave function written in terms of the coordinates of N identical particles ($x_1, y_1, ... z_N$) changes only by a factor ± 1 when two particles are interchanged. Taking the + sign gives "Bose-Einstein statistics," while the - sign leads to "Fermi-Dirac statistics." In the latter case the Pauli exclusion principle applies: no more than one particle may occupy each quantum state. In the former case it is possible for a large number of particles to condense into the lowest state.

[31] London, remarked that after the publication of Einstein's paper, "in the course of time the degeneracy of the Bose-Einstein gas has rather got the reputation of having only a purely imaginary existence." ([93], p. 644) This was presumably the result of the criticism of Uhlenbeck ([169], pp. 69-71). See also Hanle [66], pp. 232-234, 250).

[32] This is easily understandable; One "wrong" spin breaks the communication between regions of "right" spins so there is no way for a correlation to extend over finite distances, and an ordered state is unstable against thermal fluctuations. In two or more dimensions the influence can go around the wrong spin.

[33] On the strength of this conclusion Heisenberg [67] decided to work out a more complicated model for ferromagnetism. This was by no means a wasted effort, since the Heisenberg model can account for some phenomena that the Ising model can't.

[34] Onsager received the Nobel Prize in Chemistry in 1968 but the citation mentioned only his "reciprocal relations" in thermodynamics. The importance of his work on the Ising model does not seem to be generally recognized outside the physics/chemistry community.

[35] For reviews of the present situation see Stanley [156], Domb and Green [32], Fisher [53].

[36] This phrase, and a thorough discussion of the relations between the lattice gas and the magnetic system, are due to T. D. Lee and C. N. Yang [87]; for further references see Brush ([11], pp. 890-891).

[37] Alder and Wainrwright used a computer to follow the detailed motions and collisions of a few hundred elastic spheres; their method avoids statistical mechanics in favor of a numerical solution of the classical many-body problem. The main objection to this approach (aside from its expense) is that surface effects, which are significant in small systems, may hide the behavior to be expected when Avogadro's number of particles are involved.

[38] Kirzhnits [80]; Abrikosov [1]; Salpeter [137]; Brush et al. [14]. The situation is not completely symmetrical since a system with only attractive forces is not thermodynamically stable.

[39] Metz [107]; Fisher [54]; Wilson and Kogut [174]; but see Levelt Sengers and Sengers [90] for skeptical remarks. There is also considerable interest in analogies between phase transitions and instabilities in nonequilibrium systems [133].

[40] Cf. Frenkel's remarks, quoted by M. E. Fisher ([52], pp. 7-8).

[41] But one must also allow the usage "theory A reduces to theory B when a parameter in A takes on a singular value" - thus quantum statistical mechanics reduces to classical statistical mechanics when Planck's constant goes to zero. (Yaglom has given a perspicuous way of demonstrating this reduction by means of Feynman's path integral formulation of quantum mechanics - see Brush [9].) Nickles [115] has discussed this kind of reduction in relation to the more usual kind.

[42] Charles Misner (colloquium at University of Maryland, May 1976) has argued that our most reliable knowledge of the world comes from a theory (e.g., Newtonian mechanics) that has been refuted, i.e., shown to be wrong outside certain limits, since the theory that replaced it (e.g., relativity) usually confirms its accuracy within those limits. The more recent theory, whose limits of validity have not yet been determined, is less reliable. In the same way one could say that thermodynamics, after its reduction by statistical mechanics (insofar as an that has occurred) is more reliable than statistical mechanics. Failure to recognize this point has led some philosophers of science (e.g,. Smart [153], p. 159) to claim that "we cannot retain both classical and statistical thermodynamics" because, in the case of Brownian movement, the former has been falsified and "replaced" by the latter.

References

[1] Abrikosov, A. A. "Some properties of strongly compressed matter I." Soviet Physics JETP 12 (1961): 1254-1259. (Translated from Zhurnal Eksperimentalnoi i Teoreticheskoi Fiziki 39 (1960): 1797-1805.)

[2] Achinstein, Peter. Concepts of Science. A Philosophical Analysis. Baltimore: .Johns Hopkins Press, 1968.

[3] Alder, B. J. and Wainwright, T. E. "Phase Transition for a Hard Sphere System." Journal of Chemical Physics 27 (1957): 1208-1209.

[4] Andrews, Frank C. Thermodynamics. New York: Wiley-Interscience 1971.

[5] Benrath, A. "[Über die Methode der entgegengesetzten Fehler in der physikalischen Chemie." Annalen der Naturphilosophie 11 (1912): 268-288.

[6] Bergmann, G. Philosophy of Science. Madison: University of Wisconsin Press, 1957.

[7] Boltzmann, L. "Further Studies on the Thermal Equilibrium of Gas Molecules." In ref. [10], volume 2, pages 88-175. (Translated from Sitzungsberichte, K. Akademie der Wissenschaften, Wien,. Mathematisch-Naturwissenschaftliche Klasse 66 (1872): 275-370.)

[8] _____. Lectures on Gas Theory. Berkeley: University of California Press, 1964. (Translated from Vorlesungen uber Gastheorie. Leipzig: Barth, 1896-1898.)

[9] Brush, S. G. "Functional Integrals and Statistical Physics." Reviews of Modern Physics 33 (1961): 79-92.

[10] _____. Kinetic theory. Volume 1, The Nature of Gases and of Heat. Volume 2, Irreversible Processes. Volume 3, The Chapman-Enskog Solution of the Transport Equation for Moderately Dense Gases. Oxford: Pergamon Press, 1965, 1966, 1972.

[11] _____. "History of the Lenz-Ising Model." Reviews of Modern Physics 39 (1967): 883-892.

[12] Brush, S. G. The Kind of Motion we call Heat: A History of the Kinetic Theory of Gases in the 19th Century. Amsterdam: North-Holland; New York: American Elsevier, 1976.

[13] _____. "Irreversibility and Indeterminism: Fourier to Heisenberg." Journal of the History of Ideas 37 (1976): 603-630.

[14] _____, Sahlin, H. L., and Teller, E. "Monte Carlo Study of a One-Component Plasma I." Journal of Chemical Physics 45 (1966): 2102-2118.

[15] Bumstead, H. A. "Physics." In The Development of the Sciences. Edited by Woodruff. New Haven: Yale University Press, 1923. Chapter II. Pages 43-73.

[16] Bunge, M. "Problems concerning Intertheory Relations." In Induction, Physics, and Ethics. Edited by P. Weingartner and G. Zecha. Dordrecht: Reidel, 1970. Pages 285-315. discussion on pages 316-325.

[17] _____. Philosophy of Physics. Dordrecht and Boston: Reidel, 1973.

[18] Burbury, S. H. "Boltzmann' s Minimum Function." Nature 51 (1894): 78.

[19] Campbell, N. R. Physics: The Elements. Cambridge, England: Cambridge University Press, 1919. (Reprint, New York: Dover Pubs., 1957.)

[20] _____. What is Science? London: Methuen, 1921. (Reprint, New York: Dover Pubs., 1957.)

[21] Carnap, R. Philosophical Foundations of Physics. New York: Basic Books, 1966.

[22] Cassirer, E. Determinism and Indeterminism in Modern Physics. New Haven: Yale University Press 1956. (Translated from Goteborgs Hogskolas Arsskrift 42 (1936), No. 3.)

[23] _____. The Problem of Knowledge, Philosophy, Science and History since Hegel. New Haven: Yale University Press, 1950. (Translated from a German manuscript written before 1940.)

[24] Clausius, R. "The Nature of the Motion which we call Heat." In ref. [10], volume 1, pages 111-134. (Translated from Annalen der Physik, ser. 2, 105 (1857): 353-380.)

[25] _____. "On the mean Lengths of the Paths described by the separate Molecules of Gaseous Bodies on the Occurrence of Molecular Motion; together with some other Remarks upon the Mechanical Theory of Heat." In ref. [10], pages 135-147. (Translated from Annalen dec Physik. scr. 2, 105 (1858): 239-258.)

[26] Clausius, R. "Ueber verschiedene fur die Anwendung bequeme Formen der Hauptgleichungen der mechanischen Warmetheorie." Annalen der Physik, ser. 2, 125 (1856): 353-400.

[27] Clifford, W. K. "On the Aims and Instruments Of Scientific Thought." In his Lectures and Essays. London: Macmillan, 1879. Volume 1. pages 124-157.

[28] Darden, Lindley. "The Heritage of Logical Positivism." In PSA 1976 Vol. 2. Edited by F. Suppe and P. D. Asquith, East Lansing, Michigan: Philosophy of Science Association, 1977. Pages 242-258.

[29] Daub, E. E. "Probability and Thermodynamics: The Reduction of the Second Law." Isis 60 (1969): 318-330.

[30] Davies, P. C. W. the Physics of Time Asymmetry. Berkeley: University of California Press, 1974.

[31] Dingle, H. "Causality and Statistics in Modern Physics." British Journal for the Philosophy of Science 21 (1 970): 233-246.

[32] Domb, C. and Green, M. S. Phase Transitions and Critical Phenomena. Volume 1, Exact Results. New York: Academic Press, 1972.

[33] Duhem, P. L'Evolution de la Mécanique. Paris: Joanin, 1903.

[34] Dulong, P. L,. and Petit, A. T. "Recherches sur la mesure des Températures, et sur les lois de Ie Communication de la Chaleur." Annales de Chemie et de Physique, ser. 2, 7 (1817): 113-154, 225-264, 337-367.

[35] Dyson, F. J. and Lenard, A. "Stability of Matter." Journal of Mathematical Physics, 8 (1967): 423-434; 9 (1968): 698-711.

[36] Einstein, A. "On the Movement of small Particles suspended in a stationary Liquid demanded by the Molecular-Kinetic Theory of Heat." In his Investigations of the Theory

of Brownian Movement. New York: Dover Pubs., 1956. Pages 1-18. (Translated from Annalen der Physik, ser. 4, 17 (1905): 549-560.)

[37] _____. "Quantentheorie des einatomigen idealen Gases." Sitzungsberichte, Akademie der Wissenschaften, Berlin (1924): 261-267; (1925): 3-14.

[38] Eisley, L. "The intellectual Antecedents of The Descent of Man." in Sexual Selection and the Descent of Man 1871-1971. Edited by B. Campbell. Chicago: Aldine Pub. Co., 1972. Pages 1-16.

[39] Elkana, Y. "Boltzmann's Scientific Research Programme and its Alternatives." In The Interaction between Science and Philosophy. Edited by Y. Elkana. Atlantic Highlands, New Jersey: Humanities Press, 1974. Pages 234-249.

[40] Elsasser, Walter M. "A Critique of Reductionism." Preprint, Department of Earth & Planetary Sciences, Johns Hopkins University, 1976.

[41] Enriques, F. Les Concepts Fondamentaux de la Science: Leur Signification reelle et leur acquisition Psychologique. Paris; Flammarion, 1913.

[42] Feigl, H. "Notes on Causality." In Readings in the Philosophy of Science. Edited by H. Feigl and M. Brodbeck. New York: Appleton- Century Crofts, 1953.

[43] _____. The "Mental" and the "Physical." Minneapolis: University of Minnesota Press, 1967. (Reprinted with a Postscript from Minnesota Studies in the Philosophy of Science, 2 (1958).)

[44] _____."Contemporary Science and Philosophy." In Science and Contemporary Society. Edited by F. J. Crosson. Notre Dame, Indiana: University of Notre Dame Press, 1967. Pages 129-153.

[45] _____. "Beyond Peaceful Coexistence." In Historical and Philosophical Perspectives of Science (Minnesota Studies in the Philosophy of Science V). Edited by R. H. Stuewer. Minneapolis: University of Minnesota Press, 1970. Pages 3-11.

[46] Feyerabend, P. K. "Exploration, Reduction, and Empiricism." In Scientific Explanation, Space, and Time (Minnesota Studies in the philosophy of Science III). Edited by H. Feigl and G. Maxwell. Minneapolis: University of Minnesota Press, 1962. Pages 28-97.

[47] _____."Problems of Empiricism." In Beyond the Edge of Certainty (University of Pittsburgh Series in the Philosophy of Science 2). Edited by R. G. Colodny. Englewood Cliffs, N.J.: Prentice, 1965. Pages 145-260.

[48] _____."On the Possibility of a Perpetuum Mobile of the Second Kind." In Mind, Matter and Method: Essays in Honor of Herbert Feigl. Edited by P. K. Feyerabend.

Minneapolis: University of Minnesota Press, 1966. Pages 409-412.

[49] _____. "Ludwig Boltzmann." In Encyclopedia of Philosophy Edited by P. Edwards. New York: Macmillan & Free Press, 1961. Volume 1, pages 334-337.

[50] _____. "In Defence of Classical Physics." Studies in the History and Philosophy of Science 1 (1970): 59-85.

[51] Fisher, I. Z. "On the Stability of a Homogeneous Phase." Soviet Physics JETP 1 (1955): 154-160, 273-279, 280-283. (Translated from. Zhurnal Ekperimentalnoi i Teoreticheskoi Fiziki 28 (1955): 171-180, 437-446, 447-451.)

[52] Fisher, M. E. "The Nature of Critical Points." in Lectures in Theoretical Physics, Volume VIIc. Boulder: University of Colorado Press, 1965. Pages 1-159.

[53] _____. "Phase Transitions, Symmetry and Dimensionality." In Essays in Physics, Volume 4. Edited by G. K. T. Conn and G. N. Fowler. New York: Academic Press, 1972. Pages 43-89.

[54] _____."The Renormalization Group in the Theory of Critical Behavior." Reviews of Modern Physics 46 (1974): 597-616.

(55) Fourier, J. B. J. "Theorie de la Propagation de la Chaleur dens les Solides." In Joseph Fourier 1768—1830. Edited by I. Grattan-Guinness and J. R. Ravetz. Cambridge, Massachusetts: M.I.T. Press, 1972. Pages 33-440.

[56] Fox R. The Caloric Theory of Gases from Lavoisier to Regnault. Oxford: Clarendon Press, 1971.

[57] Frank, P. "The Mechanical versus the Mathematical Conception of Nature." Philosophy of Science 4 (1937): 41-74.

[58] _____. Modern Science and its Philosophy. Cambridge: Harvard University Press, 1949.

[59] _____. Philosophy of Science. Englewood Cliffs, New Jersey: Prentice-Hall, 1957.

[60] Gal-Or, B. (ed.). Modern Developments in Thermodynamics. New York: Wiley/Halsted, 1974.

[61] Gibbs, J. W. Elementary Principles in Statistical Mechanics, developed with especial reference to the Rational Foundation of Thermodynamics. New York: Scribner, 1902.

[62] Girill, T. R. "The Problem of Micro-Explanation." In PSA 1976 Vol. 1. Edited by

F. Suppe and P. D. Asquith. East Lansing, Michigan: Philosophy of Science Association, 1976. Pages 47-55.

[63] Grene, Marjorie. "Biology and the Problem of Levels of Reality." The New Scholasticism 41 (1967): 427-449.

[64] Grünbaum, A. Philosophical Problems of Space and Time. Second Edition. Boston: Reidel, 1973.

[65] Günther S. Geschichte der anorganische Naturwissenschaften in Neunzehnten Jahrhundert. Berlin: Bondi, 1901.

[66] Hanle, P. A. Erwin Schrödinger's Statistical Mechanics 1912—1925. Ph.D. Dissertation, Yale University, 1975, University Microfilms Publication Number 76-11517.

[67] Heisenberg, W. "Zur Theorie des Ferromagnetismus." Zeitschrift fur Physik 49 (1928): 619-636.

[68] Hempel, Carl. C. Philosophy of Natural Science. Englewood Cliffs, New Jersey: Prentice Hall, 1966.

[69] Hesse, M. B. Models and Analogies in Science. London: Sheed and Ward, 1963.

[70] _____. The Structure of Scientific Inference. Berkeley: University of California Press, 1974.

[71] Ising, E. "Beitrag zur Theorie des Ferromagnetismus." Zeitschrift fur Physik 31 (1925): 253-258.

[72] Johnson, M. Science and the Meanings of Truth. London: Faber and Faber, 1946.

[73] Kantor, J. R. The Logic of Modern Science. Bloomington, Indiana / Evanston, Illinois: Principia, 1953.

[74] Kattsoff, L. O. Physical Science and Physical Reality. The Hague: Nijhoff, 1957.

[75] [Kelvin] Thomson, W. "On a Universal Tendency in Nature to the Dissipation of Mechanical Energy." Philosophical Magazine, ser. 4, 4 (1852): 304-306.

[76] _____. "The Kinetic Theory of the Dissipation of Energy." Proceedings of the Royal Society of Edinburgh 8 (1874): 325-334. (Reprinted in ref. [10], Volume 2, pages 176-187.)

[77] Kelvin. Popular Lectures and Addresses, Volume 2. London: Macmillan, 1894.

[78] Kemeny, J. G. and Oppenheim, P. "On Reduction." Philosophical Studies 7 (1956): 6-19.

[79] Kirkwood, J. G. and Monroe, E. "Statistical Mechanics of Fusion." Journal of Chemical Physics 9 (1941): 514-526.

[80] Kirzhnits, D. A. "On the Internal Structure of Super-Dense Stars." Soviet Physics JETP 11 (1960): 365-368. (Translated from Zhurnal Eksperimentalnoi i Teoreticheskoi Fiziki 38 (1960): 503-508.)

[81] Klein, M. J. "Max Planck and the Beginnings of the Quantum Theory." Archive for History of Exact Sciences 1 (1962): 459-479.

[82] _____. "Mechanical Explanation at the End of the Nineteenth Century." Centaurus 17 (1972): 58-82.

[83] Kuhn, T. S. The Structure of Scientific Revolutions. Chicago: University of Chicago Press, 1962.

[84] Lakatos, I. "History of Science and its Rational Reconstructions." In PSA 1970 (Boston Studies in the Philosophy of Science, Volume 8.) Edited by R. C. Buck and R. S. Cohen. Dordrecht: Reidel, 1971. Pages 91-136.

[85] Lambert, K. and Brittan, G. G. , Jr. As an Introduction to the Philosophy of Science. Englewood Cliffs, New Jersey: Prentice-Hall, 1970.

[86] Landau, L. "The Theory of Superfluidity of Helium II." Journal of Physics, Academy of Sciences of the USSR, 5 (1941): 71-90.

[87] Lee, T. D. and Yang, C. N. "Statistical Theory of Equations of State and Phase Transitions. II. Lattice Gas and Ising Model." Physical Review, ser. 2, 87 (1952): 410-419.

[88] Lenz, W. "Beitrag zum Verständnis der magnetischen Erscheningen in festen Körpern." Physikalische Zeitschrift 21 (1920): 613-615.

[89] Levelt Sengers, J. M. H. "Liquidons and Gasons: Persistence of the Liquid State beyond the Critical Point." Colloquium at National Bureau of Standards, Gaithersburg, Maryland, 29 September 1975.

[90] _____. and Sengers, J. V. "Universality of Critical Behavior in Gases" Physical Review, ser. 3. A12 (1975): 2622-2627.

[91] _____. "Critical exponents at the turn of the century." Physics 82A (1976): 319-351.

[92] Lewis, W. C. M. Physical Chemistry and Scientific Thought. Liverpool: University of Liverpool Press, 1914.

[93] London, F. "The λ-Phenomenon of Liquid Helium and the Bose-Einstein Degeneracy." Nature 141 (1938): 643-644.

[94] _____. "Bose-Einstein Condensation" Physical Review, ser. 2, 54 (1938): 947-954.

(95) _____. Superfluids. Volume II. Macroscopic Theory of Superfluid Helium. New York: Wiley, 1954.

[96] Loschmidt, J. "Zur Grösse der Luftmolecule." Sitzungsberichte, K. Akademie der Wissenschaften, Wien, Mathematisch-Naturwissenschaftliche Klasse, 52 (1865): 395-413.

[97] Losee, J. A Historical Introduction to the Philosophy of Science. New York: Oxford University Press, 1972.

[98] MacKie, J. L. The Cement of the Universe. A Study of Causation. Oxford: Clarendon Press, 1974.

[99] Madden, E. H. The Structure of Scientific Thought. Boston: Houghton Mifflin. 1960.

[100] Margenau, Henry. The Nature of Physical Reality. New York: McGraw-Hill, 1950.

[101] _____. "The Method of Science and the Meaning of Reality." In Integrative Principles of Modern Thought. Edited by H. Margenau. New York: Gordon and Breach, 1972. Pages 3-43.

[102] Maxwell, J. C. "Illustrations of the Dynamical Theory of Gases." Philosophical Magazine, ser. 4, 19 (1860): 19-32; 20 (1860): 21-37. (Reprinted in part in ref. [10], Volume 1, pages 148-171.)

[103] _____. "On the Results of Borneol's Theory of Gases as an applied to their Internal Friction, their Diffusion, and their Conductivity for Heat." Report of the 30th Meeting of the British Association for the Advancement of Science, part 2 (1860): 15-16.

[104] _____. "Viscosity or Internal Friction of Air and other Gases." Philosophical Transactions of the Royal Society of London 156 (1866): 249-268.

[105] McGuinness, B., (ed.). Ludwig Boltzmann, Theoretical Physics and Philosophical Problems. Dordrecht and Boston: Reidel, 1974.

[106] Mehlberg, H. The Reach of Science. Toronto: University of Toronto Press, 1958.

[107] Metz, W. "Phase Changes: a Universal Theory of Critical Phenomena." Science 181 (1973): 147, 149.

[108] Meyer, O. E. "Ueber die innere Reibung der Gase." Annalen der Physik, ser. 2, 125 (1865): 177-209. 401-420, 564-599.

[109] Meyerson, E. De l'Explication dans les Sciences. Paris: Payot, 1927.

[110] _____. Identity and Reality. New York: Dover Pubs., 1962. (Reprint of the 1930 translation of the third French edition (1926).)

[111] Moser, J. (ed.). Dynamic Systems, Theory and Applications. Battelle Seattle 1974 Rencontres. New York: Springer, 1975.

(112) Nagel, E. "The Meaning of Reduction in the Natural Sciences." In Science and Civilization. Edited by R. C. Stouffer. Madison: University of Wisconsin Press, 1949. Pages 99-135.

[113] _____. "Principles of the Theory of Probability." In International Encyclopedia of Unified Science, Volume 1. Edited by O. Neurath et al. Chicago: University of Chicago Press, 1955 (reprint of 1939 edition).

[114] _____. The Structure of Science: Problems in the Logic of Scientific Explanation. New York: Harcourt, Brace and World, 1961.

[115] Nickles, T. "Two Concepts of Intertheoretic Reduction." Journal of Philosophy 70 (1973): 181-201.

[116] _____. "Theory Generalization, Problem Reduction and the Unity of Science," In PSA 1974 (Boston Studies in the Philosophy of Science XXXII). Edited by R. S. Cohen et. al. Dordrecht: D. Reidel, 1976. Pages 33-75.

[117] _____. "On some Autonomy Arguments in Social Science." In PSA 1976 Vol. 1. Edited by F. Suppe and P. D. Asquith, East Lansing, Michigan: Philosophy of Science Association, 1976. Pages 12-24.

[118] Nye, M. J. Molecular Reality: A Perspective on the Life of Jean Perrin. London: Macdonald / New York: American Elsevier, 1972.

[119] Onsager, L. "Crystal Statistics. I. A Two-Dimensional Model with as an Order-Disorder Transition.." Physical Review, ser. 2, 65 (1944): 117-149.

[120] Peierls, R. "Ising's Model of Ferromagnetism." Proceedings of the Cambridge Philosophical Society 32 (1936): 477-481.

[121] Planck, M. A Survey of Physical Theory. New York: Dover Pubs., 1960.

[122] Popper, K. R. "Irreversibility; or, Entropy since 1905." British Journal for the Philosophy of Science 8 (1957): 151-155.

[123] _____. "Normal Science and its Dangers" In Criticism and the Growth of Knowledge. Edited by I. Lakatos and A. Musgrave. London: Cambridge University Press, 1970. Pages 51-58.

[124] _____. "The Rationality of Scientific Revolutions." In Problems of Scientific Revolution. Edited by R. Harré. Oxford: Clarendon Press, 1975. Pages 72-101.

[125] Post, H. R. "Correspondence, Invariance and Heuristics: In Praise of Conservative Induction." Studies in History and Philosophy of Science 2 (1971): 213-255.

[126] Putterman, S. J. Superfluid Hydrodynamics. Amsterdam: North-Holland, 1974.

[127] Ramsay, W. "On the Critical Point." Proceedings of the Royal Society of London 31 (1881): 194-205.

[128] _____ and Young, S. "On the Nature of Liquids, as an shown by a Study of the Thermal Properties of Stable and Dissociable Bodies." Proceedings of the Physical Society of London 8 (1887): 127-137.

[129] _____. "On the Gaseous and Liquid States of Matter." Philosophical Magazine, ser. 5, 23 (1887): 547-548.

[130] Reichenbach, H. Modern Philosophy of Science. London: Routledge and Kegan Paul, 1931.

[131] _____. The Rise of Scientific Philosophy. Berkeley: University of California Press, 1951.

[132] _____. The Direction of Time. Berkeley: University of California Press, 1956.

[133] Riste, Tormod (ed.). Fluctuations, Instabilities, and Phase Transitions. New York: Plenum, 1975.

[134] Roth, Laura M. Review of "Women in Science, Illustrated Interviews" produced by Dinah L. Moche. American Journal of Physics 44 (1976): 1020.

[135] Ruelle, D. Statistical Mechanics - Rigorous Results. New York: Benjamin, 1969.

[136] Salmon, W. C. "Determinism and Indeterminism in Modern Science." In Reason and Responsibility, second edition. Edited by J. Feinberg. Encino, California: Dickenson Pub. Co., 1971. Pages 316-332.

[137] Salpeter, E. E. "Energy and Pressure of a Zero-Temperature Plasma." Astrophysical Journal 134 (1961): 669-682.

[138] Schaffner, Kenneth F. "Approaches to Reduction." Philosophy of Science 34 (1967): 137-147.

[139] _____. "Antireductionism and molecular biology." Science 157 (1967): 644-647.

[140] _____. "The Watson-Crick Model and Reductionism." British Journal for the Philosophy of Science 20 (1969): 325-348.

[141] _____. "The Peripherality of Reductionism in the Development of Molecular Biology." Journal of the History of Biology 7 (1974): 111-139.

[142] Schlegel, R. Time and the Physical World. New York: Dover Pubs., 1968. Reprint of the 1961 edition with a new preface.

[143] Schlesinger, G. Method in the Physical Sciences. London: Routledge and Kegan Paul, 1963.

[144] Sears, Francis W., and Zemansky, Mark W. University Physics. Part 1. Mechanics, Heat, and Sound 4th ed. Reading, Mass.: Addison-Wesley, 1970.

[145] Shapere, Dudley. "Scientific Theories and their Domains." In The Structure of Scientific Theories. Edited by F. Suppe. Urbana: University of Illinois Press, 1974. Pages 518-599.

[146] Shimony, A. "Carnap on Entropy." in Rudolf Carnap, Logical Empiricist. Edited by J. Hintikka. Dordrecht: Reidel, 1975. Pages 381-395. (Reprinted from the introduction to Carnap, Rudolf. Two Essays on Entropy. Berkeley: University of California Press, 1976).

[147] Siegel, Armand. "The Scientist's Personality - as an perceived, and as an perceiving." Paper given at the Center for the Study of Democratic Institutions, Santa Barbara, California, April 1, 1971 (unpublished).

[148] _____. "The Youth Culture and the Socially Concerned Scientists." Bulletin of the Atomic Scientists 28, no. 9 (November 1972): 16-21.

[149] Sklar, L. "Statistical Explanation and Ergodic Theory." Philosophy of Science 40 (1973): 194-212.

[150] _____. Space, Time and Spacetime. Berkeley: University of California Press, 1974.

[151] _____. "Thermodynamics, Statistical Mechanics and the Complexity of Reductions." In PSA 1974 (Boston Studies in the Philosophy of Science XXXII). Edited by R. S. Cohen et al. Dordrecht: D. Reidel, 1976. Pages 15-32.

[152] Smart, J. C. C. Philosophy and Scientific Realism. London: Routledge and Kegan Paul, 1963.

[153] _____. "Conflicting views about explanation." Boston Studies in the Philosophy of Science, volume II. Edited by R. Cohen and M. Wartofsky. New York: Humanities Press, 1965. Pages 157-169.

[154] _____. Between Science and Philosophy. New York: Random House, 1968.

[155] Sneed, J. The Logical Structure of Mathematical Physics. Dordrecht: Reidel, 1971.

[156] Stanley, H. E. Introduction to Phase Transitions and Critical Phenomena. New York, Oxford University Press, 1971.

[157] Stefan, J. "Über die Beziehung zwischen der Warmestrahlung und die temperatur." Sitzungsberichte K. Akademie der Wissenschaften, Wien, Mathetmatisch-Naturwissenschaftliche Kl asse 79 (1879): 391-428.

[158] Stokes, G. G. Mathematical and Physical Papers. Cambridge, England: Cambridge University Press, 1880-1905.

[159] Suppe. F. "The Search for Philosophic Understanding of Scientific Theories." In The Structure of Scientific Theories. Edited by F. Suppe. Urbana: University of Illinois Press, 1974. Pages 1-241.

[160] Suppes. P. Introduction to Logic. Princeton. New Jersey: Van Nostrand, 1957.

[161] _____. "A Comparison of the Meaning and Uses of Models In Mathematics and the Empirical Sciences." In The Concept and the Role of the model in Mathematics and Natural and Social Sciences. New York: Gordon and Breach, 1961. Pages 163-176.

[162) Theobald, D. W. "Some Considerations of the Philosophy of Chemistry," Chemical Society Revlews 5 (1976): 203-213.

[163] Thompson, C. J. Mathematical Statistical Mechanics. New York: Macmillan, 1972.

[164] Tisza, L. 'Transport Phenomena in Helium II." Nature 141 (1938), 913.

[165] _____, and Quay, P. M. "The Statistical Thermodynamics of Equilibrium." Annals of Physics 25 (1963): 48-90.

[166] Toulmin, S. Philosophy of Science. London: Hutchinson, 1953.

167] Tresca. H. "Memoire sur l'écoulement des corps solides soumis à de fortes pressions." Comptes Rendus Acad. Sci. Paris 64 (1867): 809-812.

[168] Truesdell, C. "Early Kinetic Theories of Gases." Archive for History of Exact Sciences 15 (1976): 1-66.

[169] Uhlenbeck. G. E . Over Statistische Methoden in de Theorie der Quanta. 's-Gravenhage; Nijhoff, 1927.

[170] Van Fraassen, B. C. As an Introduction to the Philosophy of Time and Space. New York: Random House, 1970.

[171] Waals. J. D. Van der. Over de Continuiteit van den gas-en Vloeistoftoestand. Leiden: Sijthoff, 1873.

[172] [Waterston, J. J.] The Collected Scientific Papers of John James Waterston. Edited, with a biography, by J. S. Haldane. Edinburgh: Oliver and Boyd, 1928.

[173] Werkmeister, W. H. A Philosophy of Science. New York: Harper, 1940.

[174] Wilson, K. G. and Kogut, J. "The Renormalization Group and the ϵ expansion." Physics Reports. Physics Letters Section C. 12C (1974): 75-200.

[175] Zwart, P. J. "The Flow of Time. Synthese 24 (1972): 133-158.

Part IV

A Guide to Historical Commentaries: Kinetic Theory of Gases, Thermodynamics, and Related Topics

A Guide to Historical Commentaries:
Kinetic Theory of Gases, Thermodynamics, and Related Topics

The bibliography below lists articles and books published from 1965 through 2001 (with occasional exceptions) on the history of kinetic theory, statistical mechanics, thermodynamics, the nature of heat, and the physical properties of gases. Reprints and edited collections of original sources are included, as well as historical commentaries. The "target period" during which the original research was done covers a quarter of a millenium: from Robert Boyle's work on gas pressure around 1660 to the beginnings of quantum theory at the end of the 19th century (a century that, in accordance with the practice of historians, is considered to end in 1914).

As it happens, both endpoints are now the subject of lively discussion among historians of science: the study of the debate between Boyle and Thomas Hobbes by Steven Shapin and Simon Schaffer (1985) is often cited in the current controversy about the social construction of scientific knowledge, and Thomas Kuhn's (1978) re-interpretation of Max Planck's 1900 papers on black-body radiation has challenged the long-held view that Planck proposed a physical quantization of energy in the second of those papers.

Author entries, cross-references, and subject entries are listed in a single alphabetical sequence. Thus editions of works by James Clerk Maxwell are followed by references to articles and books about Maxwell, and by references to discussions of the history of **Maxwell's Demon**. (Names of subjects are in bold-face.) Note that the bibliography does not in general include modern discussions of topics such as Maxwell's Demon or the Maxwell velocity distribution unless they include significant historical commentary. Book reviews are listed (and reviewers cross-referenced) for books that contain a substanial amount of material on my subject, and for biographies of scientists whose work was mostly on this subject (e.g. Boltzmann but not Maxwell).

Inclusion or exclusion of a publication does not imply any judgment about its accuracy or value.

"*Isis CB*" at the end of an entry indicates that the item itself has not been examined but the reference has been taken from the *Isis Current Bibliography of the History of Science and its Cultural Influences* or the online version of that important database. "*BISS*" means that the entry was found in *Isis CB* but originated in the Italian Bibliography of the History of Science. "*Author*" means the author of the publication has supplied the citation.

In listing my own articles, I have used an asterisk to indicate the ones reprinted in the present volume, and "*KMWCH*" for those included (with some revision) in my book *The Kind of Motion We Call Heat*.

Abbreviations:
AHES = *Archive for History of Exact Sciences*
AIHS = *Archives Internationales d'Histoire des Sciences*
BJHS = *British Journal for the History of Science*
DSB = *Dictionary of Scientific Biography*
HSPS = *Historical Studies in the Physical [and Biological] Sciences*
NRRSL = *Notes & Records of the Royal Society of London*
NTM = *Zeitschrift für Geschichte der Naturwissenschaften, Technik und Medizin*

PSA = *Proceedings of the Biennial Meeting of the Philosophy of Science Association*
SHPS = *Studies in History and Philosophy of Science*

Abbri, F. *See* Shapin & Schaffer (1985)
Abiko, S. 1997. Introduction of the Concept of Energy by Clausius, related to the problem of the Japanese Equivalent for "Verwandlung" [in Japanese]. *Kakakusi Kenkyu: Journal of History of Science, Japan*, 36: 157-161. [*Isis CB*]
Achinstein, P. 1986. Theoretical Derivations. *Stud. HPS* 17: 375-414.
> On Maxwell's derivation of his velocity distribution law.

_____ 1987a. Scientific Discovery and Maxwell's Kinetic Theory. *Philosophy of Science* 54: 409-434

_____ 1987b. (Book Review) *Maxwell on Molecules and Gases*. *Foundations of Physics* 17: 425-433.
> Review of Maxwell (1986).

_____ 1991. *Particles and Waves: Historical Essays in the Philosophy of Science*. New York: Oxford University Press.
> Part II: Maxwell and the Kinetic Theory of Gases.
> Reviewed by J. P. Halpin, *Philosophical Review* 102 (1993): 599-602; K. Hentschel, *AIHS* 44 (1994): 175-176; N. J. Nersessian, *Isis* 83: 527-528; D. F. Polis, *Modern Schoolman* 69 (1992): 156-158; R. H. Schlegel, *Review of Metaphysics* 46 (1992) 141-142 [*Isis CB*]

_____ 1994. Jean Perrin and Molecular Reality. *Perspectives on Science* 2: 396-427.

Adams, Henry: Brush (1978), Burich (1987)
Agassi, J. 1967. The Kirchhoff-Planck Radiation Law. *Science* 156: 30-37.

_____ 1977. Who Discovered Boyle's Law? *Stud. HPS* 8: 189-250.
> Townley or Hooke?

_____ 1983. The Structure of the Quantum Revolution. *Philosophy of the Social Sciences* 13: 367-381.
> Essay review of Kuhn (1978).

_____ 1993. *Radiation Theory and the Quantum Revolution*. Basel/Boston/Berlin: Birkhäuser.
> Reviewed by K. Hentschel, *AIHS* 46 (1996): 165-167; H. Kragh, *Centaurus* 37 (1994): 91-92; S. F. Mason, *Annals of Science* 51 (1994): 677-678; N. Robotti, *Nuncius* 9 (1994): 883-885 [*Isis CB*].

Air *see* **Gases**
Aiton, E. J. *See* Scott (1970)
Alekseev, I. S. 1981. K Predystorii Kvantovoi Teorii. *Voprosy Istorii Estestvoznaniia i Tekhniki* no. 2: 77-85.
> "On the prehistory of the quantum theory." On the work of Planck and W. Wien. [*Isis CB*]

Allard, G. 1983. La Termodinamica. *Storia della Scienza Contemporanea* 1: 132-147. [*BISS*]
Althin, T. *See* Cardwell (1971)

Altieri, G., and F. Sebastiani. 1985. La Teoria Cinetica dei Gas: Daniele Bernoulli. *Giornale di Fisica* 26: 323-339. [*Isis CB*]

Amano, K. 2000. Thermal Radiation Studies that led to the Genesis of Quantum Theory (I). English translation by S. Takata and S.-i. Hyodo. *Historia Scientiarum* 10: 185-210.

    From his book originally published in Japanese in 1943.

_____ 2001. Thermal Radiation Studies that led to the Genesis of Quantum Theory (II). English translation by S. Takata and S.-i. Hyodo. *Historia Scientiarum* 10: 255-280.

    From his book originally published in Japanese in 1943. "... like Boltzmann, Planck first assumed the finite Energieelement for the sake of mathematical convenience..." (p. 271).

Amano, Kiyoshi Takata (2000)

Amontons, Guillaume: Pacey (1974), Redondi (1980), Talbott & Pacey (1971)

Ampère, André Marie: Gornshtein (1966), Scheidecker-Chevalier (1997), Sebastiani (1982a, 1990a)

Andrews, D. H. *See* Donnan & Haas (1980)

Andrews, T. 1973. On the continuity of the gaseous and liquid states of matter. In *Cooperaive Phenomena near Phase Transitions*, edited by H. E. Stanley. Cambridge, MA: MIT Press. Reprinted from *Philosophical Transactions of the Royal Society of London*, 1869.

Andrews, Thomas: Levelt Sengers (1979), Rowlinson (1969)

Ångström, A. J.: James (1983)

Antman, S. *See* Truesdell (1980)

Archimedes: Baierlein (1969)

Arrhenius, Svante: Barkan (1994), Hiebert (1978b, 1982)

Assmus, A. 1992. The Molecular Tradition in early Quantum Theory. *HSPS* 21: 209-231.

    Specific heats, W. Nernst.

Atoms, reality of; Atomism: Achinstein (1994), Blackmore (1985), Brush (1968a, 1976b), Cercignani (1998), Clark (1976a), Daub (1967), Hall (1987), Heilbron (1975), Hiebert (1968b, 1974, 1996b), Jungnickel & McCormmach (1986), Kerker (1974, 1976, 1986), Kubbinga (1997a), Langevin (1971), Lindley (2001), Maddy (1994), Maiocchi (1985b, 1986, 1987a,b, 1988, 1990c, 1992), Nye (1970, 1972), Post (1968), Raman (1970), Ramunni (1986), Renn (1997)

Atoms & molecules, size of: Einstein (1989)

Avogadro, Amedeo: Mendoza (1970b), Schöpf (1984c), Sebastiani (1983)

Avogadro's Hypothesis: Causey (1971), Chevalier (1997),

Avogadro's Number Floderer (1971), Hawthorne (1970, 1973), Jech (1988), Mathieu (1984)

Bach, A. 1990. Boltzmann's Probability Distribution of 1877. *AHES* 41: 1-40.

    "What we now call Bose-Einstein Statistics actually had been introduced by Boltzmann in 1877." [*Isis CB*]

Baierlein, R. 1969. Archimedes and the Canonical Distribution. *American Journal of Physics* 37: 315-318.

    Basis of J. Perrin's analysis of the height-distribution of particles in

Brownian movement.
Bailyn, M. 1985. Carnot and the Universal Heat Death. *American Journal of Physics* 53: 1092-1099.
> Theological and historical links between Carnot's and Kelvin's formulations of the Second Law of Thermodynamics.

Baracca, A. 1987. Sadi Carnot e la Teoria del Calore: Storia e Attualità didattica. *Giornale di Fisica* 28: 52-78. [*Isis CB*]

_____ 1991. Boltzmann's Research Program and the present frontier Problems in Statistical Mechanics. In Martinás et al. (1991), 156-161. Singapore: World Scientific.

_____ *See also* Battimelli (1993)

_____ and R. Rechtman S. 1985. Einstein's Statistical Mechanics. *Revista Mexicana de Fisica* 31: 695-722.
> Einstein in 1903 established the ergodic hypothesis as a central postulate of statistical mechanics.

Barkan, D. K. 1991. Walther Nernst and Quantum Theory. In *World Views and Scientific Discipline Formation*, edited by W. R. Woodward and R. S. Cohen, 151-162. Dordrecht: Kluwer.

_____ 1994. Simply a Matter of Chemistry? The Nobel Prize for 1920. *Perspectives on Science* 2: 357-395.
> W. Nernst vs. S. Arrhenius.

_____ 1999. *Walther Nernst and the Transition to Modern Physical Science.* New York: Cambridge University Press.
> Heat Theorem ("Third Law of Thermodynamics"), quantum theory, specific heats.
> Reviewed by M. J. Nye, *Endeavour* 12 (1999): 135-136; J. S. Rowlinson, *NRRSL* 54 (2000), 123; C. Tanford & J. Reynolds, *Nature* 399 (1999): 118-119 [*Isis CB*]

_____ *See also* Kipnis et al. (1994), Medard & Tachoire (1994)

Barker, J. A. 1976. Gibbs' Contribution to Statistical Mechanics. *Proceedings of the Royal Australian Chemical Institute*, May, 131-137.

Barr, E. S. 1969. James Prescott Joule and the Quiet Revolution. *Physics Teacher* 7: 199-208.

Bartoli, Adolfo: Carazza & Kragh (1989)

Battimelli, G., M. G. Ianniello and O. Kresten, editors. 1993. *Proceedings of the International Symposium on Ludwig Boltzmann, Rome, Feb. 9-11, 1989..* Vienna: Verlag der Österreichischen Akademie der Wissenschaften.
> Papers by A. Baracca, C. Cercignani, D. Chandler, S. D'Agostino, A. Dick, D. Flamm, L. Galgani, W. Höflechner, A. Hohenester, M. G. Ianniello, G. Israel, G. Kerber, A. J. Kox, E. Oeser, J. P. Valleau, A. Wehrl, A. D. Wilson.

Baucia, G. 1973. Microfenomeni e Macrofenomeni secondo Maxwell in Relazi one alla 'Teori a Dinamica del Calore.' *Physis* 15: 333-350.

Becker, F. 1996. *Über die Natur der Wärme: Verschlungene Wege zur wissenschaftlichen Erkenntnis. (Sitzungsberichte der Wissenschaftlichen Gesellschaft an der Johann Wolfgang Goethe-Universität Frankfurt am Main,* Band 34, Nr. 1.) Stuttgart: Steiner. [*Isis CB*]

Bellone, E. 1968. L'Energia molecolare e la Velocità molecolare come Parametri discreti in alcuni Scritti di L. Boltzmann in Relazione all'Ipotesi di M. Planck ed alla Teoria della Radiazione di S. D. Poisson. *Physis* 10: 101-112.
> Comments on possible relations between Boltzmann's use of discrete energy elements and the theories of S. D. Poisson and M. Planck.

───── 1971. Osservazioni su alcuni Aspetti della Termologia del settecento, con particolare Riferimento alle Esperienze di Benjamin Thompson. *Physis* 13: 376-398.
> Criticizes the 19th-century view that 18th-century thermology was dominated by a clash between caloric-fluid explanations and models that identified heat with molecular motion: a clash that could have been resolved by certain crucial experiments. Some mechanical models, though based on fluids, also assumed that heat is molecular motion. Thompson's research should be seen in this context, in order to understand the role of caloric theories in the process leading to statistical mechanics.

───── 1972. *Aspetti dell'Approccio statistico alla Meccanica: 1849-1905.* Firenze: G. Barbera Editore.

───── 1978. *Le Leggi della Termodinamica da Boyle a Boltzmann..* Torino: Loescher Editore.

───── 1980. *A World on Paper: Studies on the Second Scientific Revolution*, translated by M. & R. Giacconi. Cambridge, MA: MIT Press.
> Reviewed by S. G. Brush, *Isis* 72 (1981): 284-286; J. D. Burchfield, *Quarterly Review of Biology* 56 (1981): 454; G. N. Cantor, *Nature* 292 (1981): 781-782; J. L. Greenberg, *American Journal of Physics* 52 (1984): 189-190; J. N. Hedgpeth, *Quarterly Review of Biology* 56 (1981): 454; C. Smith, *BJHS* 15 (1982): 197-199 [*Isis CB*]

───── 1985. Boltzmann e Lagrange: Quanti "classici" e Credenze sull' sull'Irreversibilità. In *Scienza e Filosofia: Saggi in Onore di Ludovico Geymonat*, edited by C. Mangione, 517-532. Milano: Garzanti. [*Isis CB*]

───── 1989. *I Nomi del Tempo. La seconda Rivoluzione scientifica e il Mito della Freccia temporale.* Bollati Boringhieri.
> Includes heat and atomic motion, time, Herapath, Carnot, Maxwell, Boltzmann.

───── 1996. Il "Quantum" classico e la Matematica. *Rivista di Filosofia* 87: 193-202.
> Refers especially to Boltzmann. [*Isis CB*]

───── *See also* Brush (1976b)

Benedetti, A., F. Di Carlo and G. Michelon. 1984. Aspetti storici rilevanti della Teoria Cinetica dei Gas: Un Approcio ipertestuale. In *Atti del IV Convegni Nazionale de Storia e Fondamenta della Chimica, 1983*, edited by P. Tucci, 371-382. Rozzano: Centro Stampa Rossano. [*BISS*]

Bent, H. A. 1980. Einstein and chemical Thought. Atomism extended. *Journal of Chemical Education* 57: 395-405.
> On Brownian movement, quantum, radiation.

Bérard, Jacques Étienne: Mendoza (1990a)

Beretta, M. *See* Maiocchi (1988)

Berger, J. 1999. Grenzgänge zwischen Physik und Chemie: Thermodynamik und chemische Kinetik -- kein Happy-End im 19. Jahrhundert. *Centaurus* 41: 253-279. [*Isis CB*]

Bergia, S. 1983. Lineamenti generali di un'Indagine sul Ruolo della Termodinamica Statistica nello Sviluppo della Fisica Quantistica. In *Atti del III Congresso di Storia della Fisica, Palermo, 1982*, edited by F. Bevilacqua and A. Russo, 2: 531-538. Palermo: Centro Stampa Facoltà Ingegneria. [*BISS*]

_____, P. Lugli and N. Zamboni. 1980. Zero-point Energy, Planck's Law and the Prehistory of Stochastic Electrodynamics; Part 2. Einstein and Stern's Paper of 1913. *Annales, Fondation Louis de Broglie*, 5: 39-62. [*cited by* Assmus (1992)]

_____ and L. Navarro. 1988. Recurrences and Continuity in Einstein's Research on Radiation between 1905 and 1916. *AHES* 38: 79-99.

> On the possible relationship between his 1912 paper on photochemical equivalence and his 1916 derivation of Planck's formula, and a possible origin of his dissatisfaction with the probabilistic character of quantum physics.

_____ _____ 1997. Early Quantum Concepts and the Theorem of Equipartition of Energy in Einstein's Work (1905-1925). *Physis* 34: 184-223.

Bernhardt, H. 1967. Der Umkehreinwand gegen das H-Theorem und Boltzmanns statistische Deutung der Entropie. *NTM* 4, no. 10: 35-44.

_____ 1969. Der Widerkehreinwand gegen Boltzmanns H-theorem und der Begriff der Irreversibilität. *NTM* 6, no. 2: 27-36.

_____ 1971. Über die Entwicklung und Bedeutung der Ergodenhypothese in den Anfängen der statistichen Mechanik. *NTM* 8, no. 1: 13-25.

Bernoulli, D. 1968. *Hydrodynamics*, translated from the Latin by T. Carmody and H. Kobus. Preface by H. Rouse. (First published in Latin, 1738) New York: Dover. Accompanied by a translation of *Hydraulics* by J. Bernoulli.

> Reviewed by L. Rosenfeld, *Centaurus* 15 (1970): 110. [*Isis CB*]

Bernoulli, Daniel: Altieri & Sebastiani (1985), Bowker (1971), Brush (1983), Pacey (1974), Pacey & Fisher (1967), Sheynin (1985), Truesdell (1968), Whitaker (1979)

Berthollet, C. L.: Kremer (1980)

Bevilacqua, F. 1994. Theoretical and mathematical Interpretations of Energy Conservation: The Helmholtz-Clausius Debate on central Forces 1852-5. In *Universalgenie Helmholtz*, edited by L. Krüger, 89-106. Berlin: Akademie Verlag. [*Isis CB*]

_____ *See also* Helmholtz (1994)

Bharatha, S. *See* Truesdell & Bharatha (1977)

Bierhalter, G. 1981a. Boltzmanns mechanische Grundlegung des zweiten Hauptsatzes der Wärmelehre aus dem Jahre 1866. *AHES* 24: 195-205.

_____ 1981b. Clausius' mechanische Grundlegung des zweiten Hauptsatzes der Wärmelehre aus dem Jahre 1881. *AHES* 24: 207-220.

_____ 1981c. Zu Hermann von Helmholtzens mechanischer Grundlegung der Wärmelehre aus dem Jahre 1884. *AHES* 25: 71-84.

_____ 1982. Das Virialtheorem in seiner Beziehung zu den mechanischen Grundlegungen des zweiten Hauptsatzes der Wärmelehre. *AHES* 27: 199-211.

_____ 1983a. Zu Szilys Versuch einer mechanischen Grundlegung des zweiten Hauptsatzes der Thermodynamik. *AHES* 28: 25-35.

_____ 1983b. Die v. Helmholtzschen Monozykel-Analogien zur Thermodynamik und das

Clausiussche Disgregationskonzept. *AHES* 29: 95-100.
_____ 1985. Die mechanischen Entropie- und Disgregationskonzepte aus dem 19. Jahrhundert: Ihre Grundlagen, ihr Versagen und ihr Enstehungshintergrund. *AHES* 32: 17-41.
_____ 1987. Wie erfolgreich waren die im 19. Jahrhundert betriebenen Versuche einer mechanischen Grundlegung des zweiten Hauptsatzes der Thermodynamik? *AHES* 37: 77-99.
_____ 1990. Zyklische Zeitvorstellung, Zeitrichtung und die frühen Versuche einer Deduktion des Zweiten Hauptsatzes der Thermodynamik. *Centaurus* 33: 345-367.
> On Boltzmann, Clausius.

_____ 1992. Von L. Boltzmann bis J. J. Thomson: Die Versuche einer mechanischen Grundlegung der Thermodynamik (1866-1890). *AHES* 44: 25-75.
_____ *See also* Helmholtz (1994)
Bignami, G. F. *See* Cercignani (1998)
Biology & Thermodynamics: Broda (1976, 1980, 1982); Flamm (1979)
Biot, Jean Baptiste: Yoshida (1989)
Birembaut, Arthur. 1974. A propos des Notices biographiques sur Sadi Carnot: Quelques Documents inédits. *Revue d'Histoire des Sciences* 27: 355-370.
> It is not true that Sadi died of cholera; his younger brother Hippolyte did not want to reveal the fact that Sadi died in an insane asylum.

Black, Joseph: Kubbinga (2002), Sebastiani (1985, 1990a)
Black Body Radiation: Cerreta (1992), Guidetti (1997), Kuhn (1978), Norton (1987, 1993)
Blackmore, J. 1972. *Ernst Mach.* Berkeley: University of California Press.
_____ 1985. An historical Note on Ernst Mach. *British Journal for the Philosophy of Science* 36: 299-329.
> "Mach's opposition to both the reality of atoms and to Boltzmann's kinetic theory of gases was philosophical and was not even accompanied by a serious scientific argument against either one."

_____ 1987. Ludwig Boltzmann as a major source of 20th century Philosophy. In *Logik, Wissenschaftstheorie und Erkenntnistheorie: Akten des 11. Internationale Wittgenstein-Symposiums*. Vienna: Hölder-Pichler-Tempsky. [*Isis CB*]
_____ 1995. *Ludwig Boltzmann, His Later Life and Philosophy, 1900-1906.* Book One: *A Documentary History.* Book Two: *The Philosopher.* (Boston Studies in Philosophy of Science, vols. 168 and 174). Dordrecht/Boston/London: Kluwer Academic Publishers.
> Reviewed by A. Brenner, *Revue d'Histoire des Sciences* 51 (1998): 540-541; N. De Courtenay, *AIHS* 47 (139 bis) (1997): 208-209; R. Haller, *Isis* 88 (1997): 364-365; P. Hart, *Metascience* 9 (1996): 143-145; C. Tanford & J. Reynolds, *Nature* 378 (1995): 673; L. Sklar, *British Journal for the Philosophy of Science* 47 (1996): 630-632; Book 2 reviewed by W. De Regt, *Isis* 91 (2000): 595

_____, editor. 1999. Ludwig Boltzmann: Troubled Genius as Philosopher. *Synthese* 119, nos. 1-2: 1-232.
> Articles by J. Blackmore, S. G. Brush, R. Deltete, G. Jäger, S. Meyer, J. Nabl, H. W. De Regt, M. Stöltzner, S. Tanaka, H. Visser.

_____ *See also* Cercignani (1998)

Blood, Color of: Ober (1968)
Böhm, W. 1973. Loschmidt, Johann Joseph. *DSB* 8: 507-511.
Boerhaave, Herman: Morris (1965), Sebastiani (1990a)
Bohr, N. 1972. *Collected Works*, vol. 1, *Early Work 1905-1911*, edited by J. R. Nielsen. Amsterdam: North-Holland.
        Includes letter from Bohr to C. W. Oseen, 1 December 1911, on Boltzmann's attempts to justify the Maxwell velocity distribution and the law that entropy always increases (pp. 422-426, English translation on pp. 426-431).
Bol, W. *See* Hornix & Mannaerts (2001)
Boltzmann, L. 1974. *Ludwig Boltzmann: Theoretical Physics and Philosophical Problems, Selected Writings*. Edited by B. McGuinness. Boston: Reidel.
        Reviewed by S. G. Brush in *Annals of Science* 32 (1975): 599-601; by M. Curd in *Philosophy of Science* 45 (1978): 148-149; by R. H. Stuewer in *AIHS* 26 (1976): 326-327.
_____ 1979. *Populäre Schriften*, selected and edited by E. Broda. Vieweg: Wiesbaden.
_____ 1981. *Vorlesungen über Gastheorie.* In *Ludwig Boltzmann Gesamtausgabe*, Band 1, edited by R. U. Sexl. Includes German translation of Introduction, Notes and Bibliography by S. G. Brush, originally published with the English translation of *Vorlesungen* in 1964.
_____ 1990. *Ludwig Boltzmann Principien der Naturfilosofi, Lectures on Natural Philosophy 1903-1906*, edited by I. M. Fasol-Boltzmann. Berlin: Springer-Verlag.
        Reviewed by E. N. Hiebert in *Isis* **???**
_____ 1995. *Lectures on Gas Theory*, English translation with Introduction and Notes by S. G. Brush. New York: Dover Publications. (Reprint of the 1964 edition.)
Boltzmann, Ludwig: Baracca (1991), Battimelli (1993), Bellone (1978, 1980, 1985, 1991), Blackmore (1985, 1987, 1995, 1999), Bouveresse (1991), Bowden (1972), Broda (1979, 1983), Brouzeng (1993), Brush (1970a, 1976b, 1983, 1990), Campogalliani (1991, 1992), Cercignani (1997, 1998), Curd (1978, 1982), D'Agostino (2000), De Regt (1996, 2001), Dias (1994b,c), [Einstein] Nelkowski (1979), Elkana (1974a), Feyerabend (1967, 1970), Flamm (1973, 1983), Fleischhacker & Schönfield (1977), Heilbron (1975), Helm (2000), Hiebert (1968b, 1971, 1980, 1982, 1996a, 2000), Hintikka et al. (1981), Höflechner (1994), Hörz & Laass (1989), Hoyer (1978b, 1980, 1982, 1984, 1987, 1991, 2000a), Jungnickel & McCormmach (1986), Kartsev (1981), Kastler (1983), Klein (1973a,b, 1975), Koch (1991), Jox (1990), Kubbinga (2002), Lindley (2001), Locqueneux et al. (1983), Meyenn (1982, 1984), Nairn & Kilpatrick (1972), Navarro Veguillas (1997), Nyhof (1988), Porter (1986), Post (1968), Prigogine & Stengers (1984), Ramunni (1986), Ruhla (1992), Scheibe (1995), Sexl & Blackmore (1982), Sheynin (1985), Sklar (1993b), Stiller (1988), Treder (1977), von Plato (1987, 1991), A. D. Wilson (1990), D. B. Wilson (1981), Zanarini (1996)
Boltzmann Distribution/Factor: Bach (1990), Sheynin (1971)
Boltzmann's energy elements: Amano (2001), Bellone (1968, 1996), Cerreta (1995), Hoyer (1984), Koch (1983)
Boltzmann's Equation (Transport Theory): Brush 1976b, Dresden (1995)
Boltzmann's H Theorem: Bernhardt (1967, 1969), Brush (1967c, 1976b), Dias (1994b,

1994c), Fleischhacker & Schönfield (1997) , Hjalmars (1977), Sklar (1993b), Steckline (1983)

Boltzmann's mechanical Foundation of Thermodynamics: Bierhalter (1981a, 1990), Kaiser (1982), Klein (1974a), Szumilewicz (1969)

Boltzmann's Statistical Interpretation of Entropy: Bernhardt (1967), Bohr (1972), Broda (1976), Brush (1976b, 1982), E. G. D. Cohen (1997), Daub (1969), Flamm (1979), Lestienne (1987), Renn (1997)

Bordogna, C. *See* Cotignola et al. (2002)

Bose-Einstein Statistics: Bach (1990)

Bosscha: Lunteren (2000)

Boudri, J. C. 1991. The Paradox of William Thomson: Phenomenalism and Mechanicism in the Origin of Thermodynamics (1847-1850). In Martinas (1991), 42-56.

Bouveresse, J. 1991. Hertz, Boltzmann et le Problème de la 'Vérité' des Theories. In *La Vérité est-elle Scientifique*, edited by A. Lichnérowicz & G. Gadoffie, 117-141. Paris: editions Universitaires.

Bowden, R C. 1972. *An Analysis of Lord Kelvin's scientific Career*. Ph. D. Dissertation, University of North Carolina at Chapel Hill. *Dissertation Abstracts Intermnational* 34 (1973): 240-A

    Includes equipartition theorem, correspondence with Stokes, Tait, Boltzmann, Gibbs, etc.

Bowker, Roland M. 1971. *A Translation and Discussion of Daniel Bernoulli's Work on Elastic Fluids in Hydrodynamica*. Thesis, Victoria University of Manchester.

Boyle, R. 1999. *The Works of Robert Boyle*. Volume 1, *General Introduction, Textual Note, Publications to 1660*. Volume 2, *The Sceptical Chymist and other Publications of 1661*. Volume 3, *The Usefulness of Natural Philosophy and Sequels to the Spring of the Air*. Edited by M. Hunter & E. B. Davis. London: Pickering and Chatto. [*Isis CB*]

Boyle, Robert: Bellone (1978), Brush (1976b, 1983), Chalmers (1998), Clericuzio (1998), Hall (1987), Jenkins (1997), Klever (1997), Krips (1994), Langevin (1971), Mendoza (1990b), Pinnick (1998, 1999), Redondi (1980), Sargent (1995), Shapin & Schaffer (1985, 1999)

Boyle's Law: Agassi (1977), Cohen (1964), Hawthorne (1979), Massignat (2000), Potter (2001), Webster (1963, 1969)

Brammar, W. *See* Mendelssohn (1973)

Breger, H. 1982. *Die Natur als arbeitende Maschine. Zur Entstehung des Energiebegriffs in der Physik 1840-1850*. Frankfurt & New York: Campus Verlag.

    Argues that Marx's concept of Nature as an *arbeitende Maschine* "created the necessary condition for the formulation of the conservation of energy" -- from review by K. Olesko, *Isis* 74 (1983): 589. Also reviewed by A. Lundgren, *Lychnos* (1985): 170-271; J.-P. Mathieu, *Revue d'Histoire des Sciences* 39 (1986): 87 [*Isis CB*]

Brenner, A. *See* Blackmore (1995)

Breny, H. 1992. Premières Évaluations numériques pour une 'Loi des Gaz.' *Revue d'Histoire des Sciences* 45: 345-359.

    Latter half of 17th century. [*Isis CB*]

British Scientists: Smith (1978), Wise (1983)

Brock, W. H. 1972. A Scientist among the Spirits: William Crookes and the Radiometer. In *Studies in Physics*, edited by W. H. Brock, 43-60. Amersham: Hulton Educational.

_____ *See also* Cardwell (1989), Fox (1971b)

Broda, E. 1976. Erklärung des Entropiesatzes und der Liebe aus den Prinzipien der Wahrscheinlichkeitsrechnung von Ludwig Boltzmann, Professor der theoretischen Physik an der Universität Wien. *Physikalische Blätter* 32: 337-341.

_____ 1979. Der Einfluss von Ernst Mach und Ludwig Boltzmann auf Albert Einstein. *Einstein-Symposium* 1-20

_____ 1980. Boltzmann, Einstein, Natural Law and Evolution. *Comparative Biochemistry and Physiology* 67B: 373-378.

_____ 1982. Boltzmann and Darwin. In Sexl & Blackmore (1982), 129-142.

_____ 1983. *Ludwig Boltzmann: Man, Physicist, Philosopher*. Woodbridge, CT: Ox Bow Press. Translated from the German edition (1955) by L. Gay and the author.

> Reviewed by S. G. Brush in *Technology and Culture* 26 (1985): 659-661; M. Curd, *Isis* 75 (1984): 423; M. J. Klein, *American Scientist* 72 (1984): 312-313; R. B. Lindsay, *Foundations of Physics* 14 (1984): 811-814; J. J. Roche, *Nature* 306 (1983): 128-129.

Broda, Engelbert: Heilbron (1975)

Brody, S. B. 1987. Physics in *Middlemarch*: Gas molecules and ethereal Atoms. *Modern Philology* 85: 42-53.

> References to the kinetic theory and the Heat Death by novelist George Eliot.

Brouzeng, P. 1982. Duhem et l'Irréversibilité. *Fundamenta Scientiae* 3: 189-200.

_____ 1991. Duhem's Contribution to the Development of Modern Thermodynamics. In Martinás et al. (1991), 72-80.

_____ 1993. Lubeck Conference: The Debate between Mechanism and Energeticism: Boltzmann against Ostwald. In *First European Physical Society Conference on History of Physics in Europe in the 19th and 20th Centuries*, 131-140.

_____ 1995. L'Épistémologie de l'Oeuvre Scientifique de Pierre Duhem replacée dans le Contexte du Débat Énergétisme-Mécanisme. In *Les Savants et l'Épistémologie vers la Fin du XIXe Siècle,* edited by M. Panza & J.-C. Pont, 173-180. Paris: Blanchard.

_____ *See also* Carnot (1978)

Brown, G. I. 1995. *Scientist, Soldier, Statesman, Spy -- Count Rumford: The Extraordinary Life of a Scientific Genius*. Phoenix Mill: Sutton Publishing.

> Chapter 5, The Nature of Heat.
> Reviewed by J. M. Thomas, *NRRSL* 54 (2000): 393; A. Märker, *Ambix* 47 (2000): 204-205 [*Isis CB*]

Brown, L. M. *See* Nye (1984)

Brown, Robert: Brush (1968b), Deutsch (1991)

Brown, S. C. 1967. *Benjamin Thompson--Count Rumford: Count Rumford on the Nature of Heat*. Oxford & New York: Pergamon Press.

_____ 1979. *Benjamin Thompson, Count Rumford*. Cambridge, MA: MIT Press.

> Reviewed by S. G. Brush in *Physics Today* 32, no. 11: 55, 58.

_____ *See also* Elkana (1974b), Rumford (1968-70)

Brownian Motion: Baierlein (1969), Brush (1968b, 1976b), Deutsch (1991), Dresden

(1995), Einstein (1989), Goodman (1972), Kerker (1974), Kubbinga (2002), Lavenda (1985), Lemons & Gythiel (1997), Maddy (1994), Maiocchi (1988, 1990a), Mayo (1996), Nickles (1980), Nye (1972), Pas (1971), Powles (1978), Sarkar (2000), Średniawa (1991), Teske (1969, 1972), von Plato (1994) [*Isis CB*]

Brush, S. G. *1965. *Kinetic Theory*, vol. 1. *The Nature of Gases and of Heat*. Oxford & New York: Pergamon Press.

_____ *1966. *Kinetic Theory*, vol. 2. *Irreversible Processes*. Oxford & New York: Pergamon Press.

_____ 1967a. Thermodynamics and History: Science and Culture in the 19th century. *The Graduate Journal* 7: 477-465.

_____ 1967b. Foundations of Statistical Mechanics 1845-1915. *AHES* 4: 145-183. *KMWCH*

_____ 1967c. Boltzmann's "Eta Theorem": Where's the Evidence. *American Journal of Physics* 35: 892.
      Questions the claim that the "H" in "H-Theorem" is capital Greek eta. See Hjalmars (1977).

_____ 1968a. Mach and Atomism. *Synthese* 18: 192-215. *KMWCH*

_____ 1968b. A History of Random Processes, I. Brownian Movement from Brown to Perrin. *AHES* 5: 1-36. *KMWCH*. Reprinted in *Studies in the History of Statistics and Probability*, vol. II, edited by M. Kendall & R. L. Kendall, 347-382. New York: Macmillan (1977).

_____ 1970a. Ludwig Boltzmann. *DSB* 2: 260-268.

_____ 1970b. Interatomic Forces and Gas Theory from Newton to Lennard-Jones. *Archive for Rational Mechanics and Analysis* 39: 1-29.

_____ 1970c. The Wave Theory of Heat: A Forgotten Stage in the Transition from the Caloric Theory to Thermodynamics. *BJHS* 5: 145-167.

_____ 1970d. *Kinetische Theorie, I: Die Natur der Gase und der Wärme; II: Irreversible Prozesse*. German translation of *Kinetic Theory*, vols. 1 and 2 (Oxford and New York: Pergamon Press, 1965 & 1966) Berlin: Akademie-Verlag, and Braunschweig: Vieweg.

_____ *1971a. James Clerk Maxwell and the Kinetic Theory of Gases: A Review based on Recent Historical Studies. *American Journal of Physics* 39: 631-640.

_____ *1971b. Proof of the Impossibility of Ergodic Systems: The 1913 Papers of Rosenthal and Plancherel. *Transport Theory and Statistical Physics* 1: 287-311.

_____ 1972a. *Kinetic Theory*, Vol. 3. *The Chapman-Enskog Solution of the Transport Equation for moderately dense Gases*. Oxford & New York: Pergamon Press.

_____ 1972b. John Herapath. *DSB* 6: 291-293.

_____ 1973a. J. D. van der Waals and the States of Matter. *Physics Teacher* 11: 261-270. *KMWCH*

_____ 1973b. The Development of the Kinetic Theory of Gases, VII. Heat Conduction and the Stefan-Boltzmann Law. *AHES* 11: 328-396. *KMWCH*

_____ 1974. The Development of the Kinetic Theory of Gases, VIII. Randomness and Irreversibility. *AHES* 12: 1-88. *KMWCH*

_____ 1976a. Waterston, John James. *DSB* 14: 184-186.

_____ 1976b. *The Kind of Motion We Call Heat: A History of the Kinetic Theory of Gases in the 19th Century*. Amsterdam: North-Holland Pub. Co. Reprinted, 1986.

Includes articles designated above by *KMWCH*.

Boyle and air pressure, pro- and anti-atomism, J. Herapath, J. J. Waterston, Clausius, Maxwell, Boltzmann, Van der Waals, Mach, wave theory of heat, statistical mechanics, ergodic hypothesis, interatomic forces, equation of state, viscosity, Maxwell-Boltzmann transport theory, H theorem, heat conduction, Stefan-Boltzmann law, randomness and irreversibility, 2nd Law of Thermodynamics, recurrence paradox, Planck's radiation theory, Brownian movement and the theories of Einstein and Smoluchowski, Jean Perrin and reality of atoms.

Reviewed by E. Bellone, *Scientia* 113 (1978): 703-706; J. Z. Buchwald, *Historia Mathematica* 4 (1977): 466-467; P. Clark, *British Journal for the Philosophy of Science* 33 (1982): 165-186; R. Fox, *American Scientist* 65 (1977): 646; T. L. Hankins, *Science* 196 (1977): 783-784; E. N. Hiebert, *American Journal of Physics* 45: 1130-1131; E. Mendoza in *Annals of Science* 35 (1978): 332-333; H. A. M. Snelders, *Janus* 64 (1977): 340; R. H. Stuewer, *Isis* 69 (1978): 137-138 [*Isis CB*]; L. Tisza, *Journal of Statistical Physics* 18 (1978): 409-414, with response by SGB, *ibid.* 18 (1978): 523-528.

_____ *1976c. Irreversibility and Indeterminism: Fourier to Heisenberg. *Journal of the History of Ideas* 37: 603-630.

_____ *1977. Statistical Mechanics and the Philosophy of Science. *PSA 1976, Proceedings of the Philosophy of Science Association Meeting at Chicago, October 1976*, edited by F. Suppe and P. D. Asquith, 551-584. East Lansing, MI: Philosophy of Science Association.

_____ 1978. *The Temperature of History: Phases of Science and Culture in the Nineteenth Century*. New York: Burt Franklin. German translation, *Die Temperatur der Geschichte: Wissenschaftliche und kulturelle Phasen im 19. Jahrhundert*, Braunschweig: Vieweg (1987).

Correlation between Romanticism/Realism cycles in culture and in science; 2nd Law of Thermodynamics, irreversibility & dissipation of energy; heat death; Recurrence Paradox of Poincaré and Eternal Return of Nietzsche; J. B. Stallo's critique of Kinetic Theory; mechanistic science, Comte & positivism; Henry Adams and J. W. Gibbs on thermodynamics.

Reviewed by K. Hutchison, *Annals of Science* 37 (1980): 716-718; R. Porter, *Isis* 72 (1981): 521-531; A. Romer, *American Scientist* 68 (1980): 101; C. Smith, *BJHS* 13 (1980): 272-273 [*Isis CB*]

_____ 1981a. Einleitung, Anmerkungen und Bibliographie. In Boltzmann (1981).

_____ 1981b. Nietzsche's Recurrence Revisited: The French Connection. *Journal of the History of Philosophy* 19: 235-238.

Nietzsche's "Eternal Return" and Poincaré's "recurrence paradox."

_____ 1982. Changes in the Concept of Time during the Second Scientific Revolution. In Sexl & Blackmore (1982), 305-328.

_____ 1983. *Statistical Physics and the Atomic Theory of Matter, from Boyle and Newton to Landau and Onsager*. Princeton, NJ: Princeton University Press.

Theories of Boyle, Newton, D. Bernoulli, Herapath, Clausius, maxwell, Boltzmann, Gibbs; irreversibility & indeterminism; origin of quantum

theory; interatomic forces; van der Waals theory; statistical mechanics & philosophy of science.
Reviewed by C. Domb, *Nature* 309 (1984): 731-732; E. N. Hiebert, *AIHS* 38: 342-343; J. S. Rowlinson, *Foundations of Physics* 15 (1985): 245-246; B. R. Wheaton, *American Scientist* 72 (1984): 527 *[Isis CB]*

_____ *1988. Gaseous Heat Conduction and Radiation in 19th Century Physics. In *History of Heat Transfer*, edited by E. T. Layton, Jr. & J. H. Leinhard, 25-51. New York: American Society of Mechanical Engineers.

_____ *1990. Ludwig Boltzmann and the Foundations of Natural Science. In Boltzmann (1990), 43-61.

_____ 1994. Statistical Mechanics. In *Companion Encyclopedia of the History and Philosophy of the Mathematical Sciences*, edited by I. Grattan-Guinness, 1183-1188. London & New York: Routledge.

_____ *1999. Gadflies and Geniuses in the History of Gas Theory. *Synthese* 119: 11-43.

_____ 2002. Cautious Revolutionaries: Maxwell, Planck, Hubble. *American Journal of Physics* 70: 119-127.

> They were so critical of their own ideas (kinetic theory, quantized absorption of radiation, expanding universe) that they rejected them at some point, but eventually got credit for them.

_____ *See also* Bellone (1980), Boltzmann (1974, 1995), Broda (1983), S. C. Brown (1979), Cercignani (1998), Domb (1996), Fox (1971b), Kangro (1972), Kipnis et al. (1996), Lindsay (1973, 1975), Maxwell (1986, 1995), Nye (1972), Steffens (1979), Truesdell (1980), Waals (1988), D. B. Wilson (1990) *[Isis CB]*

_____ & C. W. F. Everitt. 1969. Maxwell, Osborne Reynolds, and the Radiometer. *HSPS* 1: 105-125. *KMWCH*

Buchwald, J. Z. *See* Brush (1976b), Maxwell (1986)

Buckle: *See* Schweber (1982)

Buono, Paolo del: see **del Buono, Paolo**

Burbury, Samuel Hawkes: Dias (1994b), Hesketh (1973), Hiromasa (1975)

Burcham, W. W. *See* Cardwell (1989)

Burich, K. R. 1987. Henry Adams, the Second Law of Thermodynamics, and the Course of History. *Journal of the History of Ideas* 48: 467-482. *[Isis CB]*

Busch, L. *See* Shapin & Schaffer (1985)

Butler, J. A. V. *See* Donnan & Haas (1980)

Byrne, P. H. 1980. Statistical and Causal Concepts in Einstein's early Thought. *Annals of Science* 37: 215-228.

> Einstein's view, that statistical laws are based upon causal assumptions, originated in his earliest investigations (1901-1903) where he applied kinetic theory to phenomena of capillarity and intermolecular forces. *[Author]*

_____ 1981. The Origins of Einstein's Use of Formal Asymmetries. *Annals of Science* 38: 191-206.

> Focusing upon his earliest publications on kinetic theory and thermodynamics, the article traces the gradual evolution of Einstein's use of formal considerations from analogies through intermediate stages to a sophisticated method of formal asymmetries. *[Author]*

Cahan, D. 1990. From Dust Figures to the Kinetic Theory of Gases: August Kundt and the changing Nature of experimental Physics in the 1860s and 1870s. *Annals of Science* 47: 151-172.
_____ *See also* Helmholtz (1994, 1995), Caneva (1993)
Caloric Theory: Bellone (1971), Campogalliani (1990), Baracca (1987), Brush (1970c), Chang (2002), Costabel (1968), Dias (1996), Fox (1968b, 1971b), Morris (1965, 1972), Nakamura (1997), Nakamura & Itakura (1991), Sebastiani (1981b, 1982a,b, 1992, 1993), Watanabe (1984)
Campogalliani, P. 1990.. L'Irreversibilità nella Fisica dell'Ottocento: Il Verso privilegiato dei Fenomeni naturali: Dal Fluido Calorico al Caos Molecolare. *Scienza e Cultura*, no. 5: 53-113. English translation, pp. 89-113. [*BISS*]
_____ 1991. Irreversibilita e Principio probabilistica nella Costruzione teorica di Ludwig Boltzmann. *Science e Storia* 8: 25-30.
_____ 1992. Freccia del Tempo e Principio probabilistica nella Costruzione teorica di Ludwig Boltzmann. *Physis* 29: 447-463.
_____ 1995. Planck's Theory (1898-1906) and the Birth of Quantum Physics. In *Foundations of Quantum Mechanics*, edited by C. Garola & A. Rossi, 71-84. Boston: Kluwer.
Caneva, K. L. 1993. *Robert Mayer and the Conservation of Energy.* Princeton, NJ: Princeton University Press.
> Reviewed by J. Buchwald in *American Historical Review*, 100 (1995): 176-177; D. Cahan, *Nature* 368 (1994): 23; P. H. Harman, *Centaurus* 36 (1993): 181; F. A. J. L. James, *Ambix* 41 (1994): 52-53; P. Munday, *Physis* 31 (1994): 667-670; T. M. Porter, *Journal of Interdisciplinary History* 25 (1995): 661-662; W. D. Seufert, *Journal of the History of Medicine* 49 (1994): 125-126; R. S. Turner, *Medical History* 39 (1995): 126-127 [*Isis CB*]

_____ 1997. Colding, Ørsted, and the Meanings of Force. *HSPS* 26: 1-138.
Cantor, G. N. 1975. William Robert Grove, the Correlation of Forces, and the Conservation of Energy. *Centaurus* 19: 273-290. [*Isis CB*]
_____ *See also* Bellone (1980)
Capillarity: Byrne (1980)
Carazza, B., and G. P. Guidetti. 1997. Origini e Sviluppi della Termodinamica. *Quaderni di Storia della Fisica*, no. 1: 111-130. [*BISS*]
Carazza, B., and H. Kragh. 1989. Adolfo Bartoli and the Problem of Radiant Heat. *Annals of Science* 46: 183-194.
Cardone, G., and A. Drago. 1995. L'Omogeneità matematica nella Storia della Termodinamica. *Atti del XIV e XV Congresso Nazionale di Storia della Fisica, Udine 1993 & Lecce 1994*, edited by A. Rossi, 217-227. Lecce: Conte.
Cardwell, D. S. L. 1971. *From Watt to Clausius: The Rise of Thermodynamics in the early Industrial Age.* Ithaca, NY: Cornell University Press. Reprint, Ames, Iowa: Iowa State University Press (1989).
> Reviewed by T. Althin, *Lychnos* (1973-1974, pub. 1975): 409; E. Daub, *Nature* 235 (1972): 177-178 and *Technology & Culture* 13 (1972): 315; E. Ferguson, *Science* 175 (1972): 745-746; L. Mackenson, *Technikgeschichte* 43 (1976): 74-75; O. Mayr, *Isis* 63 (1972): 451-452;

J. Payen, *BJHS* 7 (1974): 171-175 [*Isis CB*]

___ 1972. *The Organisation of Science in England.* Revised edition. London: Heinemann.

Includes (p. 189) a graph of number of papers published on heat and thermodynamics, by year and nationality.

___ 1977. Theories of Heat and the Rise of Physics. *History of Science* 15: 138-145. Essay review of *Joseph Fourier* by J. Herivel.

___ 1978. Introduction to Symposium 6 [on "Aspects of the History of Thermodynamics: Theory and Practice"]. In Forbes (1978), pp. 289-292.

___ 1989. *James Joule: A Biography.* Manchester: Manchester University Press.

Reviewed by W. H. Brock, *Times Literary Supplement* (29 September 1989): 1070; W. E. Burcham, *Annals of Science* 49 (1992): 280-282; P. M. Harman, *BJHS* 23 (1990): 230; J. R. Hoffmann, *Isis* 82: (1991): 149-150; D. W. Hughes, *Nature* 340 (1989): 686; B. J. Hunt, *Physis* 28 (1991): 619-620; K. Hutchison, *Victorian Studies* 34 (1991): 408-409; A. Kleinert, *AIHS* 42 (1992): 408-409; J. Morrell, *English Hist. Review* 108 (1993): 237; I. R. Morus, *Stud. HPS* 21 (1990): 519-525; C. A. Russell, *NRRSL* 44 (1990): 311-313; C. Smith, *Annals of Science* 47 (1990): 196-197 [*Isis CB*]

Carnot, S. 1966. Recherche d'une Formule propre à représenter la Puissance motrice de la Vapeur d'Eau. Published by W. A. Gabbey and J. W. Herivel, Un manuscrit inédit de Sadi Carnot. *Revue d'Histoire des Sciences* 19: 151-166.

[Carnot, S.] 1976. *Sadi Carnot et l'Essor de la Thermodynamique. Table ronde du Centre National de la Recherche Scientifique, Paris: École Polytechnique, 11-13 juin 1974.* Paris: Centre National de la Recherche Scientifique. [*Isis CB*]

Reviewed by E. N. Hiebert, *Historia Mathematica* 8: 375-377; K. Hutchison, *BJHS* 13 (1980): 168; P. Redondi, *Revue d'Histoire des Sciences* 31 (1978): 90-94 [*Isis CB*]

Carnot, S. 1978. *Réflexions sur la Puissance motrice du Feu.* Edition critique par Robert Fox. Paris: Vrin.

Reviewed by E. Hiebert, *Isis* 71 (1980): 667; W. J. Hornix, *Annals of Science* 39 (1982): 399-406; P. Lervig, *Centaurus* 25 (1981): 141-143; E. Mendoza, *BJHS* 14 (1981): 75-78; P. Brouzeng, *Revue d'Histoire des Sciences* 33 (1980): 181-183 [*Isis CB*]

___ 1986. *Reflexions on the motive Power of Fire. A critical Edition with the surviving scientific Manuscripts.* Translated and edited by Robert Fox. New York: Lilian Barber Press. [*Isis CB*]

Reviewed by E. Hiebert, *Metascience* 7 (1989): 41; K. Hutchison, *Annals of Science* 45 (1987): 108-109; P. Kidwell, *Technology & Culture* 29 (1988): 144-145; F. A. J. L. James, *Ambix* 34 (1987): 49; E. Mendoza, *Isis* 78 (1987): 492-493; N. Wise, *Nature* 326 (1987): 453 [*Isis CB*]

Carnot, N. L. Sadi: Bailyn (1989), Baracca (1987), Bellone (1989), Birembaut (1974), Cimbleris (1967, 1991b), Cropper (1987a), Della Salvo & Drago (1991), Dias (1990, 1994d, 2001), Dias et al. (1995), Escudio et al. (1996), Fox (1970, 1971a, 1988), Frank (1966), Gvozdetskii (1997), Hoyer (1974a,b, 1975, 1976b,c, 1978a), Hutchison (1973, 1976a,b, 1981b) Kestin (1976), Kikuchi et al. (1996), Lervig

(1972, 1978, 1985), Locqueneux (1988-89), Ordóñez (1988), Redondi (1975, 1980), Schöpf (1984a), Sebastiani (1980, 1981b, 1990), Tansjo (1988), Truesdell (1973, 1979), Truesdell & Bharatha (1977), Velarde & Ibanez Medrano (1976), S. S. Wilson (1981), Wolff (1997c)

Carpenter, William Benjamin: Hall (1979)

Carson, C. 2000. The Origins of the Quantum Theory. *Beam Line (Stanford Linear Accelerator Center)* 30, no. 2: 6-19; also available at http://www.slac.stanford.edu/pubs/beamline/30/2/30-2-carson.pdf [*Author*]

Casanova: Sebastiani (1990a)

Casaubon, D. de. 1985. Le Rôle heuristique des Mathématiques dans la Physique: Le Cas de Planck (1894-1900). *Fundamenta Scientiae* 6: 281-297.

"The quantum hypothesis is not suggested by experiments or observations but generated by the mathematical formalism..."

Cauchy, Augustin Luis: Wolff (1995b)

Causality: Byrne (1980), Hiebert (1968b), Hörz (1975)

Causey, R. L. 1971. Avogadro's Hypothesis and the Duhemian Pitfall. *Journal of Chemical Education* 48: 365-367.

Kinetic theorists accepted the Hypothesis because they avoided the pitfall that had misled chemists.

Cercignani, C. 1992. Une Histoire de la Théorie Cinétique des Gaz. *La Vie des Sciences: Comptes Rendus de l'Academie des Sciences* 9: 275-285.

_____ 1997. *Ludwig Boltzmann e la Meccanica Statistica*. Pavia: Goliardica Pavese.

Reviewed by A. Baracca, *Isis* 90 (1999): 373-74

_____ 1998. *Ludwig Boltzmann: The Man Who Trusted Atoms*. New York: Oxford University Press.

Reviewed by G. F. Bignami, *Nature* 399 (1999): 32-33; J. Blackmore, *Physics in Perspective* 2 (2000): 108-111; by S. G. Brush, *Journal of Statistical Physics* 98 (2000): 1429-1432; by J. R. Dorfman, *Isis* 91 (2000): 595-596 [*Isis CB*]

_____ See also Battimelli (1993)

Cerreta, P. 1992. Kuhn's Analysis of History of Black Body: From 'Paradigm' to 'Group of Concepts.'" In *Italian Physical Society, Conference Proceedings* Vol. 42, *History of Physics in Europe in the 19th and 20th Centuries*, edited by F. Bevilacqua, 307-311. Bolgona: Editrice Compositori. [*Author*]

_____ 1995. Kuhn's Interpretation of Boltzmann's Statistical Heredity in Planck. In *The Foundations of Quantum Mechanics*, edited by C. Garola & A. Rossi, 139-146. Dordrecht: Kluwer.

_____ 2002. "The Birth of Quanta: A Historiographic Confrontation." In *Proceedings of the XXth International Congress of History of Science (Liège, 1997)*, Vol. XIV, *History of Modern Physics*, edited by H. Kragh, G. Vanpaernel & P. Marage, 249-259. Turnhout: Brepols.

Chalmers, A. F. 1973. On learning from our mistakes. *BJHS* 24: 164-173.

On Maxwell and the specific heats problem.

_____ 1998. Retracing the ancient steps to Atomic Theory. *Science & Education* 7: 69-84.

Atomic theory before Dalton was not scientific; critique of theories of Boyle and Newton.

Chandler, D. *See* Battimelli (1993)
Chang, H. 2001. Spirit, Air and Quicksilver: The Search for the "Real" Scale of Temperature. *HSPS* 31: 249-284. [*Author*]
_____ 2002. Rumford and the Reflection of Radiant Cold: Historical Reflections and Metaphysical Reflexes. *Physics in Perspective* 4: 127-169.
> "For Rumford heat and cold radiation consisted in sound-like undulatons in the ether ... [he pushed] the caloric theory to develop in a direction that eventually led to its downfall." Also discusses M. A. Pictet and P. Prevost.

_____ *See also* Locqueneux (1996), Smith (1998)
Channell, D. F. *See* Truesdell (1980)
Chaos, Molecular: Giorello (1972a)
Chapman, S. 1967. The Kinetic Theory of Gases Fifty Years Ago. In *Lectures in Theoretical Physics*, vol. IX-C, edited by W. E. Brittin, 1-13. New York: Gordon & Breach. Reprinted in Brush (1972a), 260-271.
Chapman, Sydney: Brush (1972a)
Charbonneau, L. 1990. Fourier et la Mécanique: Une Histoire méconnue: De la Mécanique à la Théorie de la Chaleur. In *Scientifiques et Sociétés pendant la Révolution et l'Empire*, 97-116. Paris: CTHS. [*Isis CB*]
Charles, Jacques Alexandre César: Redondi (1980)
Charles's Law: Spurgin (1987, 1989)
Chemical Constant: DeSalvo (1992)
Chemistry: Bent (1980), Berger (1999), Forrester (1975), Fullmer (1964); *See also* **Thermodynamics, Chemical**
Cheng, K. C. *See* Layton & Lienhard (1988)
Chillington, J. H. *See* Elkana (1974b)
Cimbleris, B. 1967. Reflections on the Motive Power of a Mind. *Physis* 9: 393-420.
_____ 1991a. The late Emergence of Thermodynamics. In Martinas et al. (1991), pp. 29-35.
_____ 1991b. Carnot e a Evolução das Máquinas Térmicas. *Rivista da Sociedade Brasileira de História di Ciência* 6: 39-44.
Clapeyron, B.-P.-E.: Kestin (1976), Hornix (2001), Locqueneux (1988-89), Redondi (1980)
Clark, P. 1976a. Atomism versus Thermodynamics. In *Method and Appraisal in the Physical Sciences*, edited by C. Howson, 41-105. New York: Cambridge University Press.
> Kinetic theory was a "degenerating research programme" (as defined by I. Lakatos) in the late 19th century. See Nyhof (1981).

_____ 1976b. Elkana on Helmholtz and the Conservation of Energy. *British Journal for the Philosophy of Science* 27: 167-176.
> Essay review of Elkana (1974b).

_____ 1977. *Thermodynamics and the Kinetic Theory in the 19th Century: A Case Study in the Methodology of Scientific Research Programmes*. Ph. D. Dissertation, London School of Economics. [*Isis CB*]
_____ 1982. Matter, Motion and Irreversibility. *British Journal for the Philosophy of Science* 33: 165-186.
> Essay review of Brush (1976b).

_____ 1989. Determinism, Probability and Randomness in classical Statistical Physics. In *Imre Lakatos and Theories of Scientific Change*, edited by J. Gavroglu *et al.*, 95-110. (Boston Studies in Philosophy of Science, vol. 111) Boston: Kluwer.
Clark, Peter: Maiocchi (1985), Nyhof (1981)
Clarke, Bruce. 1996. Allegories of Victorian Thermodynamics. *Configurations* 4: 67-90. On T. Pynchon's *Crying of Lot 49*, the Maxwell Demon, etc.
Clausius, Rudolf: Abiko (1997), Bevilacqua (1994), Brush (1976b, 1983), Daub (1970b, 1971a), Dias (1994a,c), Garber (1966), Hutchison (1973, 1976a,b, 1981b), Kestin (1976), Kubbinga (2002), Laidler (1993), Osietzki (1991), Schneider (1975a,b, 1988, 1995), Segers (1997), Sheynin (1985), Steffens (1979), D. B. Wilson (1981), Wolff (1993, 1995, 1997c)
Clausius on Thermodynamics: Bierhalter (1981b, 1985, 1990), Cardwell (1971), Cropper (1986), Daub (1969, 1978), Dias (2001), Dias et al. (1995), El'iashevich & Prot'ko (1984), Garber (1966, 1970, 1978), Hiebert (1968b), Hoffmann & Ebelung (1991), Hornix (2001), Kangro (1976), Kikuchi et al. (1996), Kim (1983), Locqueneux (1988-1989), Schöpf (1984b), Tansjo (1988), Yagi (1981, 1984, 2002a,b), Yagi & Tadokoro (2002a), Yagi & Hayashi (2002), Yagi, Tadokoro & Hayashi (2002), Yung (1983)
Clément, Nicolas: Costabel (1968), Fox (1970), Lervig (1978, 1985), Payen (1971)
Clericuzio, A. 1998. The Mechanical Philosophy and the Spring of the Air: New Light on Robert Boyle and Robert Hooke. *Nuncius* 13: 69-75.
_____ *See also* Klever (1997)
Cohen, E. G. D. 1997. Boltzmann and Statistical Mechanics. In *Boltzmann's Legacy 250 Years after his Birth*, 9-23. Rome: Accademia Nazionale dei Lincei. [*Isis CB*]
Cohen, I. B. 1964. Newton, Hooke, and 'Boyle's Law' (Discovered by Power and Towneley). *Nature* 204: 618.
_____ *See also* Shapin & Schaffer (1985)
Colardeau, E.: Levelt Sengers (1979)
Cold Radiation: Chang 2002
Colding, Ludvig. 1972. *Ludvig Colding and the Conservation of Energy Principle: Experimental and Philosophical Contributions.* With an Introduction, Translation, and Commentary by Per F. Dahl. New York: Johnson Reprint.
> Includes Colding's "On the Universal Forces of Nature and their mutual Dependence," translated from *Det Kgl. Danske Videnskabernes Selskabs Skrifter* (1850) and *Unified Presentation of the Mutual Dependence of the Forces of Nature with Application to the Mechanical Theory of Heat*, translated from an unpublished manuscript submitted in 1864 for the Prix Bordin of the French Academy.
> Reviewed by H. J. Steffens, *Isis* 65 (1974): 122-124.

Colding, Ludvig: Caneva (1997), Ellen (1987)
Collisions of Atoms & Molecules: Dias (1994b)
Combes, Charles: Locqueneux (1990)
Comte, August: Brush (1978), Herivel (1966)
Continuity of States: Andrews (1973)
Convection: Middleton (1968)
Cornell, J. F. *See* Elkana (1974b)

Corresponding States, Law of: Gavroglu (1990b), Hiebert (1978a)
Cortini, G. *See* Tarsitani & Vicentini (1997)
Costabel, P. 1968. Le Calorique du Vide de Clément et Desormes (1812-1819). *AIHS* 21: 3-14.
Cotignola, M. I., C. Bordogna, G. Punte and O. M. Cappannini. 2002. Difficulties in learning Thermodynamic Concepts: Are they Linked to the Historical Development of this Field? *Science & Education* 11: 279-291.
> Confusion between internal energy and heat.

Cotte, M. *See* Locqueneux (1996)
Critical Point: Andrews (1973), Domb (1996), Gel'fer (1981), Gopal & Viswanathan (1969), Goudaroulis (1994), Levelt Sengers (1976), Pineta & Velarde (1977), Rowlinson (1969)
Critical Point, Opalescence: Einstein (1993), Levelt Sengers (1983)
Cronin, J. W. *See* Mendelssohn (1973)
Crookes, William: Brock (1972), Draper (1976), D. B. Wilson (1987), Woodruff (1966)
Cropper, W. H. 1986. Rudolf Clausius and the Road to Entropy. *American Journal of Physics* 54: 1068-1074.
_____ 1987a. Carnot's function: Origins of the Thermodynamic Concept of Temperature. *American Journal of Physics* 55: 120-129.
_____ 1987b. Walther Nernst and the last Law. *Journal of Chemical Education* 64: 1-8.
Crosland, M. 1964. The Development of the Concept of the Gaseous State as a third State of Matter. In *Proceedings of the 10th International Congress on History of Science* (Ithaca, NY, 1962), 851-854. Paris: Herman.
_____ 1971. Dulong, Pierre Louis. *DSB* 4: 238-242.
_____ 1978. *Gay-Lussac: Scientist and Bourgeois*. New York: Cambridge University Press.
_____ *See also* Fox (1971b)
Crow, C. M. 1972. Paul Valéry and Maxwell's Demon: Natural Order and Human Possibility. Hull: University of Hull.
> The Demon is "an analogue for the relational and evaluative capacities of the human mind."

Cultural & Social Connections of Thermodynamics and Kinetic Theory: Brody (1987), Brush (1978), Burich (1987), Clarke (1996), Crow (1972), Gearhart (1983), Greenberg (1990), Lindquist (1998), Myers (1985), Peterfreund (1987), Porter (1981b, 1986, 1994), Potter (2001), Wise (1986)
Curd, M. V. 1978. *Ludwig Boltzmann's Philosophy of Science: Theories, Pictures and Analogies*. Ph. D. Dissertation, University of Pittsburgh. *Dissertation Abstracts International* 39 (1978): 1793-A.
> Includes Boltzmann's relation to E. Mach.

_____ 1982. Popper on Boltzmann's Theory of the Direction of Time. In Sexl & Blackmore (1982), 263-303.
_____ *See also* Boltzmann (1974)
Cuvier, G.: Kremer (1980)

D'Agostino, S. 2000. *A History of the Ideas of Theoretical Physics: Essays on the Nineteenth and Twentieth Century Physics.* (Boston Studies in Philosophy of

Science, vol. 213) Dordrecht: Kluwer.
> Includes "On Boltzmann's Mechanics and his *Bild*-Conception of Physical Theories," pp. 201-216.

_____ *See also* Battimelli (1993), Maiochhi (1988)

Dahl, P. F. 1992. *Superconductivity: Its Historical Roots and Development from Mercury to the Ceramic Oxides.* New York: American Institute of Physics.
> Includes early electron theories, kinetic theory, quantum theory of radiation. [*Author*]

Dahl, P. *See* Colding (1972)

Dalton, John: Fox (1968b), Kubbinga (2002), Mendoza (1975, 1990b), Schöpf (1984c)

Dandolo, V.: Sebastiani (1990a)

Darrigol, O. 1988. Statistics and Combinatorics in early Quantum Theory. *HSPS* 19: 17-80.

_____ 1991. Statistics and Combinatorics in early Quantum Theory. Part II. Early Symptoms of Indistinguishability and Holism. *HSPS* 21: 237-298.

_____ 1992. *From c-Numbers to q-Numbers: The Classical Analogy in the History of Quantum Theory.* Berkeley: University of California Press.
> In the context of his 1900 problem, "Planck could not possibly have understood the introduction of energy elements as a discrete selection of the admissible energy values of a resonator. Such a discontinuity would have contradicted the rest of his theory."

_____ 2000. Continuities and Discontinuities in Planck's *Akt der Verzweiflung*. *Annalen der Physik* (Leipzig) 9: 951-60.

_____ 2001a. God, Waterwheels, and Molecules: Saint-Venant's Anticipation of Energy Conservation. *HSPS* 31: 285-353.

_____ 2001b. The Historians' Disagreements over the Meaning of Planck's Quantum. *Centaurus* 43: 219-239.

Darwin, Charles: Broda (1982), Schweber (1982), Snelders (1977)

Daub, E. E. 1967. Atomism and Thermodynamics. *Isis* 58: 293-303.
> Maxwell on Rankine.

_____ 1969. Probability and Thermodynamics: The Reduction of the Second Law. *Isis* 60: 318-330.
> On Clausius and Boltzmann.

_____ 1970a. Maxwell's Demon. *Stud. HPS* 1: 213-227.

_____ 1970b. Waterston, Rankine, and Clausius on the Kinetic Theory of Gases. *Isis* 61: 105-106.

_____ 1970c. Entropy and Dissipation. *HSPS* 2: 321-354.

_____ 1971a. Clausius, Rudolf. *DSB* 3: 303-311.

_____ 1971b. Waterston's Influence on Krönig's Kinetic Theory of Gases. *Isis* 62: 512-515.

_____ 1976. Gibbs Phase Rule: A Centenary Retrospect. *Journal of Chemical Education* 53: 747-751.

_____ 1978. Sources for Clausius' Entropy Concept: Reech and Rankine. In Forbes (1978), pp. 342-358.

_____ *See also* Cardwell (1971), Elkana (1974), Lindsay (1973, 1975, 1976), Seeger (1974), Truesdell (1980), Truesdell & Bharatha (1977)

Davis, E. B. *See* Boyle (1999)
Davy, Humphry: Forgan (1980), Fullmer (1964, 1969, 2000), Vernon (1966)
Deakin, M. A. B. 1988. Nineteenth Century Anticipations of modern Theory of dynamical Systems. *AHES* 39: 183-194.
> Maxwell and others showed that Newtonian physics need not lead to a fully determined universe.

De Berg, K. C. 1995. Revisiting the Pressure-Volume Law in History: What can it teach us about the Emergence of Mathematical Relationships in Science? *Science and Education* 4: 47-64.
De Boer, J. 1974. Van der Waals in his Time and the present Revival. *Physica* 73: 1-27.
De Courtenay, N. *See* Blackmore (1995)
De Fillipo, G. *See* Tarsitani & Vicentini (1991)
Degen, H. *See* Mendelssohn (1973)
De Heen, P: Levelt Sengers (1979)
Delaroche, François: Mendoa (1990a)
del Buono, Paolo: Middleton (1969)
Della Salvo, A., and A. Drago. 1991. The Introduction of the Cycle Method in Thermodynamics. In Martinás et al. (1991), 36-41.
> On a recently-discovered manuscript of S. Carnot.

Deltete, R. J. 1983. *The Energetics Controversy in late 19th-Century Germany: Helm, Ostwald and their Critics.* Ph. D. Dissertation, Yale University. *Dissertation Abstracts International* 45 (1984): 1421-A.
_____ 1995. Gibbs and the Energeticists. In *No Truth except in the Details*, edited by A. J. Kox & D. M. Siegel, 135-169. Dordrecht: Kluwer.
_____ *See also* Blackmore (1999), Helm (2000)
Demarzo, Carlo. 1978. *Maxwell e la Fisica Classica*. Roma Bari: Laterza.
> Chap. II, Kinetic Theory of Gases.

De Regt, H. W. 1996. Philosophy and the Kinetic Theory of Gases. *British Journal for the Philosophy of Science* 47: 31-62.
> On Boltzmann and Maxwell.

_____ 2001. Erwin Schrodinger. In *Ernst Mach's Vienna 1895-1930*, edited by J. T. Blackmore et al. , 85-104. Boston: Kluwer.
> Interactions of Boltzmann, Mach, and Schrodinger.

_____ *See also* Blackmore (1995, 1999)
Desalvo, A. 1992. From the Chemical Constant to Quantum Statistics: A Thermodynamic Route to Quantum Mechanics. *Physis* 29: 465-537.
> Quantum theory of ideal monatomic gas from Sackur-Tetrode equation (1911-12) to work of Einstein and Fermi in 1920s.

Desormes, C. B.: Costabel (1968)
Determinism *see* **Indeterminism**
Deutsch, D. H. 1991. Did Robert Brown observe Brownian Motion: Probably not. *Bulletin of the American Physical Society* 36: 1374.
> News report by J. Rennie, "A small Disturbance," *Scientific American* 265, no. 2 (Aug. 1991): 20.

Dias, P. M. C. 1990. Pré-História e Histórias. *Revista da USP* [USP = Universidade de São Paulo] 7: 61-78.

Watt, 2nd Law of Thermodynamics. [*Author*]

———— 1994a. Clausius and Maxwell: The Statistics of Molecular Collisions. *Annals of Science* 51: 249-261.
    Probability, Maxwell's distribution.

———— 1994b. "Will someone say exactly what the H-Theorem proves?" A Study of Burbury's Condition A and Maxwell's Proposition II. *AHES* 46: 341-366.
    Also on Boltzmann, statistical hypotheses, number of collisions.

———— 1994c. A Hipótese Estatística do Teorema-H. *Química Nova* 17: 472-479.
    Clausius, Maxwell, Boltzmann, mean free path, statistical hypotheses. [*Author*]

———— 1994d. A Path from Watt's Engine to the Principle of Heat Transfer. In *Logic and Philosophy of Science in Uppsala*, edited by D. Prawitz & D. Westerstahl, 425-438. Dordrecht: Kluwer.
    Carnot, 2nd Law of Thermodynamics, Entropy. [*Author*]

———— 1996. William Thomson and the Heritage of Caloric. *Annals of Science* 53: 511-523.
    Kelvin, dissipation. [*Author*]

———— *See also* Locqueneux (1996)

———— 2001. A (Im)Pertinência da História ao Aprendizado da Física (Um Estudo de Caso). *Revista Brasileira de Ensiono de Física* 23: 266-236.
    "The (In)convenience of History in Learning Physics (a case study)."
    Carnot, Watt, Clausius, Kelvin, thermodynamics, heat engines. [*Author*]

————, S. P. Pinto, and D. H. Cassiano. 1995. The conceptual Import of Carnot's Theorem to the Discovery of the Entropy. *AHES* 49: 135-161.
    Clausius, 2nd law of Thermodynamics, entropy. [*Author*]

DiCarlo, F. *See* Benedetti (1984)

Dick, A. *See* Battimelli (1993)

Diffusion: Kirsch (1981), Mason (1967, 1970), Mason & Kronstadt (1967), Mason & Wright (1971), Pourprix & Locqueneux (1985)

Di Meo, A. 1997. Le Cycle du Temps, la Flèche du Temps: La Découverte d l'Irréversibilité. *Bulletin de l'Academie Royale de Belgique, Classe des Sciences* 8: 97-114.

Disgregation: Bierhalter (1983b, 1985)

Disorder: Giorello (1972a), Haitun (1991)

Dissipation: see **Irreversibility**

Domb, C. 1980. James Clerk Maxwell in London, 1860-1865. *NRRSL* 35: 67-103.
    Photo of house where viscosity experiments were done; refutes statement that Maxwell was asked to resign because he couldn't keep order in class.

———— 1995. Thermodynamics and Statistical Mechanics (in Equilibrium). In *Twentieth Century Physics*, edited by L. M. Brown, A. Pais & B. Pippard, 521-584. Bristol & Philadelphia: Institute of Physics/New York: American Institute of Physics.

———— 1996. *The Critical Point: A historical Introduction to the modern theory of critical Phenomena.* Bristol & London: Taylor & Francis.
    Chapter 2: Classical (van der Waals) theory of fluids.
    Reviewed by S. G. Brush. *NRRSL* 52 (1998): 198-200; M. Wortis, *Physics Today* 50, no. 12 (1997): 68

———— *See also* Brush (1983), Maxwell (1986)

Donnan, F. G., and A. Haas, editors. 1980. *A Commentary on the Scientific Writings of J.*

Willard Gibbs. Vol. I, *Thermodynamics*. New York: Arno Press. Reprint of the 1936 edition. For Vol. II see Haas (1980).
    Includes articles by D. H. Andrews, J. A. V. Butler, E. A. Guggenheim, H. S. Harned, F. G. Keyes, E. A. Milne, G. W. Morey, J. Rice, F. A. H. Schreinemakers, and E. B. Wilson, on thermodynamics.

Donini, E. *See* Kuhn (1978)

Dorfman, J. R.: *See* Cercignani (1998)

Dorling, J. 1970. Maxwell's Attempts to arrive at non-speculative Foundations for the Kinetic Theory. *Stud. HPS* 1: 229-248.

Dougal, R. C. 1976. The Presentation of the Planck Radiation Formula. *Physics Education* 11: 438-443.
    Includes historical outline and partial translation of article by G. Hettner (1922) on work of H. Rubens and his 1900 discussion with Planck.

———— 1979. The Centenary of the Fourth-Power Law. *Physics Education* 14: 234-238.
    Laws of Dulong-Petit and Stefan.

Drago, A. 1991. The alternative Content of Thermodynamics: The constructive Mathematics and the problematic Organization of the Theory. In Martinás et al. (1991), 329-345.

———— 1992. La Riqualificazione della Termodinamica a partire dalla Teoria chimica. *Rend. Accad. Naz. Sci. dei XL. Mem. Sci. Fis. Nat.* [ser. 5] 16, pt. 2: 279-290. [*BISS*]

———— 1995. La sfortunata Storia dei Libri di Testo di Termodinamica: L'Implicita indicita Indicazione per un Pluralismo didattico. *Atti del XIII Congresso Nazionale di Storia della Fisica, Pavia, 1992*, edited by A. Rossi, 121-136. Lecce: Conte. [*BISS*]

———— *See also* Cardone & Drago (1995), Della Salvo & Drago (1991)

———— & M. Grilli. 1991. Analisi storico critica della Formulazione della Termodinamica di Carnot-Clausius-Kelvin. *Atti del X Congresso Nazionale di Storia della Fisica, 1989*, edited by F. Bevilacqua, 139-152. Milano: Gruppo Nazionale di Coordinamento per la Storia della Fisica del CNR. [*BISS*]

———— & C. Mordillo. 1994. Perché l'Ordine logico dei Principi della Termodinamica inverte l'Ordine storico della scoperte. *Atti del IV Congresso Nazionale di Storia della Fisica*, edited by P. Tucci, 113-121. Rozzano: Centro Stampo Rozzano.

———— & P. Saiello. 1995. La Teoria Cinetica dei Gas: Una Sconfitta del Meccanicismo Newtoniano. *Atti del XIII Congresso Nazionale di Storia della Fisica, Pavia, 1992*, edited by A. Rossi. Lecce: Conte. [*BISS*]

Draper, C. W. 1976. The Crookes Radiometer Revisited. *Journal of Chemical Education* 53: 356-357.

Draper, J. W. 1973. *Scientific Memoirs, being experimental Contributions to a Knowledge of radiant Energy*. New York: Arno. Reprint of the 1878 edition [*Isis CB*]

Dresden, M. 1995. Non-Equilibrium Statistical Mechanics or *The Vagaries of Time Evolution*. In *Twentieth Century Physics*, edited by L. M. Brown, A. Pais and B. Pippard, 585-633. Bristol and Philadelphia: Institute of Physics/New York: American Institute of Physics.
    Boltzmann Equation, transport processes, Brownian motion.

Drude, Paul K. L.: Kaiser (1987), Renn (1997)

Duffy, M. C. 1983. Mechanics, Thermodynamics and Locomotive Design: The Machine-Ensemble and the Development of Industrial Thermodynamics. *History and Technology* 1: 45-78.

Dugdale, J. S. 1968. *Thermodynamics: The Men behind the Laws.* Leeds: Leeds University Press. [*Isis CB*]

Duhem, P. 1980. *Josiah-Willard Gibbs: A propos de la Publication de ses Memoires Scientifiques.* New York: Arno Press. Reprinted from *Bulletin des Sciences Mathematiques* (1907), together with commentary by Donnan and Haas on the writings of Gibbs.

Duhem, Pierre: Brouzeng (1982, 1991, 1995), Causey (1971), Klein (1990a), Maiocchi (1985a, 1988, 1990b, 1992), Miller (1966), Nye (1984)

Dulong, Pierre Louis: Crosland (1971)

Dulong-Petit Law: Dougal (1979), Fox (1968a)

Duncan, A. M. 1971. Was Heat a Chemical in the Eighteenth Century? *Actes, 13th International Congress of the History of Science, 21 August 1971,* 13: 376-398. [*Isis CB*]

    No, it was an "agent."

Dupre, F. *See* Tarsitani & Vicentini (1991)

Dutch scientists: Levelt Sengers (2002)

Dutta, M. 1968. A Hundred Years of Entropy. *Physics Today* 21, no. 1: 75-79.

Earman, J. and J. D. Norton. 1998. Exorcist XIV: The Wrath of Maxwell's Demon. Part I. From Maxwell to Szilard. *Stud. HPS* 29: 435-471.

Ebelung, W. *See* Hoffmann & Ebelung (1991)

Economic Model for Thermodynamics: Schöpf (1984b), Wise & Smith (1989-90)

Edinburgh Philosophy: Harman (1985)

Eddy, Henry Turner

Effusion: Mason & Kronstadt (1967)

Ehrenfest, Paul & Tatyana: Frenkel (1971), Klein (1970a), Nickles (1976, 1980), Norton (1993), Sklar (1993)

Einstein, A. 1989. *The Collected Papers of Albert Einstein.* Volume 2: *The Swiss Years: Writings, 1900-1909.* Edited by J. Stachel et al. Princeton, NJ: Princeton University Press.

    Includes: "Editorial Note: Einstein on the Foundations of Statistical Physics" and reprints of 3 papers on kinetic theory and thermodynamics, from *Annalen der Physik* (1902-4); Einstein's reviews for *Beiblätter zu den Annalen der Physik* of papers on thermodynamics by G. Belluzzo, G. H. Bryan, A. Fliegner, A. Giammario, W. M. Orr, N. N. Schiller, J. H. Van't Hoff, and J. J. Weyrauch; Einstein's papers on the light quantum hypothesis and on Brownian motion (*Annalen der Physik*, 1905), his dissertation on the determination of molecular sizes (University of Zurich), and Editorial Notes on these subjects; more review for the *Beiblätter* of papers by H. Birven, K. Bohlin, E. Buckingham, P. Langevin, E. Mathias, G. Meslin, M. Planck, A. Ponsot, K. F. Slotte; Einstein's papers on Brownian motion and determination of molecular sizes (*Annalen der Physik*, 1906), review of Planck on radiation (*Beiblätter*, 1906, and

Annalen der Physik, 1907); papers on thermodynamics, Brownian motion, and quantum theory of radiation (*Annalen der Physik*, 1907; *Zeitschrift für Elektrochemie*, 1907 & 1908, *Physikalische Zeitschrift*, 1909, and *Verhandlungen, Deutsche Phsyikalische Gesellschaft*, 1909). A separate volume, with a similar title, contains English translations of these papers by Anna Beck.

_____ 1993. *Collected Papers*, Volume 3: *The Swiss Years: Writings, 1909-1911*, edited by M. J. Klein et al. Princeton, NJ: Princeton University Press.

> Includes: "Editorial Note: Einstein's Lecture Notes"; lecture notes for a course on the kinetic theory of heat, 1910; papers on quantum theory (*Archives des Sciences Physiques et Naturelles*, 1910) and, with L. Hopf, on statistics of radiation (*Annalen der Physik*, 1910); "Editorial Note: Einstein on Critical Opalescence"; reprint of his paper on that subject (*Annalen der Physik*, 1910); his comments on P. Hertz's papers on thermodynamics (*Annalen der Physik*, 1911); correction on the determination of molecular sizes (*Annalen der Physik*, 1911); statement on light quantum hypothesis (*Sitzungsberichte, Naturforschende Gesellschaft in Zürich*, 1911). A separate volume, with a similar title, contains English translations of these papers by Anna Beck.

[Einstein, A.] Nelkowski, H. et al., editors. 1979. *Einstein Symposion Berlin aus Anlass der 100. Wiederkehr seines Geburtstages, 25 bis 30 März 1979*. Berlin: Springer-Verlag.

> Includes: R. Jost on Boltzmann and Planck; M. Jammer on Einstein and the quantum problem; E. Nelson on Brownian movement and quantum mechanics. [*Isis CB*]

Einstein, Albert: Bergia et al. (1980), Bergia & Navarro (1988), Broda (1980), Byrne (1980, 1981), Jech (1988), Klein (1967, 1975), Kox (1995), Kubbinga (2002), Kuhn (1980), Laidler (1993), Lewis (1973), Mayo (1996), Mehra (1999), Nickles (1976, 1980), Nugayev (2000), Nye (1984), Scheibe (1995), von Plato (1987, 1994)

Einstein and Chemistry: Bent (1980)

Einstein's Quantum Theory: Bent (1980), Bergia & Navarro (1997), Haar (1967), Mehra (2001), Mehra & Rechenberg (1982), Miller (1976), Stachel (2000)

Einstein's Statistical Mechanics: Baraaca & Rechtman S. (1985), Bergia & Navarro (1997), Ezawa (1979a,b), Gearhart (1990), Klein (1982), Landsberg (1981), Mehra (1975a, 2001), Navarro Veguillas (1988, 1991, 1994, 1998), Renn (1997)

Einstein's Theory of Brownian Motion: Bent (1980), Brush (1968b, 1976b), Sarkar (2000), Teske (1969, 1972)

Electron Gas: Dahl (1992), Kaiser (1987), Knudsen (2001), Renn (1997)

El'iashevich, M. A. and T. S. Prot'ko. 1981. Maxwell's Contribution to the Development of Molecular Physics and Statistical Methods. *Soviet Physics Uspekhi* 24: 876-903. Translated from *Uspekhi Fizicheskikh Nauk* (1981).

_____ 1984. Programma Klauziusa I Programma Maksvella v Oblasti Kinetischeskoe Teorii Gazov. *Voprosy Istorii Estestvoznaniia I Tekhniki*, no. 4: 79-88.

> "The Program of Clausius and the Program of Maxwell in the Field of the Kinetic Theory of Gases." [*Isis CB*]

Eliot, George: Brody (1987)

Elkana, Y. 1970a. The Conservation of Energy: A Case of simultaneous Discovery? *AIHS* 23: 31-60.
    Reprinted as an Appendix in Elkana (1974b).

———— 1970b. Helmholtz' 'Kraft': An Illustration of Concepts in Flux. *HSPS* 2: 263-298.

———— 1974a. Boltzmann's Scientific Research Programme and its Alternatives. In *The Interaction between Science and Philosophy*, edited by Y. Elkana, 245-279. Atlantic Highlands, NJ: Humanities.

———— 1974b. *The Discovery of the Conservation of Energy*. Cambridge, MA: Harvard University Press.
    Reviewed by S. Brown, *Physics Today* 28, no. 8 (1975):; G. N. Cantor, *BJHS* 8 (1975): 87-88; J. H. Chillington, *Synthesis: The University Journal in the History and Philosophy of Science* 2, no. 3 (1974): 56-57; P. Clark in *British Journal for the Philosophy of Science* 27: 165-176; J. F. Cornell, *Ann. Inst. Mus. Stor. Sci. Firenze* 4, no. 1 (1979): 85-87; E. E. Daub, *American Journal of Physics* 43: 468; T. Hankins in *Science* 185 (1974): 937; L. Ohlon, *Lychnos* (1975-76): 454-455; H. I. Sharlin, *American Historical Review* 81 (1976): 102; C. Smith, *Journal of Modern History* 48 (1976): 323-324; H. A. M. Snelders, *Janus* 61 (1974): 316-317; H. J. Steffens, *Isis* 67 (1976): 137-139; J. Ziman, *Minerva* 13 (1975): 124-127 [*Isis CB*]

Ellen, D. 1987. Ludvig Colding: Misrepresentation, Misunderstanding, and the Conservation of Energy. *Synthesis, The Student Journal of the History and Philosophy of Science* 6, no. 3: 3-19.
    "Colding did not understand the concept of energy in 1843. ... The minimal attention that Colding has been accorded is thus well justified ... on scientific ... grounds."

Emmerik, E. Van *See* Hornix & Mannaerts (2001)

Energetics, Energetism: Brouzeng (1993, 1995), Deltete (1983, 1995), Hakfoort (1992), Helm (2000), Hiebert (1968b, 1971, 1978c), Holt (1970), Jungnickel & McCormmach (1986), Niederson (1983), Raman (1973b), Ramunni (1986)

Energy: Abiko (1997), Breger (1982), Cotignola et al. (2002), Giorello (1972b), Greenberg (1990), Harman (1982), Hutchison (1976b, 1981a,b), Jackson (1967), Cotignola et al. (2002), Lindsay (1971, 1973, 1975, 1976, 1979), Moyer (1977), Nye (1996), Peterfreund (1987), Smith (1978, 1990, 1998), Steffens (1979), Tarsitani & Vicentini (1991), Ziggelar (1971)

Energy Conservation Law: Bevilacqua (1994), Breger (1982), Caneva (1993), Cantor (1975), Clark (1976b), Colding (1972), Darrigol (2001a), Elkana (1970a, 1974b), Ellen (1987), Forrester (1975), Hall (1978, 1979), Harman (1982, 1974a), Helmholtz (1983, 1995), Hermann (1978), Hörz (1975), James (1983, 1985), Knudsen (1995), Merleau-Ponty (1979), Monleón Pradas (1991a), Ober (1968), Ordóñez (1996), Raman (1975b), Seeger (1971), Steffens (1968), Tetens (1995), Winters (1985), Wise (1979), Wolff (1997a,b)

Energy, Electromagnetic: Knudsen (1995)

Energy, Zero Point *see* **Zero Point Energy**

Enskog, David: Brush (1972a)

Entropy: Bernhardt (1967), Bierhalter (1985), Bohr (1972), Broda (1976),Cropper (1986), Daub (1970c), Dias (1994d), Dias et al. (1995), Dutta (1968), Giorello (1972b), Haitun (1991), Hutchison (1973, 1975, 1981a,b), Jammer (1973), Leff & Rex (1990a), Maffioli (1980), Raman (1973b), Segre (1973), Snelders (1977), Tarsitani & Vicentini (1991), Wehrl (1982), Yagi et al. (2002a)

Entropy in Chemistry: Kragh & Weininger (1996)

Enz, C. P. 1974. Is the Zero-Point Energy real? In *Physical Reality and Mathematical Description*, edited by C. P. Enz and J. Mehra, 124-132. Boston: Reidel.
    Difference between Planck's and Nernst's views; quotes letter from O. Stern recalling his arguments with Pauli on the subject.

Epstein, P. S. *See* Haas (1980)

Equation of State: Brush (1976b), Garber (1978), Kubbinga (1997b, 2002)

Equipartition Theorem: Bergia & Navarro (1997), Bowden (1972)

Ergodic Hypothesis: Baracca & Rechtman S. (1985), Bernhardt (1971), Brush (1971b, 1976b), Gallavotti (1995, 1999), LoBello (1983), Von Plato (1987, 1991, 1994)

Escudié, B., M. Pennaneach, M. Tachoire & P. Gire. 1996. L'Oeuvre de Sadi Carnot dans la Perspective de Prédictions des Propriétés des Machines thermiques. *Sciences et Techniques en Perspective* 35: 119-141. [*Isis CB*]

Euler, L. 1972. Dissertation on Sound. In *Acoustics*, edited by R. B. Lindsay, 105-117. Stroudsburg, PA: Dowden, Hutchinson & Ross. Translated from *Dissertatio Physica de Sono* (1727).

Euler, Leonhard: Truesdell (1980)

Everitt, C. W. F. 1967. Maxwell's Scientific Papers. *Applied Optics* 6: 639-646.

_____ 1974. Maxwell, James Clerk. *DSB* 9: 198-230.

_____ 1975. *James Clerk Maxwell: Physicist and Natural Philosopher*. New York: Scribner. Revised and expanded version of Everitt (1974).

_____ *See also* Maxwell (1986, 1995a)

Experiments: Hilbert (1999), Sargent (1995), Shapin & Schaffer (1985), Sibum (1995, 1998a,b), Truesdell (1982)

Exner, Franz Serafin: Hiebert (2000)

Ezawa, H. 1979a. Einstein's Contribution to Statistical Mechanics, Classical and Quantum. *Japanese Studies in the History of Science* 18: 27-72. [*Isis CB*]

_____ 1979b. Einstein's Contribution to Statistical Mechanics. In *Albert Einstein, His Influence on Physics, Philosophy and Politics*, edited by P. C. Aichelburg & R. U. Sexl, 69-87. Braunschweig: Vieweg. Abridged version of Ezawa (1979a).

Faraday, Michael: Sebastiani (1990a), Smith (1976b), Spargo (1992)

Fazzari, M. *See* Klever (1997)

Feingold, M. *See* Shapin & Schaffer (1985)

Ferguson, E. *See* Cardwell (1971)

Feyerabend, P. K. 1967. Ludwig Boltzmann. *Encyclopedia of Philosophy*, edited by P. Edwards, Vol. 1, pp. 334-337. New York: Macmillan & Free Press.

_____ 1970. In Defence of classical Physics. *Stud. HPS* 1: 59-85.
    Remarks on irreversibility, randomness, Boltzmann-Zermelo debate.

Fisher, S. J. *See* Pacey & Fisher (1967)

Flamm, D. 1973. Life and Personality of Ludwig Boltzmann. *Acta Physica Austriaca Supplement* 10: 3-16.

_____ 1979. Der Entropiesatz und das Leben: 100 Jahren Boltzmannsches Prinzip. *Naturwissenschaftliche Rundschau* 32: 225-239. [*Isis CB*]

_____ 1983. Ludwig Boltzmann and his Influence on Science. *Stud. HPS* 14: 255-278.

Includes text of 1903 lecture on philosophy of science.

_____ See also Battimelli (1993), Fleischhacker & Schönfield (1997)

Flammarion, Camille: Kleinert (1992)

Fleischhacker, W. and T. Schönfield, editors. 1997. *Pioneering Ideas for the Physical and Chemical Sciences: Josef Loschmidt's Contributions and modern Developments in Structural Organic Chemistry, Atomistics, and Statistical Mechanics.* (Proceedings of the Josef Loschmidt Symposium, held June 25-27, 1995, in Vienna)

Includes H. Spohn on reversibility and the H Theorem; D. Flamm on thermal equilibrium; O. Preinung on diffusion of gases; H. Kubbinga on reality of molecules; N. Bachmayer & T. Schönfeld on size of air molecule; D. Flamm on friendship with Boltzmann.

Floderer, A. 1971. Avogadro'sche oder Loschmidt'sche Zahl? *Bohemia: Jahrbuch des Collegium Carolinum* 12: 377-384.

Includes information on Loschmidt but doesn't state his actual value for "Loschmidt's number."

Fluctuations: Klein (1982), Sredniawa (1991)

Fluids: Gavroglu (1990a)

Forbes, E. G., editor. 1978. *Human Implications of Scientific Advance: Proceedings of the Xvth International Congress of the History of Science, Edinburgh, 10-15 August 1977.* Edinburgh: Edinburgh University Press.

Includes papers by Cardwell, Daub, Hiebert, Hornix, Hoyer, Klein and Lervig [see under names of those authors].

Force: Caneva (1997), Elkana (1970b), Harman (1974a, 1974b, 1976, 1982)

Forces between Atoms and Molecules: Bevilacqua (1994), Brush (1970b, 1976b, 1983), Byrne (1980), [Harman] Heimann (1970a)

Forgan, S., editor. 1980. *Science and the Sons of Genius: Studies on Humphry Davy.* London: Science Reviews.

Forman, P., *See* Mehra & Rechenberg (1982)

Forrester, J. 1975. Chemistry and the Conservation of Energy: The Work of James Prescott Joule. *Stud. HPS* 6: 273-313.

"The mechanical equivalent of heat emerged out of a chemical programme." Joule's 1847 paper was originally supposed to have been presented to the Chemistry section of the BAAS but there was too much business there so it was moved to physics.

Fourier, J. B. J.: Brush (1976c), Cardwell (1977), Charbonneau (1990), Herivel (1966), Hiebert (1996a), Takata (1993), Yagi & Hayashi (2002), Yoshida (1989)

Fox, R. 1968a. The Background to the Discovery of Dulong and Petit's Law. *BJHS* 4: 1-22.

_____ 1968b. Dalton's Caloric Theory. In *John Dalton and the Progress of Science,* edited by D. S. L. Cardwell, 187-202. Manchester University Press; New York: Barnes & Noble.

_____ 1970. Watt's expansive Principle in the Work of Sadi Carnot and Nicolas Clement. *NRRSL* 24: 233-253.

_____ 1971a. The intellectual Environment of Sadi Carnot: A new Look. *Actes, 12th International Congress of the History of Science, 1968*, 4: 67-72.
> Influence of N. Clément and C. B. Desormes on Carnot. [*Isis CB*]

_____ 1971b. *The Caloric Theory of Gases from Lavoisier to Regnault*. Oxford: Clarendon Press.
> Reviewed by W. H. Brock, *Nature* 233 (1971): 357-358; S. G. Brush, *BJHS* 6 (1972): 218-220; M. Crosland, *Annals of Science* 28 (1972): 208-209; R. Kargon, *Science* 174 (1971): 1016-1017; O. Knudsen, *Centaurus* 16 (1972): 322; S. Pierson, *Isis* 68 (1977): 462-464; G. Schwarzenbach, *Gesnerus* 29 (1972): 104-105; D. Von Engelhardt, *Sudhoffs Archiv* 58 (1974): 98-99 [*Isis CB*]

_____ 1975. The Rise and Fall of Laplacian Physics. *HSPS* 4: 89-136.

_____ 1979. The Science of Fire: J. H. Lambert and the Study of Heat. *Colloque International Jean-Henri Lambert (1728-1777)*, 325-342. Paris: Editions Ophrys.

_____ 1988. Les "Réflexions sur la Puissance Motrice du Feu" de Sadi Carnot et la Leçon de leur Édition critique. *La Vie des Sciences: Comptes Rendus de l'Académie des Sciences* 5: 283-301.

_____ See also Brush (1976b), Carnot (1978, 1986)

Frank, F. C. 1966. Reflections on Sadi Carnot. *Physics Education* 1: 11-18. [*Isis CB*]

Franklin, Benjamin: Sebastiani (1990a)

Franz Joseph: Brody (1983)

French Science: Herivel (1966), Morris (1965)

Frenkel, V. J. 1971. *Paul Ehrenfest*. Moscow: Atomizdat.

Fujii, T. *See* Layton & Lienhard (1988)

Fullmer, J. Z. 1964. Humphry Davy's Weltaunschauung. *Actes, 10th International Congress of History of Science, Ithaca, NY, 1962*, 325-328. Paris: Hermann.
> His ideas of time and theoretical chemistry were moving toward thermodynamics.

_____ 1969. *Sir Humphry Davy's Pubpished Works*. Cambridge, MA: Harvard University Press.
> Introductory essay and annotated bibliography.

_____ 2000. *Young Humphry Davy: The Making of an Experimental Chemist*. Philadelphia: American Philosophical Society (*Memoirs*, vol. 237).
> Includes theories of heat.

Furukawa, Y. *See* Kikuchi et al. (1996)

Gabbey, W. A. *See* Carnot (1966)

Galgani, L. 1984. Sulla Storia e lo Stato attuale della Verifica sperimentale della Legge di Radiazione di Planck. *Atti del IV Congresso Nazionale di Storia della Fisica*, edited by P. Tucci, 178-181. Rozzano: Centro Stampa Rozzano.

Galison, P. 1981. Kuhn and the Quantum Controversy. *British Journal for the Philosophy of Science* 32: 71-84.
> Essay review of Kuhn (1978).

Gallavotti, G. 1995. Ergodicity, Ensembles, Irreversibility in Boltzmann and Beyond. *Journal of Statistical Physics* 7: 1571-1589.

_____ 1999. *Statistical Mechanics*. New York: Springer-Verlag.

Pp. 36-44, "A Historical Note: The Etymology of the word 'Ergodic' and the Heat Theorems."
Garber, E. 1966. *Maxwell, Clausius and Gibbs: Aspects of the Development of Kinetic Theory and Thermodynamics*. Ph. D. Dissertation, Case Institute of Technology, 1966. *Dissertation Abstracts* 28: 1030A.
_____ 1969. James Clerk Maxwell and Thermodynamics. *American Journal of Physics* 37: 146-155.
_____ 1970. Clausius and Maxwell's Kinetic Theory of Gases. *HSPS* 2: 299-319.
_____ 1972. Aspects of the Introduction of Probability into Physics. *Centaurus* 17: 11-39.
_____ 1976a. Some Reactions to Planck's Law, 1900-1914. *Stud. HPS* 7: 89-126.
_____ 1976b. Thermodynamics and Meteorology (1850-1900). *Annals of Science* 33: 51-65. [*Isis CB*]
_____ 1978. Molecular Science in Late-Nineteenth-Century Britain. *HSPS* 9: 265-297.
>Impact of kinetic theory, work of Clausius and Maxwell, specific heats problem, virial theorem, equation of state.

Garber, E. *See also* Maxwell (1986, 1995a), Steffens (1979)
Gaseous State of Matter: Brush (1976b), Crosland (1964), Gavroglu (1990a)
Gases, Laws and Spring of: Breny (1992), Brush (1965, 1970d, 1976b), Clericuzio (1998), De Berg (1995), Klever (1997), Kubbinga (2002), Massignat (2000), Meerkulova (1978), Middleton (1969), Raman (1973a), Webster (1965 (*See also* **Equation of State**)
Gassendi, Pierre: Massignat (2000)
Gavroglu, K. 1990a. From Gases and Liquids to Fluids: The Formation of new Concepts during the Development of Theories of Liuquids. In *Greek Studies in the History and Philosophy of Science*, edited by P. Nicolacopoulos, 251-277. Dordecht: Kluwer.
_____ 1990b. The Reaction of the British Physicists and Chemists to van der Waals' early Work and to the Law of Corresponding States. *HSPS* 20: 199-237.
_____ *See also* Kamerlingh Onnes (1991)
Gay-Lussac, Joseph Louis: Crosland (1978), Kubbinga (2002), Redondi (1981), Schöpf (1984c), Snelders (1965), Spurgin (1987)
Gearhart, C. A., Jr. 1983. Gibbs, Liouville's Theorem, and the American Frontier. *American Journal of Physics* 51: 81-82.
_____ 1990. Einstein before 1905: The early Papers on Statistical Mechanics. *American Journal of Physics* 58: 468-480.
_____ 2002. Planck, the Quantum, and the Historians. *Physics in Perspective* 4: 170-215.
Gel'fer, I. M. 1981. *Istoriia I Metodologiia Termodinamiki I statisticheskoi Fizike*. 2nd ed. Moscow: Vysshaia Shkola.
>18th-century ideas of heat, thermodynamics, kinetic theory of gases, critical point, statistical methods, quantum theory.

Gender: Potter (2001)
German science: Hoffmann & Ebelung (1991)
Geymonat, L. *See* Redondi (1980)
Gibbs, J. Willard: Bowden (1972), Brush (1978, 1983), Deltete (1995), Donnan & Haas (1980), Duhem (1980), Garber (1966), Gearhart (1983), Gol'dberg & Naziuta (1969), Haas (1980), Hiebert (1968b, 1971, 1978b, 1996a), Hornix (1978, 2001),

Kastler (1983), Kestin (1976), Khatib (1992), Kipnis (1991, 2001), Klein (1969, 1978, 1983, 1987, 1990a,b), Knudsen (1987), Kox (1990), Laidler (1993), Levelt Sengers (2002), Moulines (1989, 1991), Moyer (1988), Navarro Veguillas (1994, 1998), Seeger (1974)

Gibbs Ensemble: Hiromasa (1975)
Gibbs Phase Rule: Daub (1976)
Gibbs Statistical Mechanics: Barker (1976), Mehra (1998, 2001), Sklar (1993)
Giffard's Injector: Kranakis (1982)

Gillmor, C. S. *See* Redondi (1980)

Giorello, G. 1972a. Le "Ipotesi del Disordine" nell'Opera di Max Planck: Caos molecolare e Radiazione naturale. In *Alcuni Aspetti dello Sviluppo delle Teori Fisiche 1743-1911, Saggi di Piero Delsedime et al.*, 239-342. Pisa: Domus Galilaeana.

_____ 1972b. I Concetti di Energia, Temperatura, Entropia in Max Planck (Periodi: 1880-1913). *Physis* 14: 211-214. [*Isis CB*]

Gire, P. *See* Escudié et al. (1996)

Gol'dberg, M. N., and N. P. Naziuta. 1969. Teoriia geterogennogo Ravnovesiia Gibbsa, ee Istoki I Razvitie. *Voprosy Istorii Estestvoznaniia i Tekhniki* 26: 39-46. [*Isis CB*]
"Gibbs' theory of heterogeneous equilibrium, its sources and development."

Goldberg, S. *See* Kuhn (1978), Mendelssohn (1973)

Goldfarb, S. J. 1977. Rumford's Theory of Heat - a Reassessment. *BJHS* 10: 25-36.

Gooding, D. C. *See* Lindsay (1975), Steffens (1979)

Goodman, D. C. 1972. The Discovery of Brownian Motion. *Episteme, Rivista Critica della Storia delle Scienze, Mediche e Biologiche* 6: 12-29.

Gopal, E. S. R., and B. Viswanathan. 1969. One Hundred Years of Critical Point Phenomena. *Journal of Scientific and Industrial Research (New Delhi)* 28: 204-214.

Gorham, G. 1991. Planck's Principle and Jeans's Conversion. *Stud. HPS* 22: 471-497.
Jeans's "conversion from classical mechanics to quantum theory stands as a remarkable exception to Planck's Principle."

Gornshtein, T. N. 1966. Razvitie Predstavleniia o edinoi Prirode Teplovogo Izlucheniya I Sveta. *Voprosy Istorii Estestvoznaniia i Tekhniki* no. 20: 55-58.
"Development of ideas of identical nature of heat and light." Ampère, W. Herschel, Melloni, Seebeck.

Gosiewksi, Wladislaw. 1962. Various Theories of Pressure in Gases. Translated from *Pamiet-nik Towarzystwa Nauk Scislych w Paryzu* (1874). Livermore, CA: Lawrence Livermore National Laboratory, report UCRL-Trans-885(L).

Goudaroulis, Yorgos. 1994. Searching for a Name: The Development of the Concept of the Critical Point (1822-1869). *Revue d'Histoire des Sciences* 47: 353-379.
C. Cagniard de la Tour, M. Faraday, T. Andrews.

_____ *See also* Kamerlingh Onnes (1991)

Gouy, G.: Levelt Sengers (1979)

Graffi, D. *See* Truesdell (1980)

Graham, Thomas: Munro (1972)

Graham's Laws: Kirsch (1981), Mason (1967), Mason & Kronstadt (1967), Mason & Wright (1971), Pourprix & Locqueneux (1985)

Grattan-Guinness, I. *See* Truesdell (1980)

Greenberg, D. 1990. Energy, Power, and Perceptions of social Change in the early Nineteenth Century. *American Historical Review* 95: 693-714.
> "[A]nalyzes the emergence of an energy mystique during the early stages of Anglo-American industrialization, a perception of power as the primary source of well-being and beneficial social changes. Misunderstanding of the relation of biological and nonbiological power produced an exaggerated view of the importance of energy ..."

Greenberg, J. L. *See* Bellone (1980)

Gregory, F. *See* Caneva (1993)

Grilli, M., and F. Sebastiani. 1982. Le Origini della Fisica del Calore: Le Teorie sulla Natura del Calore da Galileo a Newton. *Physis* 24: 301-356. [*Isis CB*]

Gross, M. A. *See* Mehra & Rechenberg (1982)

Grove, William Robert: Cantor (1975), Vernon (1966)

Guggenheim, E. A. *See* Donnan & Haas (1980)

Guidetti, G. P. 1985. I primi "Corpi neri" utilizzati in Laboratorio. *Physis* A27: 157-162. [*BISS*]

_____ *See also* Carazza & Guidetti (1997)

Gurikov, V. A. 1977. K Voprosy Razvitiia Teorii Teplovogo Izlucheniia. *Voprosy Istorii Estestvoznaniia I Tekhniki*, no. 56-57: 69-72.
> "On the Development of Heat Radiation Theory." [*Isis CB*]

Gvozdetskii, V. L. 1997. Sadi Carnot and Heat Energy Production [in Russian]. *Voprosy Istorii Estestvoznaniia I Tekhniki* no. 3: 96-114.

Gythiel, A. *See* Lemons & Gythiel (1997)

H: Brush (1967c), Hjalmars (1977)

Haar, D. ter. 1967. *The Old Quantum Theory*. Oxford & New York: Pergamon Press.
> Includes translations of papers of Planck (1900) and Einstein (1905).

Haas, A., editor. 1980. *A Commentary on the Scientific Writings of J. Willard Gibbs*, Vol. II. *Theoretical Physics*. New York: Arno Press, 1980. Reprint of the 1936 edition. For Vol. I see Donnan & Haas (1980).
> Includes articles by A. Haas and P. S. Epstein on thermodynamics and statistical mechanics.

Haber, Fritz: Hiebert (1978b)

Hackmann, W. *See* Shapin & Schaffer (1985)

Hadfield, E. C. R. 1967. *Atmospheric Railways: A Victorian Adventure in silent Speed*. Newton Abbot (Devon, UK): David & Charles.
> Several references to John Herapath.

Haitun, S. D. 1991. Entropy and Disorder: The Evolution of Views concerning their Connection. In Martinas et al. (1991), 220-227.

Hakfoort, C. 1992. Science deified: Wilhelm Ostwald's Energeticist World-View and the History of Scientism. *Annals of Science* 49: 525-544. [*Isis CB*]

Hall, M. B. 1987. Boyle's Method of Work: Promoting his Corpuscular Philosophy. *NRRSL* 41: 111-143.

_____ *See also* Shapin & Schaffer (1985)

Hall, V. 1978. *Some Contributions of Medical Thought in the late 18th and early 19th Centuries to the Emergence of the Idea of Conservation of Energy.* Ph. D. Thesis, University College London. [*Isis CB*]

_____ 1979. The Contribution of the Physiologist, William Benjamin Carpenter (1813-1885), to the Development of the Principles of the Correlation of Forces and the Conservation of Energy. *Medical History* 23: 129-155. [*Isis CB*]

Haller, R. *See* Blackmore (1995)

Halpin, J. P. *See* Achinstein (1991)

Hankins, T. L. *See* Scott (1970), Elkana (1974b)

Hannaway, O. *See* Shapin & Schaffer (1985)

[Harman] Heimann, P. M. 1970a. Molecular Forces, Statistical Representation and Maxwell's Demon. *Stud. HPS* 1: 189-211.

_____ 1970b. Maxwell and the Modes of Consistent Representation. *AHES* 6: 171-213.

_____ 1974a. Conversion of Forces and the Conservation of Energy. *Centaurus* 18: 147-161.

_____ 1974b. Helmholtz and Kant: The metaphysical Foundations of *Über die Erhaltung der Kraft. Stud. HPS* 5: 205-238.

_____ 1976. Mayer's Concept of "Force": The "Axis" of a new Science of Physics. *HSPS* 7: 277-296.

Harman, P. M. [formerly Heimann, P. M.] 1982. *Energy, Force, and Matter. The Conceptual Development of Nineteenth-Century Physics.* New York: Cambridge University Press.

Energy conservation, thermodynamics, kinetic theory.

_____ 1985. Edinburgh Philosophy and Cambridge Physics: The Natural Philosophy of James Clerk Maxwell. In *Wranglers and Physicists*, edited by P. M. Harman, 202-224. Dover, NH: Manchester University Press.

_____ 1987. Mathematics and Reality in Maxwell's Dynamical Physics. In *Kelvin's Baltimore Lectures and Modern Theoretical Physics*, edited by R. Kargon & P. Achinstein, 267-297. Cambridge, MA: MIT Press.

_____ 1988. Newton to Maxwell: The *Principia* and British Physics. *NRRSL* 42: 75-96.

_____ 1992. Maxwell and Saturn's Rings: Problems of Stability and Calculability. In *Investigation of Difficult Things. Essays on Newton and the History of the Exact Sciences in Honour of D. T. Whiteside*, edited by P. M. Harman & A. E. Shapiro, 477-502. Cambridge: Cambridge University Press.

Link between problem of Saturn's rings and that of the gas system. [*Isis CB*]

_____ 1998. *The Natural Philosophy of James Clerk Maxwell.* Cambridge: Cambridge University Press.

Harman [Heimann], P. M. *See also* Caneva (1993), Cardwell (1989), Everitt (1975), Maxwell (1990, 1995a,b, 2002), Steffens (1979)

Harned, H. S. *See* Donnan & Haas (1980)

Hart, P. *See* Blackmore (1995)

Hartsoeker: Sebastiani (1984)

Hawthorne, R. M., Jr. 1970. Avogadro's Number: Early Values by Loschmidt and others. *Journal of Chemical Education* 47: 451-455.

Doubts about Loschmidt's 1865 paper.

_____ 1973. The Mole and Avogadro's Number: A forced fusion of Ideas for teaching purposes. *Journal of Chemical Education* 50: 282-284.
    History of these ideas and their use in chemistry textbooks.
_____ 1979. Boyle's/Hooke's/Towneley and Power's/Mariotte's Law. *Journal of Chemical Education* 56: 741-742.
Hayashi, H. *See* Yagi & Hayashi (2002), Yagi, Tadokoro & Hayashi (2002)
Hayles, N. Katherine. 1990a. *Chaos Bound: Orderly Disorder in Contemporary Literature*. Ithaca, NYM: Cornell University Press.
    Includes Maxwell's Demon.
_____ 1990b. Self-reflexive Metaphors in Maxwell's Demon and Shannon's Choice. Finding the Passages. In *Literature and Science: Theory & Practice*, edited by S. Peterfreund, 209-237. Boston: Northeastern University Press.
Heat Conduction: Brush (1973b, 1976b, 1988)
Heat Death: Bailyn (1985), Brody (1987), Brush (1978)
Heat, Nature of: Becker (1996), Bellone (1971, 1980, 1989), G. I. Brown (1995), S. C. Brown (1967), Brush (1965, 1970c, 1970d), Cardwell (1972, 1977), Duncan (1971), Fox (1979), Fullmer (2000), Cotignola et al. (2002), Gel'fer (1981), Goldfarb (1977), Grilli & Sebastiani (1982), D. N. Jones (1969), Kranakis (1982), Kremer (1980), Liszi (1991), Locqueneux (1990, 1996), Mach (1987), Maxwell (1970, 1995a, 2001), Olson (1970), Rumford (1968-70, Schöpf (1984c), Sebastiani (1980, 1981a, 1982c, 1983, 1984, 1985, 1987a,b, 1990a,b, 1992), Tarsitani & Vicentini (1991).
Heat Radiation: Amano (2000, 2001), Brush (1970c, 1988), Carazzi & Kragh (1989), Draper (1973), Gurikov (1977), Hilbert (1999), Lovell (1968), McRae (1969), Olson (1969, 1970), Palik (1977), Takata (1993), Watanabe (1978)
Heat Transfer: Layton & Lienhard (1988)
Hedgpeth, J. N. *See* Bellone (1980)
Heilbron, J. L. 1975. Clarification. *Science* 187: 792.
    On E. Broda's account of the battle between Boltzmann and the anti-atomists.
_____ *See also* Kangro (1976b), Mehra & Rechenberg (1982), Shapin & Schaffer (1985)
Helm, G. 2000. *The Historical Development of Energetics*. (Boston Studies in the Philosophy of Science, vol. 209). Boston: Kluwer. Translated by R. J. Deltete from *Die Energetik nach ihrer geschichtlich Entwickelung* (1898).
    Includes his response to criticism from Boltzmann and Planck.
Helm, Georg: Hiebert (1968b, 1971)
[Helmholtz, H. von] 1973. Hermann von Helmholtz' philosophische und naturwissenschaftliche Leistungen aus der Sicht des dialektischen Materialismus und der modernen Naturwissenschaftlichen. *Wissenschaftliche Zeitschrift der Humboldt Universität zu Berlin, Mathematische-Naturwissenschaft Reihe* 22: 277-361.
    Includes a paper by H. Bernhardt on 2nd Law of Thermodynamics. [*Isis CB*]
Helmholtz, H. von. 1983. *Über die Erhaltung der Kraft*. 2 vols. Weinheim: Physik-Verlag.
    Vol. 1 contains an introduction by H.-J. Treder and a transcription of the manuscript text by C. Kirsetn. Vol. 2 contains a facsimile of Helmholtz's manuscript. [*Isis CB*]

Reviewed by R. L. Kremer, *Isis* 76 (1985): 281-282.

_____ 1994. *Hermann von Helmholtz and the Foundations of Nineteenth-Century Science.* Edited, introduced, and with an essay by D. Cahan. Berkeley, Los Angeles & Oxford: University of California Press.

> Contains essays by F. Bevilacqua, H. Kragh, and G. Bierhalter relating to Helmholtz and kinetic theory. [*Editor*]

_____ 1995. *Science and Culture: Popular and Philosophical Essays,* edited, introduced, and in part translated by D. Cahan. Chicago and London: University of Chicago Press.

> Contains his 1854 and 1862-63 addresses on the conservation of force (energy). [*editor*]

Helmholtz, Hermann von: Bevilacqua (1994), Bierhalter (1981c, 1983b), Clark (1976b), Elkana (1970b), Harman (1974b), Hiebert (1978b), Hoffmann & Ebelung (1991), Knudsen (1995), Lloyd (1970), Nye (1984), Ordóñez (1996), Winters (1985), Wolff (1995b, 1997a,b)

Hendry, J. *See* Mehra & Rechenberg (1982)

Hentschel, K. *See* Agassi (1993), Achinstein (1991)

Herapath, J. 1972. *Mathematical Physics and Selected Papers.* Edited and with an Introduction and Bibliography by S. G. Brush. New York: Johnson Reprint Corp.

Herapath, John: Bellone (1989), Brush (1972b, 1976b, 1983), Hadfield (1967), Knight (1968), Knudsen (1971), Mendoza (1975, 1982), Scott (1970), Talbott & Pacey (1966), Truesdell (1968), Whitaker (1979)

Herivel, J. W. 1966. Prerequisites for Creativity in Theoretical Physics. *Scientia* [ser. 7] 60: 1-6.

> Notes the "striking absence of any contribution to the development of the kinetic theory of gases by French theoretical physicists during a period (c. 1850-1880) when French physical thought was largely dominated by the positivist attitudes of Comte and Fourier."

_____ *See also* Carnot (1966)

Hermann, A. 1969. *Frühgeschichte der Quantentheorie (1899-1913).* Mosbach in Baden: Physik. [*Isis CB*].

> Translated into English, by C. W. Nash, as *The Genesis of Quantum Theory (1899-1913),* Cambridge, MA: MIT Press (1971). Author's preface says "the extensive literature survey that was originally published in the German version ... has here been omitted."

_____ 1978. Die Entdeckung des Energie-Prinzip: Wie der Arzt Julius Robert Mayer die Physiker belehrte. *Bild der Wissenschaft* 15, no. 4: 140-148. [*Isis CB*]

_____ *See also* Planck (1969)

Hermann, Jacob: *See* Middleton (1965), Truesdell (1968)

Herschel, John: James (1985)

Herschel, William: Hilbert (1999), Lovell (1968), Gornshtein (1966)

Hertz, Heinrich: Bouveresse (1991), Hiromasa (1975)

Hesketh, R. 1973. The conspicuous Merit of a Victorian Scientist. *Times Higher Education Supplement,* 20 April, p. 11.

> S. H. Burbury.

Hiebert, E. N. 1966. The Uses and Abuses of Thermodynamics in Religion. *Daedalus* 95:

1046-1080.
> On using the 1st & 2nd laws of thermodynamics to examine questions of a fundamentally religious cast. [*Author*]

___ 1967. Thermodynamics and Religion. In *Science and Contemporary Society*, edited by F. J. Crosson, 57-104. Notre Dame: University of Notre Dame Press.

___ 1968a. The Role of Mechanics in Chemistry. *Proceedings of the XI International Congress of History of Science, Warsaw*, vol. 3, pp. 402-405.
> Prior to 1920 the efforts of chemists to represent chemical reactions in the thought and language of the mechanical notion of force came to naught. By contrast, the mechanical notion of energy put chemists on the right track in the development of a chemical thermodynamics. [*Author*]

___ 1968b. *The Conception of Thermodynamics in the Scientific Thought of Mach and Planck*. (Wissenschaftlicher Bericht, Nr. 5/68) Freiburg I. Br.: Ernst-Mach Institut der Fraunhofer-Gesellschaft zur Förderung der angewandten Physik. Expanded 104-page version of a lecture delivered at the Ernst Mach Symposium, Frieburg I. Br., March 1966
> On mechanics, thermochemistry, energetics, causality, realism, and Planck's equivocation on the role of kinetic theory, probability, and atomic-molecular considerations. On G. Helm, W. Ostwald, R. Clausius, J. W. Gibbs, L. Boltzmann. [*Author*]

___ 1971. The Energetics Controversy and the New Thermodynamics. In *Perspectives in the History of Science and Technology*, edited by D. H. D. Roller, 67-86. Norman: University of Oklahoma Press.
> On the meeting of the German Society of Scientists and Physicians in Lübeck in September 1895: W. Ostwald, G. Helm, L. Boltzmann, M. Planck, E. Mach, J. W. Gibbs. [*Author*]

___ 1974. Ernst Mach. *DSB* 8: 595-607.
> Biographical resumé of Mach's scientific contributions to heat theory, thermodynamics, atomism, kinetic theory and other topics. [*Author*]

___ 1976. An Appraisal of the Work of Ernst Mach: Scientist-Historian-Philosopher. In *Motion and Time, Space and Matter*, edited by P. K. Machamer & R. G. Turnbull, 360-388. Columbus: Ohio State University Press.
> On Mach's perspective of the origins, nature, methods, and functions of the exact sciences. [*Author*]

___ 1978a. Nernst, Hermann Walther. *DSB* 15: 432-453.
> Biographical resumé of Nernst's scientific contributions to thermochemistry, the 3rd law of thermodynamics, the heat theorem, absolute zero considerations, zero-point energy, corresponding states, and other topics. [*Author*]

___ 1978b. Chemical Thermodynamics and the Third Law: 1884-1914. *Proceedings of the 15th International Congress of History of Science, Edinburgh, 1977*, 305-313. Edinburgh: Edinburgh University Press.
> Chemical origins, analysis and spin-off of Nernst's line of reasoning in the enunciation of the 3rd law. On S. Arrhenius, J. W. Gibbs, H. Von Helmholtz, F. Haber, H. Le Châtelier, L. Boltzmann, M. Planck. [*Author*]

___ 1978c. Wilhelm Friedrich Ostwald. *DSB* 15: 432-453.

On the Leipzig school of physical chemistry, energetics, atomism and other topics. [*Author*]

_____ 1980. Boltzmann's Conception of Theory Construction: The Promotion of Pluralism, Provisionalism, and Pragmatic Realism. In Hintikka et al. (1981), 2: 175-198.

Examination of Boltzmann's philosophy of science, and his historical analysis of major advances in 19th century theoretical physics, statistical mechanics, and the kinetic theory of gases. [*Author*]

_____ 1982. Developments in Physical Chemistry at the Turn of the Century. In *Science, Technology and Society in the Time of Alfred Nobel*, edited by C. G. Bernhard, E. Crawford, and P. Sorbom, 97-118. Oxford.

Physical chemistry pioneers: W. Ostwald, J. H. Van't Hoff, S. Arrhenius. The physics of chemistry: L. Boltzmann, M. Planck, W. Nernst. [*Author*]

_____ 1983. Walther Nernst and the Application of Physics to Chemistry. In *Springs of Scientific Creativity*, edited by R. Aris, H. T. Davis, and R. H. Stuewer, 203-231. Minneapolis: University of Minnesota Press.

Nernst's scientific career as a study in the creative integration of physics and chemistry in the domain of thermodynamics.

_____ 1984. The Influence of Mach's Thought on Science. *Philosophia Naturalis* 21: 598-615.

Survey of Mach's legacy in physics, the physiology and psychology of sensations, and his critical exposé of the history and philosophy of science. His status, in the 19th century, as "the scientists' philosopher." [*Author*]

_____ 1991. Reflections on the Origin and Verification of the Third Law of Thermodynamics. In Martinas et al. (1991), 90-138.

Examination of Nernst's line of reasoning in the search for the criteria of chemical equilibrium and chemical spontaneity.

_____ 1996a. The Reduction of Thermodynamics to Mechanics: Historical-Philosophical Problems. In *Memorial Symposium for Lorenz Krüger [1995]*, 43-60. Berlin: Max-Planck-Institut für Wissenschaftsgeschichte, preprint 38.

On the classical 19th-century formulation of a theory of heat. Krüger's intertheoretic relations approach to the problem of reducing thermodynamics to mechanics. He supports reduction without adopting a reductionist philosophy. J. B. J. Fourier, L. Boltzmann, J. W. Gibbs, L. Krüger. [*Author*]

_____ 1996b. The Macro- and Microstructures of Matter: Historical Reflections. In *Das Ganze und seine Teile: Europäische Forum Alpbach 1995*, edited by H. Pfusterschmid-Hartenstein, 66-87. Vienna: Ibera Verlag.

Historical comparison of the particulate nature of matter: from wholes to parts and from parts to whole. Mechanics as a paradigm model. Classical mechanics subverted as quantum mechanics takes over. [*Author*]

_____ 1997. Walther Hermann Nernst (1864-1941). In *Die Grossen Physiker*, 2. Band, *Von Maxwell bis Gell-Mann*, edited by K. Von Meÿenn, 178-195, 437, 469-472. Munich: Verlag C. H. Beck.

Significance of his work in establishing the common frontiers of chemistry and physics that led to the genesis of chemical physics as a major discipline. [*Author*]

_____ 2000. Common Frontiers of the Exact Sciences and the Humanities. *Physics in Perspective* 2: 6-29.

    A study of Franz Serafin Exner (1849-1926) and his cross-disciplinary discussions on the common and the divergent frontiers of the exact sciences and the humanities in relation to concept transfer, laws of nature, probability, and the notion of chance. On the Exner Kreis, L. Boltzmann, the Austrian revolt in classical mechanics, and thermodynamics. Exner's critique of Oswald Spengler's pessimistic rejection of man's ability to acquire objctive truths in the exact sciences. [*Author*]

_____ *See also* Boltzmann (1990), Brush (1976b, 1983), Carnot (1976, 1978, 1986), Hintikka et al. (1981), Mach (1987)

Hilbert, D. 1972. "Foundations of the Kinetic Theory of Gases," translated from *Mathematische Annalen*, 1912, by J. Kopp. In Brush (1972), 89-101.

Hilbert, M. 1999. Herschel's Investigation of the Nature of Radiant Heat: The Limitations of Experiment. *Annals of Science* 56: 357-378.

Hill, C. *See* Shapin & Schaffer (1985)

Hindle, B. *See* Rumford (1968-1970)

Hintikka, J., D. Gruender, and E. Agazzi, editors. 1981. *Proceedings of the 1978 Pisa Conference on the History and Philosophy of Science*, Vol. II: *Probabilistic Thinking, Thermodynamics and the Interaction of the History and Philosophy of Science*. Dordrecht: Reidel.

    Includes papers by L. Krüger on statistical mechanics and the reduction problem; E. N. Hiebert on Boltzmann; V. Kartsev on Mach, Boltzmann, and Maxwell; O. A. Lezhneva on Boltzmann, Mach, and Russian physicists; C.-U. Moulines on equilibrium thermodynamics as a theory-frame.

Hiromasa, N. 1975. Formation of the concept of the Gibbs Ensemble. *Proceedings of the 14th International Congress on History of Science, 1974*, 2: 265-268.

    On S. H. Burbury and Hertz.

Hirosige, T., and S. Nisio. 1970. The Genesis of the Bohr Atom Model and Planck's Theory of Radiation. *Japanese Studies in the History of Science*, 9: 35-47.

Hjalmars, S. 1977. Evidence for Boltzmann's H as a Capital Eta. *American Journal of Physics* 45: 214-215.

Hobbes, Thomas: Pinnick (1998, 1999), Schaffer (1988), Shapin & Schaffer (1985, 1999)

Hochkirchen, T. *See* von Plato (1994)

Höflechner, W., editor. 1994. *Ludwig Boltzmann, Leben und Briefe.* (*Ludwig Boltzmann Gesamtausgabe*, vol. 9; *Publicationen aus dem Archiv der Universität Graz*, vol. 30) Graz: Akademische Druck- und Verlagsanstalt. [*Isis CB*]

_____ *See also* Battimelli (1993)

Hörz, H. 1975. Zom Verhältnis von Kausalität und Energieerhaltungssatz in der Physik des 19. Jahrhunderts. *Proceedings of the 14th International Congress of History of Science, 1974*, 2: 273-276. [*Isis CB*]

Hörz, H., and A. Laass. 1989. *Ludwig Boltzmanns Wege nach Berlin. Ein Kapitel österreichisch-deutscher Wissenschaftsbeziehungen.* Berlin: Akademie-Verlag.

    Reviewed by C. Jungnickel, *Centaurus* 33 (1990): 271-273; A. Wilson, *Isis* 82 (1991): 754-755.

Hoffmann, D., and W. Ebelung. 1991. Thermodynamics in Berlin. In Martinas et al.

(1991), 262-275.
> R. Clausius, H. von Helmholtz, A. K. Krönig, H. G. Magnus, W. Nernst, M. Planck and others.

Hoffmann, J. R. *See* Cardwell (1989)
Hohenester, A. *See* Battimelli (1993)
Holt, N. R. 1970. A Note on Wilhelm Ostwald's Energism. *Isis* 61: 386-389.
Hooge, F. N. *See* Hornix & Mannaerts (2001)
Hooke, Robert: Agassi (1977), Clericuzio (1998), Cohen (1964), Hawthorne (1979)
Hornix, W. J. 1978. The Thermostatics of J. Willard Gibbs and 19th-Century Physical Chemistry. In Forbes (1978), pp. 314-329.
_____ *See also* Carnot (1978)
_____ and S. H. W. M. Mannaerts, editors. 2001. *Van't Hoff and the Emergence of Chemical Thermodynamics*. Delft: Delft University Press/DUP Science.
> Includes translation of his 1885 paper on chemical equilibrium, with comments by Mannaerts, E. Van Emmerik, F. H. Hooge, W. Bol, Hornix, and invited papers by H. Kragh, A. Kipnis and others.

Horstmann, August Friedrich: Kipnis (1997)
Horz, H. *See* Hörz
Howard, D. A. *See* Sexl & Blackmore (1982)
Hoyer, U. 1974a. Über den Zusammenhang der Carnotschen Theorie mit der Thermodynamik. *AHES* 13: 359-375.
_____ 1974b. Carnot's "Réflexions" -- Zur Entstehung der Thermodynamik vor 150 Jahren. *Physikalische Blätter* 30: 385-393.
_____ 1975. How did Carnot Calculate the Mechanical Equivalent of Heat? *Centaurus*: 207-219.
_____ 1976a. Theoriewandel und Strukturerhaltung: Das Biespiel der Thermodynamik. *Philosophia Naturalis* 16: 421-436.
_____ 1976b. Das Verhältnis der Carnotschen Theorie zur klassischen Thermodynamik. *AHES* 15: 149-197.
_____ 1976c. La Théorie de Carnot -- Première et seconde Approximations de la Thermodynamique. In [Carnot] (1976), pp. 221-228. [*Author*]
_____ 1977. Eine folgerichtige Begründung der phänomenologischen Thermodynamik. *Praxis der Naturwissenschaften* 26: 40-47. [*Author*]
_____ 1978a. Considerations on Carnot's mechanical Equivalent of Heat. In Forbes (1978), pp. 359-367.
_____ 1978b. Kinetische Gastheorie und Boltzmannsches Prinzip. *Praxis der Naturwissenschaften*, 27, Heft 2: 29-34.
_____ 1978c. Über Waterstons mechanische Wärmeäquivalent. *AHES* 19: 371-381. [*Isis CB*]
_____ 1980. Von Boltzmann zu Planck. *AHES* 23: 47-86.
_____ 1982. Boltzmanns Verhältnis zum Positivismus. *Rivista di Filosofia* 73: 275-289.
_____ 1984. Ludwig Boltzmann und das Grundproblem der Quantentheorie. *Zeitschrift für allgemeine Wissenschaftstheorie* 15: 201-210.
> Planck's theory of radiation can be derived from Boltzmann's statistics without abandoning classical physics. [*Author*]

_____ 1987. Thermodynamics and Philosophy -- Ludwig Boltzmann. *Journal of Non-*

Equilibrium Thermodynamics 12: 11-26. [*Author*]
_____ 1991. From Boltzmann to Schrödinger. In Martinás et al. (1991), pp. 162-175.
_____ 1994. Klassische Naturphilosophie und moderne Physik. *Existentia* III/IV: 57-83. [*Author*]
_____ 2000a. Meine Begegnungen mit Ludwig Boltzmann. *Wege zur Wissenschaft: Gelehrte erzählen aus ihren Leben. Pathways to Science: Scientists tell of their Life and Work*, edited by W. Schröder, 147-153. Bremen-Rönnebeck & Potsdam: Arbeitskreis Geschichte der Geophysik und Kosmischen Physik.
_____ 2000b. Hundert Jahre Quantentheorie. *Existentia* X: 225-237. [*Author*]
Hudson, R. G. 1997. Classical Physics and early Quantum Theory: A legitimate Case of theoretical Underdetermination. *Synthese* 110: 217-256.
Hughes, D. W. *See* Cardwell (1989)
Hund, F. *See* Klein (1970a)
Hunt, B. J. *See* Cardwell (1989)
Hunter, M. *See* Boyle (1999)
Hutchison, K. 1973. Der Ursprung der Entropiefunktion bei Rankine und Clausius. *Annals of Science* 30: 341-364.
> Thomson, irreversibility, vortex atom, Carnot, energy, 2nd Law of Thermodynamics.

_____ 1975. W. J. M. Rankine and the Entropy Function. *Proceedings of the 14th International Congress of History of Science, 1974*, 2: 281-284.
_____ 1976a. *W. J. M. Rankine and the Rise of Thermodynamics*. D. Phil. Thesis, Oxford University.
> Thomson, irreversibility, vortex atom, Carnot, Clausius, energy, specific heat, 2nd Law of Thermodynamics.

_____ 1976b. Mayer's Hypothesis: A Study of the Early Years of Thermodynamics. *Centaurus* 20: 279-304.
> Temperature, Thomson, irreversibility, Carnot, energy, specific heat, 2nd Law of Thermodynamics, Clausius, Joule.

_____ 1981a. Rankine, Atomic Vortices, and the Entropy Function. *AIHS* 31: 72-134.
> Irreversibility, energy, specific heat, 2nd Law of Thermodynamics

_____ 1981b. W. J. M. Rankine and the Rise of Thermodynamics. *BJHS* 14: 1-26.
> Thomson, irreversibility, vortex atom, Clausius, Carnot, energy, 2nd Law of Thermodynamics, Forbes, Regnault, MacCullagh, temperature, entropy

_____ 2002. Miracle or Mystery? Hypotheses and Predictions in Rankine's Thermodynamics. In *Recent Themes in the Philosophy of Science*, edited by S. Clarke & T. D. Lyons, 91-120. Dordrecht: Kluwer.
_____ *See also* Brush (1978), Cardwell (1989), Carnot (1976, 1986), Redondi (1980), Steffens (1979)Truesdell & Bharatha (1977)
Hutton, James: Watanabe (1978, 1980)
Huxley, T. H.: Myers (1985)
Hydrodynamics: Bernoulli (1968), Pourprix & Locqueneux (1988), Szabo (1975), Wolff (1994)
Hylozoism: Potter (2001)

Ianiello, G. *See* Battimelli (1993), Tarsitani & Vicentini (1991)
Ibanez Medrano, J. L. *See* Velarde & Ibanez Medrano (1976)
Indeterminism: Brush (1976c, 1983), Clark (1982), Deakin (1988); *see also* **Randomness**
Indistinguishability: Kastler (1983)
Infrared Radiation *see* **Heat Radiation**
Instrumentalism: Nyhof (1981)
Irreversibility and Dissipation: Bellone (1985), Bernhardt (1969), Brouzeng (1982), Brush (1974, 1976b, 1976c, 1978, 1983), Campogaliiani (1990, 1991), Clark (1982), DiMeo (1997), Feyerabend (1970), Hutchison (1973, 1976a,b, 1981a,b), Dias (1996), Needell (1980), Niedersen (1983, 1986), Prigogine & Stengers (1984), Smith (1998), Tyapkin (1991), von Plato (1994)
Irreversible Processes *see* **Transport Processes**
Israel, G. *See* Battimelli (1993)
Itakura, K. *See* Najamura & Itakura (1991)
Italian scientists: Maiocchi (1985), Sebastiani (1993)
Ito, K. *See* Kikuchi et al. (1996)

Jackson, S. W. 1967. Subjective Experiences and the Concept of Energy. *Perspectives in Biology and Medicine* 10: 602-626. [*Isis CB*]
Jacob, M. C. *See* Shapin & Schaffer (1985)
Jäger, G. *See* Blackmore (1999)
James, F. A. J. L. 1982. Thermodynamics and Sources of Solar Heat, 1846-1862. *BJHS* 15: 155-81.
───── 1983. The Conservation of Energy, Theories of Absorption and Resonating Molecules, 1851-1854: G. G. Stokes, A. J. Ångström and W. Thomson. *NRRSL* 38: 79-107. (*Author*)
───── 1985. Between two Scientific Generations: John Herschel's Rejection of the Conservation of Energy in his 1864 Correspondence with William Thomson. *NRRSL* 40: 53-62.
───── *See also* Caneva (1993), Carnot (1986)
James, P. J. *See* Shapin & Schaffer (1985)
Jammer, M. 1973. Entropy. *Dictionary of the History of Ideas*, edited by P. P. Wiener, Vol. II, pp. 112-120. New York: Scribner.
───── *See also* [Einstein] Nelkowski (1979)
Jeans, J. H.: Gorham (1991)
Jech, B. 1988. Autopsie d'une Erreur: Á Propos de la Thèse de Doctorat d'Einstein. *Fundamenta Scientiae* 9: 55-95.
> "Discusses an error in Einstein's thesis on fluid mechanics which prevented him from finding the correct value of the Avogadro-Loschmidt number." [*Isis CB*]

Jedlik, Ányos: Liszi (1991)
Jenkins, Jane E. 1996. *Matter and Vacuum in Robert Boyle's Natural Philosophy*. Dissertation, University of Toronto. *Dissertation Abstracts International* 57 (1997): 3650-A. [*Isis CB*]
Jennings, R. C. *See* Shapin & Schaffer (1985)
Job, G. 1969. Der Zwiespalt zwischen Theorie und Anschauung in der heutigen

Wärmelehre und seine geschichtlichen Ursachen. *Sudhoffs Archiv* 53: 378-396.

Jones, D. N. 1969. *M. L. Lomonosov: The formative Years, 1711-1742.* Ph. D. Dissertation, University of North Carolina, Chapel Hill. *Dissertation Abstracts International* 30: 3402-A.

> Includes his ideas about atomic structure and motion in fluids, and about heat.

Jones, G. 1968. Joule's early Researches. *Centaurus* 13: 198-219.

> His interest in electromagnetic power production as a motivation for his work.

Jones, H. W. *See* Shapin & Schaffer (1985)

Jones, R. V. 1973. James Clerk Maxwell at Aberdeen 1856-1860. *NRRSL* 28: 57-81.

> Evidence that Maxwell has been completely forgotten in Aberdeen; his 1856 inaugural lecture is reprinted.

_____ 1980. The complete Physicist: James Clerk Maxwell, 1831-1879. *Yearbook of the Royal Society of Edinburgh [for 1978-79]*, 5-23.

Jost, R. *See* [Einstein] Nelkowski (1979)

Joule, James Prescott (including Mechanical Equivalent of Heat): Barr (1969), Cardwell (1989), Forrester (1975), Hutchison (1976b), G. Jones (1968), Lloyd (1970), Mendoza (1981, 1982), Merleau-Ponty (1979), Rossi (1977), Sebastiani (1987b, 1990a), Smith (1976b, 1998), Spargo (1992), Steffens (1968, 1979), Truesdell (1968), D. B. Wilson (1981)

Jungnickel, C. *See* Hörz & Laass (1989)

_____ and R. McCormmach. 1986. *Intellectual Mastery of Nature, Theoretical Physics from Ohm to Einstein*, Vol. 2. *The Now Mighty Theoretical Physics, 1870-1925.* Chicago: University of Chicago Press.

> Boltzmann at Graz and Vienna; thermodynamics, atomism, energetics.

Kac, M. 1974. The Emergence of Statistical Thought in Exact Sciences. In *The Heritage of Copernicus*, edited by J. Neyman, 433-444. Cambridge, MA: MIT Press.

Kaiser, W. 1982. Boltzmanns mechanische Darstellung von Thermodynamik und Elektrodynamik. In Sexl & Blackmore (1982), 207-230.

_____ 1987. Early Theories of the Electron Gas. *HSPS* 17: 271-297.

> E. Riecke (1898), P. Drude (1900), H. A. Lorentz (1905).

_____ *See also* Schöpf (1978)

Kamerlingh Onnes, H. 1991. *Through Measurement to Knowledge: The Selected Papers of Heike Kamerlingh Onnes, 1853-1926*, edited with an introductory essay by K. Gavroglu and Y. Goudaroulis. (Boston Studies in the Philosophy of Science, vol. 124) Dordrecht/Boston/London: Kluwer.

Kamerlingh Onnes, Heike: Levelt Sengers (1979, 2002), Van Helden (1991)

Kangro, H. 1970. *Vorgeschichte der Planckschen Strahlungsgesetzes.* Wiesbaden: F. Steiner.

_____ 1972. *Planck's Original Papers in Quantum Physics.* New York: Halsted/Wiley; London: Taylor & Francis. Translation of introduction and notes by S. G. Brush.

_____ 1976a. Le Developpement de la Thermodynamique de Clausius a Planck. In [Carnot] (1976), 229-245.

_____ 1976b. *Early History of Planck's Radiation Law.* New York: Crane, Russak. Translation/revision of Kangro (1970).

Reviewed by J. L. Heilbron, *Annals of Science* 35 (1978): 214-216; L. Ohlon, *Lychnos* (1977-78): 359-361; S. B. Sinclair, *BJHS* 11 (1978): 88; T. J. Trenn, *AIHS* 27 (1977): 324; M. N. Wise, *Science* 199 (1978): 1331-1332. [*Isis CB*]

Kant, Immanuel: Harman (1974b)

Kargon, R. *See* Fox (1971b), Shapin & Schaffer (1985)

Kartsev, V. 1981. The Mach-Boltzmann Controversy and Maxwell's Views on Physical Reality. In Hintikka et al. (1981), 199-205. Dordrecht & Boston: Reidel.

Kastler, A. 1983. On the Historical Development of the Indistinguishability Concept for Microphysics. In *Old and New Questions in Physics, Cosmology, Philosophy, and Theoretical Biology*, edited by A. Van der Merwe, 607-623. London: Plenum.
> Boltzmann, Gibbs, Planck; L. Natanson's 1911 papers on the quantization of radiation. See also Stachel (2000).

Kelvin, Lord (William Thomson) Bailyn (1985), Bellone (1980), Boudri (1991), Bowden (1972), Dias (1996), Hutchison (1973, 1976a,b, 1981b), James (1983, 1985), Kestin (1976), Knudsen (1971, 1995), Lloyd (1970), Moyer (1977), Myers (1985), Nye (1984), Smith (1977, 1980, 1998), Smith & Wise (1989), Steffens (1979), Tansjo (1988), Truesdell (1979), Tunbridge (1971), D. B. Wilson (1981, 1987, 1990), Wise (1979)

Kerber, G. *See* Battimelli (1993)

Kerker, M. 1974. Brownian Movement and Molecular Reality prior to 1900. *Journal of Chemical Education* 51: 764-768.

_____ 1976. The Svedberg and Molecular Reality. *Isis* 67: 190-216.

_____ 1986. The Svedberg and Molecular Reality. An autobiographical postscript. *Isis* 77: 278-282.
> On the circumstances of his Nobel Prize.

Kestin, J., editor. 1976. *The Second Law of Thermodynamics*. (Benchmark Papers on Energy, 5) Stroudsburg, PA: Dowden, Hutchinson & Ross.
> Extracts from works of S. Carnot, E. Clapeyron, Lord Kelvin, R. Clausius, J. W. Gibbs, C. Caratheodory, and others.

_____ *See also* Truesdell & Bharatha (1977)

Keyes, F. G. *See* Donnan & Haas (1980)

Khatib, G. K. 1992. *Change of Phase: The Transformation of 19th-Century Thermodynamics: Josiah Willard Gibbs (1873-1878)*. Ph. D. Dissertation, Cornell University. *Dissertation Abstracts International* 53(1993):3348-A.

Kidwell, P. *See* Carnot (1986)

Kikuchi, Y., K. Ito, and Y. Furukawa. 1996. Rudolf Clausius' Copy of Sadi Carnot's *Réflexions*: A new Look at Clausius' Access to Carnot's Work. *Historia Scientiarum* 6: 31-36. [*Isis CB*]

Kilpatrick, J. E. *See* Nairn & Kilpatrick (1972)

Kim, Y. S. 1983. Clausius' Endeavor to Generalize the Second Law of Thermodynamics, 1850-1865. *AIHS* 33: 256-273.

Kinetic Theory of Gases: Achinstein (1987a, 1991), Benedetti et al. (1984), Blackmore (1985), Boltzmann (1981, 1995), Brody (1987), Brush (1965, 1966, 1970b,d, 1971a, 1972a, 1973b, 1974, 1976b, 1978, 2002), Byne (1981), Cahan (1990), Causey (1971), Cercignani (1992), Chapman (1967), Clark (1977), Daub (1970b),

Demarzo (1978), De Regt (1996), Dorling (1970), Drago & Saiella (1995), Einstein (1989), El'iashevich & Protko (1994), Garber (1966, 1970, 1978), Gel'fer (1981), Harman (1982), Helmholtz (1994), Herivel (1966), Hiebert (1968b), Hilbert (1972), Hoyer (1978b), Knudsen (2001), Koizumi (1985), Kox (1990), Kubbinga (2002), Laidler (1993), Locqueneux et al (1983), Mason (1970), Maxwell (1986), Mendoza (1975, 1982), Middleton (1965), Morrison (1988b), Nakamura (1997), Nakamura & Itakura (1991), Nyhof (1981, 1988), Pourprix & Locqueneux (1988), Sebastiani (1992), Sklar (1993), Talbott & Pacey (1966), Tisza (1991a), Truesdell (1968), Whitaker (1979), D. B. Wilson (1981), Wolff (1994, 1995b, 1997b) [*Isis CB*]

Kipnis, A. Ya. 1964. *Razvitie khimicheskoi Termodinamiki v Rossii*. Moscow & Leningrad: Nauka. [*Author*]

_____ 1991. J. W. Gibbs and Chemical Thermodynamics. In Martinas et al. (1991), 492-507.

_____ 1997. *August Friedrich Horstmann und die physikalische Chemie*. Berlin: ERS Verlag. [*Author*]

_____ 2001. Early Chemical Thermodynamics: Its Duality Embodied in Van't Hoff and Gibbs. In Hornix & Mannaerts (2001), 212-242.

_____, B. E. Yavelov, and J. S. Rowlinson. 1996. *Van der Waals and Molecular Science*. Oxford: Clarendon Press/New York: Oxford University Press.

Reviewed by D. Barkan, *Studies in History and Philosophy of Modern Physics* 30 (1999): 433-435; J. M. H. Levelt Sengers and S. G. Brush in *Journal of Statistical Physics* 89 (1997): 1099-1103; M. Yamalidou, *BJHS* 33 (2000): 239 [*Isis CB*]

Kirchhoff Radiation Law: Agasssi (1967)

Kirsch, A. S. 1981. A prekinetic Explanation of Graham's Law. *American Journal of Physics* 49: 1076.

Thomas Thomson.

Klein, M. J. 1966. Thermodynamics and Quanta in Planck's Work. *Physics Today* 19, no. 11: 23-32.

_____ 1967. Thermodynamics in Einstein's Thought. *Science* 157: 509-516.

_____ 1969. Gibbs on Clausius. *HSPS* 1: 127-149.

_____ 1970a. *Paul Ehrenfest*, Vol. 1: *The Making of a Theoretical Physicist*. New York: American Elsevier.

Includes his work on foundations of statistical mechanics and early quantum theory.

Reviewed by F. Hund, *Centaurus* 16 (1972): 323-324; T. S. Kuhn, *American Scientist* 60 (1972): 98

_____ 1970b. Maxwell, his Demon, and the Second Law of Thermodynamics. *American Scientist* 58: 84-97.

_____ 1972. Mechanical Explanation at the End of the Nineteenth Century. *Centaurus* 17: 58-82.

_____ 1973a. The Development of Boltzmann's Statistical Ideas. In *The Boltzmann Equation, Theory and Applications*, edited by E. G. D. Cohen & W. Thirring, 53-106. Vienna & New York: Springer-Verlag.

_____ 1973b. The Maxwell-Boltzmann Relationship. In *Transport Phenomena-1973*, edited by J. Kestin. New York: American Institute of Physics.

———— 1974a. Boltzmann, Monocycles and Mechanical Explanation. *Boston Studies in the Philosophy of Science* 11: 155-175.

———— 1974b. Carnot's Contributions to Thermodynamics. *Physics Today* 27, no. 8: 23-28.

———— 1974c. The Historical Origins of the van der Waals equation. *Physica* 73: 28-47.

———— 1975. Einstein, Boltzmann's Principle, and the Mechanical World View. *Proceedings of the 14th International Congress of History of Science, 1974*, 1: 183-194.

———— 1977. The Beginnings of the Quantum Theory. In *History of Twentieth Century Physics*, edited by C. Weiner, 1-39. (Varenna International School of Physics "Enrico Fermi," Course LVII, 1972) New York: Academic Press.

———— 1978. The early Papers of J. Willard Gibbs: A Transformation of Thermodynamics. In Forbes (1978), pp. 330-341.

———— 1982. Fluctuations and Statistical Physics in Einstein's early Work. In *Albert Einstein, Historical and Cultural Perspectives: The Centennial Symposium in Jerusalem*, 39-58. Princeton, NY: Princeton University Press.

———— 1983. The Scientific Style of Josiah Willard Gibbs. In *Springs of Scientific Creativity*, edited by R. Aris et al., 142-162. Minneapolis: University of Minnesota Press.

———— 1987. Some Historical Remarks on the Statistical Mechanics of Josiah Willard Gibbs. In *From Ancient Omens to Statistical Mechanics: Essays on the Exact Sciences presented to Asger Aaboe*, edited by J. L. Berggren & B. R. Goldstein, 281-289. Copenhagen: University Library. [*Isis CB*]

———— 1990a. Duhem on Gibbs. In *Beyond the History of Science*, edited by E. Garber, 52-66. Bethlehem, PA: Lehigh University Press.

———— 1990b. The Physics of J. Willard Gibbs in his Time. *Physics Today* 43, no. 9: 40-48.

———— *See also* Broda (1983), Einstein (1993), Kuhn (1978), Mach (1987), Mendelssohn (1973), Sexl & Blackmore (1982)

Kleinert, A. 1992. Camille Flammarion und der zweite Hauptsatze der Thermodynamik. *Berichte zur Wissenschaftsgeschichte* 15: 243-249.

———— *See* Cardwell (1989)

Klever, W., editor. 1997. *Die Schwere der Luft in der Diskussion des 17. Jahrhunderts.* Wiesbaden: Harrassowitz.

>Includes papers by A. Clericuzio on Boyle's pneumatic experiments; M. Fazzari on Boyle's elasticity of air and philosophical polemics; H. Zehe on a thermometer "qui marquera toujours la véritable agitation des parties de l'air." [*Isis CB*]

Knight, D. M., editor. 1968. *Classic Scientific Papers -- Chemistry*. New York: American Elsevier.

>Includes papers by J. Herapath and J. C. Maxwell.

Knudsen, O. 1971. From Lord Kelvin's Notebook: Ether Speculations. *Centaurus* 16: 41-53.

>Includes an 1859 reference by Kelvin to Herapath's kinetic theory.

———— 1987. The Influence of Gibbs's European Studies on his later Work. In *From Ancient Omens to Statistical Mechanics*, edited by J. L. Berggren & B. R. Goldstein, 271-280. Copenhagen: University Library.

_____ 1995. Electromagnetic Energy and the Early History of the Energy Principle. In *No Truth except in the Details*, edited by A. J. Kox & D. M. Siegel, 55-78. Dordrecth: Kluwer.

> Helmholtz, W. Thomson, and Maxwell on electromagnetic work and energy. [*Author*]

_____ 2001. O. W. Richardson and the Electron Theory of Matter, 1901-1916. In *Histories of the Electron*, edited by J. Z. Buchwald & A. Warwick, 227-253. Cambridge, MA: MIT Press.

> Kinetic theory and thermodynamics applied to thermionic phenomena.

_____ *See also* Fox (1971b), Steffens (1979)

Kober, F. 1980. Die Geschichte des ersten und zweiten Hauptsatzes der Thermodynamik. *Chemiker Zeitung* 104: 195-200. [*Isis CB*]

Koch, M. 1983. Ludwig Boltzmann: Ein Vorläufer der Quantentheorie? Bemerkungen zum Verhältnis von klassischer und nichtklassischer Physik. *Wissenschaftliche Zeitschrift der Humboldt Universität, Mathematisch-Naturiwissenschaftliche Reihe* 32: 317-320. [*Isis CB*]

_____ 1991. From Boltzmann to Planck: On Continuity in Scientific Revolutions. In *World Views and Scientific Discipline Formation*, edited by W. R. Woodward & R. S. Cohen, 141-150. Boston: Kluwer.

> Rejects the view of Kuhn (1978) that Planck's 1900 paper was not a jump to a new paradigm.

Koizumi, K. 1985. The Birth of a new Mathematical Physics: The Establishment of the Kinetic Theory of Gases. [In Japanese] *Kagakusi Kenkyu: Journal of History of Science, Japan* 24: 65-75. [*Isis CB*]

Kokjowski, M. 1997. O usilowanich Wladyslawa Natansona zbudowania Termodynamiki Procesów nicodwracalnych. *Kwartalnik Historii Nauki I Techniki* 42, no. 2: 23-68.

> "On Natanson's Attempts to create Thermodynamics of Irreversible Processes." [*Isis CB*]

Komkov, V. *See* Truesdell (1980)

Kopperl, S. *See* Nye (1972)

Kox, A. J. 1990. H. A. Lorentz's Contributions to Kinetic Gas Theory. *Annals of Science* 47: 591-606.

> Influence of Boltzmann and Gibbs.

_____ 1995. Einstein, Specific Heats, and residual Rays: The History of a retracted Paper. In *No Truth except in the Details*, edited by A. J. Kox & D. M. Siegel, 245-257. Dordrecht: Kluwer.

Kragh, H. 2000. Max Planck: The Reluctant Revolutionary. *Physics World* 13, no. 12: 31-35.

> Planck did not in 1900 intend a physical quantization of energy.

_____ 2001. Van't Hoff and the Transition from Thermochemistry to Chemical Thermodynamics. In Hornix & Mannaerts (2001), 191-211.

_____ 2002. The Vortex Atom: A Victorian Theory of Everything. *Centaurus* 44: 32-114.

> Includes its relations (as a possible alternative) to kinetic theory.

_____ *See also* Agassi (1993), Brush (1983), Maxwell (1995a)

Kragh, H., and S. J. Weininger. 1996. Sooner Silence than Confusion: The tortuous Entry of Entropy into Chemistry. *HSPS* 27: 91-130.

Kranakis, E. F. 1982. The French Connection: Giffard's Injector and the Nature of Heat. *Technology and Culture* 23: 3-38.
>On H. Poincaré's attempt to explain an apparent violation of the laws of thermodynamics. [*Isis CB*]

Kranzberg, M. *See* Shapin & Schaffer (1985)

Kremer, R. L. 1980. Defending Lavoisier: The French Academy's Prize Competition of 1821. *History and Philosophy of the Life Sciences* 7: 41-65.
>G. Cuvier and C. L. Berthollet favored Lavoisier's theory of heat. [*Isis CB*]

——— *See also* Helmholtz (1983)

Krips, H. 1994. Ideology, Rhetoric, and Boyle's *New Experiments. Science in Context* 7: 53-64.

Kritsman, V. A. 1991. The Influence of the Ideas of Chemical Thermodynamics on the Formation of the Principles of Chemical Kinetics in the XIX Century. In Martinas et al. (1991), 480-491.

Krönig, A. K.: Daub (1971b), Hoffmann & ebelung (1991), Truesdell !968)

Kroes, P. A., and A. Sarlemijn. 1989. Fundamental Laws and Physical Reality. In *Physics in the Making*, edited by A. Sarlemijn & M. J. Sparnay, 303-328. Amsterdam: Elsevier.
>On J. D. van der Waals' theory.

Krüger, L. 1980. Reduction as a Problem. Some Remarks on the History of Statistical Mechanics from a Philosophical Point of View. In Hintikka et al. (1981), vol. II, pp. 147-174.

Krüger, Lorenz: Hiebert (1996a), Hintikka et al. (1981)

Krug, K. 1981. Zur Herausbildung der technischen Thermodynamik am Beispiel der wissenschaftlichen Schule der G. A. Zeuner. *NTM* 18, no. 2: 79-97. [*Isis CB*]

Kruse, W. *See* Porterfierield & Kruse (1995)

Krylov, N. S.: Tyapkin (1991)

Kubbinga, H. 1988. Newton's Theory of Matter. In *Newton's Scienific and Philosophical Legacy* (Proceedings of Congress, Nijmegen, 9-12 June 1987), edited by P. B. Scheurer & G. Debrock, 321-341. Dordrecht: Kluwer. [*Author*]

——— 1991. Thermodynamics, Molecularism, and Positivism. In Martinás (1991), 404-415.

——— 1996. Some Aspects of the Status of Molecularism in Physics 1900-1915. In *The Emergence of Modern Physics*, edited by D. Hoffmann et al., 253-265. Pavia: Goliardica Pavese. [*Author*]

——— 1997a. Josef Loschmidt and the Reality of Molecules. In Fleischhacker & Schönfeld (1997), 217-22. [*Author*]

——— 1997b. J. D. Van der Waals: achtergronden van de toestandsvergelijking. *Nederlands Tijdschrift voor Natuurkunde* 63, no. 3: 65-68.
>"J. D. Van der Waals: Background of the Equation of State." Assesses the *mathematical* reasons for the controversial correction term a/v^2. [*Author*]

——— 1999. Jean Perrin ou le triomphe du molécularisme. In *De la Diffusion des Sciences à l'Espionage Industriel. XVe-XXe Siècle* (Proceedings of congress at Lyon, 30-31 May 1996), edited by A. Guillerme, 115-131. Fontenay Saint-Cloud: ENS Editions.

2002. *L'Histoire du Concept de "Molécule."* Paris: Springer-Verlag France.
>Theory of 3 states of aggregation (Black, Turgot, Lavoisier, Laplace), study of gases (Gay-Lussac and Dalton), thermometry; shift from static lattice gas theory (Newton, Dalton, Laplace) to kinetic model (Clausius) and introduction of statistics (Maxwell, Boltzmann); equation of state (Van der Waals); rise of quantum physics (Wien, Planck) and role of Brownian motion in confirming the molecular theory in the kinetic setting (Einstein, Smoluchowski, Perrin). [*Author*]

Kubo, R. 1979. Statistical Mechanics: A Survey of its One Hundred Years. In *Scientific Culture in the Contemporary World*, edited by V. Mathieu & P. Rossi, 131-157. Milan: Scientia.

Kuenen, J. P. : Levelt Sengers (1979)

Kuhn, T. S. 1978. *Black-Body Theory and the Quantum Discontinuity: 1894-1912.* Oxford & New York: Oxford University Press.
>Reviewed by J. Agassi (1983), E. Donini, *Testi Contesti* (1979), no. 2: 122-129; A. D. Franklin, *Annals of Science* 37 (1980): 713-714; P. Galison (1981); S. Goldberg, *American Journal of Physics* 48 (1980): 327-331; by M. J. Klein, A. Shimony, and T. J. Pinch, *Isis* 70 (1979), 430-434, 434-437, 437-440 resp.; P. T. Landsberg, *Nature* 282 (1979):180-182; R, McCormmach, *Science* 203 (1979): 1100-1102; D. F. Moyer, *Centaurus* 25 (1981): 146-148; J. Nicholas, *Philosophy of Science* 49 (1982): 295-297; R. H. Stuewer, *American Scientist* 67 (1979): 623-624; J. M. Ziman, *Minerva* 17 (1979): 321-327. [*Isis CB*]. See also Koch (1991).

_____ 1980. Einstein's Critique of Planck. In *Some Strangeness in the Proportion*, edited by H. Woolf, 186-191. Reading, MA: Addison-Wesley.
>See E. Wigner's comment, p. 194.

_____ 1984. Revisting Planck. *HSPS* 14: 231-252.

_____ 1987. *Black-Body Theory and the Quantum Discontinuity: 1894-1912.* Chicago: University of Chicago Press. Reprint of Kuhn (1978) with a "New Afterword" which is a reprint of Kuhn (1984).
>Reviewed by A. A. Needell, *Isis* 78 (1987): 604-605; D. R. Topper, *Annals of Science* 45 (1988): 547-548 [*Isis CB*]

Kuhn, Thomas S.: Cerreta (1992, 1995), Tetens (1995)

Kundt, August: Cahan (1990), Wolff (1992a)

Kuznetsova, O. V. 1988. *Istoriia Obosnovaniia statisticheskoi Mekhaniki.* Moscow: Nauka. [*Isis CB*]

Laass, A. *See* Hörx & Laass (1989)

Laidler, K. J.   1993. *The World of Physical Chemistry.* Oxford, New York & Toronto: Oxford University Press.
>Semi-historical account including chapters on thermodynamics, kinetic theory, statistical mechanics, etc.; contributions to Clausius, Gibbs, Einstein, Nernst.

Lakatos, Imre: Clark (1976a)

Lambert, J. H.: Fox (1979)

Landsberg, P. T. 1981. Einstein and Statistical Thermodynamics. *European Journal of Physics* 2 (1981): 203-219.
     See also Kuhn (1978), Truesdell (1980)
Langevin, L. 1971. Apport de Lomonosov (1711-1765) au Développement de la Théorie Corpusculaire. In *Actes, XII Congrès Internationale d'Histoire des Sciences, Paris, 1968*, 55-58.
          Influence of Boyle on Lomonosov; his views on conservation of matter and motion.
Langevin, Paul: Einstein (1989), Lemons & Gythiel (1997), Navarro Veguillas (1997)
Laplace, Pierre Simon de: Fox (1975), Kubbinga (2002), Saraiva (1997), Sebastiani (1982a, 1990a), Sheynin (1985)
Lattice Theory of Gases: Kubbinga (2002), Mendoza (1990b)
Latour, B. See Shapin & Schaffer (1985)
Lavenda, B. H. 1985. Brownian Motion. *Scientific American* 252, no. 2: 70-85.
Lavis, D. 1977. The Role of Statistical Mechanics in Classical Physics. *British Journal for the Philosophy of Science* 28: 255-279.
Lavoisier, Antoine Laurent: Kremer (1980), Kubbinga (2002), Morris (1965, 1972), Saraiva (1997), Schöpf (1984c), Sebastiani (1990a)
Layton, E. T., Jr., and J. H. Lienhard, editors. 1988. *History of Heat Transfer. Essays in Honor of the 50th Anniversary of the ASME Heat Transfer Division.* New York: American Society of Mechanical Engineers.
          Includes Brush (1988); K. C. Cheng and T. Fujii, "Review and some Observations on the Historical Development of Heat Transfer Theory from Newton to Eckert -- 1700-1960. An Annotated Bibliography," 213-260.
Le Chatelier, H.: Hiebert (1987b)
Lee, G.: Pacey (1974)
Leff, H. S., and A. F. Rex, editors. 1990a. *Maxwell's Demon: Entropy, Information, Computing.* Princeton, NJ: Princeton University Press.
          Collection of reprinted articles.
     1990b. Resource Letter MD-1: Maxwell's Demon. *American Journal of Physics* 58: 201-209.
          Survey and bibliography.
Leicester, H. M. See [Lomonosov] (1970)
Lemons, D. S., and A. Gythiel. 1997. Paul Langevin's 1908 Paper "On the Theory of Brownian Motion." *American Journal of Physics*, 65: 1079-1081.
          Includes a translation of the paper.
Lenker, T. D. 1979. Carathéodory's Concept of Temperature. *Synthese* 42: 167-171.
Lervig, P. 1972. On the Structure of Carnot's Theory of Heat. *AHES* 9: 222-239.
     1978. Sadi Carnot and Nicolas Clément. In Forbes (1978), 293-304.
     1982 What is Heat? C. Truesdell's View of Thermodynamics. A Critical Discussion. *Centaurus* 26: 85-122.
     1985. Sadi Carnot and the Steam Engine: Nicolas Clément's Lectures on Industrial Chemistry 1823-28. *BJHS* 18: 147-196.
     See also Carnot (1978), Redondi (1980)
Leslie, John: Olson (1969, 1970)
Lestienne, R. 1987. A la Mémoire de Ludvig Boltzmann: L'Entropie est-elle objective?

Fundamenta Scientiae 8: 173-184.

Levelt Sengers, J. M. H. 1974. From Van der Waals' Equation to the Scaling Laws. *Physica* 73: 73-106.

    Also on J. Verschaffelt who anticipated later work on critical exponents.

_____ 1976. Critical Exponents at the Turn of the Century. *Physica* 82A: 319-351.

    In 1900 J. Verschaffelt "established precise nonclassical values for the critical exponents beta and delta."

_____ 1979. Liquidons and Gasons: Controversies about the Continuity of States. *Physica* 89A: 363-402.

    Are liquid and gas molecules different? Theories and experiments of T. Andrews, L. Cailletet, E. Colardeau, P. De Heen, G. Gouy, H. Kamerlingh Onnes, J. P. Kuenen, W. Ramsay, J. D. Van der Waals.

_____ 1983. Physics Nobel Prize. *Science* 219: 1172.

    Role of Ornstein-Zernike theory of critical opalescence in history of theories of critical point phenomena.

_____ 2002. *How Fluids Unmix. Discoveries by the School of van der Waals and Kamerlingh Onnes.* Amsterdam: Edita, Publishing House of the Royal Netherlands Academy of Arts and Sciences.

    Includes applications of Gibbs' thermodynamics. Dutch scientists.

_____ *See also* Kipnis et al. (1996)

Levere, T. H. *See* Steffens (1979)

Lewis, H. R. 1973. Einstein's Derivation of Planck's Radiation Law. *American Journal of Physics* 41: 38-44.

Lezhneva, O. A. *See* Hintikka et al. (1981)

Lindley, David. 2001. *Boltzmann's Atom: The Great Debate that Launched a Revolution in Physics.* New York: Free Press.

Lindquist, B. L. 1998. *Literature, Popular Science, and Gender: Thermodynamics in 19th and early 20th Century American Culture and Thought.* Ph. D. Dissertation, University of Wisconsin-Milwaukee. *Dissertation Abstracts International* 59 (1998): 2024-A.

Lindsay, R. B. 1971. The Concept of Energy and its early historical Development. *Foundations of Physics* 1: 383-393.

_____ 1973. *Julius Robert Mayer: Prophet of Energy.* Oxford & New York: Pergamon Press.

    Reviewed by S. G. Brush, *American Journal of Physics* 42: 920-921; E. E. Daub, *BJHS* 8 (1975): 88-89; H. J. Steffens, *Isis* 66 (1975): 145-146 [*Isis CB*]

_____, editor. 1975. *Energy: Historical Development of the Concept.* Stroudsburg, PA; Dowden, Hutchinson & Ross.

    Anthology of extracts.

    Reviewed by S. G. Brush, *Annals of Science* 33 (1976): 611-612; E. E. Daub, *Physics Teacher* 14 (1976): 583-584; D. C. Gooding, *Isis* 68 (1977): 464-465. [*Isis CB*]

_____, editor. 1976. *Applications of Energy: Nineteenth Century.* New York: Wiley/Halsted.

    Reviewed by E. E. Daub, *Isis* 69 (1978): 310-311.

_____, editor. 1979. *Early Concepts of Energy in Atomic Physics*. New York: Academic Press.
_____ *See also* Broda (1983)
Linus, Franciscus: Reilly (1969), Potter (2001), Lützen (1989)
Liouville's Theorem: Gearhart (1983), L
Liquids: Gavroglu (1990a), Levelt Sengers (1979), Lunteren (2000)
Liszi, J. 1991. Ányos Jedlik: Theory of Heat. A Hungarian Manuscript from the Middle of the 19th Century. In Martinas et al. (1991): 255-261.
Lloyd, J. T. 1970. Background to the Joule-Mayer Controversy. *NRRSL* 25: 211-225.
"Based on newly available letters of J. P. Joule, P. G. Tait, J. C. Maxwell, and H. Helmholtz to Lord Kelvin" [*Isis CB*]
Lo Bello, A. 1983. On the Origin and History of Ergodic Theory. *Bollettino di Storia delle Scienze Matematiche* 3: 37-75. [*Isis CB*]
Lockyer, Norman: Myers (1985)
Locqueneux, R. 1988-89. Le Mathématisation dans les Travaux der Carnot, Clapeyron et Clausius sur le 'Puissance motrice de la Chaleur.' *Sciences et Techniques en Perspective* 16:135-159. [*Isis CB*]
_____ 1990. Charles Combes (1801-1872): Les Principes de la 'Theorie de la Chaleur' fondés sur les Principes metaphysiques. *AIHS* 40: 11-29. [*Isis CB*]
_____ 1996. *Préhistoire et Histoire de la Thermodynamique Classique: Une Histoire de la Chaleur*. Paris: Sociéte Française d'Histoire des Sciences et des Techniques. [*Isis CB*]
Reviewed by H. Chang, *Ambix* 45 (1998): 37; M. Cotte, *AIHS* 48 (1998): 423-424; P. M. C. Dias, *Centaurus* 41 (1999): 309-310; P. Redondi, *Isis* 89 (1998): 550-551.
_____ *See also* Pourprix & Locqueneux (195, 1988)
_____, B. Maitte, and B. Pourprix. 1983. Les Statuts épistémologiques des Modèles de la Théorie des Gaz dans les Oeuvres de Maxwell et Boltzmann. *Fundamenta Scientiae* 4: 29-54.
[Lomonosov, M. V.] 1970. *Mikhail Vasil'evich Lomonosov on the Corpuscular Theory*, translated with an introduction by H. M. Leicester. Cambridge, MA: Harvard University Press.
Lomonosov, Mikhail Vasilyevich: D. N. Jones (1969), Langevin (1971), Schöpf (1984c), Sebastiani (1984)
Lorentz, Hendrik Antoon: Kaiser (1987), Kox (1990), Navarro Veguillas (1997)
Loschmidt, Josef: Böhm (1973), Fleischhacker & Schönfeld (1997), Floderer (1971), Kubbinga (1997a), Porterfield (1995)
Loschmidt's Number: Hawthorne (1970), Jech (1988)
Lovell, D. J. 1968. Herschel's Dilemma in the Interpretation of Thermal Radiation. *Isis* 59: 46-60.
Lützen, J. 1989. Joseph Liouville and die nach ihm benannten Sätze. *NTM* 26, no. 2: 5-17.
There are at least 6 theorems named for him including the one on volume in phase space.
Lugli, P. *See* Bergia (1980)
Lundgren, A. *See* Breyer (1982)
Lunteren, F. Von. 2000. Bosscha's Leerboek en Van der Waals' Proefschrift: Aantrekkende

Krachten in Den Haag. *Gewina: Tijdschrift voor de Geschiedenis der Geneeskunde* 23: 247-265.
   Genesis of van der Waals's work on physics of liquids. [*Isis CB*]

Mach, E. 1987. *Principles of the Theory of Heat: Historically and Critically Elucidated.* Edited by B. McGuinness, translated by T. J. McCormack, P. E. B. Jourdain and A. E. Heath from the German edition of 1900. Introduction by M. J. Klein. Dordrecht: Reidel. [*Isis CB*]
   Review by E. Hiebert in *Isis* 80 (1989): 159-161.

Mach, Ernst: Blackmore (1972, 1985), Broda (1979), Brush (1968a, 1976b), Curd (1978), De Regt (2001), Hiebert (1968b, 1971, 1974, 1976, 1984), Hintikka et al. (1981), Kartsev (1981), Maiocchi (1988), Mladjenovich (1991), Nye (1984), Takata (1981, 1987), Trues-dell (1979)

Mackenson, L. *See* Cardwell (1971)

Maddy, P. 1994. Taking Naturalism seriously. In *Logic, Methodology and Philosophy of Science IX*, edited by D. Prawitz et al., 383-407.
   On the use of historical evidence in philosophy of science, using J. Perrin's experiments on Brownian Motion and other examples from atomism.

Märker, A. *See* G. I. Brown (1995)

Maffioli, C. 1980. La Genesi del Concetto di Entropia. *Giornale di Fisica* 21: 3-15.
_____ *See also* Redondi (1980)

Magnetism: Navarro Veguillas (1997)

Magnus, Gustav: Hoffmann & Ebelung (1991), Wolff (1995a)

Maiocchi, R. 1985a. *Chimica e Filosofia, Scienza, Epistemologia, Storia e Religione nell'Opera di Pierre Duhem.* Firenze: La Nuova Italia. [*Author*]

_____ 1985b. La Fisica Italiana e la Vittoria dell'Atomismo (1890-1914). In *Scienza e Filosofia*, edited by C. Mangione, 697-711. Milano: Garzanti. [*Author*]

_____ 1986. La "Scoperta" dell'Atomo. Un Problem Storiografico. *Nuova Civiltá delle Macchine* 4, no. 2: 52-59. [*Author*]

_____ 1987a. Il Segreto di Pulcinella: La Vittoria dell'Atomismo attraverso la manualistica Fisica. *Societá e Storia* 10: 17-52, 301-331. [*Author*]

_____ 1987b. Volere e Vedere. Bolle di Sapone e Vittoria dell'Atomismo nella Fisica Contemporanea. *Nuova Civiltá delle Macchine* 5: 111-124. [*Author*]

_____ 1988. *La "Belle Epoque" dell'Atomo. Ricerche sulla Vittoria dell'Atomismo nella Fisica del primo Novecento.* Milan: Franco Angeli.
   Includes sections on Brownian motion, Mach, Duhem, Ostwald. Disagrees with P. Clark (1976) on alleged failure of kinetic-atomic programme in late 19th century.
   Reviewed by M. Beretta, *Nuncius* 4, no. 1 (1989): 309-310; S. D'Agostino, *Centaurus* 33 (1990): 96; M. A. Morselli in *BJHS* 22 (1989): 243-245; D. Palladino, *Epistemologia* 12 (1989): 359-362 [*Isis CB*]

_____ 1990a. The Case of Brownian Motion. *BJHS* 23: 257-283.

_____ 1990b. Pierre Duhem's "The Aim and Structure of Physical Theory": A Book against Conventionalism." *Synthese* 83: 385-400. [*Author*]

_____ 1990c. Mach nel Dibattito sull'Atomismo tra Ottocento e Novecento. *Nuova Civiltá delle Macchine* 8: 55-65. [*Author*]

_____ 1995. Duhem et l'Atomisme. *Revue Internationale de Philosophie* 46: 376-389. [*Author*]

Maitte, B. *See* Locqueneux et al. (1983)

Majumdar, C. K. *See* Mehra & Rechenberg (1982)

Mannaerts, S. H. W. M. *See* Hornix & Mannaerts (2001)

Marat, J. P.: Sebastiani (1990a)

Marić, Mileva: Renn (1997)

Mariotte, Edme: Hawthorne (1979)

Martinás, K., L. Ropolyi, and P. Szegedi, editors. 1991. *Thermodynamics: History and Philosophy. Facts, Trends, Debates.* Proceedings of a Conference at Veszprém, Hungary, 23-28 July 1990. Singapore/Teaneck, NJ/London: World Scientific.
    Includes several papers listed here under authors' names, where this volume is cited as Martinás et al. (1991).

Martinich, A. P. *See* Shapin & Schaffer (1985)

Marx, Karl: Breger (1982)

Mason, E. A. 1967. Equal Pressure Diffusion and Graham's Law. *American Journal of Physics* 35: 434

_____ 1970. Thomas Graham and the Kinetic Theory of Gases. *Philosophical Journal* 7: 99-115.

_____, and Kronstadt, B. 1967. Graham's Laws of Diffusion and Effusion. *Journal of Chemical Education* 44: 740-744.

_____, and P. G. Wright. 1971. Graham's Laws. *Contemporary Physics* 12: 179-186.

_____ *See also* Agassi (1993)

Massignat, C. 2000. Gassendi et l'Élasticite de l'Air: Une Étape entre Pascal et la Loi de Boyle-Mariotte. *Review d'Histoire des Sciences* 53: 179-204.
    P. Gassendi's study of air compressibility in 1648. [*Isis CB*]

Materialism, Dialectical: [Helmholtz] (1973)

Mathieu, J.-P. 1984. *Histoire de la Constante d'Avogadro.* Paris: Centre de Documentation Sciences Humaines.

_____ *See also* Breger (1982), Sexl & Blackmore (1982)

Mattioli, C. *See* Truesdell (1980)

Maxwell, J. C. 1965. *The Scienific Papers of James Clerk Maxwell.* New York: Dover. Reprint of the 1890 edition.

_____ 1970. *Theory of Heat.* Westport, CT: Greenwood Press. Reprint of the 3rd edition, 1872.

_____ 1986. *Maxwell on Molecules and Gases*, edited by E. Garber, S. G. Brush, and C. W. F. Everitt. Cambridge, MA: MIT Press.
    Includes published and unpublished papers, letters from and to Maxwell. Reviewed by P. Achinstein, *Foundations of Physics* 17 (1987): 425-433; J. Z. Buchwald, *Annals of Science* 45 (1988), 207-208; C. Domb, *Nature* 326 (1987): 26; I. Schneider, *Isis* 80 (1989): 535-536; B. R. Wheaton, *HSPS* 17 (1986): 186 [*Isis CB*]

_____ 1990. *The Scientific Letters and Papers of James Clerk Maxwell*, Vol. 1: *1846-1862*, edited by P. M. Harman. New York: Cambridge University Press.

_____ 1995a. *Maxwell on Heat and Statistical Mechanics. On "Avoiding all Personal Enquiries" of Molecules*, edited by E. Garber, S. G. Brush, and C. W. F. Everitt.

Bethlehem, PA: Lehigh University Press; London: Associated University Presses. Reviewed by P. M. Harman, *BJHS* 29 (1996), 107-109; H. Kragh, *Lychnos* (1997): 324-325; D. Siegel, *Isis* 87 (1996): 511-516; D. B. Wilson, *Annals of Science* 53 (1996): 649-650 [*Isis CB*]

_____ 1995b. *The Scientific Letters and Papers of James Clerk Maxwell*, Vol. 2: *1862-1873*, edited by P. H. Harman. New York: Cambridge University Press.

_____ 2001. *Theory of Heat*. With a new Introduction and Notes by P. Pesic and the 1891 Index by Lord Rayleigh. Mineola, NY: Dover.

_____ 2002. *The Scientific Letters and Papers of James Clerk Maxwell*, Vol. 3: *1873-1879*. Edited by P. M. Harman. New York: Cambridge University Press.

Maxwell, James Clerk: Achinstein (1987a, 1987b, 1991), Baucia (1973), Bellone (1989), Brush (1971a, 1976b, 1983, 2002), Brush & Everitt (1969), Chalmers (1973), Daub (1967), Deakin (1988), DeMarzo (1978), DeRegt (1996), Dias (1994a,b,c), Domb (1980), Dorling (1970), El'iashevich & Protko (1981, 1984), Everitt (1967, 1974, 1975), Garber (1966, 1969, 1970, 1978), Harman (1970b, 1985, 1987, 1988, 1992, 1998), R. V. Jones (1973, 1980), Kartsev (1981), Knight (1968), Knudsen (1995), Lloyd (1970), Locqueneux et al. (1983), Moyer (1977), Myers (1985), Navarro Veguillas (1997), Nye (1984), Porter (1981b, 1986, 1994), Ruhla (1992), Sheynin (1985), Smith (1998), Theerman (1980), Tolstoy (1981), Toyoda (1997), D. B. Wilson (1981), Wise (1983), Yamalidou (2001)

Maxwell Demon: Clarke (1996), Crow (1972), Daub (1970a), Earman & Norton (1998), Harman (1970a), Hayles (1990a,b), Klein (1970b), Leff & Rex (1990a,b), Porter (1986), Schweber (1982)

Maxwell Velocity Distribution: Achinstein (1986), Bohr (1972), Dias (1994a)

Mayer, J. R. 1978a. *Die Mechanik der Wärme: Sämtliche Schriften*. Edited by P. Münzenmayer. Heilbronn: Stadtarchiv. [*Isis CB*]

_____ 1978b. *Robert Mayer: Die Idee aus Heilbronn, Umwandlung und Erhaltung der Energie. Magazin und Katalog zur Ausstellung anlässlich des 100. Todestage von Robert Mayer*. (Kleine Schriftenreihe des Archives der Stadt Heilbronn, 11) Heilbronn: Stadtarchiv. [*Isis CB*]

Mayer, Julius Robert: Caneva (1993), Harman (1976), Hermann (1978), Hutchison (1976b), Lloyd (1970), Ober (1968), Schütz (1972), Sebastiani (1987a), Walter (1995)

Mayer, M. *See* Tarsitani & Vicentini (1991).

Mayo, D. G. 1996. *Error and the Growth of Experimental Knowledge*. Chicago: University of Chicago Press.

>Chapter 7, Brownian motion -- Jean Perrin's test of the Einstein-Smoluchowski theory.

Mayr, O. *See* Cardwell (1971)

McCormmach, R. *See* Jungnickel & McCormmach (1986)

McGuinness, B. *See* Boltzmann (1974), Mach (1987)

McRae, R. J. 1969. *The Origin of the Conception of the Continuous Spectrum of Heat and Light*. Ph. D. Dissertation, University of Wisconsin

Mean free path: Dias (1994c)

Mechanical Explanation/Philosophy/World View: Boudri (1991), Breger (1982),

Brouzeng (1993, 1995), Brush (1978), Clericuzio (1998), Drago & Saiello (1995), Klein (1972, 1974a, 1975)

Medard, L., and H. Tachoire. 1994. *Histoire de la Thermochimie: Prelude a la Thermodynamique Chimique.* Provence: University of Provence. [*Isis CB*]
    Reviewed by D. Barkan, *Isis* 87 (1996): 147-148.

Medical Thought: Hall (1978)

Medicus, H. A. *See* Mendelssohn (1973)

Mehra, J. 1975a. Einstein and the Foundation of Statistical Mechanics. *Physica* 79A: 447-477.

──── 1975b. *The Solvay Conferences on Physics. Aspects of the Development of Physics since 1911.* With a Foreword by Werner Heisenberg. Dordrecht-Holland/Boston: D. Reidel.
    Includes an account of the 1st Conference (1911) on Radiation Theory and the Quanta.

──── 1987. Erwin Schrödinger and the Rise of Wave Mechanics. I. Schrödinger's Scientific Work before the Creation of Wave Mechanics. *Foundations of Physics* 17: 1051-1112.

──── 1998. Josiah Willard Gibbs and the Foundations of Statistical Mechanics. *Foundations of Physics* 28: 1785-1815.

──── 1999. *Einstein, Physics and Reality.* River Edge, NJ: World Scientific.

──── 2001. *The Golden Age of Theoretical Physics*, Vol. 1. Singapore: World Scientific.
    Includes "Max Planck and the Law of Blackbody Radiation," 19-55; "Planck's Half-Quanta: A History of the Concept of Zero-Point Energy," 56-93; "Josiah Willard Gibbs and the Foundations of Statistical Mechanics," 94-122; "Einstein and the Foundaion of Statistical Mechanics," 123-152; "Albert Einstein and Marian von Smoluchowski: Early History of the Theory of Fluctuation Phenomena," 153-209; "Albert Einstein and the Origin of the Light-Quantum Theory," 326-350.

──── & Rechenberg, H. 1982. *The Hisorical Development of Quantum Theory.* Vol. 1, Part 1. *The Quantum Theory of Planck, Einstein, Bohr, and Sommerfeld: Its Foundation and the Rise of its Difficulties, 1900-1925.* New York: Springer-Verlag.
    Reviewed by P. Forman, *Science* 220 (1983): 824-827; M. A. Gross, *American Scientist* 71 (1983): 551; J. L. Heilbron, *Isis* 76 (1985): 388-393; J. Hendry, *BJHS* 19 (1986): 206-208; C. K. Majumdar, *Indian Journal History of Science* 19 (1984): 406-412; C. Nording, *Lychnos* (1983): 175-187; I. Prigogine, *Foundations of Physics* 14 (1984): 275-277; E. Rüdinger, *Centaurus* 28 (1985): 81-83 [*Isis CB*]
    See also Stachel (2000)

──── ──── 1999. Planck's Half-Quanta: A History of the Concept of the Zero-Point Energy. *Foundations of Physics* 29: 91-132.

Melloni, Macedonio: Gornshtein (1966)

Mendelssohn, K. 1973. *The World of Walther Nernst: The Rise and Fall of German Science, 1864-1941.* Pittsburgh: University of Pittsburgh Press.
    Reviewed by W. Brammar, *Nature* 245 (1973): 107 J. W. Cronin in

Synthesis, The University Journal in the History and Philosophy of Science, 3, no. 1 (1975): 59-61; H. Degen, *Naturwissenschaftliche Rundschau* 27 (1974): 381; P. Forman, *American Scientist* 63 (1975): 482-483; S. Goldberg in *Physics Today* (March 1974); by M. J. Klein in *Science* 186 (1974): 342-343; H. A. Medicus, *American Journal of Physics* 42 (1974): 1135-1136; Süsskind, *Technology and Culture* 17 (1976): 151-153; L. Suhling, *Physikalische Blätter* 32 (1976): 527-528 [*Isis CB*]

Mendoza, E. 1975. A critical Examination of Herapath's Dynamical Theory of Gases. *BJHS* 8: 155-165.

    It is similar to Dalton's model.

____ 1981. On a Suggestion concerning the Work of J. P. Joule. *BJHS* 14: 177-180.

    On Steffens' (1979) contention that Joule may have known of Mayer's work earlier than is usually suggested. [*Isis CB*]

____ 1982. The Kinetic Theory of Matter, 1845-1855. *AHES* 32: 184-220.

    J. Herapath, J. P. Joule, J. J. Waterston, C. C. Person.

____ 1990a. Delaroche and Bérard and Experimental Error. *BJHS* 23: 285-292.

    Their measurements of the specific heat of gases. [*Isis CB*

____ 1990b. The Lattice Theory of Gases: A Neglected Episode in the History of Chemistry. *Journal of Chemical Education* 67: 1040-1042.

    Boyle, Dalton, Avogadro.

____ *See also* Brush (1976b), Carnot (1978, 1986)

Merkulova, N. M. 1978. *Istoriia Mekhanikhi Gaza: Do nach XX v.* Moscow: Nauka. [*Isis CB*]

Merleau-Ponty, J. 1978. Thèmes cosmologiques chez les Fondateurs de la Thermodynamique classique. *Proceedings of the 15th International Congress of History of Science*, 559-566. [*Isis CB*]

____ 1979. La Découverte des Principes de l'Energie de Joule. *Revue d'Histoire des Sciences* 32: 315-331. [*Isis CB*]

Metcalfe, S. L.: Watanabe (1982, 1984)

Meteorology: Garber (1976b)

Meyenn, K. von. 1982. Boltzmann als Kritiker und Rezensent. In Sexl & Blackmore (1982), 97-127.

____ 1994. Boltzmann y la Mecánica Estadistica. *Arbor: Ciencia, Pensamiento y Cultura* 148, no. 581: 51-79.

____ *See also* Redondi (1980)

Meyer, Oscar Emil: Wolff (1994)

Meyer, S. *See* Blackmore (1999)

Michaelson, R. *See* Nye (1984)

Middleton, W. E. K. 1965. Jacob Hermann and the Kinetic Theory. *BJHS* 2: 247-250.

____ 1968. Carlo Rinaldini and the Discovery of Convection in Air. *Physis* 10: 299-305.

____ 1969. Paolo del Buono on the Elasticity of Air. *AHES* 6: 1-28.

    Text and partial translation of 1657 letter that suggests an experiment identical to the one Boyle made in 1661.

Miller, A. I. 1976. On Einstein, Light Quanta, Radiation, and Relativity in 1905. *American Journal of Physics* 44: 912-923.

Miller, D. G. 1966. Pierre Duhem. *Physics Today* 19, No. 12: 47-53.
Milne, E. A. *See* Donnan & Haas (1980)
Milonni, P. W., and M.-L. Shih. 1991. Zero-Point Energy in early Quantum Theory. *American Journal of Physics* 59: 684-698.
> Its roots were in black-body research of Planck and Einstein in 1912-1913 [*Isis CB*].

Mladjenović, M. 1991. Mach's "Principien der Wärmelehre." In Martinás et al. (1991), 63-71.
Molecularism: Kubbinga (1991, 1996, 1999, 2002)
Molecular Reality, *see* **Atoms**
Monleón Pradas, M. 1991a. Analysis of a trivial Example and critical Considerations following from it regarding the Historiography of Energy Conservation. In Martinás et al. (1991), 81-89.
Monleón Pradas, M. 1991b. Thermodynamics as a Physics of Qualities: The Evolution of the Concept of State. In Martinás et al. (1991)
Monocycles: Klein (1974a)
Morey, G. H. *See* Donnan & Haas (1980)
Morrell, J. *See* Cardwell (1989)
Morris, R. J., Jr. 1965. *Eighteenth-Century Theories of the Nature of Heat.* Ph. D. Dissertation, University of Oklahoma. University Microfilms 65-13,888.
> H. Boerhaave, Lavoisier, caloris theory, French scientists.

_____ 1972. Lavoisier and the Caloric Theory. *BJHS* 6: 1-38.
_____ *See also* Rumford (1968-1970)
Morrison, M. 1988. Reduction and Realism. *PSA 1988* (Proceedings of the Philosophy of Science Association meeting, 1988), 1: 286-293.
> Van der Waals model of gas.

_____ 1988. Unification, Realism and Inference. *British Journal for the Philosophy of Science* 41: 305-332.
> Includes kinetic theory and van der Waals equation.

Morselli, M. A. *See* Maiocchi (1988)
Morus, I. R. *See* Cardwell (1989)
Moulines, C. U. 1989. The Emergence of a Research Programme in Classical Thermodynamics. In *Imre Lakatos and Theories of Scientific Change*, edited by K. Gavroglu et al., 111-121. Dordrecht: Reidel.
> J. W. Gibbs.

_____ 1991. The Classical Spirit in J. Willard Gibbs's Classical Thermodynamics. In Martinás et al. (1991), 7-28.
_____ *See also* Hintikka et al. (1981)
Moyer, A. E. 1988. Josiah Willard Gibbs. *The World & I* 3, no. 6: 192-197.
Moyer, D. F. 1977. Energy, Dynamics, hidden Machinery: Rankine, Thomson and Tait, Maxwell. *Stud. HPS* 8: 251-268. [*Isis CB*]
_____ *See also* Kuhn (1978)
Munday, P. *See* Caneva (1993)
Munro, A. C. 1972. Thomas Graham, 1805-1869. *Philosophical Journal* 9: 30-42. [*Isis CB*]
Myers, G. 1985. Nineteenth-Century Popularizations of Thermodynamics and the Rhetoric

of Social Prophecy. *Victorian Studies* 29: 35-66. [*Isis CB*] Reprinted in *Energy & Entropy*, edited by P. Brantlinger, 307-338. Bloomington: Indiana University Press (1989).

  T. H. Huxley, Kelvin, N. Lockyer, Maxwell, B. Stewart, P. G. Tait, J. Tyndall.

Nabl, J. *See* Blackmore (1999)

Nairn, J. H., and J. E. Kilpatrick. 1972. Van der Waals, Boltzmann, and the fourth Virial Coefficient of Hard Spheres. *American Journal of Physics* 40: 503-515.

Nakamura, K. 1997. Process through Disaffirmance of "Material Theory of Heat" to Introduction and Diffusion of "Kinetic Theory of Heat" in Japan. *Historia Scientiarum* 6: 187-208. [*Isis CB*]

_____ and K. Itakura. 1991. From the "Material Theory of Heat" to the Introduction and Diffusion of the 'Kinetic Theory of Heat' in translated Physics Texts in Meiji Japan. [in Japanese] *Kagaku Kenkyu: Journal of History of Science, Japan* 30: 107-119. [*Isis CB*]

Natanson, L.: Kastler (1983), Kokjowski (1997), Stachel (2000)

Navarro Veguillas, L. 1988. El Papel de la Mecánica Estadistica en la Evolutión del Concepto de Quantum de Radiación en Einstein (1905-1916). In *Història de la Fisica*, edited by L. Navarro Veguillas, 119-129. Barcelona: CIRIT. [*Isis CB*]

Navarro [Veguillas], L. 1991. On Einstein's Statistical-Mechanical Approach to the early Quantum Theory. *Historia Scientiarum* [2nd ser.] 1: 39-58.

_____ 1994. Gibbs y Einstein: Una o dos Formulaciònes de la Mecánica Estadistica del Equilibrio? *Arbor: Ciencia, Pensamiento y Cultura* 148 no. 581: 109-129.

_____ 1997. On the Nature of the Hypotheses in Langevin's Magnetism. *AIHS* 47: 316-345.

  His theory should not be regarded as proto-quantum, rather it was based entirely on classical physics (Lorentz, Maxwell, Boltzmann).

_____ 1998. Gibbs, Einstein and the Foundations of Statistical Mechanics. *AHES* 53: 147-180.

_____ *See also* Bergia & Navarro (1988)

Navier, Claude-Louis-Marie-Henri: Szabo (1975)

Needell, A. A. 1980. *Irreversibility and the Failure of Classical Dynamics: Max Planck's Work on the Quantum Theory*. Ph. D. Dissertation, Yale University. *Dissertation Abstracts International* 41 (1980): 2742-A.

_____ *See also* Planck (1988).

Nelson, E. *See* [Einstein] Nelkowski (1979)

Nernst, Walther: Assmus (1992), Barkan (1991, 1994, 1999), Cropper (1987b), Enz (1974), Hiebert (1978a, 1978b, 1983, 1991, 1997), Hoffmann & Ebelung (1991), Laidler (1993), Mendelsson (1973),

Nernst Heat Theorem *see* **Thermodynamics, 3rd Law**

Newton, Isaac: Brush (1970b, 1983), Chalmers (1998), Cohen (1964), Grilli & Sebastiani (1982), Harman (1988), Kubbinga (1988, 2002), Layton & Lienhard (1988)

Nicholas, J. M. 1988. Planck's Quantum Crisis and Shifts in Guiding Assumptions. In *Scrutinizing Science*, edited by A. Donovan, L. Laudan, and R. Lauden, 317-335. Boston: Kluwer.

_____ *See also* Kuhn (1978)

Nickles, T. 1976. Theory Generalization, Problem Reduction, and the Unity of Science. *PSA 1974* (Boston Studies in the Philosophy of Science, vol. 32), 31-74. Dordrecht & Boston: Reidel.

> Ehrenfest (also Einstein, Debye, and others) on thermodynamic and statistical mechanical approahces to the old quantum theory. [*Author*]

_____ 1978. Scientific Problems and Constraints. *PSA 1978, Proceedings of the 1978 Meeting of the Philosophy of Science Association* 1: 134-148.

> On the black-body radiation problem, 1859-1900. [*Isis CB*]

_____ 1980. Can Scientific Constraints be Violated Rationally? In *Scientific Discovery, Logic, and Rationality*, edited by T. Nickles, 285-315. Dordrecht: Reidel.

> Planck, Einstein on Brownian motion, P. Ehrenfest.

Niedersen, U. 1983. Die Energetik und der Irreversibilitätsgedanke bei Wilhelm Ostwald. *Wissenschaftliche Zeitschrift der Humboldt Universität, Mathematisch-Naturiwissenschaftliche Reihe* 32: 325-329.

_____ 1986. Zu den Problemen von Reversibilität und Zeit im Schaffen Wilhelm Ostwalds. *NTM* 23, no. 1: 47-59.

Nielsen, J. R. *See* Bohr (1972)

Nietzsche, Friedreich: Brush (1978, 1981b)

Nisio, S. *See* Hirosige & Nisio (1970)

Nording, C. *See* Mehra & Rechenberg (1982)

North, J. D. *See* Shapin & Schaffer (1985)

Norton, J. 1987. The Logical Inconsistency of the Old Quantum Theory of Black Body Radiation. *Philosophy of Science* 54: 327-350.

_____ 1993. The Determination of Theory by Evidence: The Case of Quantum Discontinuity 1900-1915. *Synthese* 97: 1-31.

> P. Ehrenfest and H. Poincaré showed in 1911 & 1912 that the black-body radiation evidence entails quantization; this is a counter-example to the "underdetermination thesis."

Norton, J. *See* Earman (1998)

Nugayev, R. M. 2000. Einstein's Revolution: Reconciliation of Mechanics, Electrodynamics and Thermodynamics. *Physis* 37: 181-208. [*Isis CB*]

Nye, M. J. 1970. *Jean Perrin and Molecular Reality*. Ph. D. Dissertation, University of Wisconsin. *Dissertation Abstracts International* 31: 715-A. [*Isis CB*]

_____ 1972. *Molecular Reality: A Perspective on the Life of Jean Perrin.* New York: American Elsevier.

> Includes Perrin's research on Brownian motion.
> Reviewed by S. G. Brush in *Centaurus* 17 (1972) 174-175; S. Kopperl, *Isis* 64 (1973): 135-136; C. A. Russell, *Nature* 240 (1972): 55-56; *Scientific American* 227, no. 1 (1972): 118-119; S. B. Sinclair, *BJHS* 7 (1974): 300-301; T. Shinn, *HSPS* 16 (1986): 353-369 [*Isis CB*]

_____, editor. 1984. *The Question of the Atom: From the Karlsruhe Congress to the First Solvay Conference, 1860-1911. A Compilation of Primary Sources.* Los Angeles: Tomash.

> Includes papers by Boltzmann, Duhem, Einstein, Helmholtz, Kelvin, Mach, Maxwell, Ostwald, Perrin, Poincaré.

Reviewed by L. M. Brown & R. Michaelson, *Isis* 76 (1985): 102-104; D. R. Topper, *Annals of Science* 47 (1986): 104-105. [*Isis CB*]

____ 1996. *Before Big Science: The Pursuit of Modern Chemistry and Physics, 1800-1940*. New York: Twayne. Reprinted by Harvard University Press (1999).

Chapter 4: "Thermodynamics, Thermochemistry, and the Science of Energy."

____ *See also* Barkan (1999)

Nyhof, J. 1981. *Instrumentalism and Beyond*. Ph. D. Thesis, University of Otago, New Zealand.

Uses the history of kinetic theory to argue that scientific developments cannot rightly force instrumentalism or positivism on scientists. Critique of Clark (1976).

____ 1988. Philosophical Objections to the Kinetic Theory. *British Journal for the Philosophy of Science* 39: 81-109.

The objections (in the late 19th century) were due primarily to philosophical rather than scientific difficulties. Discusses Boltzmann and others on the specific heats problem.

Ober, W. B. 1968. Robert Mayer, M. D. (1814-1878) and the Mechanical Equivalent of Heat. *New York State Journal of Medicine* 68: 2447-2454.

On the story that Mayer's idea of energy conservation was inspired by his observation of the color of venous blood in the tropics.

Oersted, Hans Christian: Caneva (1997)

Oeser, E. *See* Battimelli (1993)

Ohlon, L. *See* Elkana (1971b), Kangro (1976b), Truesdell (1980)

Oldroyd, D. *See* Shapin & Schaffer (1985)

Olesko, K. *See* Breger (1982)

Olson, R. G. 1969. A Note on Leslie's Cube in the Study of Radiant Heat. *Annals of Science* 25: 203-208.

____ 1970. Count Rumford, Sir John Leslie, and the Study of the Nature and Propagation of Heat at the Beginning of the Nineteenth Century. *Annals of Science* 26: 273-304.

____ *See also* Steffens (1979)

Opalescence, Critical *see* **Critical Point**

Ordónez, J. 1988. La Recepción di Sadi Carnot: El Significado e la Aportación de la Termodinámica. *Historia de la Fisica*, edited by L. Navarro Veguillas, 267-279. Barcelona: CRIT. [*Isis CB*]

____ 1996. The Story of a Non-Discovery: Helmholtz and the Conservation of Energy. In *Spanish Studies in the Philosophy of Science*, edited by G. Munévar, 1-18. Dordrecht: Kluwer. [*Isis CB*]

Ornstein-Zernike Theory: Levelt Sengers (1983)

Osietzki, M. 1991. Rudolf Clausius -- Entropy and Environment. In Martinás et al. (1991), 57-62.

Ostwald, Wilhelm: Brouzeng (1993), Deltete (1983), Hakfoort (1992), Hiebert (1968b, 1971, 1978c, 1982), Holt (1970), Maiocchi (1988), Niedersen (1983, 1986), Nye (1984), Post (1968), Ramunni (1980)

Pacey, A. J. 1974. Some early Heat Engine Concepts and the Conservation of Heat. *BJHS* 7: 135-145.
        On G. Amontons, D. Bernoulli, J. Smeaton, G. Lee.
_____ *See also* Scott (1970), Talbott & Pacey (1966)
Pacey, A. J., and S. J. Fisher. 1967. Daniel Bernoulli and the *Vis Viva* of Compressed Air. *BJHS* 3: 388-392.
Palik, E. D. 1977. History of Far-Infrared Research, I: The Rubens Era. *Journal of the Optical Society of America* 67: 857-865.
Palladino, D. *See* Maiochhi (1988)
Pas, P. W. van der. 1971. The early History of the Brownian Motion. *Actes, 12th Congres Internationale d'Histoire des Sciences, 1968*, 8: 143-158.
        On 18th-century predecessors of R. Brown [*Isis CB*]
Pascual, M. J. *See* Shapin & Schaffer (1985)
Pauli, Wolfgang: Enz (1974)
Payen, J. 1971. Clément, Nicholas. *DSB* 3: 315-317.
_____ *See also* Cardwell (1971)
Peirce, Charles Sanders: Porter (1986), Reynolds (1996)
Pennaneach, M. *See* Escudié et al. (1966)
Perrin, F. *See* J. Perrin (1970)
Perrin, J. 1970. *Les Atomes*. Présentation et compléments par Francis Perrin. Paris: Gallimard. [*Isis CB*]
_____ 1990. *Atoms*. Reprint of the 1923 edition of D. Ll. Hammick's translation of *Les Atomes* (1913). Woodbridge, CT: Ox Bow Press.
Perrin, Jean: Achinstein (1994), Baierlein (1969), Brush (1968b, 1976b), Kubbinga (1999, 2002), Maddy (1994), Mayo (1996), Nye (1970, 1972, 1984), Raman (1970).
Person, C. C.: Mendoza (1982)
Pesic, P. *See* Maxwell (2001)
Pestre, D. *See* Shapin & Schaffer (1985)
Peterfreund, P. 1987. Organicism and the Birth of Energy. In *Approaches to Organic Form*, edited by F. Burwick, 113-152. (Boston Studies in the Philosophy of Science, vol. 105) Boston: Reidel.
        Usages of word "energy" back to 17th century.
Phenomenalism: Boudri (1991)
Philosophy: Blackmore (1987, 1995, 1999), De Regt (1996), Nyhof (1988), Schöpf (1984a)
Philosophy of Science: Brush (1977, 1983), Curd (1978), Krüger (1980), Maddy (1994)
Pictet, M. A.: Chang (2002)
Pierson, S. *See* Fox (1971b)
Pinch, T. J. *See* Kuhn (1978), Shapin & Schaffer (1985)
Pineda, C. F., and M. G. Velarde. 1977. *Cuestiones de Termodinamica y de Fisica Estadistica*. (Memorias de la Real Academia de Ciencias de Madrid, Serie de Ciencias Fisico-Quimicas, v. 7, no. 2).
        Chronology of major events in the history of statistical thermodynamics, 1679-1973; evolution of ideas about phase transitions and critical phenomena.
Pinnick, C. L. 1998. What is wrong with the Strong Programme's Case Study of the "Hobbes-Boyle Dispute"? In *A House Built on Sand*, edited by N. Koertge, 227-

239. New York: Oxford University Press.
> The history. Critique of Shapin & Schaffer (1985); see their reply and rejoinder (1998).

_____ 1999. Caught in a Sandy Shoal of the Shallow: Reply to Shapin and Schaffer. *Social Studies of Science* 29 (1999): 253-257.

Pippard, B. *See* Ruhla (1992)

Plancherel, M.: Brush (1971b)

Planck, M. 1969. *Die Quantenhypothese,* edited by A. Hermann. München: Battenberg Verlag.
> Includes his two 1900 papers and three from 1901-2.

_____ 1988. *The Theory of Heat Radiation.* Translation by M. Masius (1914) and original German text (1906). Introduction by A. A. Needell, xi-xlv. New York: AIP/Tomash.

Planck, Max: [Einstein] Nelkowski (1979), Enz (1994), Giorello (1972b), Helm (2000), Hiebert (1968b, 1971, 1978b, 1982), Hoffmann & Ebelung (1991), Kangro (1976), Scheibe (1995)

Planck's Principle: Gorham (1991)

Planck's Radiation Law/Quantum Hypothesis: Agassi (1967), Alekseev (1981), Amano (2000, 2001), Bellone (1968< Bergia et al. (1980), Brush (1983, 2002), Carson (2000), Casaubon (1985), Cerrita (1995), Darrigol (1992, 2000, 2001b), Einstein (1989), Galgani (1984), Galison (1981), Garber (1976a), Gearhart (2002), Giorello (1972b), Haar (1967), Hirosige & Nisio (1970), Hoyer (1984, 2000b), Kangro (1970, 1972, 1976b), Klein (1966), Koch (1991), Kragh (2000), Kubbinga (2002), Kastler (1983), Kuhn (1978, 1980, 1984), Lewis (1973), Mehra (2001), Needell (1980), Nicholas (1988), Nickles (1976, 1980), Norton (1987, 1993), Stachel (2000),

Plato, J. von *See* von Plato, J.

Poincaré, Henri: (*See also* **Recurrence Paradox**) Brush (1978, 1981b), Kranakis (1982), Norton (1993), Nye (1984), Ruhla (1992)

Poisson, S. D.: Bellone (1968), Sebastiani (1982a,c), Wolff (1995b)

Polis, D. F. *See* Achinstein (1991)

Popper, Karl: Curd (1982)

Porter, T. M. 1981a. *The Calculus of Liberalism: The Development of Statistical Thinking in the Social and Natural Sciences of the Nineteenth Century.* Ph. D. Dissertation, Princeton University. *Dissertation Abstracts International* 42 (1981): 2827A.

_____ 1981b. A Statistical Survey of Gases: Maxwell's Social Physics. *HSPS* 12: 77-116.
> Argues that Maxwell's understanding of probability in his kinetic theory came from Quetelet's social statistics.

_____ 1986. *The Rise of Statistical Thinking 1820-1900.* Princeton, NJ: Princeton University Press.
> Kinetic theory, Maxwell and his Demon, Boltzmann, C. S. Peirce, time's arrow and statistical uncertainty.

_____ 1994. From Quetelet to Maxwell: Social Statistics and the Origins of Statistical Physics. In *The Natural Sciences and the Social Sciences*, edited by I. B. Cohen, 345-362. Dordrecht: Kluwer.

_____ *See also* Brush (1978), Caneva (1993)

Porterfield, W. W., and W. Kruse. 1995. Loschmidt and the Discovery of the Small. *Journal of Chemical Education* 72: 870-875.
>   Includes translation of 1865 paper.

Positivism: Brush (1978), Hoyer (1982), Kubbinga (1991), Nyhof (1981)

Post, H. R. 1968. Atomism 1900. *Physics Education* 3: 225-232, 307-312.
>   Ostwald vs. Boltzmann.

Potter, E. 2001. *Gender and Boyle's Law of Gases*. Bloomington: Indiana University Press.
>   Argues that Boyle rejected the funiculus theory, proposed by F. Linus as an alternative to Boyle's own theory that the elasticity or weight of air adequately explains his experiments, not because it failed to account for the data but because it was based on an animistic (hylozoic) worldview associated with the radical feminist movement that Boyle disliked.

Pourprix, B. *See* Locqueneux et al. (1983)

Pourprix, B., and R. Locqueneux. 1985. Thomas Graham (1805-1869) et la Diffusion, Gazeuse et Liquide: Une Contribution au Débat sur la Structure de la Matière. *Fundamenta Scientiae* 6: 179-207.

―――― ―――― 1988. Josef Stefan (1835-1893) et les Phenomènes de Transport dans les Fluides: La Jonction entre l'Hydrodynamique Continuiste et la Théorie Cinetique des Gaz. *AIHS* 38: 86-118.

Power, Henry: Cohen (1964), Hawthorne (1979), Webster (1965, 1966)

Powles, J. G. 1978. Brownian motion -- June, 1827. *Physics Education* 13: 310-312. [*Isis CB*]

Prandtl, L.: Szabo (1975)

Prevost, P. : Chang (2002), Sebastiani (1982a)

Prigogine, I. *See* Mehra & Rechenberg (1982)

―――― and I. Stengers. 1984. *Order out of Chaos: Man's New Dialogue with Nature*. London: Heinemann.
>   Includes historical comments on thermodynamics, irreversibility, Boltzmann.

Probability, Probability Theory: Clark (1989), Daub (1969), Garber (1972), Gel'fer (1981), Harman (1970a), Hiebert (1968b), Dias (1994a,b,c), Kubbinga (2002), Ruhla (1992), Schneider (1975a,b, 1985, 1988, 1999), Schweber (1982), Sheynin (1971, 1985, 1990, 1995), von Plato (1987, 1994)

Prot'ko, T. S. *See* El'iashevich (1981)

Punte, G. *See* Cotignola et al. (2002)

Purs, J., editor. 1984. *Energy in History*. Prague: Institute of Czechoslovak and World History of the Czechoslovak Academy of Sciences.
>   Steam engines.

Pynchon, Thomas: Clarke (1996)

Quantum Theory (*see also* **Planck's Radiation Law**): Agassi (1993), Amano (2000, 2001), Barkan (1991, 1999), Bergia (1983), Brush (1983), Campogalliani (1995), DeSalvo (1992), Cerreta (2002), Dahl (1992), Darrigol (1988, 1991, 1992), Einstein (1989, 1993), Gel'fer (1981), Hermann (1969), Hudson (1997), Klein (1977), Kuhn (1978), Mehra (1975b, 2001), Mehra & Rechenberg (1982), Takata

(2000)
Quetelet, Adolphe: Porter (1981b, 1994), Schweber (1982), Toyoda (1997)

Radiant Heat *see* **Heat Radiation**
Radiation, Black Body (*see also* **Planck's Radiation Law**): Agassi (1993), Brush (1976b), Giorello (1972a)
Radiometer: Brock (1972), Brush & Everitt (1969), Draper (1976), D. B. Wilson (1981), Woodruff (1966, 1968)
Raman, V. V. 1970. Jean Perrin: Advocate for the Atoms. *Physics Teacher* 8: 380-386.
_____ 1973a. Where Credit is Due: The Gas Laws. *Physics Teacher* 11: 419-424.
_____ 1973b. William John Macquorn Rankine 1820-1872. *Journal of Chemical Education* 50: 274-276.
> Theory of molecular vortices, "potential energy," energetics, entropy.

_____ 1975a. The Permeation of Thermodynamics into Nineteenth Century Chemistry. *Indian Journal of History of Science* 10: 16-37.
_____ 1975b. Where Credit is due: The Energy Conservation Principle. *Physics Teacher* 3: 80-86. [*Isis CB*]
Ramsay, William: Levelt Sengers (1979)
Ramunni, G. 1986. Peut-on Analyser le Langage Scientifique sans se soucier de l'Outil mathématique? Le Cas de Boltzmann confronté à celui de Ostwald. *Documents pour l'Histoire du Vocabulaire Scientifique* 8: 121-132.
> On the controversy between energists and atomists. [*Isis CB*]

Randomness: Bergia & Navarro (1988), Brush (1968b, 1974, 1976b), Clark (1989), Feyerabend (1970), Sheynin (1995)
Rankine, William John Macquorn: Daub (1967, 1970b, 1978), Hutchison (1973, 1975, 1976a, 1981a,b, 2002), Moyer (1977), Raman (1973b), Steffens (1979)
Rayleigh, *see* Maxwell (2001)
Realism: Brush (1978), Hiebert (1968b), Morrison (1988a,b), Scheibe (1995)
Rechtman S., R. *See* Baracca & Rechtman S. (1985)
Recurrence Paradox: Bernhardt (1969), Brush (1976b,m 1978), Feyerabend (1970), Steckline (1983)
Redondi, P. 1975. Contributo alla Conoscenza di Sadi Carnot e dei Principi della Termodinamica a nell'Opera di Paolo Saint-Robert. *Atti della Academia delle Scienze di Torino, Classe di Scienze Morali Storiche e Filologiche* 109: 281-318. [*Isis CB*]
_____ 1980. *L'Accueil des Idées de Sadi Carnot et la Technologie Française de 1820 à 1860: De la Legende à lHistoire*. Paris: Vrin.
> Also Boyle, Townley, Amontons, Charles, Gay-Lussac, Clapeyron.
> Reviewed by L. Geymonat, *Scientia* 117 (1982): 183-187; C. S. Gillmor, *Technology & Culture* 23 (1982): 102-104; I. Grattan-Guinness, *Annals of Science* 39 (1982): 521; K. Hutchison, *Isis* 73 (1982): 145-146; P. Lervig, *Centaurus* 26 (1982-83): 227-230; C. Maffioli, *Janus* 69 (1982): 141-143; K. Von Meyenn, *Technikgeschichte* 51 (1984): 227-228; A. Yoshida, *Historia Scientiarum* 23 (1982): 129-130 [*Isis CB*]

_____ *See also* [Carnot] (1976), Locqueneux (1996)
Reech, F.: Daub (1978), Truesdell & Bharatha (1977)

Regt, H. W. de, *see* de Regt, H. W.
Reilly, C. 1969. *Francis Line, S. J. An exiled English Scientist 1595-1675*. Rome: Institutum Historicum S. J.
Renn, J. 1997. Einstein's Controversy with Drude and the Origin of Statistical Mechanics: A new Gimpse from the "Love Letters." *AHES* 51: 315-354. Also published in *Einstein: The Formative Years, 1879-1909*, edited by D. Howard & J. Stachel, Boston: Birkhäuser (2000).
>Correspondence with M. Marić; misinterpretation of Boltzmann;s results; atomism; dispute with P. Drude on electron theory of metals.

Reversibility Paradox: Bernhardt (1967), Fleischhacker & Schönfield (1997)
Rex, A. F. *See* Leff & Rex (1990a,b)
Reynolds, A. 1996. Peirce's Cosmology and the Laws of Thermodynamics. *Transactions of the Charles S. Peirce Society* 32: 403-423.
Reynolds, J. *See* Blackmore (1995), Barkan (1999)
Reynolds, Osborne: Brush & Everitt (1969), Szabo (1975)
Rice, J. *See* Donnan & Haas (1980)
Richardson, O. W. *See* Knudsen (2001)
Riecke, Eduard: Kaiser (1987)
Rinaldini, Carlo: Middleton (1968)
Robotti, N. *See* Agassi (1993)
Roche, J. J. *See* Broda (1983)
Romer, A. *See* Brush (1978)
Rosenfeld, L. *See* Bernoulli (1968)
Rosenthal, Artur: Brush (1971b)
Rossi, A. 1977. L'Esperimento di Joule sull-Equivalente Meccanico del Calore e il secondo Principio della Termodinamica. *Physis* 19: 337-353. [*Isis CB*]
Rouse, H. *See* Bernoulli (1968)
Rowlinson, J. S. 1969. Thomas Andrews and the Critical Point. *Nature* 224: 541-543.
_____ 1973. The Legacy of van der Waals. *Nature* 244: 414-417.
_____ 1980. Van der Waals revisited. *Chemistry in Britain* 16: 32-35.
>"In 1893 van der Waals published the first adequate treatment of inhomogeneous systems. His results were rediscovered in 1958 and later derived by the methods of statistical mechanics..." [*Isis CB*]

_____ *See also* Barkan (1999), Brush (1983), Kipnis et al. (1996), Smith (1998)
Rubens, H.: Palik (1977)
Rüdinger, E. *See* Mehra & Rechenberg (1982)
Ruhla, C. 1992. *The Physics of Chance: From Blaise Pascal to Niels Bohr*. Translated by G. Barton. New York: Oxford University Press.
>Maxwell, Boltzmann, Poincaré
>Reviewed by J. R. Hofmann, *Isis* 85 (1994): 680-681; B. Pippard, *Nature* 362 (1993): 216; C. A. Whitney, *American Journal of Physics* 63 (1995): 190-191 [*Isis CB*]

Rumford, Count (Benjamin Thompson). 1968-1970. *Collected Works*, edited by S. C. Brown, 5 vols. Cambridge, MA: Belknap Press of Harvard University Press.
>Vol. 1 (*The Nature of Heat*) reviewed by B. Hindle, *Technology & Culture* 12 (1971): 640-642; R. J. Morris, *Isis* 60 (1969): 410-411; R. E.

Schofield, *Science* 163 (1969): 462-463; P. Swinbank, *BJHS* 5 (1971): 403-404 [*Isis CB*]
Rumford, Count (Benjamin Thompson): Bellone (1971), G. I. Brown (1995), S. C. Brown (1967, 1979), Chang (2002), Goldfarb (1977), Olson (1970), Schagrin (1994), Schöpf (1984c), Sebastiani (1990a), Yoshida (1989)
Russell, C. A. *See* Cardwell (1989), Nye (1972)
Russian scientists: Hintikka et al. (1981), Kipnis (1964), Volkov (1991)

Sackur-Tetrode Equation: DeSalvo (1992)
Saint-Venant, A. J. C. Barre de: Darrigol (2001), Szabo (1975)
Saint-Robert, Paolo: Redondi (1975)
Saraiva, L. M. R. 1997. Laplace, Lavoisier, and the Quantification of Heat. *Physis* 34: 99-137.
Sargent, R.-M. 1988. Explaining the Success of Science. *PSA 1988*, 1: 55-63.
        Critique of Shapin & Schaffer (1985).
_____ 1995. *The Diffident Naturalist: Robert Boyle and the Philosophy of Experiment.* Chicago: University of Chicago Press.
_____ *See also* Shapin & Schaffer (1985)
Sarkar, S. 2000. Physical Approximations and Stochastic Processes in Einstein's 1905 Paper on Brownian Motion. In *Einstein: The Formative Years, 1879-1909*, edited by D. Howard & J. Stachel, 203-229. Boston: Birkhäuser.
Sarlemijn, A. *See* Kroes & Sarlemijn (1989)
Saturn's Rings: Harman (1992)
Schaffer, S. 1988. Wallifaction: Thomas Hobbes on School Divinity and Experimental Pneumatics. *Stud. HPS* 19: 275-298.
Schagrin, M. L. 1994. More Heat than Light: Rumford's Experiments on the Materiality of Light. *Synthese* 99: 111-121. [*Isis CB*]
Scheibe, E. 1995. L'Origin du Réalisme Scientifique: Boltzmann, Planck, Einstein. In *Les Savants et l'Epistémologie vers la Fin du XIXe Siècle*, edited by M. Panza & J.-C. Pont, 157-172. Paris: Blanchard. [*Isis CB*]
Scheidecker-Chevalier, M. 1997. L'Hypothèse d'Avogadro (1811) et d'Ampère (1814); La Distinction Atome/Molécule et la Théorie de la Combinaison Chimque. *Revue d'Histoire des Sciences et de leurs Applications* 50: 159-194. [*Isis CB*]
Schlagel, R. H. *See* Achinstein (1991)
Schneider, I. 1975a. Clausius' erste Anwendung der Wahrscheinlichkeitsrechnung im Rahmen der atmosphärischen Lichtstreuung. *AHES* 14: 143-158. [*Author*]
_____ 1975b Rudolph Clausius' Beitrag zur Einführung wahrscheinlichkeitstheoretischer Methoden in die Physik der Gase nach 1856. *AHES* 14: 237-261.
_____ 1985. Physics, Statistics in (Early History). In *Encyclopedia of Statistical Sciences*, edited by Kotz & Johnson, vol. 6, pp. 718-724. New York: Wiley. [*Author*]
_____ 1988. *Die Entwicklung der Wahrscheinlichkeitstheorie von den Anfängen bis 1933 -- Einführungen und Texte*. Darmstadt: Wissenschaftliche Buchgesellschaft/Berlin: Akademie Verlag.
        Chapter 7: applications of probability theory to physics, texts of Clausius, Maxwell, Boltzmann, Planck and Einstein concerning the kinetic theory of gases, statistical mechanics and thermodynamics. [*Author*]

_____ 1995. Rudolf Clausius. In *Deutsche Biographische Enzyklopädie*, edited by W. Killy, vol. 2, pp. 335f. Munich/New Providence/London/Paris: K. G. Saur.

_____ 1999. Stochastik von Laplace bis Poincaré. In *Fuzzy Theory und Stochastik. Modelle und Anwendungen in der Diskussion*, edited by R. Seising, 86-128. Wiesbaden/Braunschweig.

    Refers to kinetic theory of gases, background of David Hilbert's program for the axiomatization of statistical mechanics. [*Author*]

_____ *See also* Maxwell (1986)

Schöpf, H.-G. 1978. *Von Kirchhoff bis Planck: Theorie der Wärmestrahlung in historisch-kritisch Darstellung*. Braunschweig: Vieweg. [*Isis CB*]

    Reviewed by W. Kaiser, *Ber. Wissenschaftsgeschichte* 3 (1980): 238-239

_____ 1984a. Die Carnotsche "Paradigm" und seine erkenntnistheoretische Implikationen. *Annalen der Physik* 41: 151-160.

_____ 1984b. Rudolf Clausius. Ein Versuch, ihn zu verstehen. *Annalen der Physik* 41: 185-207.

    Second Law of Thermodynamics; quasi-economic model

_____ 1984c. Frühe Ansichten und Einsichten über die Wärme und die Gase. *NTM* 21, no. 2: 35-47.

    Lomonosov, Lavoisier, Gay-Lussac, Rumford, Dalton, Avogadro.

Schreier, W. *See* Scott (1970)

Schreinemakers, F. A. H. *See* Donnan & Haas (1980)

Schrödinger, Erwin:De Regt (2001), Mehra (1987)

Schütz. W. 1972. *Robert Mayer* 2nd edition. Leipzig: Teubner. [*Isis CB*]

Schweber, S. S. 1982. Demons, Angels, and Probability: Some Aspects of British Science in the 19th Century. In *Physics as Natural Philosophy: Essays in Honor of Laszlo Tisza on his 75th Birthday*, edited by A. Shimony & H. Feshbach, 319-363. Cambridge, MA: MIT Press.

    Maxwell Demon's relation to ideas of Darwin, Buckle, Quetelet.

Scientism: Hakfoort (1992)

Scott, W. L. 1970. *The Conflict between Atomism and Conservation Theory 1644-1860*. New York: Elsevier.

    Discusses views of J. Herapath and others.

    Reviewed by E. J. Aiton, *Isis* 63 (1972): 110-111; A. Gabbey, *Stud. HPS* 3 (1973): 373-385; T. L. Hankins, *History of Science* 9 (1970): 119-128; A. J. Pacey, *BJHS* 5: 191; W. Schreier, *NTM* 9, no. 1 (1972): 105-106 [*Isis CB*]

Sebastiani, F. 1980. Sadi Carnot e le Teorie sulla Natura del Calore. *Giornale di Fisica* 21: 61-71. [*Isis CB*]

_____ 1981a. La Memoria Voltiana Intorno al Calore. *Physis* 23: 89-113. [*Isis CB*]

_____ 1981b. La Teorie Caloricistiche di Laplace, Poisson, Sadi Carnot, Clapeyron et la Teoria dei Fenomeni Termici nei Gas formulata da Clausius nel 1850. *Physis* 23: 397-438. [*Isis CB*]

_____ 1982a. La Teorie Microscopico-Caloricistiche dei gas di Laplace, Ampère, Poisson e Prèvost. *Physis* 24: 197-236. [*Isis CB*]

_____ 1982b. Alcune Considerazione sulla Teoria Caloricistica dei Fenomeni termici nei gas. *Physis* A24: 519-527. [*BISS*]

_____ 1982c. Poisson e le Teorie sulla Natura del Calore. *Giornale di Fisica* 23: 221-238. [*BISS*]
_____ 1983. Avogadro e le Teorie sulla Nature del Calore. *Giornale de Fisica* 24: 211-230. [*BISS*]
_____ 1984. La Fisica dei Fenomeni Termici nella prima meta del Settecento: Le Teorie sulla Natura del Calore da Hartsoeker a Lomonosov. *Physis* 26: 29-127. [*Isis CB*]
_____ 1985. La Fisica dei Fenomeni Termici nella Seconda Metà del Settecento: Le Teorie sulla Natura del Calore da Black a Volta. *Physis* A27: 45-126.
_____ 1987a. Mayer e la Teoria sulla Natura del Calore. *Giornale di Fisica* 28: 141-157. [*Isis CB*]
_____ 1987b. Joule e la Teoria sulla Natura del Calore. *Giornale di Fisica* 28: 209-235, 279-300. [*Isis CB*]
_____ 1990a. *I Fluidi Imponderabili: Calore ed Elettricità da Newton a Joule.* Bari: Dedalo.
> Also discusses H. Boerhave, B. Franklin, Casanova, J. Black, A. Volta, J.-P. Marat, A. Lavoisier, P. S. De Laplace, V. Dandolo, Rumford, S. Carnot, Ampère, Faraday.

_____ 1990b. Quantification of Heat. In *Scritti di Storia della Scienza, Rend.* Accad. Naz. Sci. dei. XL Mem. Sci. Fis. Nat. Ser. 5, 14, no. 2, part 2: 53-68. [*BISS*]
_____ 1992. Dal Calore come Sostanza al Calore come Forma di Energia: L'Affermazione della Concezione Cinetica. *Giornale di Fisica* 33, no. 2: 83-109. [*BISS*]
_____ 1993. La Diffusione della Teoria del Calorico in Italia. *Giornale di Fisica* 34: 131-150. [*BISS*]

Sebastiani, F. *See also* Altieri & Sebastiani (1985)

Seebeck: Gornshtein (1966)

Seeger, R. J. 1971. Development of the Conservation of Energy during the first Half of the 19th Century. *Physis* 13: 218-224.
_____ 1974. *Josiah Willard Gibbs: American Mathematical Physicist par Excellence.* Oxford & New York: Pergamon Press.
> Reviewed by L. Tisza, *Physics Today* 28, no. 12 (1975): 56; E. E. Daub, *Isis* 67 (1976): 655-657.

Segers, J. G. 1997. Contribution à l'Histoire du Viriel de Clausius. *Scientiarum Historia* 23: 21-26. [*Isis CB*]

Segre, E. 1973. Otto Stern. *Biographical Memoirs of the National Academy of Sciences* 43: 215-236.
> Includes his early work on statistical thermodynamics, calculation of entropy constant.

Sengers, J. M. H. Levelt, *see* Levelt Sengers, J. M. H.

Sexl, R., and J. Blackmore, editors. 1982. *Ludwig Boltzmann, Internationale Tagung anlässlich des 75. Jahrestages seines Todes, 5.-8. September 1981, Ausgewählte Abhandlungen.* (Ludwig Boltzmann Gesamtausgabe, Band 8) Graz: Akademische Druck-u. Verlagsanstalt.
> Reviewed by D. A. Howard, *Isis* 75 (1984): 621; M. J. Klein, *Centaurus* 28 (1985): 72-79; J. P. Mathieu, *Revue d'Histoire des Sciences* 37 (1984): 172; B. R. Wheaton, *Annals of Science* 41 (1984): 500-501 [*Isis CB*]

Shapin, S., and S. Schaffer. 1985. *Leviathan and the Air-Pump. Hobbes, Boyle, and the*

Experimental Life. Princeton, NJ: Princeton University Press.
> Reviewed by F. Abbri, *Nuncius* 2, no. 1 (1987): 241-244; L. Busch, *Science & Technology Studies* 5, no. 1: 39-40 (1987); I. B. Cohen, *American Historical Review* 92 (1987): 658-659; M. Feingold, *English Historical Review* 106 (1991): 187-188; I. Hacking, *BJHS* 24 (1991): 235-241; W. D. Hackmann, *Nature* 321 (1986): 480; M. B. Hall, *Ambix* 33 (1986): 157-158 and *Annals of Science* 43 (1986): 575-576; T. L. Hankins, *Science* 232 (1986): 1040-1042; O. Hannaway, *Technology & Culture* 29 (1988): 291-293; P. M. Harman, *History* 72 (1987): 176; J. L. Heilbron, *Medical History* 33 (1989): 256-257; C. Hill, *Social Studies of Science* 16 (1986):726-735; M. C. Jacob, *Isis* 77 (1986): 719-720; P. J. James, *History and Philosophy of the Life Sciences* 12 (1990): 134-137; R. C. Jennings, *British Journal Philosophy of Science* 39 (1988): 403-410; H. W. Jones, *BJHS* 20 (1987): 122-123; R. Kargon, *Albion* 18 (1986): 665-666; M. Kranzberg *American Scientist* 75 (1987): 216-217; B. Latour, *Stud. HPS* 21 (1990): 145-171; A. P. Martinich, *Journal of the History of Philosophy* 27 (1989): 308-309; J. D. North, *American Scientist* 75 (1987): 216; M. J. Pascual, *Sylva Clius* 4, no. 2 (1988): 81-82; D. Oldroyd & W. Lynch, *Social Epistemology* 3 (1989): 355-372; D. Pestre, *Revue d'Histoire des Sciences* 43 (1990): 109-116; R.-M. Sargent, *PSA 1988*, 1(1988): 53-63; L. Stewart, *HSPS* 19 (1988): 193-197; T. Pinch, *Sociology* 20 (1986): 653-654; J. G. Traynham, *Journal of Interdisciplinary History* 18 (1987): 351-353; C. Webster, *Times Literary Supplement* (13 March 1987): 281; R. S. Westfall, *Philosophy of Science* 54 (1987): 128-130; P. B. Wood, *History of Science* 26 (1988): 103-109 [*Isis CB*]
>
> See also Pinnick (1998); Sargent (1988, 1995).

_____ 1999. Response to Pinnick. *Social Studies of Science* 29: 249-253, and "On Bad History. Reply to Pinnick," *ibid.* 257-259.

Shapin & Shaffer: Pinnick (1998, 1999), Sargent (1988)

Sharlin, H. 1975. William Thomson's Dynamical Theory. *Annals of Science* 32: 133-147.
_____ *see also* Elkana (1974b)

Sheynin, O. B. 1971. On the History of some Statistical Laws of Distribution. *Biometrika* 58: 234-236.
> Describes results of J. Herschel, J. C. Maxwell, L. Boltzmann

_____ 1985. On the History of the Statistical Method in Physics. *AHES* 33: 351-382.
> Daniel Bernoulli, Clausius, Maxwell, Boltzmann; why Laplace did not promote the theory. [*Author*]

_____ 1990. On the History of Statistical Methods in Natural Science [in Russian]. *Istoriko-Matematicheskie Issledovania* 32-33: 384-408. [*Author*]

_____ 1995. The Concept of Randomness in Natural Science [in Russian]. *Istoriko-Matematicheskie Issledovania* 36, no. 1: 85-105. [*Author*]

Shih, M.-L. *See* Milonni & Shih (1991)

Shinn, T. *See* Nye (1972)

Shimony, A. *See* Kuhn (1978)

Sibum, H. O. 1995. Reworking the Mechanical Value of Heat. *Stud. HPS* 26: 73-106.

Importance of Joule's brewery experience in developing precision temperature measurements.

____ 1998a. An old Hand in a new System. In *The invisible Industrialist: Manufactures and the Production of Scientific Knowledge*, edited by J.-P. Gaudillière & I. Löwy, 23-57. Houndsmill, Eng.: Macmillan.

Joule's experimental practice and brewing skills. [*Isis CB*]

____ 1998b. Les Gestes de la Mesure: Joule, les Pratiques de la Brasserie et la Science. *Annales: Économies, Sociétés, Civilisations* 53: 745-774. [*Isis CB*]

Siegel, D. *See* Maxwell (1995a)

Simões, A. 1990. L'Interpretation Statistique de la Second Loi de la Thermodynamique. In *Scientific and Philosophical Controversies*, edited by F. Gil, 203-212. Lisbon: Fragmentos. [*Author*]

Sinclair, S. B. *See* Kangro (1976b), Nye (1972)

Sklar, L. 1993a. *Physics and Chance: Philosophical Issues in the Foundations of Statistical Mechanics*. Cambridge & New York: Cambridge University Press.

Includes historical sketch of thermodynamics, kinetic theory of gases, Gibbs statistical mechanics, the Ehrenfests' 1912 article.

____ 1993b. Idealization and Explanation: A Case Study from Statistical Mechanics. In *Philosophy of Science*, edited by P. A. French et al., 258-270. Notre Dame, IN: University of Notre Dame Press.

Boltzmann's attempt to describe the approach to equilibrium in a low density gas and recent efforts to improve it.

Smeaton, J.: Pacey (1974)

Smith, C. W. 1975. *Natural Philosophy and Thermodynamics: Patterns of Thought in 19th Century Physics*. Ph. D. Dissertation, Cambridge University.

____ 1976a. Natural Philosophy and Thermodynamics: William Thomson and 'the Dynamical Theory of Heat.' *BJHS* 9: 293-319.

____ 1976b. Faraday as Referee of Joule's Royal Society Paper "On the Mechanical Equivalent of Heat." *Isis* 67: 444-449.

____ 1977. William Thomson and the Creation of Thermodynamics: 1840-1855. *AHES* 16: 231-288.

____ 1978. A new Chart for British Natural Philosophy: The Development of Energy Physics in the Nineteenth Century. *History of Science* 16: 231-279.

____ 1980. Engineering the Universe: William Thomson and Fleeming Jenkin on the Nature of Matter. *Annals of Science* 37: 387-412.

____ 1990. Energy. In *Companion to the History of Modern Science*, edited by R. C. Olby et al., 326-341. New York: Routledge.

____ 1998. *The Science of Energy: A Cultural History of Energy Physics in Victorian Britain*. Chicago: University of Chicago Press.

Joule, steam engines, thermodynamics, Thomson & Tait book, Maxwell, dissipation.

Reviewed by H. Chang, *Ambix* 47 (2000): 50-51; J. S. Rowlinson, *NRRSL* 53 (1999): 281-282; J. M. Thomas, *Nature* 400 (1999): 329-330; R. M. Yost, *BJHS* 33 (2000): 118 [*Isis CB*]

____ *see also* Bellone (1980), Brush (1978), Cardwell (1989), Elkana (1974b), Steffens (1979) Wise (1989-1990)

_____ and M. N. Wise. 1989. *Energy and Empire: A Biographical Study of Lord Kelvin.* Cambridge, Eng.: Cambridge University Press.

Biography of William Thomson, Lord Kelvin.

Smoluchowski, Marian von Smolan: Brush (1976b), Kubbinga (2002), Mayo (1996), Sredniawa (1991), Teske (1969, 1970, 1972, 1977), Tyapkin (1991), Wolff (1998)

Snelders, H. A. M. 1965. De Opvattingen van Gay-Lussac over de soortelijke Warmte van Gassen en de Betekenis ervan voor de Ontwikkeling van de Thermdynamika. *Scientiarum Historia* 7: 16-32.

"The Conceptions of Gay-Lussac of the Specific Heat of Gases and their Influence on the Development of Thermodynamics."

_____ 1977. Dissociation, Darwinism and Entropy. *Janus* 64: 51-75.

_____ 1982. Negentiende-eeuwse Theorieën over de Materie. *Tijdschrift voor de Geschiedenis der Geneeskunde, Natuurwetenschappen, Wiskunde en Techniek* 4: 168-187.

"Theories of Matter in the 19th Century."

_____ *See also* Brush (1976b), Elkana (1974b)

Sound: Euler (1972)

Spargo, P. E. 1992. Faraday, Joule and the Mechanical Equivalent of Heat. *Transactions of the Royal Society of South Africa* 48, no. 1: 47-53. [*Isis* CB]

Specific Heats: Barkan (1999), Chalmers (1973), Garber (1978), Kox (1995), Hutchison (1976b, 1981a), Mendoza (1990a)

Spohn, H. *See* Fleischhacker & Schönfield (1997)

Spurgin, C. B. 1987. Gay-Lussac's Gas-Expansivity Experiments and the traditional Mis-Teaching of 'Charles's Law.' *Annals of Science* 44: 489-505.

_____ 1989. Charles's Law -- the Truth. *School Science Review* 71, no. 254: 47-59. [*Author*]

Średniawa, B. 1991. Collaboration of Marian Smoluchowski and Theodor Svedberg in the Investigation of Brownian Movement and Density Fluctuations. Exchange of Letters. In Martinás et al. (1991), 176-189.

Stachel, J. 2000. Einstein's Light-Quantum Hypothesis, Or Why didn't Einstein Propose a Quantum Gas a Decade-and-a-Half Earlier? In *Einstein: The Formative Years*, edited by D. Howard & J. Stachel, 231-251. Boston: Birkhäuser. Also published (with a slightly different title) in Stachel, *Einstein from 'B' to 'Z'*, 427-444. Boston: Birkhäuser (2002).

On Einstein's connection with the work of W. Wien, M. Planck, and L. Natanson's comments on the interpretations of Mehra & Rechenberg (1982) and Kastler (1983).

Stachel, J. *See also* Einstein (1989)

Stallo, J. B.: Brush (1978)

State Concept: Brush (1973a), Crosland (1964), Kubbinga (2002), Levelt Sengers (1979), Monleon Pradas (1991)

Statistical Mechanics: Baracca (1991), Baracca & Rechtman S. (1985), Barker (1976), Bellone (1971, 1972), Bernhardt (1971), Brush (1967b, 1977, 1983, 1994), E. G. D. Cohen (1997), Domb (1995), Haas (1980), Krüger (1980), Kubo (1979), Kuznetsova (1988), Laidler (1993), Lavis (1977), Maxwell (1995a), Mehra (1975a, 1998, 2001), Meyenn (1994), Navarro Veguillas (1988, 1991, 1994, 1998)

Statistical Thought: Burich (1987), Kac (1974), Porter (1981a,b, 1986, 1994), Toyoda (1997), Wise (1986). (*See also* **Probability**)

Steam or Heat Engines: Cardwell (1971), Carnot (1966), Cimbleris (1991b), Dias (2001), Duffy (1983), Escudie et al. (1996), Lervig (1985), Pacey (1974), Purs (1984), Smith (1998), Truesdell & Bharatha (1977), Velarde & Ibanez Medrano (1976)

Steckline, V. S. 1983. Zermelo, Boltzmann, and the Recurrence Paradox. *American Journal of Physics* 51: 894-897. [*Isis CB*]

Stefan, Josef: Pourprix & Locqueneux (1988)

Stefan-Boltzmann Law: Brush (1973b, 1976b), Dougal (1979)

Steffens, H. J. 1968. *James Prescott Joule and the Development of the Principle of the Conservation of Energy*. Ph. D. Dissertation, Cornell University. *Dissertation Abstracts* 28 (1968): 4107-A.

———— 1979. *James Prescott Joule and the Concept of Energy*. New York: Science History Publications.

    Includes Joule's influence on W. Rankine, R. Clausius, Kelvin.

    Reviewed by S. G. Brush, *Nature* 281 (1979): 714; J. D. Burchfield, *Isis* 71 (1980): 183; E. Garber, *Science* 205 (1979): 1371-1372; D. S. L. Cardwell, *Technology & Culture* 22 (1981): 321-322; D. Gooding, *BJHS* 14 (1981): 217-219; P. M. Heimann [Harman], *Annals of Science* 37 (1980): 247-248; K. Hutchison, *Centaurus* 25 (1981): 143-144; O. Knudsen, *AIHS* 32 (1982): 124-125; T. H. Levere, *American Scientist* 68 (1980): 102-103; R. G. Olson, *American Historical Review* 85 (1980): 600-601; C. Smith, *Ambix* 27 (1980): 142 [*Isis CB*]

———— *See also* Colding (1972), Elkana (1974b), Lindsey (1973)

Stern, Otto: Enz (1974), Segre (1973)

Stewart, Balfour: Myers (1985)

Stewart, L. *See* Shapin & Schaffer (1985)

Stiller, W. 1988. *Ludwig Boltzmann: Altmeister der Klassischen Physik. Wegbereiter der Quantenphysik und Evolutionstheorie*. Leipzig: Barth.

    Reviewed by A. Wilson, *Isis* 82 (1991): 754.

Stöltzner, M. *See* Blackmore (1999)

Stokes, George Gabriel: Bowden (1972), Szabo (1975), James (1983), D. B. Wilson (1981, 1990)

Strnad, J. 1984. The Second Law of Thermodynamics in a Historical Setting. *Physics Education* 19: 94-100.

    How to teach it to 16-year-olds.

Strong Programme (in Sociology of Science): Pinnick (1998, 1999)

Stubbs, P. *See* Mendelssohn (1973)

Stuewer, R. H. *See* Boltzmann (1974), Brush (1976b), Kuhn (1978)

Süsskind, C. *See* Mendelssohn (1973)

Suffczvnski, M. *See* Teske (1977)

Suhling, L. *See* Mendelssohn (1973)

Sun, Solar Heat: James 1982

Svedberg, The: Kerker (1976, 1986), Sredniawa (1991)

Swinbank, P. *See* Rumford (1968-1970)

Szabo, I. 1975. Die Geschichte der Theorie der zähen Flüssigkeiten. Zur Geschichte der

Hydramechanik. *Humanismus und Technik* 19: 1-22.
      Theory of viscous fluids: Saint-Venant, Navier, Stokes, Reynolds, Prandtl.
Szily von Nagy-Szigeth, Coloman (Kálmán): Bierhalter (1983a)
Szumilewicz, I. 1969. Mechanicyzm Ludwika Boltzmann a Postulat Mikroredukcji. *Rozprawy Filozoficzne* 21, no. 2: 357-368.
      "The Mechanics of Ludwig Boltzmann and his Postulate of Microreduction" [*Isis CB*]

Tachoire, M. *See* Escudié et al. (1996)
Tadokoro, R. *See* Yagi & Tadokoro (2002a), Yagi, Tadokoro & Hayashi (2002)
Tait, Peter Guthrie: Bowden (1972), Lloyd (1970), Moyer (1977), Myers (1985), Smith (1998)
Takata, S. 1981. ["A comparative Study of the Editions of Mach's *Die Principien der Wärmelehre*" -- in Japanese] *Kagakusi Kenkyu* 20: 110-116. [*Isis CB*]
_____ 1987. A textual Comparison of all the four Editions of Mach's "Wärmelehre." *Historia Scientiarum* 32: 63-73. [*Isis CB*]
_____ 1993. J. B. Fourier in the History of Thermal Radiation Research. *Historia Scientiarum* 2: 203-221. [*Isis CB*]
_____ 2000. Kiyoshi Amano's Pioneering Studies on the History of Quantum Theory. *Historia Scientiarum* 10: 112-119.
      See Amano (2000, 2001) for partial translations.
Talbot, G. R., and A. J. Pacey. 1966. Some early Kinetic Theories of Gases: Herapath and his Predecessors. *BJHS* 3: 133-149.
_____ _____ 1971. Antecedents of Thermodynamics in the work of Guillaume Amontons. *Centaurus* 16: 20-40.
Tanaka, S. *See* Blackmore (1999)
Tanford, C. *See* Blackmore (1995)
Tansjo, L. 1988. Comment on the Discovery of the Second Law. *American Journal of Physics* 56: 179-182.
      Carnot, Clausius, and Kelvin.
Tarsitani, C. 1991. History of and Education in Thermodynamics: Comments and Problems related to Mental Representations and Semantics. In Martinás et al. (1991), 440-446.
_____ 1995. Le Trame della Ricerca in Storia della Fisica: Il Caso della Termodinamica. *Atti del XIII Congresso Nazionale di Storia della Fisica, Pavia, 1992*, edited by A. Rossi, 137-152. Lecce: Conte.
_____ 1997. La seconde Legge della Termodinamica tra Storia e Didattica. *Giornale di Fisica* 38: 151-162. [*BISS*]
_____ and M. Vicentini, editors. 1991. *Calore, Energia, Entropia: Le Basi concettuali della Termodinamica e il loro Sviluppo storico.* Milano: Angeli.
      Papers by G. Cortini, F. Duprè, G. De Filippo, M. G. Ianniello, M. Mayer, F. Sebastiani, C. Tarsitani, M. Vicentini. [*BISS*; for detailed contents see *Isis CB* record XBIS208631-H]
Temperature, Thermometry: Chang (2002), Cropper (1987a), Giorello (1972b), Hutchison (1976b), Klever (1997), Kubbinga (2002), Lenker (1979), Sibum (1995), Truesdell (1979)

Teske, A. 1969. Einstein und Smoluchowski. Zur Geschichte der Brownsche Bewegung und der Opaleszenz. *Sudhoffs Archiv* 53: 192-305.
        Includes correspondence.
_____ 1970. An Outline Account of the Work of Marian Smoluchowski 1872-1917. *Monogr. Dziej. Nauki Techn. Pol.* 51: 14-20.
_____ 1972. *The History of Physics and the Philosophy of Science: Selected Essays.* Wroclaw: Ossolineum.
        Includes Einstein and Smoluchowski on Brownian movement. [*Isis CB*]
_____ 1977. *Marian Smoluchowski, Leben und Werk.* Translated from Polish by A. Teske and R. Ulbrich; bibliography by M. Suffczvński. Wrroclaw: Ossolineum. [*Isis CB*]

Tetens, H. 1995. Natur und Erhaltungssätze. Exemplarische Überlegungen am Beispiel des Energieerhaltungssatzes. In *Naturauffassungen in Philosophie, Wissenschaft, Technik*, edited by L. Schäfer & E. Ströker. Band III: *Aufklärung und späte Neuzeit*, 13-40. Freiburg & München: Verlag Karl Alber.
        Discusses T. S. Kuhn's 1959 paper on energy conservation.

Textbooks: Drago (1995), Nakamura & Itakura (1991)

Theerman, P. H. 1980. *James Clerk Maxwell: Physicist and Intellectual in Victorian Britain.* Ph. D. Dissertation, University of Chicago. *Dissertation Abstracts International* 41 (1981): 4146-A. [*Isis CB*]

Thermionic Phenomena: Knudsen (2001)

Thermodynamics, Chemical, and Thermochemistry: Berger (1999), Drago (1992), Hiebert (1968b, 1978a,b), Hornix (1978, 2001), Kipnis (1964, 1997, 2000) , Kritsman (1991), Medard & Tachoire (1994), Nye (1996), Raman (1975a), Hornix & Mannaerts (2001), Kragh (2001)

Thermodynamics, 1st Law: Hiebert (1966), Sibum (1995), Walter (1995) (*See also* **Energy Conservation Law; Joule)**

Thermodynamics, 2nd Law: Bailyn (1985), Bierhalter (1981a,b, 1982, 1983a, 1987, 1990), Brush (1976b, 1978), Burich (1987), Daub (1969), Dias (1990, 1994d), Dias et al. (1995), Drago & Grilli (1991), [Helmholtz] 1973, Hiebert (1966), Hutchison 1973, 1976a,b, 1981a,b), Kestin (1976), Kim (1983), Klein (1970b), Kleinert (1992), Kober (1980), Rossi (1977), Schöf (1984b), Simoes (1990), Strnad (1984), Tansjo (1988), Tarsitani (1997), Tarsitani & Vicentini (1991), Yung (1983) (see also **Irreversibility; Time, Direction of)**

Thermodynamics, 3rd Law: Barkan (1999), Cropper (1987b), Hiebert (1978a,b,1991)

Thermodynamics, General/Miscellaneous: Allard (1983), Bauci (1973), Bellone (1978), Boudri (1991), Brouzeng (1991), Brush (1967a, 1970c, 1978), Byrne (1981), Carazzi & Guidetti (1997), Cardone & Drago (1995), Cardwell (1971, 1972, 1978), Carnot (1976), Cimbleris (1967), Clark (1976a, 1977), Daub (1967), Della Salvo & Drago (1991), Dias (2001), Domb (1995), Donnan & Haas (1980), Drago (1991), Drago & Mordello (1994), Dugdale (1968), Einstein (1989, 1993), Garber (1966, 1969, 1976b), Gel'fer (1981), Haas (1980), Harman (1982), Hiebert (1967, 1968b, 1971), Hintikka et al. (1981), Hoffmann & Ebelung (1991), Hoyer (1974a,b, 1976a,b,c, 1977, 1987), Hutchison(1976b, 2002), James (1982), Job (1969), Jungnickel & McCormmach (1986), Kangro (1976a), Klein (1966, 1967, 1974b, 1978), Knudsen (2001), Kranakis (1982), Krug (1981), Kubbinga (1991), Laidler (1993), Locqueneux (1996), Martinás et al. (1991), Merleau-Ponty (1978),

Monleon Pradas (1991b), Myers (1985), Nugayev (2000), Nye (1996), Prigogine & Stengers (1984), Redondi (1975), Reynolds (1996), Smith (1975, 1976a, 1977, 1998), Sklar (1993), Snelders (1965), Talbott & Pacey (1971), Tarsitani (1991, 1999), Tisza (1966, 1991a,b), Truesdell (1973b, 1975, 1980, 1982), Truesdell & Bharatha (1977), Volkov (1991), Yagi (1994), Yagi & Tadokoro (2002b) (See also **Cultural Connections of Thermodynamics & Kinetic Theory**)

Thermodynamics of Irreversible Processes: Kokjowski (1997)

Thermodynamics, Reduction to Mechanics: Bierhalter (1981a,b,c, 1982, 1983a,b, 1987, 1992), Hiebert (1996a)

Thermodynamics, Statistical: Bergia (1983), Daub (1969), Landsberg (1981), Pineda & Velarde (1977), Segre (1973)

Thermology, *see* **Heat, Nature of**

Thiele, J. 1967. Briefe von Gustav Theodor Fechner und Ludwig Boltzmann an Ernst Mach. *Centaurus* 11: 222-235.

Thomas, J. M. *See* G. I. Brown (1995), Smith (1998)

Thompson, Benjamin, *see* **Rumford**

Thomson, J. J.: Bierhalter (1992)

Thomson, Thomas: Kirsch (1981)

Thomson, William, *see* **Kelvin**

Thomson Effect: Tunbridge (1971)

Time, Concept of: Bellone (1989), Brush (1982), Fullmer (1964)

Time, Direction: Bierhalter (1990), Curd (1982), Porter (1986)

Tisza, L. 1966. *Generalized Thermodynamics*. Cambridge, MA: MIT Press.
    Includes "Evolution of the Concepts of Thermodynamics," 3-52.

_____ 1991a. A new look at the Relation of Thermodynamics to the Kinetic Theory. In Martinás et al. (1991), 141-155.

_____ 1992a. Concluding Remarks. In Martinás et al. (1991), 515-522.
    On thermodynamics.

_____ *see also* Brush (1976b), Seeger (1974)

Tolstoy, I. 1981. *James Clerk Maxwell: A Biography*. Chicago: University of Chicago Press.

Topper, D. R. *See* Kuhn (1987), Nye (1984)

Towneley, Richard: Agassi (1977), Cohen (1964), Hawthorne (1979), Redondi (1980), Webster (1963, 1965, 1966)

Toyoda, T. 1997. Essay on Quételet and Maxwell: From *la Physique Sociale* to Statistical Physics. *Revue des Questions Scientifiques* 168: 279-302.

Transport Processes, Irreversible Processes: Brush (1966, 1970d, 1972a), Dresden (1975), Pourprix & Locqueneux (1988)

Traynham, J. G. *See* Shapin & Schaffer (1985)

Treder, H.-J. 1977. Boltzmanns Kosmogonie und die räumliche und zeitliche Unendlichkeit der Welt. *Wissenschaftliche Zeitschrift der Humboldt Universität, Matematische-Naturwissenschaftliche Reihe* 26: 13-15. [*Isis CB*]

_____ *See also* Helmholtz (1983)

Trenn, T. J. *See* Kangro (1976b)

Truesdell, C. 1968. Early Kinetic Theories of Gases. In his *Essays in the History of Mechanics*, 272-304. New York: Springer-Verlag. Also published in *AHES* 15

(1975): 1-66.

J. Hermann, L. Euler, D. Bernoulli, J. Herapath, J. P. Joule, A. K. Krönig, J. J. Waterston.

———— 1973a. Theoria de Effectibus Mechanicis Caloris pridem ab ill^{mo} Sadi Carnoto verbis Physicis promulgata nunc primum Mathematica Enucleata. *Atti della Accademie delle Scienze dell'Istituto di Bologna, Classe di Scienze Fisiche, Rendiconti* [ser. XII] 10: 29-41.

———— 1973b. *The Tragicomedy of Classical Thermodynamics.* New York: Springer-Verlag.

———— 1975. How to Understand and Teach the Logical Structure and the History of Classical Thermodynamics. *Proceedings of the International Congress of Mathematicians, Vancouver* 2: 577-586.

———— 1979. Absolute Temperatures as a Consequence of Carnot's general Axiom. *AHES* 20: 357-380.

Also discusses Kelvin, Mach.

———— 1980. *The Tragicomical History of Thermodynamics, 1822-1854.* New York: Springer-Verlag.

Reviewed by S. Antman, *American Mathematical Monthly* 90 (1983): 343-346; S. G. Brush, *Isis* 72 (1981): 284-286; D. F. Channell, *Technology & Culture* 23 (1982): 104-106; E. E. Daub, *Science* 212 (1981): 783-784; D. Graffi, *Scientia* 117 (1982): 117; I. Grattan-Guinness, *Bulletin of the American Mathematical Society* 7 (1982): 640-643; V. Komkov, *Mathematical Intelligencer* 3, no. 3 (1981): 135-136; P. T. Landsberg, *Nature* 292 (1981): 782-783; C. Mattioli, *Janus* 68 (1981): 311-313; L. Ohlon, *Lychnos* (1981-82): 296; B. R. Wheaton, *AIHS* 32 (1982): 319-320. [*Isis CB*]

See also Lervig (1982).

———— 1982. The disastrous Effects of Experiment upon the early Development of Thermodynamics. In *Scientific Philosophy Today: Essays in Honor of Mario Bunge,* edited by J. Agassi & R. S. Cohen, 415-423. Dordrecht: Reidel. [*Isis CB*]

———— and S. Bharatha. 1977. *The Concepts and Logic of Classical Thermodynamics as a Theory of Heat Engines: Rigorously Constructed upon the Foundation laid by S. Carnot and F. Reech.* New York: Springer-Verlag.

Reviewed by E. E. Daub, *Isis* 70 (1979): 478; K. Hutchison, *Annals of Science* 36 (1979): 660-663; J. Kestin, *American Scientist* 67 (1979): 118. [*Isis CB*]

Truesdell, C.: Lervig (1982)

Tunbridge, P. A. 1971. A Letter by William Thomson, F.R.S., on the 'Thomson Effect.' *NRRSL* 26: 229-232. [*Isis CB*]

Turgot: Kubbinga (2002)

Tyapkin, A. A. 1991. On the Problem of Mechanical Representation of Irreversibility in Statistical Physics. In Martinás et al. (1991), 190-219.

On M. Smoluchowski and N. S. Krylov.

Tyndall, John: Myers (1985)

Ulbrich, R. *See* Teske (1977)

Umkehreinwand *see* **Reversibility Paradox**
Underdetermination Thesis: Norton (1993)

Valéry, Paul: Crow 1972
Valleau, J. P. *See* Battimelli (1993)
Van der Waals, *see* **Waals, van der**
Van Helden, A. C. 1991. Kamerlingh Onnes three-dimensional Graphical Method. In Martinás et al. (1991), 455-461.
Van't Hoff, J. H.: Hornix & Mannaerts (2001), Kipnis (2001), Kragh (2001)
Velarde, M. G., and J. L. Ibanez Medrano. 1976. Sadi Carnot y el Desarrollo de la Máquina de Vapor. *Arbor: Ciencia, Pensamiento y Cultura*, no. 361: 57-66.
Vernon, K. D. C. 1966. The Royal Institution's Manuscripts -- Humphry Davy and W. R. Grove. *Proceedings of the Royal Institution of Great Britain* 4: 241-258.
Verschaffelt, J.: Levelt Sengers (1976)
Vicentini, M. *See* Tarsitani & Vicentini (1991)
Virial Coefficients: Navin & Kilpatrick (1972)
Virial Theorem: Bierhalter (1982), Garber (1978), Segers (1997)
Viscosity: Domb (1980), Brush (1976b)
Visser, H. *See* Blackmore (1999)
Viswanathan, B. *See* Gopal & Viswanathan (1969)
Volkov, V. A. 1991. The first Russian Scientists in the field of Thermodynamics of the XIXth Century. In Martinás et al. (1991), 276-282.
Volta, A.: Sebastiani (1981a, 1985, 1990a)
Von Engelhardt, D. *See* Fox (1971b)
Von Plato, J. 1987. Probabilistic Physics the Classical Way. In *The Probabilistic Revolution*, edited by L. Kruger et al., Vol. 2, 379-407. Cambridge, MA: MIT Press.
    Boltzmann, Einstein, ergodic theory.
_____ 1991. Boltzmann's Ergodic Hypothesis. *AHES* 42: 71-89.
_____ 1994. *Creating Modern Probability: Its Mathematics, Physics, and Philosophy in Historical Perspective.* New York: Cambridge University Press.
    Chapter 3, "Probability in Statistical Physics" discusses irreversibility, ergodic theory, Brownian motion, Einstein.
    Reviewed by S. G. Brush, *Journal of Statistical Physics* 77 (1994): 1105-1107; T. Hochkirchen, *Historia Mathematica* 23 (1996): 203-207; C. Howson, *Philosophical Quarterly* 47 (1997): 122-125; L. Sklar, *Journal of Philosophy* 91 (1994): 622-626; S. L. Zabell, *Isis* 86 (1995): 671-672 and *Studies in History and Philosophy of Modern Physics* 31B (2000): 109-116.
Vortex Atom: Hutchison (1973, 1976a, 1981a,b), Raman (1973b), Kragh (2002)

Waals, J. D. van der. 1988. *On the Continuity of the Gaseous and Liquid States.* Edited with an Introductory Essay by J. S. Rowlinson. Amsterdam: North-Holland.
    Reviewed by S. G. Brush, *Journal of Statistical Physics* 53 (1988): 1337-1339.
Waals, Johannes Diderik van der: Brush (1973a, 1976b, 1983), De Boer (1974), Domb

(1996), Gavroglu (1990b), Kipnis et al. (1996), Kroes & Sarlemijn (1989), Kubbinga (1997b, 2002), Levelt Sengers (1979, 2002), Lunteren (2000), Morrison (1988a,b), Nairn & Kirkpatrick (1972), Rowlinson (1973, 1980)

Warburg, Emil: Wolff (1992b, 1998)

Walter, J. 1995. Über die Mathematik des Ersten Hauptsatzes der Thermodynamik. 150 Jahre "Bemerkungen über die Kräfte der unbelebten Nature" von J. R. Mayer. *Results in Mathematics* 28: 15-32.

Watanabe, M. 1978. James Hutton's "Obscure Light": A Discovery of Infrared Radiation predating Herschel's. *Japanese Studies in the History of Science* 17: 97-104. [*Isis CB*]

_____ 1980. ["James Hutton's Theories of Light, Heat, and Matter as Expounded chiefly in his *A Dissertation* (1794)," in Japanese] *Kagakusi Kenkyu* 19: 35-44, 73-82. [*Isis CB*]

_____ 1982. ["On the Revised Edition of S. L. Metcalfe's *Caloric* and the Appendix to it," in Japanese] *Kagakusi Kenkyu* 21: 46-49. [*Isis CB*]

_____ 1984. The Caloric Theory of S. L. Metcalfe. *BJHS* 17: 210-212.

Metcalfe's work was published in 1843.

Waterston, John James: Brush (1976a,b), Daub (1970b, 1971b), Hoyer (1978c), Mendoza (1982), Truesdell (1968), Whitaker (1979)

Watt, James: Cardwell (1971), Dias (1990, 2001), Fox (1970),

Wave Theory of Heat: Brush (1970c, 1976b)

Webster, C. 1963. Richard Towneley and Boyle's Law. *Nature* 197: 226-228.

_____ 1965. The Discovery of Boyle's Law, and the Concept of the Elasticity of Air in the Seventeenth Century. *AHES* 2: 441-502.

R. Towneley and H. Power.

_____ 1966. Richard Towneley (1629-1707), the Towneley Group and Seventeenth-Century Science. *Transactions of the Historical Society of Lancashire & Cheshire* 118: 51-76.

Also Henry Power (1623-1668).

_____ See also Shapin & Schaffer (1985)

Wehrl, A. 1982. Entropy seit Boltzmann. In Sexl & Blackmore (1982), 329-339.

_____ See also Battimelli (1993)

Westfall, R. S. *See* Shapin & Schaffer (1985)

Wheaton, B. R. *See* Brush (1983), Maxwell (1986), Sexl & Blackmore (1982), Truesdell (1980)

Whitaker, R. D. 1979. The early Development of Kinetic Theory. *Journal of Chemical Education* 56: 315-318.

D. Bernoulli, J. Herapath, J. J. Waterston. [*Isis CB*]

Whitney, C. A. *See* Ruhla (1992)

Wiederkehreinwand: *see* **Recurrence Paradox**

Wien, Wilhelm: Alekseev (1981)

Wigner, E. *See* Kuhn (1980)

Wilson, A. D. 1990. *Representing Reality: Ludwig Boltzmann and the Nature and Purpose of Theoretical Physics*. Ph. D. Dissertation, Cornell University. *Dissertation Abstracts International* 50 (1990): 4078-A.

_____ *see also* Battimelli (1993), Hörz & Laass (1989), Stiller (1988), Tolstoy (1981)

Wilson, D. B. 1981. Kinetic Atom. *American Journal of Physics* 49: 217-222.
> Joule, Kelvin, Maxwell, Clausius, Boltzmann.
_____ 1987. *Kelvin and Stokes: A Comparative Study in Victorian Physics.* Bristol: Adam Hilger.
> Includes W. Crookes and the Radiometer.
_____ 1990. *The Correspondence between Sir George Gabriel Stokes and Sir William Thomson, Baron Kelvin of Largs.* 2 vols. New York: Cambridge University Press.
_____ *See also* Maxwell (1995a)
Wilson, E. B. *See* Donnan & Haas (1980)
Wilson, S. S. 1981. Sadi Carnot. *Scientific American* 245, no. 2: 134-145.
Winters, S. M. 1985. *Hermann von Helmholtz's Discovery of Force Conservation.* Ph. D. Dissertation, Johns Hopkins University. *Dissertation Abstracts International* 46 (1985): 1722-A.
Wise, M. N. 1979. William Thomson's Mathematical Route to Energy Conservation: A Case Study of the Role of Mathematics in Concept Formation. *HSPS* 10: 49-83.
_____ 1983. The Maxwell Literature and British Dynamical Theory. *HSPS* 13: 175-201.
_____ 1986. How do Sums Count? On the cultural Origins of Statistical Causality. In *The Probabilistic Revolution, 1800-1930: Dynamics of Scientific Development*, vol. 1, *Ideas in History*, edited by L. Daston, M. Heidelberger & L. Krüger, 395-425. Cambridge, MA: MIT Press.
_____ *See also* Carnot (1986), Kangro (1976b)
_____ with the collaboration of C. Smith. 1989-90. Work and Waste: Political Economy and Natural Philosophy in Nineteenth Century Britain. *History of Science* 27 (1989): 263-301, 391-449; 28 (1990): 221-261.
Wolff, S. L. 1992a. August Kundt (1839-1894): Die Karriere eines Experimentalphysikers. *Physis* 29: 403-446.
_____ 1992b. Emil Warburg -- mehr als ein halbes Jahrhundert Physik. *Physikalische Blätter* 48: 275-279. [*Author*]
_____ 1993. Origins of Theoretical Physics in Germany in the 19th Century: A Case Study. In *I Beni Culturali Scientifici nella Storia e Didattica*, edited by F. Bevilacqua, 161-176. Pavia: Universita degli Studi Pavia.
> On the career of R. Clausius between 1840 and 1858 [*Isis CB*]
_____ 1994. Von der Hydrodynamik zur kinetische Gastheorie -- Oskar Emil Meyer. *Centaurus* 37: 321-348.
_____ 1995a. Gustav Magnus -- ein Chemiker prägt die Berliner Physik. In *Gustav Magnus und sein Haus*, edited by D. Hoffmann, 11-31. Stuttgart: Verlag f. Gesch. Naturw. u. Technik.
_____ 1995b. Clausius' Weg zur kinetischen Gastheorie. *Sudhoffs Archiv* 79: 54-72.
> Influence of Cauchy, Poisson, Helmholtz.
_____ 1997a. Ein Militärarzt in der Physik vor 150 Jahren: Die Formulierung Energieerhaltung durch Hermann Helmholtz. *Kultur & Technik: Zeitschrift des Deutschen Museums* no. 3: 41-45. [*Isis CB*]
_____ 1997b. Zwischen Wärmestoff und kinetischer Gastheorie: Die Behandlung der Physik der Wärme durch Hermann Helmholtz. *NTM* 5: 90-103. [*Isis CB*]
_____ 1997c. Rumford, Carnot und Clausius. In *Klassiker der Physik*, edited by K. V. Meyenn, Band I, 289-302, 467-468, 515-517. München: Beck-Verlag. [*Author*]

_____ 1998. Emil Warburg und Marian von Smoluchowski. *Physikalische Blätter* 54: 65. [*Author*]
Wood, P. B. *See* Shapin & Schaffer (1985)
Woodruff, A. E. 1966. William Crookes and the Radiometer. *Isis* 57: 188-198.
_____ 1968. The Radiometer and how it does not Work. *Physics Teacher* 6: 358-364.
Wortis, M. *See* Domb (1996)

Yagi, E. 1981. Analytical Approach to Clausius's first Memoir on Mechanical Theory of Heat (1850). *Historia Scientiarum* 20: 77-94.
_____ 1984. Clausius's Mathematical Method and the Mechanical Theory of Heat. *HSPS* 15: 177-195.
_____ 1994. Thermodynamics. In *Companion Encyclopedia of the History and Philosophy of the Mathematical Sciences*, edited by I. Grattan-Guinness, 1171-1182. London and New York: Routledge.
_____ et al. 2002a. *A Historical Approach to Entropy. Collected Papers of Eri Yagi and her Coworkers, at the Occasion of Her Retirement, 2002*. Tokyo: International Publishing Institute.
      Includes autobiographical notes, reprints of her papers, list of Clausius's publications on Mechanical Theory of Heat
_____ _____ 2002b. *A Supplement of the Collected Papers of Eri Yagi and her Coworkers. A Database from R. Clausius's Abhandlungen I-XVI*. Tokyo: Eri Yagi Institute for History of Science, Rm 404 Honkawagoe 2nd LM, 30-4, Renjaku-cho, Kawagoe-shi, Saitama, 350-0066, Japan. [*Author*]
_____ and H. Hayashi. 2002. Clausius's First and Second Laws of Thermodynamics with Fourier's Influence. *Proceedings of the 20th International Congress of History of Science, Liege, 1997*, vol. XIV, *History of Modern Physics*, edited by H. Kragh, G. Vanpaemel & P. Marage, 133-141. Turnhout (Belgium): Brepols. Also in Yagi et al. (2002), 81-95. [*Author*]
_____ and R. Tadokoro. 2002a. Theory of Electricity by R. Clausius in the Development of Thermodynamics. In Yagi et al. (2002), 113-160.
_____ _____ 2002b. Studies on the History of Thermodynamics through a Database. *Keizai-ronshu* (The Economic Review of Tokyo University) 27; reprinted in Yagi et al. (2002), 99-112.
_____ _____ and H. Hayashi. 2002. Studies on R. Clausius through various useful Methods. In Yagi et al. (2002), 15-37.
Yamalidou, M. 1997. *Problems in Molecular Science in the 19th Century*. Ph. D. Thesis, University of Lancaster. [*Isis CB*]
_____ 2001. Molecular Representations: Building Tentative Links between the History of Science and the Study of Cognition. *Science & Education* 10: 423-451.
      Maxwell.
_____ *See also* Kipnis et al. (1996)
Yavelov, B. E. *See* Kipnis et al. (1996)
Yoshida, H. 1989. The Growth of Fourier's Theory of Heat Conduction in Relation to the Experimental Studies by Count Rumford and Biot [in Japanese]. *Kagakusi Kenkyuu: Journal of History of Science, Japan* 28: 89-98. [*Isis CB*]
_____ *See also* Redondi (1980)

Yost, R. M. *See* Smith (1998)

Yung[,] Sik Kim. 1983. Clausius's Endeavor to generalize the Second Law of Thermodynamics 1850-1865. *AIHS* 33: 256-273.

Zabell, S. L. 2000. The Rise of Modern Probability Theory. *Studies in History and Philosophy of Modern Physics* 31B: 109-116.
      Essay review of Von Plato (1994).

Zamboni, N. *See* Bergia (1980)

Zanarini, G. 1996. *Ludwig Boltzmann: Una Passione Scientifica*. Napoli: CUEN. [*Isis CB*]

Zehe, H. *See* Klever (1997)

Zermelo, Ernst: Feyerabend (1970), Steckline (1983)

Zero Point Energy: Bergia et al. (1980), Enz (1974), Hiebert (1978a), Mehra & Rechenberg (1999), Milonni & Shih (1991)

Zeuner, G. A.: Krug (1981)

Ziggelar, A. 1971. From the Prehistory of Energy before 1850. In *Seminar on the Teaching of Physics in Schools, 1969*, edited by S. Sikjaer, 29-48. Copenhagen: Gyldendal.

Ziman, J. M. *See* Elkana (1974b), Kuhn (1978)

Index to Parts I, II, III[1]

Abrikosov, A. A. 540
Academia del Cimento 2
Academie des Sciences 483
Achinstein, P. 536
Adair, J. F. 466n
Adams, H. 187
Air, physical nature of 1-5
Air pressure 422-424
Alder, B. J. 467n, 476, 478, 535, 540
Alexander, H. G. 453n
Amdur, I. 475n
Andler, C. 192n
Andrade, E. N. da C. 187n
Andrews, F. C. 537
Andrews, T. 459
Annals of Philosophy 427
Asquith, P. D. 546
Atkinson, C. F. 503
Atomism and atomic theory 1, 31-34, 109, 112, 149-150, 451-479
Amontons, G. 62, 64, 460
Avogadro 427, 443, 540

Badash, L. 498
Baillie, J. 191n
Bailly, J. S. 482
Baire, R. 518
Bak, T. 511
Baker, G. A. 477
Barometer, D. 2, 64
Bates, D. R. 475n
Battimelli 444
Beez, R. 293n
Benfey, O. 498
Benrath A. 537
Bergmann, G. 536, 537
Bernoulli, D. 7, 8, 57-65, 149, 424-426, 428, 432, 457, 460-461, 485

Bernoulli, J. 7n
Bernoulli, L. 7n
Bernstein, H. T. 161n
Berthollet, C. L. 11, 107
Beyer, R. 499
Bird, R. B. 473, 474
Black, J. 10
Blackmore, J. viii, 444
Blaisse, B. S. 473n
Boas, M. 453n
Boer, J. de 473
Boerhaave, H. 68n
Bogoliubov 478
Bohm, D. 502
Böhm, W. 430
Bohr, N. 495, 497
Boltwood, B. B. 187n
Boltzmann, L. 20, 25, 26, 30, 35, 180-182, 189, 190, 194, 260n, 388, 391, 403, 406-409, 422, 429, 431-445, 463, 478, 481, 486, 487-493, 494, 496, 497, 513, 524, 526, 527, 532, 539
Boltzmann factor 182, 340, 341, 431
Boltzmann *H*-curve 194n, 392-394, 403, 406-408, 413, 418, 440-442
Boltzmann *H*-theorem 182, 189, 194, 263, 280-291, 299-302, 332, 334, 339, 393, 433, 436-438, 440, 488, 489, 490, 505-512, 526, 531, 539
Boltzmann transport equation 181, 262, 278, 307, 338, 433
Borel-Lebesgue measure theory 509
Bork, A. M. 499
Born, M. 468, 470n, 472, 481, 492-497, 505
Bosanquet, R. H. M. 30
Bosch, F. von 475n

---

1. Part IV is self-indexed.

Boscovich, R. 451, 454, 457n, 458, 469
Bose-Einstein gas 533, 540
Bose-Einstein statistics 472, 540
Bothe 497
Boyle, R. 1-4, 6, 13, 28, 34, 43-51, 164, 422-424, 427, 443, 453, 455, 481, 525
British Association for the Advancement of Science 18, 19, 436, 457n, 463, 477, 487, 489, 506, 530
Brittan, G. G. 536
Brittanica 488
Broda 444
Brodbeck, M. 544
Broglie, de 497
Brouwer, L. E. J. 509, 510, 523
Brown, R. 492
Brown, S. C. 35, 180n
Brownian movement 22, 26, 183, 492, 493, 526, 532, 541
Brück, H. 471n
Brunhes, B. 492
Brush, S. G. 35, 180n, 425, 431, 432, 434, 435, 439, 440, 442, 443, 444, 445, 452n, 454n, 455n, 460n, 462n, 464n, 467n, 477n, 478n, 480, 505, 524*ff*
Bryan, G. H. 189n
Buck, R. C. 547
Buckingham, R. A. 473, 474n, 475n, 476
Buffon 482
Bumstead, H. A. 525
Bunge, M. 525
Burbury, S. H. 182n, 390, 436-437, 439, 442, 444, 487, 489, 490, 539
Burstyn 428
Burtt, E. 453n
Buys-Ballot, C. H. D. 24, 136, 147, 422, 427-429, 430, 529

Cajori, F. 34, 52, 452n, 456n

Caloric theory 9-13, 34, 66-70, 107, 425, 484, 527, 528
Campbell, L. 30n, 500
Campbell, N. 526, 536
Cantor, G. 439, 508, 518
Carathéodory, C. 193n
Carnap, R. 499, 537
Carnot, S. 16, 20, 93, 183, 483, 484
Casimir, H. G. B. 475
Cassirer, E. 498, 537
Cavendish, H. 453
Cercignani 445
Chapman, S. 25, 181, 182n, 468-470, 474, 478
Chapman, T. 502
Charles 460, 525
Chemical reactions 73, 109, 124
Clapeyron, E. 16, 93, 110
Clausius, R. 23-26. 28-30, 111-147, 149, 160, 161n, 166, 172-178, 184, 201, 202, 207, 233, 261, 358, 392, 422, 427-430, 433, 463, 466, 483-486, 488, 497, 505-506, 529
Clifford, W. K. 536
Cohen, E. G. D. 478, 511
Cohen, I. B. 52n
Cohen, R. S. 550
Colding, L. A. 20, 527
Collisions 3, 81, 105, 114, 149, 150-152, 166-168
Collisions of molecules 189, 197, 200, 201, 205-222, 266, 269-279, 294-298, 319-327, 336-346
Colodny, R. G. 544
Compton, A. 497
Conant, J. B. 34, 443
Conn, G. K. T. 545
Conservation of energy 19-22, 35, 64, 71-110, 462, 527; in fluid flow 7-8
Cook, W. R. 470, 471
Cooling by expansion 246
Copeland, D. A. 476n
Copernicus 480
Corner, J. 473, 474n

Coulomb 475
Coulson, T. 11
Cournot 481
Craig, H. 449
Creation 186, 192
Crombie, A. C. 499
Crosson, F. J. 544
Culverwell, E. P. 390, 422, 436-437, 439, 443, 489
Curd, M. 444

Dalton, J. 34, 237, 457
Darden, L. 536
Darwin, C. 185, 186, 188, 421, 480, 481
Daub, E. E. 428, 444, 537
Davies, G. 498, 537
Davy, H. 6, 10, 14, 15, 17, 107n, 426-427, 462
Debus, A. 421
DeGroot, S. R. viii, 537
De Luc, J. A. 8n
Democritus 199
Dent, B. M. 470n
Descartes, R. 6, 45, 424, 453
Determinism 11; see Indeterminism
Devonshire, A. F. 473
Diffusion (mixing) of gases 24, 27, 135, 161n, 166
Diffusion 179, 181, 197, 201, 237, 241-243, 311, 312, 356-361, 416
    thermal 181n; thermal effects of 248-249, 353
Dingle, H. 537
Dissipation of energy 184, 185, 187, 188, 190, 193, 350-361
Domb, C. 540
Dorfman, J. R. 478
Dorling, J. 511
Dueren, R. 475n
Dugas, R. 189n
Duhamel, 64
Duhem, P. 183, 193n
Dulong-Petit radiation law 463, 538
Dyson, F. J. 537

Earth, and thermal history of 184-187
Eddington, A. S. 178, 480, 495
Edwards, P. 545
Ehrenfest, P. 189n, 194n, 441, 443, 444, 445, 491, 492, 505, 507, 509-510, 513
Ehrenfest, T. 189n, 194n, 441, 443, 445, 492, 507, 505-510, 513
Einstein, A. 421, 480, 481, 492-494, 497, 539
Einstein-Podolsky-Rosen paradox 421
Eisenschitz, R. 467, 471
Eisley, L. 186n
Elasticity 200, 201n, 205-207, 252,
Electricity 19, 21, 107-110
Electromagnetism (Maxwell's theory) 25
Elsasser, W. M. 536
Eliade, M. 191n
Elkana, Y. 481, 526
Energy 182, 184, 185, 188, 189, 246, 247, 249, 262, 405, 408, 412-415; and probablity 366n
Enriques, E. 544
Enscog, D. 25, 468-470, 474, 478
Entropy 422, 433-435, 438, 444, 485-491, 531, 532, 538
Epicurus 199
Equation of state 23, 31, 172
Equipartition theory 17, 29, 125, 149, 170, 445, 506, 509, 529
Equivalent volumes, law of 198, 199, 201, 206, 244, 245
Ergal 175
Ergodic hypothesis 445, 505-523, 528
Ernst, M. H. 478n
Eternal return 191
Ether 425, 427, 436, 443, 527, 528, 538
Euler, L. 7
Evaporation 118-120
Everett, J. D. 199n
Everitt, C. W. F. viii, 19n, 30n, 429, 431, 432, 434
Euler, L. 457
Evolution 185-187, 506

Exner, F. 496

Faraday, M. 20, 21, 454n
Feigl, H. 536, 537
Feinberg, J. 551
Fermi-Dirac statistics 540
Feyerabend, P. K. 525, 536, 537
Feynman, R. 541
Fichte, I. H. 454n
Fisher, I. Z. 535
Fisher, M. E. 477, 535, 540, 541
Floderer, A. 445
Fluctuations 180, 198, 199, 350, 354, 395, 396
Fludd, R. 421
Fluid velocity 7-8
Forbes 260
Forces 19, 52-56, 71-77, 79, 80, 92, 93, 97, 102, 116, 138, 150; see also *Vis viva*
Ford, K. 480
Forman, P. 496, 504
Foucault- Panthéon pendulum experiment 379, 438
Fourcroy, A. F. de 11
Fourier, J. B. J. 480-483, 538
Fourier theory of heat conduction 185, 186, 353
Fowler, G. N. 545
Fowler, R. H. 470n, 472, 473n, 534
Fox, R. 527
Fraassen, B. van 536, 537
Franck, J. 497
Frank, P. 526, 537
Freeman, A. 498
Frenkel 541
Fresnel 104, 486, 527
Frieman, E. A. 478
Frost, A. A. 475n
Funiculus 5, 423-424
Furry, W. 180n

Galilei, G. 1, 421, 422, 480
Gal-Or, B. 537
Gamow, G. 499
Garber, E. viii, 431, 432, 434

Gardner, M. 504
Garnett, W. 500
Gas laws 3-4, 34, 90, 116, 428
Gassendi 421
Gay, L. 446
Gay-Lussac 34, 90, 115, 129, 137, 145, 460-461
Geiger 497
Ghieslin, M. 498
Gibbs, J. W. 20, 182n, 398, 406n, 444, 490, 491, 497, 507, 524, 526
Gillispie, C. C. 30n, 185n, 499
Giorello, G. 501
Girill, T. R. 535
Goethe 454n
Goldberg, S. 501
Goldman, R. 478
Gouy, L. 502
Grad, H. 478n
Graham 197, 238-242, 253, 256
Grattan-Guinness, I. 498
Gravity 14, 53, 71, 74, 79, 81, 89, 102
  effect on distribution of molecules 190, 226, 237, 258, 259
Green, M. S. 477n, 540
Greenough, G. 482
Gregory, G. 11, 66-70
Grene, M. 532
Gross, E. P. 478n
Grove, W. R. 20
Grünbaum, A. 537
Guericke, O. von 2, 422, 455
Guggenheim, E. A. 473n
Günther, S. 545
Guthrie, F. 422, 429-432
Guyton de Morveau, L. B. 11

H-curve, H-theorem *see* Boltzmann
Haar, D. ter vii, 189n, 194n, 511
Haber, F. C. viii, 498
Hahn, R. viii
Haines, L. K. 478n
Halicioğlu, T. 476n
Hall, A. R. 453n

Hall, F. 5
Hall, M. B. 453n
Hall, N. S. viii
Hanle, P. 504, 540
Hann, J. 448
Hardy, W. B. 459n
Harré, R. 550
Hartley, H. viii, 17n
Hassé, H. R. 470, 471
Heat 9-16, 19, 21-23, 62, 66-71, 74-79, 83-88, 104-115, 131-134, 173
Heat conduction 150, 161n, 166, 179, 180n, 181, 185, 197-201, 234, 242, 257-261, 307, 313-315, 352, 356, 412
Heat death 184, 187, 190, 193, 380, 433, 439, 487
Heine 192
Heisenberg, W. 470, 480-483, 486, 493, 495, 497, 505, 540
Heitler, W. 499
Helm, G. 193n
Helmholtz, H. von 20-22, 89-110, 184, 379, 383, 390n, 398, 527
Hemmer, P. C. 467, 468n
Hempel, C. 536, 537
Henry, J. 11
Henry, W. 107
Herapath, J. 14-17, 22, 149, 204, 426-427, 431, 457, 462-463, 485
Herschel, J. 30n, 459, 486, 527
Hertz 390
Hertz's experiments on electromagnetic waves 25
Hess, H. 110
Hesse, M. B. 454n, 537
Hiebert, E. viii
Hilbert, D. 478, 510
Hintikka, J. 551
Hirschfelder, J. O. 473, 474, 476
Hobbes, T. 5, 424, 443
Holism 525
Holton, G. viii
Holtzmann, K. 20, 110
Hooke, R. 3, 4, 6, 423

Hugoniot equation 476
Hunter, J. 482
Hutton 185, 482
Huxley, T. H. 186
Huygens, C. 3, 424
Hydrodynamics 198
Hypothetico-deductive method 451, 530

Ianello 444
Ikenberry, E. 478n
Impact, *see* Collisions
Indeterminism 480-497, 526, 532
Integral variant 369-371
Integrals, replacement by sums 291-303, 332, 333
Irreversibility 182, 183, 200, 251, 377-380, 388, 398, 400, 404, 405, 409, 415, 480-497, 524-526, 531, 533'
Ising, E. 477, 533-535, 540

Jackson, E. A. 478n
Jacobi's principle of last multiplier 279
Jäger, G. 180
James, W. S. 4n, 34
Jammer, M. 454n, 497
Jancel, R. 478n
Jauch. J. M. 496
Jeans, J. H. 180, 192, 509
Jevons, W. S. 499
Johnson, M. 537
Jones, J. E. *see* Lennard-Jones
Jones, W. 68n
Jordan, J. E. 475n
Joule, J. P. 14, 19, 21, 23, 78-88, 89, 106, 110, 113n, 135, 136, 149, 201, 251, 427, 465, 473, 485, 527, 528
Jourdain, P. E. B. 512

Kac, M. 444, 467, 468n
Kahan, T. 478n
Kangro, H. 500
Kantor, J. R. 537
Kattsoff, L. O. 536

Kaufman, W. A. 192n
Keegan, F. 512
Kelvin (William Thomson) 20, 32, 161n, 184-186, 188, 190, 192, 259, 430, 434, 436, 454n, 457n, 458, 468, 484, 489, 497, 528, 530, 531, 538, 539
Kemeny, J. G. 536
Kendall, M. S. 500, 503
Kennard, E. 35, 180n
Kepler, J. 47, 421, 480, 486
Kestner, N. R. 476n
Kihara, T. 473, 476
Kinetic energy, see *Vis viva*
King, W. J. viii
Kingston, A. E. 474
Kirchoff, G. 490
Kirkwood, J. G. 471n, 535
Kirwan, R. 482
Kirzhnits, D. A. 540
Klein, M. J. viii, 183n, 431, 439, 497, 537
Klein, M. L. 476
Klein, O. 503
Knott, C. G. 499
Kogut, J. 540
Kohler 442
Kohlrausch 204
Körner, S. 502
Kotani, M. 473
Koyré, A. 453n, 454n
Krajewski, W. 499
Kramers, H. A. 497
Kranendonk, J. van 473
Krause 4454
Kresten 444
Krönig, A. K. 14, 22-23, 111-113, 115, 122, 123, 126, 129, 135, 136, 149, 485
Kruse, W. 449
Kubrin, D. 498
Kuhn, T. S. 20, 34, 35, 455, 500, 527
Kundt, A. 29

Lagrange 185, 293
  equations of motion 399
Lakatos, I. 547
Lambert, K. 536
Langevin, P. 495
Landau, L. D. 512, 533
Landé, A. 496
Laplace, P. S. 11-13, 15, 185, 426, 451, 459-460, 485
Larmor, J. 390n
Latent heats 10, 93, 109, 121
Laue, M. von 503
Lavoisier, A. 11
Lebesgue, H. 193, 508, 510, 523
Lebowitz, J. L. viii
Lee, T. D. 540
Leibniz 453n
Lenard, A. 537
Lennard-Jones, J. E. 451, 468-476
Lenz, W. 477, 533
Le Sage, G.-L. 8n, 200, 426, 454n
Levelt Sengers, J. H. M. 539, 540
Lewis, W. C. M. 548
Liebig, J. 20
Lifshitz, E. N. 512
Light, reflexion and refraction of 90, 104
Liley, P. E. viii
Lilley, S. 34
Linus, F. 5, 422-424
Liouville's theorem 369, 384, 406, 417, 440
Lissajous figures 506
Living force, see *Vis viva*
Loeb, L. B. 35, 180n
Lomonosov, M. V. 8n
London, F. 467, 471, 533, 540
Lorentz, H. A. 391, 492
Lorenz, L. 32, 530
Loschmidt, J. 31-32, 190, 362, 363, 422, 429-430, 433-436, 443, 488, 489, 530, 531, 539
Losee, J. 537
Lotka, A. J. 498
Lubbock, J. W. 17, 18n
Lucretius 199, 200

Ludwig 504
Lyell, C. 185, 482

Mach, E. 20, 22, 183, 193n, 444, 454n, 495, 526
MacKie, J. L. 545
Madden, E. H. 536
Mairan, D. de 482
Margenau, H. 466n, 476n, 532, 537
Mariotte, E. 4, 90, 116, 129, 137, 145, 164
Masius, M. 501
Mason, E. A. viii, 474, 475n
Massey, H. S. W. 472, 473
Materialism 192, 193, 251
Maxwell, G. 544
Maxwell, J. C. 5, 19n, 25, 30, 31, 148-171, 179-181, 188, 190, 263, 265, 266, 279, 280, 308, 309, 380, 391, 392, 412, 421, 422, 429-436, 464-466, 478 480, 481, 485-488, 489, 494, 497, 505-511, 513, 524, 527, 528-531, 535
    demon 188, 350, 352-354, 377, 379, 380, 421, 422, 434, 436, 439, 487, 488, 539
    electromagnetic theory 183
    transfer equations 180-182, 212, 224-236
    velocity distribution 181, 182, 189, 197, 221-223, 259, 262, 265, 266, 280, 291, 305, 306, 347, 392-396, 402, 406, 417, 430
Maxwellian (r^{-5}) molecules 181, 216-218, 253, 312, 313, 451, 464-465, 478
Mayer, J. E. 444, 472
Mayer, J. Robert 19, 21, 22, 71-77, 527
McCormmach, R. 453
McGuinness, B. 449, 537
Mean free path 24, 26, 32, 135-147, 149, 159, 160, 164, 180, 201, 428-430, 529

Measure of a set of points 193, 194
Mechanico-corpuscular view of nature 1, 21, 53, 422
Mechanistic viewpoint 192, 193, 377, 378, 311, 318, 319, 326, 330, 333, 334, 344, 421, 451, 485, 526
Mehra, J. 444
Mehlberg, H. 537
Mendelssohn, K. 499
Mendoza, E. viii, 499
Mersenne 421
Merz, J. T. 35
Metz, W. 540
Meyer, L. 32, 475n
Meyer, O. E. 28, 180, 202, 253, 464n, 465, 530
Meyerson, E. 536, 537
Michels, A. viii
Miller, S. L. 449
Mises, R. von 504
Misner, C. 541
Moche, D. L. 550
Mohr, C. B. O. 472
Mohr, K. F. 20
Molecular theories 198
Monocyclic systems 20
Monroe, E. 535
Moravcsik, M. J. 501
Morse 476
Moser, J. 549
Motte-Cajori 452n, 456n
Munn, R. J. 476
Murray, J. 482
Musgrave, A. 550

Nagel, E. 526, 536, 537
Nature 431, 432, 436-437
Naturphilosophie 21, 22
Navier-Stokes equation 198, 251
Nernst, W. 496
Neumann, J. von 499
Neurath, O. 550

Newton, I. 1, 3, 11, 13, 21, 34, 421, 424, 433-434, 436, 438, 480, 481, 483, 487, 488, 489, 492, 496, 541
  repulsion theory of gases 6, 12, 52-56, 424-425, 451-462
Nickles, T. 532, 541
Nietzsche, F. 191-193
Niven, W. D. 160n-161n, 166, 432, 500
Nye, M. J. 502, 539

Obermayer, A. v. 465n
Occult forces 1, 53
Oersted's discovery of electromagnetism 21, 22
Olson, R. viii, 457n
Onsager, L. 534, 540
Oppenheim, P. 536
Osteen, C. W. 504
Ostwald, W. 20, 22, 183, 526, 539

Pappus 53
Pascal, B. 1, 2, 49, 422, 455
Pauli, W. 442, 540
Pauling, L. 471n
Peano 508
Peckhaus 439
Peirce, B. 481, 499
Peierls, R. 534
Peltier, J. C. A. 21
Perier, F. 2, 49n
Perrin, J. 22, 539
Phase of motion 248
Phlogiston 73, 536
Plancherel, M. 445, 505, 510, 520-523
Planck, M. 183, 443, 444, 480, 481, 486, 490-492, 494, 495, 497, 509, 527, 537, 538, 541
Playfair 482
Pledge, H. T. 35
Poggendorff, J. C. 22, 35, 71

Poincaré, H. 191-193, 260n, 383, 392-396, 399, 401, 404, 405, 415, 419, 438-439, 489, 539
Poisson, S. D. 13, 16, 185, 251, 368, 370, 460
Polder, D. 475
Polyatomic gas molecules 216-246, 463
Popper, K. 455, 527, 536, 537
Porterfield 445
Positivism 495, 524-525, 536
Post, H. R. 537
Powell, B. 17, 18n
Power, H. 455
Pressure 179, 202, 203, 205, 232
  of air 1-2, 4-6, 43-51
Preston, T. 19n
Prevost, P. 200
Probability theory 11-12, 264, 357-361, 365, 392, 395-397, 405, 409, 410, 412, 414, 415
Puluj, J. 465n
Putterman, S. J. 533

Quantum mechanics 421
Quantum theory 30, 33, 183, 480, 490, 494, 496, 497, 527
Quay, P. M. 537
Quetelet 486

Radiant heat 10, 12
Radioactivity 187
Railway Magazine 427, 462
Ramsay, W. 29, 539
Randall, J. H., Jr. 498
Rankine, W. J. M. 19, 19n, 115, 463
Ravetz, J. R. 545
Rayleigh, J. W. S. 18, 29, 463n, 509, 527, 538
Recurrence 191-193, 368-376, 380, 382-415, 438-439, 489, 490, 509, 531, 539
Reduction 524-526
Regnault 117, 130, 145
Reichenbach. H. 194, 442, 526, 537

Reilly 423, 424
Relaxation 180, 204, 205, 251, 254
Reversibility paradox 190, 350-355,
    362-367, 389, 290, 433-436,
    444, 509, 531, 538
Rey, A. 191n
Riccioli, G. 48
Rice, W. E. 474
Riemann, B. 293
Rigidity 204, 205, 253, 254
Riste, T. 550
Robinson, N. H. viii, 17n
Roget, P. M. 18n
Romanticism 21
Romer, A. 502
Rootselaar 439
Rosenfeld, L. 504
Rosental, S. 504
Rosenthal, A. 445, 505, 510, 513-519
Ross, M. 476

Rotation of gas molecules 19, 28-30,
    45, 111, 113-114, 150, 167-171
Roth, L. M. 525
Rowlinson, J. S. viii, 473
Royal Society of London 2, 17, 18, 35,
    424, 426-427, 431, 462, 528-529
Rüdinger, E. 504
Ruelle, D. 537
Rumford, B. T. 10, 89
Rutherford, E. 494

Sahlin, H. L. 467n, 542
Salmon, W. C. 537
Salpeter, E. E. 540
Sarton, G. 35
Scattering angle 209-211, 215-217
Schaffner, K. 535, 536
Schelling, F. 21, 454n
Scheutz 437
Schiller, F. 444
Schlegel, R. 551
Schlesinger, G. 536

Schlier, C. 475n
Schneider, W. G. 474n
Schoenflies, A. 520
Schofield. R. 453
Schrödinger, E. 470, 480, 483, 496,
    497, 536
Schwartz, J. viii
Schweidler, E. von 494
Scott, W. L. 450
Sears, F. 551
Second law of thermodynamics
    (Carnot-Clausius principle)
    183-185, 188-190, 194, 259,
    263, 291, 347, 362, 363, 367,
    389-393, 397, 403, 404, 407,
    412, 416, 417, 431, 433-435,
    438-443, 480, 483-487, 489,
    490, 492, 496, 497, 525, 531,
    535, 538-539
Seebeck, T. J. 21
Seguin, M. 20
Sengers, J. V. 477n, 478n, 540
Sexl 444
Siemens, W. 113n
Shapere, D. 527
Shapin and Schaffer 423, 443
Sharlin, H. I. viii
Shimony, A. 552
Shostak, A. 476n
Siegel, A. 525
Silliman, R. 454n
Simon 497
Sinai, J. G. 511
Sinanoğlu, O. 476
Sirovich, L. 478n
Sklar, L. 525, 537
Slater, J. C. 467, 470, 471n, 472, 497
Smart, J. C. C. 536, 537, 541
Smith, B. B. 500
Smith, E. J. 476
Sneed, J. 553
Snelders, H. A. M. 428
Socrates 421
Soddy, F. 494
Sorokin, P. 191n
Sound, velocity of 13, 427

Specific heats 13, 28, 115-116, 133, 134, 150, 171, 207, 246-248, 315, 463, 506, 509, 528, 529, 530
Spencer, H. 187
Spencer, J. B. 454n
Spengler, O. 503
Spotz, E. L. 473
Stability of mechanical systems 368, 370, 371
Stamper, J. 476m
Stanley, H. E. 540
Statistical mechanics 20, 27, 509, 510, 524-536
Stefan, J. 180, 183, 293, 315, 465n, 538
Stiller 445
Stockmayer 476
Stokes, G. G. 26, 27, 251, 254, 464, 530
Stoney, G. J. 32, 444, 530
Stouffer, R. C. 550
Strutt, R. J. 187n
Stuewer, R. H. 544
Suppe, F. 524, 536
Suppes, P. 526, 537
Swiss Physical Society 523

Tait, P. G. 6n, 180, 186, 188, 434, 486
Taylor, B. A. 470n
Teller, E. 467n
Temperature distinguished from heat 10
Theobold, D. W. 536
Thermal diffusion 25
Thermodynamics 16, 19, 23, 24, 112, 525, 537, 540
Thermoelectricity 21, 110
Thompson, B. *see* Rumford
Thompson, C. J. 536
Thompson, S. P. 186n
Thomson, W. *see* Kelvin
Threfall, R. 466n
Tisza, L. 533, 537
Tolman, R. C. 14n, 442, 444

Torricelli, E. 1-2, 4, 422-423, 427, 455
Toulmin, S. 536, 537
Tour, C. de la 459
Towneley, R. 3, 4, 455
Transpiration 253, 255, 256
Trembley, J. 62n
Tresca, H. 539
Truesdell, C. viii, 452n, 453, 453n, 456n, 478n, 479

Uhlenbeck, G. E. 444, 467, 468n, 478n, 540
Ulam, S. 498
Ultraviolet catastrophe 491, 527
Uniformitarian geology 185, 186, 482
Unsöld, A. 471n

Vacuum 422-424
Vanderslice, J. T. 475n
Van Melson, A. 35
Velocity distribution 25, 26, 30, 127, 148, 153-155
Velocity-distribution function (*see also* Maxwell) 181, 197, 218-222, 262, 265, 267, 391
Vienna Academy 434, 526
Virial theorem 23, 30-31, 172-178, 466
Vis viva (living force, kinetic energy) 31, 74, 80-88, 89, 93, 94, 95, 98, 100, 106, 108, 111, 115, 121, 125, 129, 131, 152, 172-178, 432
Viscosity (internal friction) 5, 26-28, 31-32, 149, 150, 164-166, 179, 181, 197, 199, 201-203, 205, 216, 243, 250, 251, 253-255, 307, 308, 313, 429-431, 528-530
Volta's invention of battery 21
Vortex atoms 6, 19, 20

Waals, J. D. van der viii, 31, 179, 451, 465-468, 532-533

Waerden, B. L. van der 502
Wainright, T. E. 467n, 478, 535, 540
Wald, F. 193n
Walker, J. 512
Wallace, A. R. 185
Wallis, J. 3, 5
Wang, S. C. 467, 470, 471
Wang Chang, C. S. 478n
Warburg, E. 29
Wartofsky, M. 552
Wasserburg, G. J. 449
Waterston, J. J. 17-19, 463, 485, 505, 510, 528-529, 535, 538
Wave motion 103-104, 527
Weber 30
Webster, C. 422, 455n
Weimar Germany 481, 496
Weinstock, J. 478
Werkmeister, W. H. 537
Whyte, L. L. 454n, 457n
Wilberforce 506
Wiener, P. P. 498
Wightman, A. S. 511
Williams, L. P. 454n
Wilson, K. G. 540
Wolf, A. 185n
Woodson, J. H. 475n
Work 21, 89, 121, 122, 123, 173
Wren, C. 3

Yaglom 541
Yang, C. N. 540
Yntema, J. L. 474n
Young, S. 539
Young, T. 457

Zemansky, M. 551
Zermolo, E. 193, 194, 392-396, 400, 412, 413, 417, 422, 438-443, 489, 490, 531
Zwart, P. J. 553